ALLE ZEIT WACH
1842

Progress in Sensory Physiology

Editors: H. Autrum, D. Ottoson,
E.R. Perl, R.F. Schmidt, H. Shimazu, W.D. Willis
Editor-in-Chief: D. Ottoson

Volume 8
1987. 56 figures. V, 207 pages
ISBN 3-540-16300-X

Contents: *W. Skrandies:* The Upper and Lower Visual Field of Man: Electrophysiological and Functional Differences. – *J. Le Magnen:* Central Processing of Sensory Information in the Control of Feeding. – *A. Faurion:* Physiology of the Sweet Taste.

Volume 9
1989. 71 figures. V, 227 pages
ISBN 3-540-50282-3

Contents: *R.D. Foreman:* Organization of the Spinothalamic Tract as a Relay for Cardiopulmonary Sympathetic Afferent Fiber Activity. – *G.N. Akoev, G.N. Andrianov:* Synaptic Transmission in the Mechano- and Electroreceptors of the Acousticolateral System. – *J. Syka:* Experimental Models of Sensorineural Hearing Loss – Effects of Noise and Ototoxic Drugs on Hearing. – *Y. Oomura:* Sensing of Endogenous Chemicals in Control of Feeding. – *J.E. Sánchez-Criado, O.A. Mora, A. Gallego:* Structure and Function of the Vomeronasal System – The Vomeronasal Organ as a Priming Pheromone Receptor in Mammals.

Volume 10
M. Steriade, D. Paré, B. Hu, M. Deschênes
The Visual Thalamocortical System and Its Modulation by the Brain Stem Core
1990. 27 figures. VII, 124 pages.
ISBN 3-540-51380-9

Contents: Prologue. – Basic Circuits in the Visual Thalamocortical Systems and Their Physiological Aspects. – Regulatory Systems of the Brain Stem Core, Basal Forebrain, and Hypothalamus. – Neurotransmitters. – State Dependency of Visual Thalamic and Cortical Activities.

Progress in

Sensory Physiology 11

P.J. Snow P. Wilson

Plasticity in the Somatosensory System of Developing and Mature Mammals – The Effects of Injury to the Central and Peripheral Nervous System

With 139 Figures

Springer-Verlag
Berlin Heidelberg NewYork London
Paris Tokyo HongKong Barcelona

Editor-in-Chief

Professor Dr. David Ottoson
Karolinska Institutet, Fysiologiska Institutionen II
Solnavägen 1, 10401 Stockholm 60, Sweden

Editors

Professor Dr. Hansjochem Autrum
Zoologisches Institut der Universität München
Luisenstraße 14, 8000 München 2, Germany

Professor Dr. Edward R. Perl
Department of Physiology
University of North Carolina at Chapel Hill
Chapel Hill, NC 27514, USA

Professor Dr. Robert F. Schmidt
Physiologisches Institut der Universität
Röntgenring 9, 8700 Würzburg, Germany

Professor Dr. Hiroshi Shimazu
Department of Neurophysiology
University of Tokyo, Institute of Brain Research
7.3.1. Hongo, Bunkyo Ku, Tokyo, Japan

Professor Dr. William D. Willis
The Marine Biomedical Institute
University of Texas Medical Branch
Galveston, TX 77550, USA

Authors

Peter J. Snow, BSc, PhD
Peter Wilson, BSc, PhD

Department of Anatomy, Mammalian Neurobiology Laboratories
University of Queensland, St. Lucia 4067, Queensland, Australia

ISBN-13:978-3-642-75703-7 e-ISBN-13:978-3-642-75701-3
DOI: 10.1007/978-3-642-75701-3

LC card number: 85-647743 (ISSN 0721-9156)

Softcover reprint of the hardcover 1st edition 1991

Phototypesetting: Thomson Press (India) Ltd, New Delhi, India
21/3130-543210 – Printed on acid-free paper

Acknowledgements

We wish to thank Dr. M.B. Plenderleith for his valuable comments on all parts of this manuscript and Ms D. Crook for her excellent technical assistance. We also wish to thank Dr. P. Waite for her expert criticism of Chapter 3, those members of the Anatomy Department who read specific chapters and Ms L.A. Snow for her assistance with the art work.

Contents

Abbreviations

Development Times:

P	postnatal day
G	gestation day

Drugs and Chemicals:

ACh	acetylcholine
CO	cytochrome oxidase
GABA	γ-aminobutryric acid
HRP	horseradish peroxidase
NGF	nerve growth factor
SDH	succinic dehydrogenase
2-DG	2-deoxyglucose
WGA-HRP	Wheat germ agglutinin – horseradish peroxidase conjugate

Regions of Parts of the Nervous Systems:

DCN	dorsal column nuclei
DRG	dorsal root ganglion
FCN	femoral cutaneous nerve
HN	hypogastric nerve
IMN	inferior mesenteric nerve
ION	infraorbital nerve
LCN	lateral cervical nucleus
PFCN	posterior femoral cutaneous nerve
PMBSF	posterior medial barrel subfield
PrV	principal sensory nucleus of the trigeminal complex
PSDC	postsynaptic dorsal column (cell or unit)
SI cortex	first somatosensory area of the cerebral cortex
SII cortex	second somatosensory area of the cerebral cortex
SCT	spinocervical tract
SpVo	subnucleus oralis of the spinal nucleus of the trigeminal complex
SpVi	subnucleus interpolaris of the spinal nucleus of the trigeminal complex

SpVc	subnucleus caudalis of the spinal nucleus of the trigeminal complex
TNC	trigeminal nuclear complex
VPI	ventral posterior inferior nucleus
VPL	ventral posterior lateral nucleus
VPM	ventral posterior medial nucleus
VPO	ventral posterior oralis nucleus
VPS	ventral posterior superior nucleus
VB complex	ventrobasal complex

1 Introductory Remarks

The *Concise Oxford Dictionary* defines the adjective *plastic* as applicable for describing something which is pliant, supple or capable of being moulded. That object or entity thus has the property known as *plasticity*. This definition implies change in physical structure but when we apply this term to dynamic phenomena such as behaviour or nervous function we need to consider more carefully its meaning. For instance, does our ability to learn and to experience changes in mood and perception suggest that the nervous system is plastic or must we consider only recovery from a lesion as diagnostic? Alternatively, do we accept changes in the efficacy of a reflex, brought about by a conscious decision, as indicative of neural plasticity? In each case changes have occurred in the functional connectivity of the nervous system and in neither case do we necessarily have information concerning structural changes. Fortunately, there is one outstanding criterion which perhaps transcends all others – that of *permanency*. This aspect emerges for it carries with it the covenant that the clay, once moulded, shall never again be the same. It permits us to discount the transient modulation of connections and to focus on changes that can never be fully reversed. It facilitates definition without embroiling us unnecessarily in premature discussion of mechanisms.

Anyone who reflects upon the relatively enormous changes which occur in behaviour during the first 15 years of the life of a human may well consider that any malleability of the mature nervous system is, in some way, intimately linked with the processes that control normal development. Again it is the permanency of most of these developmental changes that catches our attention and persuades us that the immature nervous system is highly plastic. It must, however, be emphasized that on this basis such a conclusion is groundless, for it assumes that merely repeatedly observing a set of stereotyped, though lasting, changes reflects the potential for more diverse changes within the system. Hence emerges another aspect of plasticity – that of *potential for change*. Both during development and in maturity, this potential can only be explored by the imposition of abnormal circumstances. Thus in studying neural plasticity at any age we disrupt the normal and we observe the outcome. Such an approach soon confirms our suspicion that plasticity is an age-related process, for while the mature nervous system may be capable of adaptation, it is, by any measure, tonically inflexible when compared with its predecessor (see Chap. 3).

Research is driven by curiosity. What then has stimulated our curiosity in plasticity in the somatosensory system? The answer to this must be more than the practical demands of neurology, even though the derived knowledge is crucial to many aspects of that discipline. Most certainly it lies in the frequency of relatively stable somatosensory illusions. Fascinating descriptions of distorted perceptions of the body, and its surface, in response to fever, local anaesthesia, schizophrenia, psychoactive drugs and sensory deprivation are rife throughout the professional and lay literature. However, of all somatosensory illusions one in particular stands out for both its frequency of occurrence and the permanency of its creation. That illusion is the *phantom limb.*

Phantom sensations usually begin immediately after amputation of an arm or a leg, the blockade of nerves that supply sufficiently large areas of the body or section of the spinal cord (Wall 1981; Berger and Gerstenbrand 1981). After limb amputation they characteristically involve the distal parts of the limb although the thigh and upper arm are present in some cases. In contrast, spinal cord section can produce phantom sensations that involve the entire limb. Thus the size of the phantom roughly equates the size of the area that is isolated from the central nervous system. Primarily phantom limb sensations indicate the *presence* of an appendage. However, especially in the early postoperative period, the phantom may give rise to quite specific sensations which are often associated with sensations prevailing preoperatively. Pain is one such sensation that may persist for years and which seems to be related to preoperative trauma in the amputated limb. Another less disturbing illusion is that of position, which often reflects the preoperative position (Fig. 1.1) (Avenarius and Gerstenbrand 1967). With "great concentration" some patients are, in fact, able to perform movements of the phantom (Berger and Gerstenbrand 1981). While the existence of a phantom confirms overwhelmingly that conscious awareness of the parts of our body is indelibly written in our nervous system, the persistence, in the phantom, of specific sensations that existed preoperatively indicates that this is also the case for the recent history of those parts. Perhaps the cortical locus for conscious awareness of a stimulus and the location of simple sensory memory are one and the same (see Sect. 8.6).

With time the phantom usually gets shorter so that in cases of limb amputation the digits approach the stump – a phenomenon known as *telescoping* (Fig. 1.2) (Siegfried and Zimmermann 1981). Indeed the phantom may often be absent. In amputees the phantom can be restored or exaggerated by stimulation of the stump and on this basis Wall (1981) has suggested that regulation of the phantom results from alterations in functional connectivity in ascending pathways. In addition, however, mental activity or concentration can exaggerate or recall or, as mentioned above, manipulate a phantom. To account for this we must clearly consider the involvement of higher neural functions such as directed attention and cognitive strategies (see Sect. 8.6). Perhaps it is not unrelated that relaxation-encouragement (reward) methods and acupuncture have been implicated as among the most successful strategies in treatment of phantom pain (Sherman et al. 1980).

Research into plasticity within the somatosensory system has also been stimulated by the clinical observations following peripheral nerve injuries. It has been known for a long time that following nerve damage many sensory axons will regenerate connections with peripheral structures. The degree to which a particular axon reinnervates the area it previously innervated (specific reinnervation) depends on the type of nerve injury (see Chap. 2). Following nerve section, specific reinnervation does not occur and the patient's ability to localize stimuli on the reinnervated area is grossly impaired. That this impairment can be corrected by retraining is direct evidence for plasticity somewhere along the somatosensory axis (Fig. 8.4) (Wynn-Parry 1981). Moreover, like the strategies involved in the alleviation of phantom

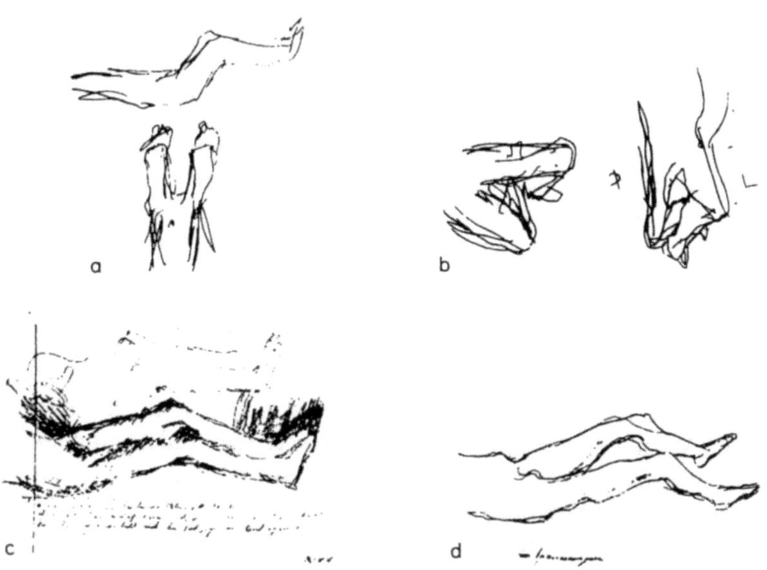

Fig. 1.1 a–d. Drawing of phantom legs by a sculptor who suffered spinal transection in a fall. **a** and **b** show the perceived change in the position of the phantom while the patient was in hospital. **c** With great concentration the patient was able to perform phantom movements and a decrease in the degree of flexion was achieved 2.5 weeks after the accident which was sometimes associated with a feeling of tension in the thighs **d**. (From Berger and Gerstenbrand 1981, p. 68)

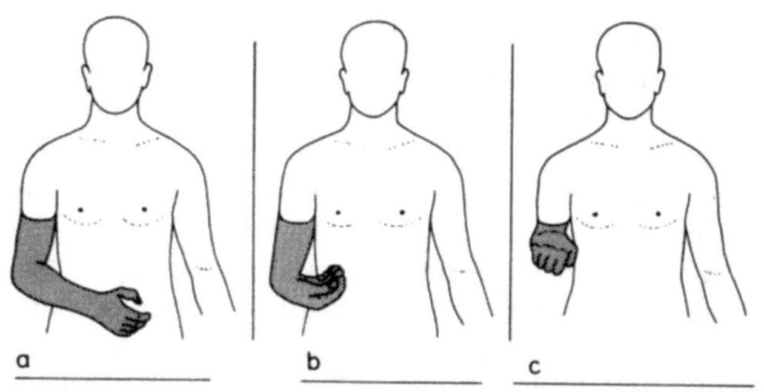

Fig. 1.2 a–c. The phenomenon of telescoping of a phantom limb in a 34-year-old man who suffered severe nerve damage that totally paralysed the right upper extremity presumably totally denervating the right arm below the level of the humerus (see *hatching*). The injuries were received when the man drove his motor bike into a lamp post. After the accident his paralysed arm felt as though it was still holding the handlebars of the motor bike in the position shown in **a**. After a few weeks the patient developed pains referred to his right hand and fingers and distal one-third of the forearm. The pains were squeezing and of increasing intensity during the following 3 months. At the same time the phantom began to shrink and 6 months after the accident the hand felt as though it were localized at the elbow joint **b**. Amputation about 8 months after the accident had no effect on the pain and shrinking continued until the hand felt as if it were placed directly on the amputation stump and clenched more tightly **c**. (From Jensen and Rasmussen, p. 404)

pain, these retraining experiments challenge us to consider the mechanism by which neural systems associated with reward, attention and cognition might mould connections within the somatosensory system.

An enormous number of papers contain information relevant to plasticity in the somatosensory system. Therefore, in reviewing what is known about this area, it has been necessary for us to establish some guidelines. We have found this to be a difficult task, for, as will be apparent from the preceding paragraphs, plasticity is not a subject that is easily confined to a particular part of the nervous system or to a small set of nervous phenomena. Nevertheless, such confinement has been necessary. Consequently we have regarded the somatosensory system as comprising those ascending pathways that deliver information to the somatosensory cortex. Thus we believe that we have covered the systems that convey information from the receptor to a site where neural activity can give rise to stimulus awareness (Libet 1965). We have given minimal attention to systems outside this somatosensory axis, selecting only those that we could implicate in some way in the interpretation of experiments on plasticity (see Sect. 8.6).

We have chosen to deal with the somatosensory system in a hierarchical manner. This has the disadvantage of de-emphasizing mechanisms and general principles, but the advantage of enabling a critical assessment and comparison of a variety of experiments conducted on the same tissue. In addition, we hope that it has facilitated a more objective emergence of those principles that are applicable to several, or all, levels of the neuraxis. Dealing with the cerebral cortex (see Chap. 7) posed particular problems in relation to interspecies differences that are less acute at lower levels. Consequently we have subdivided this section in relation to the species concerned. Finally, we have placed in a separate chapter all material dealing with the vibrissae system of rodents. To offset the difficulties arising from this system of organization we have sought in Chap. 8 to emphasize the mechanisms and principles relating to plasticity in the somatosensory system and to fashion our conclusions into questions and proposals that may help to direct future research.

Many experiments on plasticity in the somatosensory system have been conducted over the last decade – a period which has also seen an exponential growth of our knowledge of the basic anatomy and physiology of this system. In order to assess experimental data relating to plasticity it soon became clear to us that we would have to cite many data recently derived from experiments on normal animals. We therefore resolved to begin each chapter with a brief review of what is currently known about the relevant part of the somatosensory system in normal animals. We hope the reader will find this information useful in the subsequent discussion of experiments designed to show plasticity or elucidate the underlying mechanisms.

In selecting relevant information we have focused our attention on experimentally induced plasticity. We have dealt broadly with this matter, entertaining ideas that relate to pharmacological and anatomical, as well as physiological changes. The literature on such a fascinating subject tends to contain many anecdotal reports which reduce very quickly to phenomenology. In these cases we have tried to extract relevancy on the basis of what the report might tell us about underlying mechanisms and particularly those mechanisms that have been clearly demonstrated in experiments on intact animals. Finally, we have sought to bring the conclusions of experimental studies to bear on clinical observations such as phantom sensations and sensory retraining after nerve section, an endeavour which has drawn us irresistably towards the cerebral cortex.

Strategically our approach has been to critically evaluate the specific and general conclusions reached in published research papers on plasticity. We have purposely avoided the rote citation of the opinions of the authors. For the same reason we have tried to avoid copious citations from the many smaller and excellent reviews on the subject of plasticity. Writing this review has been an enlightening and thus enjoyable process, a reflection on the imagination and skills of all those cited within.

2 Plasticity in the Peripheral Somatosensory Nervous System

2.1 Aspects of Plasticity in the Peripheral Nervous System

Nerve injury or disease in man is commonly associated with a variety of somatosensory perceptual abnormalities, including mislocalization of tactile stimuli (Hawkins 1948; Sunderland 1978), reduction of tactile acuity, paresthesiae, hypo- and hyperesthesia, hyperalgesia and often chronic intractable pain (Sunderland 1978). The clinical consequences of nerve injury and their possible mechanisms have been reviewed many times and will not be discussed in detail in this review. However, lesions of primary afferent fibres brought about either by crush or transection of peripheral nerves or dorsal roots or by application of neurotoxins have been used extensively to investigate the influence of changes of input on connectivity in somatosensory pathways. The peripheral sensory apparatus (consisting of receptor sheet and peripheral nerves) reacts in various ways to such interventions, and it is therefore important, in interpreting changes of function in different central regions of the somatosensory system, to bear in mind the changes occurring in the periphery. The most important peripheral changes from the point of view of long-term central plasticity are (1) degeneration and other responses of axotomized afferents (see Sect. 2.1), (2) peripheral collateral sprouting (see Sect. 2.2) and (3) peripheral nerve regeneration (see Sect. 2.3).

2.2 Survival and Loss of Sensory Neurons After Lesions of the Peripheral Nervous System

It has been known for many years that injury to peripheral nerves is rapidly followed by Wallerian degeneration of the distal terminals of the injured axons and sensory loss in the affected area of skin (see, e.g. Ramon y Cajal 1928). However, a variable degree of functional recovery later occurs and this is thought to be accomplished in two stages: first, collateral sprouting by intact fibres innervating skin adjacent to the denervated region, and second, regenerative outgrowth of the fibres injured by the nerve lesion. In addition, some sensory neurons do not regenerate and a variable number may even die following axotomy.

The degree and time-course of any loss of sensory cells after peripheral nerve injury will obviously have an impact on the final outcome of nerve regeneration and any central reorganization of afferent connections occurring as a result of the lesion. Thus, for example, if cell death were to occur rapidly and extensively after a peripheral lesion there would be far fewer sensory neurons available for regeneration, and furthermore the types of surviving neurons might be unrepresentative of the original population. Accordingly, in Sect. 2.2.1 we discuss the effects of crush or transection of peripheral nerves on the survival of neurons in sensory ganglia of both adult and neonatal mammals.

2.2.1 Effect of Crush or Transection of Peripheral Nerve on Neurons of Sensory Ganglia

Cajal's classical histopathological studies of degeneration and regeneration in the nervous system led him to the following conclusion concerning the effect of peripheral nerve lesions on sensory neurons of the dorsal root ganglion (DRG). "For degeneration of the tubes and sensory cells to take place it is not necessary to operate on the ganglion itself; it is sufficient for the traumatism to act on the nerves at a distance, subject to the condition that the mechanical commotion is propagated, like an oscillatory movement, to the neurone of origin. In this way the extirpation of the sciatic nerve at some distance from the ganglia brings about the mortification of some cells, as well as the primary retrograde degeneration of their expansions" (Ramon y Cajal 1928). From a perusal of the many studies of sensory cell loss after peripheral nerve section it would appear that, when applied to neonatal animals, Cajal's view was substantially correct, but that he probably overstated the case for the extent of cell death occurring in the adult.

Sensory neuron loss in adults. It has been known for over 100 years that axotomized neurons undergo retrograde changes and that some may actually degenerate (Lieberman 1971). Quantitative studies of sensory cell death following nerve section began at the turn of the century (Ranson 1906), and the results of a comprehensive, though not exhaustive, range of these are shown in Table 2.2. The most obvious feature of these results is the enormous variability from system to system. Proximity of the lesion to the cell bodies in the sensory ganglion has long been thought to be an important factor enhancing the rate and extent of retrograde neuronal changes following axotomy (Lieberman 1971). Accordingly, the first quantitative report of retrograde sensory cell death in the adult rat DRG after nerve section showed that about 50% of neurons in the C2 DRG disappear by 120 days after section of the dorsal ramus of the spinal nerve (Ranson 1906). A similar magnitude of cell loss was observed by Cavanaugh (1951) in rat thoracic DRGs. In both studies the nerves were cut close to the ganglia in order to axotomize as many of the sensory neurons as possible. A more modest loss of 30%–40% has recently been reported for rat thoracic DRGs following intercostal nerve section a little further (20 mm) from the ganglion (Ygge and Aldskogius 1984).

Of those studies in which the nerve lesion has been made more peripherally, nerve cell loss has tended to be much less dramatic, around 10%–15% in cat L7 after sciatic nerve transection (Tessler 1985) and in rat trigeminal ganglion after infraorbital nerve transection (Aldskogius and Arvidsson 1978). On the other hand, a greater loss of about 34% has been reported in the cat L7 dorsal root ganglion 90 days after sciatic nerve transection (Risling 1983b). The detailed numerical results in these studies may be subject to great variability owing to at least two factors: (1) a lack of normal left/right symmetry (in most cases the contralateral DRG was used as a

Table 2.1 a. Neuron loss in sensory ganglia after nerve lesions in the rat

Ref.	Lesioned nerve	Ganglion	Age at operation	Survival time	Mean neuron loss (%)	Comments
1	DR	C2	12 d	2 mo	52 ($n=5$)[b]	Cut nerve ends remained approximated. Controls from 4 normal animals
				4 mo	49 ($n=2$)[a,b]	
			140 d	4 mo	50 ($n=2$)[a,b]	
2		T7 to T10	adult	9 d	6 ($n=1$)	Three main branches cut and central ends capped 15 mm from DRG Counts are for 3 DRGs
				20 d	21 ($n=1$)	
				49 d	43 ($n=1$)	
				50 d	58 ($n=1$)	
				175 d	65 ($n=1$)	
				291 d	52 ($n=1$)	
				292 d	12 ($n=1$)	
3	IO	trigeminal	adult	60–70 d	14 ($n=7$)	Nerve sectioned and 2 mm of distal stump resected
4	S	L4	0 d	84 d	60 ($n=10$)[a,b]	Nerve crushed in upper thigh less than 16 h after birth
		L5	0 d	84 d	62 ($n=10$)[a,b]	
		L6	0 d	84 d	59 ($n=8$)[a,b]	
5	IC	T5 to T11	adult	20–30 d	27 ($n=4$)	Nerve transected 20 mm distal to DRG and a few mm resected. Counts are for 3 consecutive DRGs.
				40 d	33 ($n=2$)	
				70 d	42 ($n=7$)	
6	S	L4 plus L5	adult	Loss of axotomized neurons calculated to be 8% per 100 d ($n=23$)		Nerve ligated and 5 mm of distal stump resected
			0 d	Loss of axotomized neurons almost total at 232–461 d ($n=2$)		
			9–10 d	Like adult ($n=2$)		

Table 2.1 b. Neuron loss in sensory ganglia after nerve lesions in the cat

Ref.	Lesioned nerve	Ganglion	Age at operation	Survival time	Mean neuron loss (%)	Comments
7	S	L7	1 week	80 d	46 ($n=3$)[a]	Nerve segment resected
				200 d	42 ($n=2$)[a]	
8	S	L7	1 week	80–90 d	53 ($n=4$)	15 mm of distal stump resected. After 80–90 d neuroma resected and animals survived 6–14 d more
9	S	L7	8 d	48 d	16 ($n=3$)[a]	Nerve crushed just distal to hamstring branch
				220 d	22 ($n=3$)[a]	
10	S	L7	adult	35 d	14 ($n=2$)[a]	Nerve resected from 10 mm below hamstring branch to popliteal fossa
				90 d	34 ($n=5$)[a]	
				190 d	10 ($n=3$)[a]	
11	S	L7	adult	30–40 d	0 ($n=3$)	Proximal stump ligated, 10 mm removed distally. Lesion 15 cm from DRG
				60 d	16 ($n=2$)[a]	
				90–109 d	5 ($n=7$)[a]	

References: 1, Ranson (1906); 2, Cavanaugh (1951); 3, Aldskogius and Arvidsson (1978); 4, Bondok and Sansone (1984); 5, Ygge and Aldskogius (1984); 6, Devor et al. (1985); 7, Aldskogius and Risling (1981); 8, Aldskogius and Risling (1983); 9, Risling et al. (1983a); 10, Risling et al. (1983b); 11, Tessler et al. (1985).

Abbreviations: h, hours; d, days; mo, months; C, cervical; T, thoracic; L, Lumbar; IO, Infraorbital; IC, Intercostals; DR, Dorsal ramus.

[a]Means calculated from authors' original data.

[b]Data uncorrected for split nucleoli.

control, and left/right asymmetry in counts of DRG neurons has been estimated to be as high as 50% for individual pairs of thoracic DRGs in the rat (Ygge et al. 1981)) and (2) large variations between individuals in the proportion of neurons in the DRG having an axon in the transected nerve. The effect of this variability is clearly evident in the study of Risling et al. (1983b), where one cat examined 190 days after sciatic nerve transection had a greater number of sensory neurons in the L7 DRG on the operated side than on the unoperated side. As only 3 cats were examined at this time interval, the mean loss of sensory neurons at this time was reduced to only 10% (Table 2.1b).

In attempting to overcome the problem of asymmetry and variability Devor et al. (1985) evaluated the loss of peripherally axotomized sensory neurons in the rat by first estimating the proportion of neurons in the L4 and L5 DRGs combined which project an axon into the sciatic nerve. They concluded that only about 8% of axotomized neurons are lost for every 100 days which elapse postoperatively. This study also pointed out that the number of neuronal perikarya in the DRG may actually *increase* after birth. Thus in the combined L4 and L5 ganglia of the rat there appears to be an increase of about 18 neurons per day throughout life, resulting in a final count about 150% higher in aged than in young rats. According to Devor et al. (1985) the net neuron loss following axotomy could be explained simply as a result of a cessation of this constant slow accretion of new neurons to the "sciatic compartment" of the ganglia, rather than an actual loss of already existing neurons. Death of the newly formed neurons would be ensured by their inability to obtain trophic support from the periphery (see Sect. 2.2.2). This, of course, implies that the newly formed neurons normally do send a process out to the periphery, but there is no available information on this point. It is known that the number of myelinated fibres is stable in rat peripheral nerves (Birren and Wall 1956), but in Devor's experiments the newly formed neurons apparently belong to the small cell population of the ganglia (Devor et al. 1985) and would therefore be more likely to have unmyelinated axons (Harper and Lawson 1985). Clearly, further studies are required to reconcile the findings of Devor et al. (1985) with those of other groups because the question of whether existing sensory neurons die and degenerate, even at a slow rate, following peripheral axotomy, may have important consequences in relation to the mechanisms underlying reorganization of somatotopic maps, particularly in the spinal dorsal horn (see Sect. 4.5.5).

Effect of Amputation of a Limb. Since many more neurons are axotomized, and there is a much reduced chance of fibres regenerating back to the skin, one might reasonably expect the effect of amputation of the hindlimb on sensory neuron loss to be more severe than the effect of sciatic nerve transection alone. This appears to hold true for the adult cat, where, following hindlimb amputation, there was a loss of about 23%–25% of neurons from the L7 DRG by 3 months (Table 2.1) (Carlson et al. 1979). However, there may be large species differences in the effect of limb amputation on sensory cell loss. High amputation of the hindlimb in the rat was found to cause only minor loss of sensory neurons (Feringa et al. 1985). In contrast, electrophysiological recordings in man have indicated that peripheral nerves in the amputation stumps of long-term forearm amputees may have drastically reduced numbers of conducting axons (McComas et al. 1978). Presumably this indicates that the numbers of DRG neuron are similarly depleted. Since the limbs of these human subjects had been amputated at least 2 years earlier, there might be continuing gradual decline in the number of surviving axotomized sensory neurons which would not be apparent at the much shorter survival times generally used in animal experiments.

Sensory Neuron Loss After Nerve Crush or Transection in Neonates. If peripheral nerve transection is performed in immature animals the resulting loss of sensory neurons is considerably greater, ranging from 42%–53% in the cat L7 DRG after sciatic nerve transection at 1 week of age (Aldskogius and Risling 1981, 1983) to an estimated complete elimination of axotomized sensory neurons from the L4 and L5 DRGs of the rat after sciatic nerve transection on postnatal day 0 (P 0) (Devor et al. 1985). In agreement with the latter result, Bondok and Sansone (1984) noted the elimination of about 60% of neurons from the L4, L5 and L6 DRGs 12 weeks after crushing the sciatic nerve on P 0 (Table 2.1a). According to the calculations of Devor et al. (1985), this would represent the loss of the whole sciatic compartment of these ganglia in the rat. A substantial loss of neurons, amounting to 36.5% (n = 3), has also been noted in the opthalmic-maxillary part of the trigeminal ganglion of the mouse 60 days after coagulation of all mystacial vibrissae follicles on P 0 (Savy et al. 1981). Unfortunately it is not possible to determine what proportion of axotomized neurons die in this situation. However, the simultaneous loss of about 60% of myelinated axons from the maxillary nerve (Waite and Cragg 1979), only part of which innervates the whisker pad (see Sect. 3.3.1), would tend to indicate again that perhaps all the injured neurons die. This early postnatal period of sensitivity to nerve transection appears to terminate by P 9–10 in the rat (Devor et al. 1985). In the week-old kitten, crushing the sciatic nerve causes the loss of fewer neurons than does transection (Risling et al. 1983a). Although the bounds of any sensitive period have not been determined for the kitten, the effect on sensory cell death of sciatic nerve transection at 1 week of age is substantially greater than it is in the adult cat (Table 2.1b). The effect of nerve injury on sensory neurons at earlier postnatal times has not been studied in this species.

Retrograde Labelling of DRG Cells After Peripheral Nerve Lesions. In addition to those studies where the perikarya of all surviving sensory cells have been counted (Table 2.1), loss of sensory neurons has also been observed by workers who have evaluated surviving cell numbers from retrograde labelling with HRP from the damaged nerve. For example, 150 days after cutting the superficial peroneal nerve in the adult cat only about 50% of the normal number of DRG cells may be labelled following application of HRP to the nerve (Jänig and McLachlan 1984). However, the extent of neuron loss may have been overestimated in these studies because retrograde axonal transport appears to be severely depressed in chronically axotomized neurons (Goshgarian et al. 1983; Peyronnard et al. 1986). The unreliability of the retrograde HRP technique for estimating numbers of DRG neurons is illustrated by the fact that whereas Yip et al. (1984) found 82% (n = 3) of L5 sensory neurons to be labelled from the sciatic nerve in the rat, Devor et al. (1985) reported that only 62.7% (n = 9) were so labelled. Despite these limitations the technique has been used ostensibly to show that, in the rat, 45% of L5 DRG neurons are lost only 1 day after crushing the sciatic nerve on P 0, at a time when no cell death is occurring in the normal ganglion. Following this there is no further loss of neurons from the ganglion up to 2 weeks postnatally (Yip et al. 1984). According to the estimate of Devor et al. (1985) of the size of the sciatic compartment of the L5 DRG, this degree of sensory cell loss would actually represent the death of all or nearly all those neurons injured by the nerve crush. Finally it is interesting to note that the rapid degeneration and death of DRG cells after axotomy in the neonatal rat appears to occur without an intervening stage of chromatolysis (Hess 1956; La Velle and La Velle 1958; Savy et al. 1981; Bondok and Sansone 1984).

Spontaneous Activity in Axotomized Afferent Fibres. It is argued elsewhere (Sect. 8.5.1) that spontaneous activity in sensory neurons is important in the stabilization of central representations of the body. Thus it is important to note that following peripheral nerve transection, some surviving DRG cells may become chronically active for a considerable time. Wall and Gutnick (1974a,b) first noted an unusually large spontaneous discharge in L4 and L5 dorsal rootlets of rats with a chronic sciatic nerve neuroma. The discharge was increased by pressure on the neuroma, and decreased by local anaesthesia of this structure, indicating that the origin of much of the spontaneous discharge was in the neuroma itself. Recordings by Govrin-Lippmann and Devor (1978) in rats, and Scadding (1981) in mice, from the nerve proximal to the neuroma, found that the spontaneous discharge in myelinated fibres was a transient phenomenon, beginning after about 2 days and reaching a peak at 2 weeks after nerve transection. Following this, the discharge fell to a low, but maintained, level. The discharge was enhanced by adrenergic compounds (Korenman and Devor 1981; Wall and Gutnick 1974b) and by sympathetic stimulation (Devor and Janig 1981), and was blocked by alpha antagonists or, after catecholamine depletion by guanethidine (Korenman and Devor 1981; Wall and Gutnick 1974b). Spontaneous firing occurs in both large and small afferent fibres (Devor and Janig 1981; Devor and Raber 1983; Devor and Wall 1976a; Govrin-Lippmann and Devor 1978; Korenman and Devor 1981; Wall and Gutnick 1974b). However, activity in C fibres is rare until about 20–30 days after nerve section, after which it is claimed to show a considerable increase (Devor and Raber 1983). In contrast to these findings in rats and mice, it has been reported that injured nerves in the cat produce very low levels of spontaneous discharge (Blumberg and Janig 1981). After nerve injury, additional ectopic nerve impulse generation may occur in DRG cells of rats, cats and rabbits (Kirk 1974; Howe et al. 1977; De Santis and Duckworth 1982; Wall and Devor 1984) but relatively little is known about this phenomenon.

More recently, Burchiel and Russell (1987) found that a purely sensory nerve neuroma rarely produced spontaneous impulses in myelinated fibres and never in unmyelinated fibres. These workers ascribed earlier reports of spontaneous discharge in neuromas to the use of gallamine, a potent K^+ channel blocking agent, as a paralysing agent. Furthermore, they found that C fibre activity, presumed to originate within the neuroma, may have been from intact surrounding receptive fields. It is not clear, however, how these intact C fibres come to be in the nerve proximal to the transection neuroma. Clearly, further basic investigations of spontaneous discharge by axotomized afferents are required before any convincing role can be ascribed to the phenomenon in the changes of central receptive fields that follow peripheral nerve lesions.

2.2.2 Trophic Dependence of Immature Sensory Neurons on the Periphery

The reason for the rapid and profound loss of sensory neurons after nerve lesions in the neonatal rat is undoubtedly related to the cessation of a supply of trophic material from the periphery at a critical time during early postnatal development. One likely candidate for a trophic substance is NGF, the trophic effect of which appears to be particularly important for neurons that are still actively growing and differentiating (Thoenen and Barde 1980). As well as enhancing the survival *in vitro* of DRG neurons taken from embryonic chicks (Levi-Montalcini and Angeletti 1963), embryonic rodents (Crain and Peterson 1974), human embryos (Crain et al. 1980) and newborn rodents (Varon et al. 1973), NGF has also been shown to be essential for the normal embryonic development of mammalian sensory neurons *in vivo*. This has been demonstrated by raising antibodies to NGF in adult rats and guinea-pigs.

In the rat, small amounts of maternal antibodies are transferred to offspring prenatally, and large amounts are transferred postnatally in milk, whereas in the guinea-pig large amounts of antibodies are transferred to the offspring prenatally. Antibodies to NGF, transferred to the offspring *in utero*, cause profound destruction of DRG neurons. In the rat the number of neurons in the C8 DRG was decreased by 70% in animals exposed to maternal NGF antibodies prenatally, but was normal in animals exposed only postnatally by fostering with immunized mothers. In the guinea-pig, where the prenatal transfer of maternal antibodies is much greater, the C8 DRG of offspring of immunized mothers suffered an 80%–85% loss of sensory neurons (Gorin and Johnson 1979; Johnson et al. 1980). Direct exposure of rat fetuses to NGF antibodies by subcutaneous injection also caused severe damage to sensory neurons, including hypotrophy and degeneration, and the affected animals lacked normal responsiveness to noxious stimuli. In contrast, similar injections given postnatally did not cause any morphological or behavioural changes. (Aloe et al. 1981). However, larger doses of NGF antibodies given just postnatally may cause a loss of up to 20% of L5 sensory cells in the rat, and administration of NGF may at least temporarily prevent the neuron loss following neonatal sciatic nerve crush in this species (Yip et al. 1984). Thus, in the rat at least, a residual of trophic dependence of NGF is still present in somatosensory neurons in the early postnatal period. A recent clear demonstration of the continuing dependence on NGF of sensory neurons in adult mammals comes from the work of Diamond et al. (1987), who showed that NGF is necessary for the collateral sprouting of nociceptive afferents into denervated skin of the adult rat (Sect. 2.3.6). By controlling sprouting, the peripheral supply of NGF may thus be responsible for an ongoing regulation of the peripheral receptive fields of nociceptive afferents throughout life.

Finally, the effect of NGF on the survival of axotomized sensory neurons in the adult rat has been investigated directly by Rich et al. (1987). In these experiments it was found that cell death in the L4 and L5 DRGs combined, was reduced to zero, from control levels of 17% and 13% at 3 and 6 weeks, respectively, after nerve injury, by the chronic application of NGF to the proximal cut end of the sciatic nerve. Owing to atrophy of surviving cell bodies in the control animals the authors were unable to determine whether neuron loss preferentially involved small or large neurons (see also Sect. 2.2.1). However, the degree of atrophy of surviving neurons was also significantly reduced by NGF application. Unfortunately, the authors did not comment on the large difference between the levels of cell death in their control animals (calculated to be 22% and 16% of *axotomized* neurons at 3 and 6 weeks, respectively), and the results of Devor et al. (1985) (calculated to be only 8% of *axotomized* neurons per 100 days postoperatively). This difference would indicate that cell death might be precipitated simply by the application of a chamber to the cut nerve. Moreover, since Devor et al. (1985) were of the opinion that the loss of sensory cells after sciatic nerve section may have represented not true cell-loss, but rather a *deficit of newly generated neurons* unable to make contact with the periphery (see Sect. 2.2.1), one might further suppose that the provision of NGF in the experiment of Rich et al. (1987) would merely have allowed any newly generated neurons to survive. These considerations clearly weaken the case for dependence on NGF for survival of *mature* sensory neurons.

2.2.3 Effect of Peripheral Nerve Transection on Different Types of Sensory Neurons in Dorsal Root Ganglia

When the diameter of the surviving DRG neurons was measured after chronic peripheral nerve section in the adult rat it first appeared that there was a preferential loss of small cells (Ranson 1909; Cavanaugh 1951). This view was later stated for sciatic nerve transection in the cat in a brief report by Humbertson (1963). Since it is now known that the population of small neurons in the DRG is mainly related to the unmyelinated primary afferent fibres (Harper and Lawson 1985), these observations imply that unmyelinated afferents are more sensitive to axotomy than myelinated fibres. However, the results of more recent studies of transection of the sciatic nerve in the cat (Aldskogius and Risling 1981; Risling et al. 1983b), and rat (Devor et al. 1985) or of section of intercostal nerves in the rat (Ygge and Aldskogius 1984) are not in agreement with these earlier findings. Bondok and Sansone (1984) have even claimed cell loss to be greater among the larger cells (diameter $> 40\,\mu m$) than the small cells (diameter $< 25\,\mu m$) in the rat L4 to L6 DRGs after neonatal sciatic nerve crush. However, selective loss of either large or small neurons after neonatal nerve lesions in the rat is unlikely if it is true that almost all axotomized sensory ganglion cells die (see Sect. 2.2.1). After sciatic nerve transection in 1-week-old kitten, Aldskogius and Risling (1981) noted a consistent shift in cell size histograms towards smaller cell sizes in all L7 DRGs (see Table 2.1) they examined. In addition, the overall size range of the neurons was slightly reduced. Although the proportion of large neurons was markedly reduced, the proportion of medium-sized cells was actually increased, while the *relative* proportion of small cells remained unchanged, suggesting that the observed shift in cell-size histograms might be largely due to atrophy of the surviving large cells.

The reasons for the above disparities are not clear, but are probably related to size changes of axotomized neurons. These may initially undergo swelling as part of the chromatolytic response but, if functional reinnervation does not occur, they may later tend to atrophy (Lieberman 1974). Such changes of cell size will obviously tend to complicate any attempt to determine a possible size-related preferential loss of neurons from measurements of cell size alone.

In contrast to the apparently non-selective effect of axotomy, the loss of sensory neurons in rats exposed to maternal NGF antibodies *in utero* primarily involves those cells having diameters $< 30\,\mu m$ (Johnson et al. 1980). One possible reason which has been suggested for the sparing of large neurons is that because these are formed earlier in gestation than the small cells (Lawson et al. 1974) they might be through the stage of NGF dependence before the maternal antibodies reach the fetus (Johnson et al. 1980). However, there is an 18% loss of sensory neurons from the rat DRG following neonatal treatment with large doses of NGF antiserum, and this also appears to be due primarily to a loss of small cells (Yip et al. 1984). This indicates that the small cells might at *all* stages have a greater dependence on NGF than large cells. No doubt this question could be best addressed in cell culture experiments.

2.2.4 Effect of Peripheral Nerve Section on Fibre Composition of Dorsal Roots

The population of axons in dorsal roots has also been studied after peripheral nerve transection, both in adult and immature animals. The results of the major quantitative studies are summarized in Table 2.2. The loss of fibres from the L7 dorsal root following sciatic nerve section in both the adult cat and the week-old kitten is apparently less severe than the loss of neurons from the ganglion (Table 2.1). Since dorsal root afferents outnumber DRG cells by about 2:1 (Coggeshall 1986), one possible explanation is that neurons supporting only one central process are more liable to die than those supporting two or more processes. Centrally directed axons may also be formed by sprouting of uninjured neurons in the DRG similar to that which occurs in a spared dorsal root (Hulsebosch and Coggeshall 1983a) (see Sect. 4.5.5). A third possibility, and one for which there is some direct evidence, is that recurrent unmyelinated sprouts of injured fibres formed at the site of the transection (Ramon y Cajal 1928) (Sect. 2.4) contribute to the axon count in dorsal roots. The relative proportions of unmyelinated and myelinated fibres in the L7 dorsal root are largely normal 2–3 months after sciatic nerve transection in the kitten (Risling et al. 1980), indicating that there is no preferential loss of either type of afferent. However, recurrent sprouts are known to invade the dorsal root after sciatic nerve transection in the kitten (Aldskogius and Risling 1983). If these are caused to degenerate by subsequent removal of the neuroma, the loss of unmyelinated fibres in dorsal roots in then seen to be more than twice as severe as

Table 2.2. Effect of peripheral nerve transection on axon numbers in dorsal roots

Ref.	Nerve lesioned (species)	Ganglion	Age at operation	Survival time	Loss of dorsal root axons (%)		
					Myelinated	Unmyelinated	All
1	DR (rat)	C2	12 d	2 mo	17	—	—($n=4$)
				4 mo	15	—	—($n=3$)
2	S (cat)	L7	6–9 d	8 d	8[b]	16[b]	14[b]($n=2$)
				12 d	9[b]	10[b]	–4[b]($n=2$)
				24 d	21[b]	23[b]	23[b]($n=2$)
				49 d	30[b]	36[b]	34[b]($n=2$)
				80 d	22[b]	35[b]	31[b]($n=9$)
				200 d	28[b]	32[b]	31[b]($n=2$)
3	S (cat)	L7	1 week	80–90 d[a]	20[b]	45[b]	33[b]($n=5$)
4	S (cat)	L7	8 d	14 d	20[b]	28[b]	26[b]($n=3$)
				48 d	13[b]	6[b]	10[b]($n=3$)
				220 d	10[b]	4[b]	7[b]($n=3$)
5	S (cat)	L7	adult	35 d	4[b]	7[b]	6[b]($n=2$)
				90 d	6[b]	5[b]	6[b]($n=4$)
				190 d	–3[b]	5[b]	2[b]($n=2$)

With the exception of Ref. 1 all studies used contralateral ganglia as controls.
Abbreviations: C, Cervical; DR Dorsal ramus; L, Lumbar; S Sciatic.
[a] In this study the neuroma was resected after the stated time and the animals survived a further 6–14 d to allow degeneration of any recurrent axon sprouts.
[b] Means calculated from the authors' original data.
References: 1, Ranson (1906); 2, Risling et al. (1980); 3, Aldskogius and Risling (1983); 4, Risling et al. (1983a); 5, Risling et al. (1983b).

the loss of myelinated fibres (Aldskogius and Risling 1983). Nevertheless, the total loss of dorsal root fibres, at around 33% following removal of recurrent sprouts, is still substantially lower than the 53% loss of neurons from the DRG, indicating that some other factor, such as mentioned above, must also be involved.

2.2.5 Fate of the Lost Neurons

The fate of the sensory neurons which disappear after nerve transection is not well understood. Degeneration-like argyrophilia (Arvidsson and Grant 1979) and electron-dense myelinated axons (Arvidsson 1979) have been observed in sections of the spinal trigeminal root and tract following infraorbital nerve transection, and an electron-dense type of degeneration has also been observed in fasciculus gracilis (FG) after sciatic nerve transection (Knyihar and Csillik 1976). Argyrophilic degeneration has also been found in the spinal dorsal horn and DCN of the adult rat after sciatic nerve transection (Moradian and Rustioni 1977). However, Devor and Wall (1981b), using the Fink-Heimer technique, could not find any trace of degeneration in the lumbosacral dorsal horn 13–105 days after saphenous and sciatic nerve transection in adult cats, and Devor et al. (1985) were unable to find more than an occasional trace of degeneration argyrophilia in the L4–L5 spinal cord of rats in which the sciatic nerve had been transected at various ages, and for various times (Table 2.1). Only very moderate argyrophilia is present in the trigeminal nuclei after trigeminal nerve injury in kittens (Grant and Arvidsson 1975). These observations support the evidence from cell and fibre counts showing that degeneration of primary afferent myelinated fibres occurs after peripheral nerve injury, but that the retrograde cell death occurs gradually. Moreover, the degeneration must be accompanied by efficient and rapid clearing of degeneration products. Degeneration may be more extensive following damage to some nerves than to others. For example, it has been observed in the cat sensory trigeminal nuclei after tooth extractions or pulpectomies (Johnson and Westrum 1980; Westrum et al. 1976) and after the natural exfoliation of the deciduous teeth (Westrum and Canfield 1979).

A curious feature of transganglionic degeneration argyrophilia is its non-uniform laminar distribution. In the spinal dorsal horn of the rat 13–53 days after intercostal nerve transection, argyrophilic material was present in laminae I, III and IV, but absent from lamina II (Grant and Ygge 1981). This is in contrast to the degeneration argyrophilia which occurs following dorsal rhizotomy and classical Wallerian degeneration, which is found throughout laminae I–III of the dostal horn in the rat at relatively short times (1–4 days) after dorsal root transection (Heimer and Wall 1968). This difference of distribution presumably reflects the termination of different classes of primary afferent fibres having varying degeneration rates following peripheral axotomy (Sect. 4.2.2). The characteristic distribution, as well as the paucity of degeneration argyrophilia after peripheral nerve injury might be due also to a more gradual loss of trophic support from the cell body for central processes than that which occurs after dorsal rhizotomy.

For a relatively long time ($>$ 12 weeks) after sciatic nerve transection in the adult rat very few degenerating myelinated fibres can be found with the Marchi technique in the L5 dorsal root. After this period many scattered groups of small Marchi-positive structures (MPS) are found. Central to the root entry zone the incidence of MPS is clearly increased by 2 weeks, and there is no further increase thereafter. Anatomically, the borderline between these two types of reaction corresponds with the change from peripheral to central type myelination as afferent fibres course centrally (Aldskogius et al. 1985). Since the loss of myelinated dorsal root axons and sensory neurons is only slight at this time (see Sects. 2.2.3 and 2.2.4), the morphological basis for the pattern of MPS in the peripheral part of the root is unknown. However, it has been suggested that it could be related to the late degeneration of the newly formed axons, some of which are thinly myelinated, that appear in dorsal roots as a result of the recurrent sprouting following peripheral nerve transection (Ramon y Cajal 1928; Aldskogius and Risling 1983) (see Sect. 2.2.4). Since the Marchi method does not distinguish between primary demyelination (i.e. demyelination without axonal degeneration) and secondary demyelination (i.e. demyelination secondary to axonal degeneration) it is also possible that some of the MPS in both central and peripheral portions of the dorsal root are due to primary demyelination of atrophied neurons, rather than degeneration of dying cells.

2.2.6 Sensory Cell Loss After Chemical Lesions of Afferent Fibres

In addition to peripheral nerve section or crush, various chemical agents have been used either to destroy all sensory neurons in a particular peripheral nerve, or to selectively damage or destroy a particular type or types of primary afferent fibre, while leaving others relatively unaffected. To date, three substances, ricin, capsaicin and pronase, have been utilized in this way, and their effects on sensory neurons are briefly described below.

Ricin. The cytotoxic plant lectin ricin appears to be a particularly useful tool for those interested in plasticity of somatosensory pathways, because it may be used to selectively destroy all the sensory neurons whose axons run in a particular peripheral nerve, without affecting those whose axons run in other nerves. Thus, unlike nerve transection, which destroys relatively few sensory cells in the short term in adult mammals (see Sect. 2.2.1) and is beset with problems of persistent regeneration (see Sect. 2.4) the use of ricin should allow the complete and irreversible elimination of sensory input to contiguous regions of somatosensory maps.
Ricin is a highly toxic plant lectin with a molecular weight of 60 000, obtained from castor beans. Unfortunately, the extreme toxicity of the lectin, which has an intravenous LD_{50} of 2.5 μg/kg in mice (Fodstad et al. 1979), demands that extreme care be taken in its use as a selective neurotoxin (Wiley and Oeltmann 1986). To destroy axons in peripheral nerves, the lectin has either been applied to the proximal cut ends of the nerves, or microinjected into the intact nerve. The latter method has been reported to reduce the incidence of systemic intoxication, but in order to destroy all neurons separate microinjections must be placed in each fascicle of the nerve (Wiley and Oeltmann 1986). The lectin is transported retrogradely to the cell body of the neuron, where it acts on the large subunit of ribosomes to inhibit neuronal protein synthesis (MacConnell et al. 1982). Thus, the first sign of ricin action is chromatolysis, or destruction of the Nissl substance, which occurs very rapidly once the lectin reaches the cell body (Wiley 1985). Complete degeneration and disappearance of all motor and sensory neurons and their axons of the sciatic nerve in the rat has been observed to occur within 20 days of microinjection of a minimum of 2 μg of ricin into the nerve (Wiley and Oeltmann 1986). Interestingly, while the lectin appears to be transported by and to destroy all kinds of peripheral neurons, including sensory neurons, motoneurons and autonomic neurons, it is apparently not transported by neurons whose afferent fibres run entirely within the CNS (Wiley et al. 1983).

Capsaicin. The chemical agent most widely used to date in studies of somatosensory plasticity is the neurotoxin capsaicin, the pungent, "hot" active principle of red pepper. Since the properties and peripheral neurotoxic actions of capsaicin have

been reviewed in detail (Fitzgerald 1983), only a brief outline will be given here. In relation to somatosensory plasticity, capsaicin has been used primarily as a tool for the elimination of unmyelinated sensory neurons (see Sect. 4.5.4).
A single subcutaneous injection of 50 mg/kg of capsaicin into 2-day-old rats destroys more than 90% of unmyelinated (C) fibres and about 25% of thinly myelinated (A∂) fibres in the dorsal roots (Nagy et al. 1983). Other workers have confirmed that about 10% of C fibres survive neonatal capsaicin treatment in rats, and that the surviving afferents include significant numbers of nociceptors (Carpenter and Lynn 1983; Lynn 1984; Cervero et al. 1984; Welk et al. 1984). Recent studies have also shown that capsaicin affects the uptake and retrograde transport of NGF by some types of afferent fibres (Miller et al. 1982; Otten et al. 1983; Taylor et al. 1985), the fibres in question originating mainly from the small dark B cell population of the DRG (Harper and Lawson 1985). As noted above, the small DRG cells appear to require NGF up to a late stage of prenatal development, and even immediately postnatally, for their survival (Yip et al. 1984). Thus while there are several theories concerning the mechanism of capsaicin's fairly specific action on C fibres (Fitzgerald 1983), one strong possibility is that capsaicin administered neonatally may act by killing only those sensory neurons which still retain a developmental trophic dependence on NGF (see Sect. 2.2.2).

Pronase. Recently, a novel neurotoxic material, pronase, has been introduced into studies of somatosensory plasticity (LaMotte and Kapadia 1987; LaMotte et al. 1989). Pronase consists of a combination of proteolytic enzymes, which, when injected into a peripheral nerve, produces massive cell death among DRG neurons, together with degeneration of their central terminal fields. Although this material would appear to be less toxic than ricin, it nevertheless requires great care in its application. For example, LaMotte et al. (1989) state that it was necessary to enclose the injected region of a peripheral nerve in an impervious sheath, such as polyethylene tubing, to prevent the escape of pronase into surrounding tissues. Despite this disadvantage, pronase would seem to be a more acceptable and safer neurotoxic tool than ricin for the permanent removal of normal peripheral input to selected regions of somatotopic maps (see Sect. 4.5.5).

2.3 Collateral Sprouting of Primary Afferent Fibres in the Periphery

A second major feature of plasticity in the response of the peripheral somatosensory system to nerve injury is the reinnervation of denervated regions of skin by sprouting of intact afferent fibres. The sprouting of intact cutaneous sensory fibres into nearby denervated regions of skin, like the sprouting of motor nerve axons into denervated muscle, has been termed collateral reinnervation. That cutaneous nerves can sprout new collaterals has been widely accepted since Weddell and his colleagues obtained behavioural and anatomical evidence that at least high-threshold mechanoreceptive nerve fibres could invade and functionally innervate adjacent denervated skin (Weddell et al. 1941; Weddell 1942) and cornea (Zander and Weddell 1951) in adult rabbits. Such sprouting following peripheral nerve section with or without regeneration clearly may affect the outcome of any reorganization of the receptive fields of central neurons which receive primary afferent input from the sprouted fibres. It is therefore necessary to examine the evidence concerning peripheral sprouting in order to determine (1) the extent and time-course of the phenomenon, (2) which kinds of cutaneous afferents are involved in sprouting and (3) how sprouting is affected by the developmental stage of the animal.

2.3.1 Collateral Reinnervation of the Skin in Adult Mammals

There is now an impressive body of evidence showing that collateral sprouting of at least some types of primary afferent fibres occurs in the skin of adult mammals. Weddell and his colleagues, in their original studies, followed the progress of collateralization by testing the response of awake rabbits to pricking of the dorsal skin of the ear at various times after partial denervation (Weddell et al. 1941; Weddell 1942). This region of skin is supplied by two nerve trunks, the greater auricular nerve supplying the lateral part of the ear, and the occipital nerve supplying the medial part, with a small region of overlap. Section of the greater auricular nerve produced a region of skin along the lateral edge of the ear within which pin-prick failed to evoke a characteristic withdrawal response. During the following 2–3 weeks the area of skin which remained sensitive to such noxious stimulation expanded to include areas which had previously been insensitive. These observations have been confirmed (Jackson and Diamond 1983). In the absence of further evidence, it is not possible to know whether the spread of sensitivity is due to collateral reinnervation, or whether it is due to a strengthening within the CNS of normally ineffective afferent projections from pre-existing innervation. However, the border between regions of skin sensitive and insensitive to prick was found always to correspond within 1 mm with the border between skin respectively containing and lacking a demonstrable subepidermal innervation in silver or methylene blue stained sections (Weddell et al. 1941; Weddell 1942). Thus the behavioural observations may be explained by intact nociceptive afferents of the occipital nerve expanding their cutaneous receptive fields by collateral sprouting into the adjacent denervated skin. Similar behavioural signs of the sprouting of intact high-threshold mechanoreceptive afferents in adult mammalian skin have since been obtained for the saphenous innervation of the hindpaw following transection of the sciatic nerve in the rat (Devor et al. 1979; Pomeranz et al. 1984), and for the dorsal cutaneous nerve of the rat following isolation of an innervated island of skin on the back. In this preparation the response in the anaesthetized animal to pinching the skin consists of a bilateral reflex contraction of the cutaneus trunci muscle (Nixon et al. 1984). In humans also, collateral sprouting has been proposed to explain the restoration of cutaneous sensation in the absence of regeneration of an injured nerve (Livingston 1947; Leonard 1973; Inbal et al. 1987).

Extent and Time-Course of Collateral Sprouting. It is generally agreed that the spread of sensitivity into a denervated region of skin does not proceed indefinitely, nor does the whole of the denervated region usually become reinnervated. In the

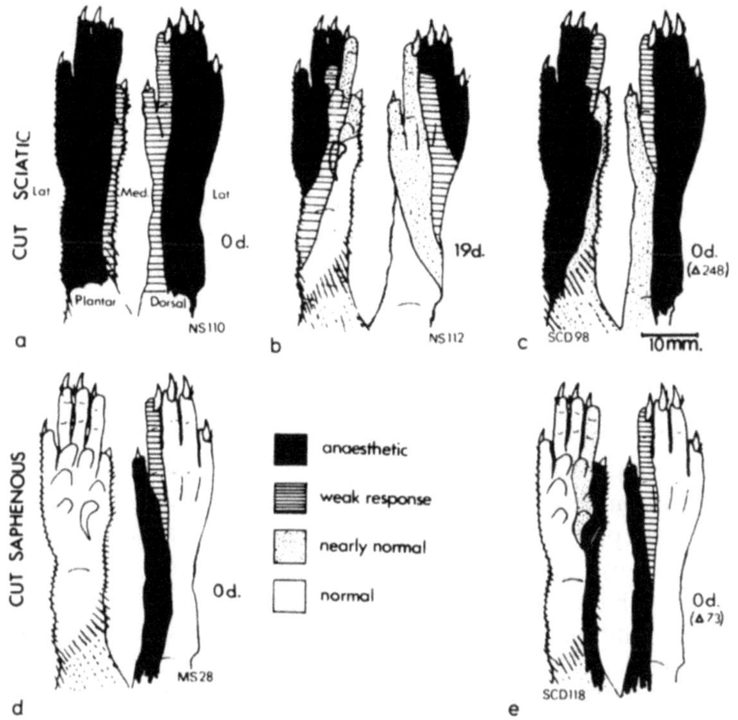

Fig. 2.1 a–e. Collateral reinnervation of denervated skin in the hindpaw of the adult rat. Expansion and contraction of the peripheral distribution of the saphenous nerve after transection of the sciatic nerve. Each pair of figures indicates the strength of flexion reflex evoked by noxious pinching in different regions of the paw. **a** Normal rat immediately after section of the sciatic nerve. A weak response is evoked from a thin strip of skin along the medial border of the paw, innervated by the saphenous nerve. The peripheral field of the sciatic nerve, which occupies the remainder of the paw, is totally anaesthetic. **b** 19 days after sciatic nerve section. The denervated (i.e. non-responsive) region has shrunk due to collateral sprouting of afferents from the saphenous nerve. **c** Immediately after section of a sciatic nerve which had been crushed 248 days earlier and allowed to regenerate. The extent of the innervation by the saphenous nerve is less than its maximum extent in **b**, indicating that regeneration of the crushed sciatic nerve caused the saphenous nerve to withdraw its collateral sprouts. **d** Normal rat immediately after section of the saphenous nerve. **e** Immediately after section of the saphenous nerve in a rat whose sciatic nerve had been crushed 73 days earlier and allowed to regenerate. Expansion of the sciatic nerve field to innervate the saphenous nerve field (not shown) has been reversed by regeneration of the saphenous nerve. (Devor et al. 1979, p. 213)

hindpaw of the rat, for example, transection with ligation of the sciatic nerve leaves only a small medial region of the paw dorsum, innervated by the saphenous nerve, sensitive to noxious pinching (Fig. 2.1a). Within 4 days of the lesion the sensitive region begins to expand, and there is good evidence that the expansion is due to collateral sprouting of saphenous nerve afferents into denervated skin (see below). After 13–30 days the spread of sensitivity comes to a halt (Fig. 2.1b), and even after 200–470 days responsiveness does not return to stimulation of the most lateral three toes (Devor et al. 1979). In the case of the isolated field of the medial branch of the dorsal cutaneous nerve in the back skin of the rat, expansion of the area sensitive to pinch is first detectable at 13–14 days and ceases at about 24 days after the surgery (Nixon et al. 1984). The initial rate of sprouting may be affected by factors such as activity in the afferents (see Sect. 2.2.5), but the reason for the cessation of sprouting in the presence of still-denervated skin in not known. Certainly the collateral sprouts of high-threshold mechanosensory fibres are able to cross dermatomal boundaries in the back skin of the rat with ease (Nixon et al. 1984). Presumably then, sprouting stops because an equilibrium is reached between the neuron's capacity to extend more processes, with the consequent increased metabolic demands this places upon it, and the fixed supply of some hypothetical trophic factor present in the denervated skin which stimulates the cell to sprout (see Sect. 2.3.6). In relation to this hypothesis it is interesting to note that regenerating afferents exhibit a greater ability than intact afferents to sprout into denervated skin of adult mammals. Following sciatic nerve section in the adult rat WGA-HRP injections into the L3 and L4 ganglia labelled thin saphenous nerve afferents slightly more laterally than normal in the glabrous skin of the paw. However, if the saphenous nerve was crushed at the same time as the sciatic transection, regenerated saphenous afferents extended further into sciatic territory (Kinnman and Alskogius 1986). The degree of collateral sprouting under these conditions was, in fact, comparable to that seen after simple sciatic nerve transection in neonates (Sect. 2.3.2). The enhanced sprouting by regenerating saphenous afferents may be explained by the recent findings of Diamond et al. (1987) that the extension of uninjured, but not regenerating, sensory fibres into denervated skin is dependent on NGF (see also Sect. 2.3.6)

Sprouting of Myelinated High-Threshold Mechanoreceptive Afferents. Behavioural evidence for the spread of sensitivity to noxious mechanical stimulation is insufficient in itself to show that collateral sprouting occurs in denervated skin of the adult mammal, as such changes might equally be due to the functional strengthening, within the CNS, of a pre-existing but ineffective innervation. Thus several studies have attempted to obtain direct anatomical and electrophysiological evidence for collateral sprouting, along with information concerning the different types of afferent fibres which sprout. First, Devor et al. (1979) extended their behavioural study of the expansion of the saphenous innervation of the rat hindpaw by recording from single afferent units in the saphenous nerve. Normally only the most medial edge of the plantar skin in the rat is innervated by saphenous afferents, but after sciatic nerve transection the area of sensitivity to noxious stimulation spreads more laterally. Devor et al. (1979) therefore looked particularly for afferent fibres with high-threshold mechanoreceptive fields on plantar skin following (1) acute transection of the sciatic nerve, (2) chronic (11–217 days) transection and ligation of the sciatic nerve (if there was clear behavioural evidence of sprouting) and (3) crush of the sciatic nerve 161 and 163 days earlier. In each case they found saphenous afferents which responded to maintained pressure with a graded, slowly adapting discharge. These fibres, which had conduction velocities in the A∂ range, resembled the A∂ high threshold mechanoreceptive afferents (HTMs) found in normal rats (Lynn and Carpenter 1982), cats (Hunt and McIntyre 1960) and primates (Perl 1968). Although there was no significant difference between the mean sizes of the receptive fields of fibres in their samples, 11 out of 22 of the fibres in the chronically denervated paw had receptive fields which extened further laterally than any of a sample of 30 normal fibres, and these fibres had statistically significantly larger receptive fields than normal. Interestingly, the sensitivity of the fibres with expanded receptive fields was uniform over their whole receptive field, indicating that the sprouted terminals have normal mechanical thresholds. Examination of silver stained sections of the foot pad 60 days after transection of the sciatic nerve showed the presence of intact, fine, apparently unmyelinated axons running parallel to the skin surface just below the epidermis, where at 14 days only degenerating fibres had been present (Brenowitz and Devor 1981). Similar anatomical findings were made in expanded regions of pinch sensitivity around innervated islands of back skin in the rat (Nixon et al. 1984). In addition, injection of WGA-HRP into the L3 and L4 ganglia of the adult rat after chronic sciatic nerve section labels saphenous nerve endings more laterally than normal in the plantar skin of the hindpaw (Kinnman and Aldskogius 1986). Since it is unlikely that unmyelinated afferents could be visualized by these means, these anatomical findings point to the sprouting of A∂ fibres being responsible for the spread, into denervated skin, of sensitivity to strong mechanical stimulation. That A∂ fibres are involved in the spread of pinch sensitivity in the experiments of Nixon et al. (1984) was also shown by direct electrical stimulation of the skin, which evokes a reflex contraction of the cutaneus trunci muscle at a latency corresponding to A∂ activation when applied either within or up to 1–2 mm outside the region of pinch sensitivity both before and after expansion.

Sprouting of Myelinated Low-Threshold Mechanoreceptive Afferents. A decade ago it was commonly thought that low-threshold mechanosensory myelinated cutaneous afferent fibres were also capable of sprouting into denervated skin in mature mammals. At the time there was, in fact, very little direct evidence for this other than a few clinical studies indicating recovery of the sensation of "touch" after nerve injury without regeneration. When, however, a region of thigh skin in the cat was denervated by section of its nerve supply (a small branch of the posterior femoral nerve) it was found that the specialized sensory endings called touch domes persisted near the borders of the denervated region, whereas few survived in the central region (Burgess et al. 1974). Touch domes are specialized low-threshold mechanosensory structure characterized by 20–50 specialized epithelial cells (Merkel cells) in a small dome-like elevation, and each is supplied by 1–3 (usually only 1) type I sensory afferent fibre. Since regenerating type I fibres appear to reinnervate domes specifically in the cat, it was assumed that the persisting domes survived as a result of reinnervation by collateral sprouts of type I fibres that supplied surrounding skin. However, electrophysiological recording from whole nerves has since been used to demonstrate conclusively that low-threshold mechanoreceptive myelinated afferents do not sprout into adjacent denervated skin of adult mammals. For example Jackson and Diamond (1981, 1984) did not find any expansion of the low-threshold field of the dorsal cutaneous nerve in the adult rat up to 85 days after denervation of the surrounding skin (Fig. 2.2d). In the rabbit these workers also examined the low-

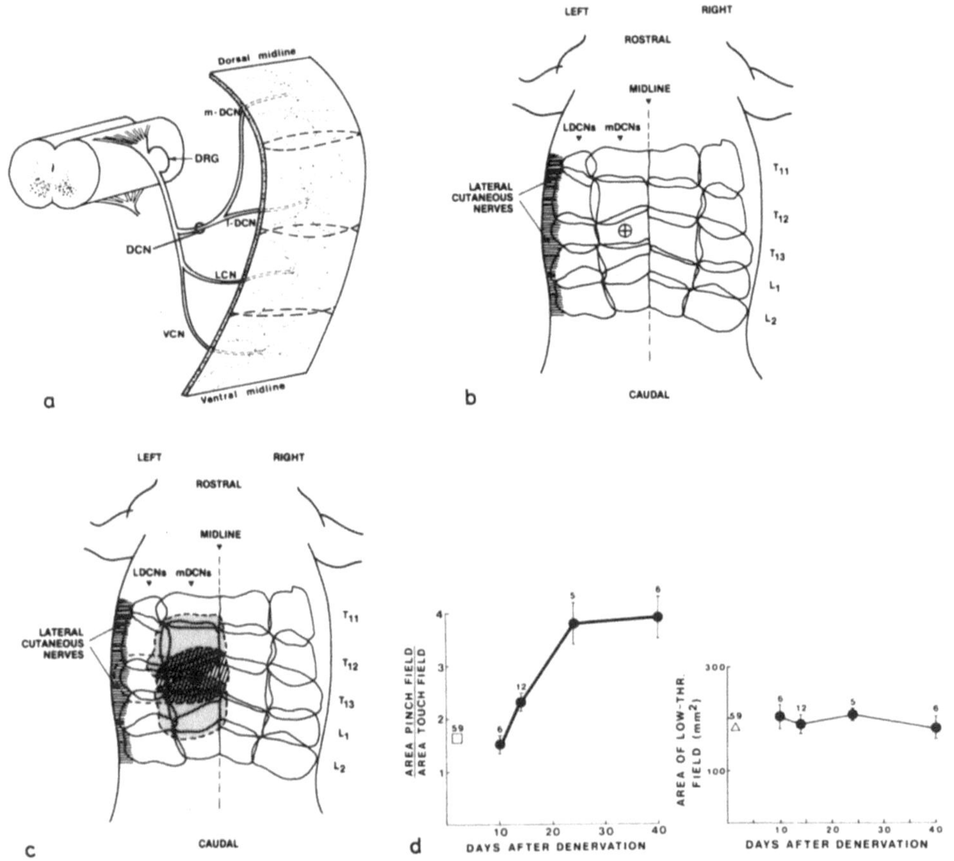

Fig. 2.2 a–d. Collateral reinnervation of denervated back skin by high-threshold mechanoreceptive afferents in the adult rat. **a** Arrangement of low-threshold mechanosensory dermatome in the thoracic region of the rat. The peripheral receptive field of a single DRG extends as a band from the ventral to the dorsal midline. The dermatome is divided into several regions, each supplied by a different cutaneous nerve. Overlap between the fields is shown by the *broken lines*. (Not to scale.) **b** Dorsal view of rat, showing typical banding pattern of low-threshold tactile receptive fields of segmental DCNs determined by electrophysiological recording from whole nerve. *Hatched area* at left indicates skin innervated by LCNs. *Cross within circle* indicates low-threshold receptive field of the mDCN-T13. **c** High-threshold (pinch) receptive fields of mDCN-T13 (*diagonal hatching*) and of mDCN-T12 and mDCN-L1 (*stippled*) are superimposed on the low-threshold fields shown in **b**. High-threshold fields are determined by mapping points on the skin from which reflex contractions of the underlying cutaneus trunci muscle are evoked by a pinch. **d** Areas of high- (left) and low- (right) threshold receptive fields determined for mDCN-T13 in different groups of animals at different times after denervation of surrounding skin by transection of DCNs T10–T12, L1–L3 and the lateral branch of DCN T13 on the same side, and of DCNs T10–L3 on the opposite side. To compensate for inter-animal variation in the initial size of the receptive fields, the area of the high-threshold field is expressed as a proportion of the low-threshold field, which does not change after denervation of surrounding skin. Note the initial delay of 13–14 d before any detectable expansion of the high-threshold field. Expansion ceases at about 24 d postoperatively.

threshold field of the occipital nerve of the ear following section of the greater auricular nerve. In addition they mapped the low-threshold field of the sural nerve up to 60 days after section of the lesser sural nerve, the low-threshold field of which it slightly overlaps (Fig. 2.3). In neither case was there any sign of expansion of the low-threshold field of the intact nerve (Jackson and Diamond 1983). Similarly, Devor et al. (1979) were unable, using whole nerve recordings, to find any unequivocal spread of the low-threshold field of the saphenous nerve onto plantar skin of the hindpaw after sciatic nerve transection in the rat. Neither was there any spread of the receptive fields of single myelinated low-threshold mechanoreceptive afferents isolated from the saphenous nerve. In the light of this evidence Horch (1981) reinvestigated the effect of cutting the femoral cutaneous nerve in the cat, and found, by electrophysiological recording, that, in fact, none of the domes denervated by nerve transection became innervated by type I fibres in the surrounding nerves (Fig. 2.4). The evidence for lack of sprouting of large myelinated afferent fibres does

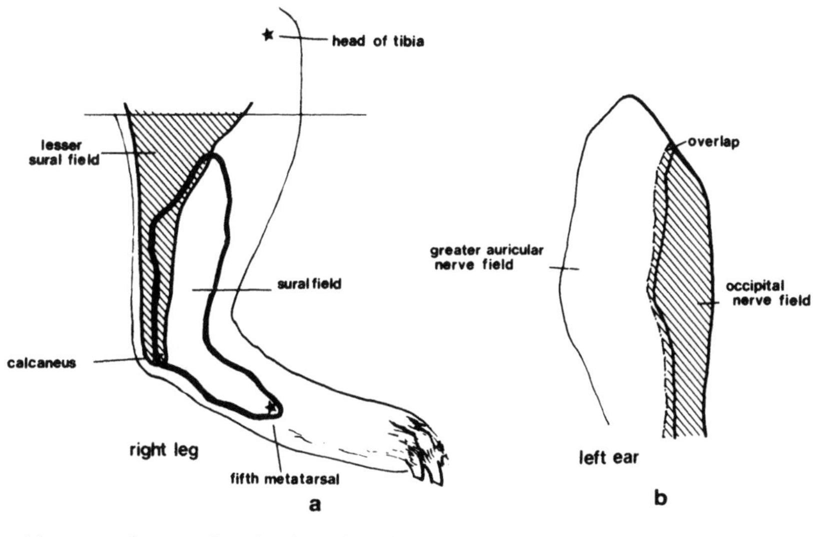

Fig. 2.3 a, b. Absence of sprouting by low-threshold mechanoreceptive afferents into denervated areas of skin in the adult rabbit. The low-threshold mechanosensory receptive fields of the sural and lesser sural nerves in the leg (**a**) and of the greater auricular and occipital nerves in the ear (**b**) were mapped by electrophysiological recording from the whole nerve. The fields in **a** are as they would appear if the skin were removed from the leg and flattened out. Note the small region of overlap between the two nerve fields in each case. The area of the lesser sural field was measured below the horizontal line in **a**. Chronic (60 day) section of the sural nerve had no effect on the area of the lesser sural nerve field. Similarly, chronic (65 day) section of the greater auricular nerve did not cause expansion of the low-threshold field of the occipital nerve. From Jackson and Diamond (1983, p. 279)

◀ **Fig. 2.2 a–d** (*continued*). *DRG*, dorsal root ganglion; *DCN*, dorsal cutaneous nerve; *m-DCN*, medial branch of dorsal cutaneous nerve; *l-DCN*, lateral branch of dorsal cutaneous nerve; *LCN*, lateral cutaneous nerve; *VCN*, ventral cutaneous nerve, (**a** From Jackson and Diamond 1984, p. 337; **b–d** from Nixon et al. 1984, pp. 105, 107)

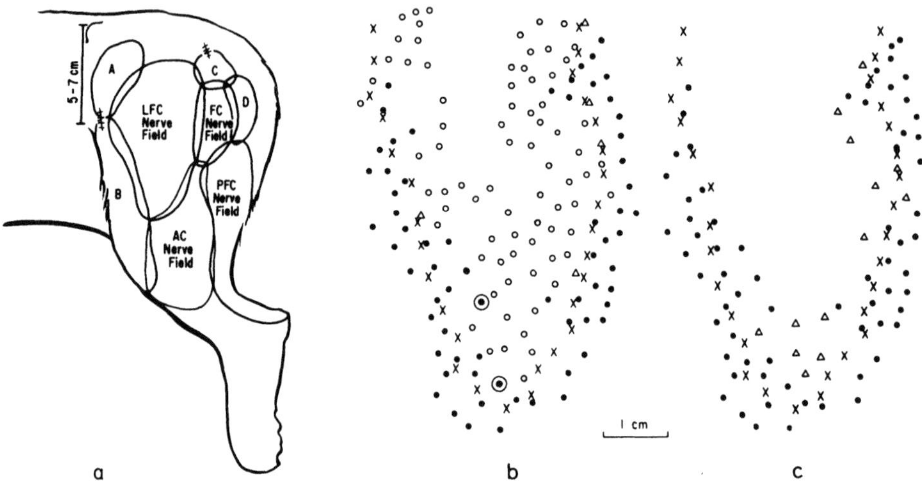

Fig. 2.4 a–c. Absence of sprouting by type I mechanoreceptive afferents into denevated skin of the adult cat. **a** Drawing of typical sensory nerve distributions on the thigh of the cat determined by electrophysiological recording from whole nerves. *FC*, femoral cutaneous nerve field; *PFC*, posterior femoral cutaneous nerve field; *LFC*, lateral femoral cutaneous nerve field; *AC*, accessory sural nerve field; *A,B,C,D*, fields of unnamed nerves. **b, c** Maps of the touch domes and their innervation determined by electrophysiological recording from the FCN and surrounding nerves, on the border of the peripheral field of the FCN. The positions of domes were noted in relation to fixed tattoo marks placed on the skin at the edge of the FCN nerve field. **b** Before and **c** 9 months after permanent transection of the FCN. Note that surrounding nerves have failed to reinnervate touch domes along the border of the FCN field denervated by the FCN transection. *Symbols*: ×, tattoo marks placed on the skin when the area was mapped originally; ○, domes innervated by the FCN; ●, domes innervated by the surroundings nerves; ⊙, domes innervated by both the FCN and the surrounding nerves; Δ, domes not producing activity in any of the recorded nerves. (**a** From English 1977, p.138; **b, c** from Horch 1975, p. 315)

not, of course, exclude the possibility that the peripheral terminals of these fibres sprout without developing normal physiological function, or that they sprout within regions they initially share with fibres from other nerves.

In a recent study Kinnman (1987) obtained clear anatomical evidence for the collateral sprouting of both small and large diameter myelinated afferents in hairy skin of the adult rat. In the first of these experiments, the cutaneous territory of the T10 intercostal nerve of one side was isolated from the surrounding dermatomes and from the territory of the dorsal cutaneous nerve by transection of intercostal nerves T7–T9 and T11–T13 together with dorsal cutaneous nerves T7–T13. Between 1 and 6 weeks later, WGA-HRP was injected into the ipsilateral T10 DRG in order to label the cutaneous branches of T10 afferents. The medial extension of axons of the T10 lateral cutaneous nerve in the skin was then compared with that in control rats in which the DRG was injected at the time of nerve section. In the chronic animals, fine labelled epidermal and dermal and coarse dermal fibres were found between 2 and 2.4 mm medial to the normal (control) T10 lateral cutaneous nerve territory, although the density of labelled fibres was somewhat less than that in controls. Interestingly, labelled nerve fibres of normal appearance were found around hair follicles in the collaterally reinnervated skin. To investigate the ability of collateral sprouts to cross dermatomal barriers, in a second experiment, the T10 dermatome of one side was isolated by resection of the intercostal and dorsal cutaneous nerves T7–T9 and T11–T13. Between 1 and 6 weeks later axons of T10 afferents were found in the skin up to 4 mm caudal to the normal T10 segmental innervation area. However, relatively few labelled axons were present in this case, and none were seen terminating around hair follicles or in any other kind of specialized sensory nerve endings. In both cases, sprouting was first detectable between 1 week and 2 weeks after the lesion. Kinnman (1987) suggested that the conflict between his anatomical findings of collateral sprouting by coarse cutaneous myelinated afferents and the various electrophysiological studies, described above, in which no evidence of sprouting by low threshold cutaneous afferents was found, could be resolved if the sprouted axons were either non-conducting or possessed, like free nerve endings in normally innervated skin, an elevated threshold to mechanical stimuli. In comparing these findings with the results of anatomical (Kinnman and Aldskogius 1986) and physiological (Devor et al. 1979; Markus et al. 1984; Pomeranz et al. 1984) studies of collateral sprouting in the adult rat forepaw, Kinnman (1987) further suggested that collateral sprouting of sensory axons might occur more extensively in hairy skin than in glabrous skin.

Sprouting of C Polymodal Nociceptors. More recently, evidence has been accumulating that at least some unmyelinated, or C fibres, are able to sprout into adjacent denervated skin. Initially Brenan (1983, 1986) showed that peptide-containing C fibres of the rat saphenous nerve undergo a limited sprouting into adjacent denervated skin within about 3 weeks of transection of the sciatic nerve. The presence of these fibres was detected by the method of plasma extravasation in response to their antidromic activation. It is well known that C fibres contribute to the nociceptive innervation of skin (Bessou and Perl 1969; Van Hees and Gybels 1972), and behavioural evidence for the sprouting of thermal nociceptors into denervated skin on the rat's back has now been obtained by mapping the zone from which noxious thermal stimulation elicited reflex contraction of the cutaneus trunci muscle (Doucette and Diamond 1987). Using the electron microscope it was found that sensory C fibres of the horizontal sub-epidermal unmyelinated fibre system disappeared from insensitive skin, leaving axon-free sub-epidermal Schwann tubes. However, after the return of thermal nociception a few of these tubes always contained C fibres, though many also remained empty. The return of noxious thermal sensitivity to adjacent denervated skin was also related to the presence of plasma extravasation in response to antidromic activation of the intact nerve (Doucette and Diamond 1987). In contrast, Jancso and Kiraly (1983) were unable to obtain any evidence for sprouting of sciatic nerve C fibres in the rat hindpaw following transection to the saphenous nerve. These workers used the extravasation which occurs following application of mustard oil to the skin as an index of polymodal nociceptor innervation. It is particularly difficult to explain these negative findings in view of the results of Brenan (1983, 1986) showing that sprouting of saphenous C fibres occurs in the rat paw. One possible explanation is that Jancso and Kiraly (1983) did not examine rats in the same time-window after nerve transection as other workers. They noted that the inflammatory reaction was absent from the sciatic nerve field 5–7 days after transection, and that it was still absent 8 months later. It is not clear from their account whether they looked at intermediate times. However, 5–7 days might be too short a time for a collateral innervation to be established (Doucette and Diamond 1987) and there is some indication from the data of Brenan (1983, 1986) that the expansion of C fibre innervation may peak at about 2 months and then possibly regress. An alternative, and perhaps more likely, explanation is that the sprouted nociceptor terminals lack the chemical sensitivity of normal, or of regenerated, polymodal nociceptors, but this has not been determined. It may be relevant, however, that an absence of responsiveness to irritant chemicals, but not to noxious thermal or mechanical stimuli has been noted in neonatal rats (see Sect. 4.6.2).

In relation to the mechanism which might give A∂ HTMs and C fibres an advantage in responding to peripheral denervation, it has been pointed out that a feature common to both types of afferent fibre is the characteristic delay, due to activation of a voltage sensitive Ca^{++} channel, in the repolarization phase of the somatic action potential. It has been suggested that this feature, which is not shared by A∂ D hair afferents, at least in the cat (Rose et al. 1986), may be characteristic of developmentally immature sensory neurons (Yoshida and Matsuda 1979). While it is true that collateral sprouting occurs more readily and in a wider range of afferents in the neonatal animal, there is no good evidence that D hair afferents do not also undergo collateral sprouting in *adult* mammals.

2.3.2 Collateral Sprouting in Neonates

In contrast to the lack of collateral sprouting of low-threshold myelinated afferents in the skin of adult mammals, there is good evidence that some such fibres will sprout if adjacent skin is denervated during a relatively restricted postnatal period. For example, whole nerve recordings from intact nerves supplying islands of innervated back skin in rats have shown that expansion of the low-threshold receptive field occurs when surrounding skin is denervated before P 19 and that expansion ceased at about this age. When the denervation was carried out at P 9, there was apparently no sprouting of the intact innervation until P 14. Thus there appears to be a developmental window between P 14 and P 19 in the rat, during which collateral sprouting of low-threshold myelinated cutaneous afferents will take place, although it remains possible that sprouts are formed earlier than P 14, but have not yet acquired normal sensory function at this time (Jackson and Diamond 1981, 1984). Expansion seems to occur equally in rapidly adapting field and hair-follicle afferents and type I and type II afferents. Like the sprouting of low-threshold cutaneous afferents in the salamander (Diamond et al. 1976), but unlike the sprouting of C and A∂ fibres in adult mammals, the sprouting of A∂ fibres in immature rats appears to be restricted to the area within the denervated skin of their own dermatome.

Sprouting of the saphenous innervation of the rat hindpaw is also initially more extensive when the sciatic nerve is sectioned neonatally. For example, at 6–10 weeks after sciatic nerve transection on P 1–2 in rats, saphenous nerve fibres, including coarse myelinated axons, are found throughout the whole mediolateral extent of the plantar skin. By 14 weeks, however, the extent of this collateral innervation is reduced almost to that occurring after nerve lesions in the adult (Kinnman and Aldskogius 1986). In this case a similar result would, of course, be obtained if the initial innervation of hindpaw skin in the neonate were more extensive than in the adult, and if withdrawal of this extra innervation were delayed by the nerve lesion. Unfortunately it is not known whether the initial innervation of the skin in the rat is non-specific and exuberant. Comparison with the chick embryo, where the initial innervation fields of cutaneous nerves appear to be similar to those in the adult, renders this unlikely, at least for large myelinated fibres (Scott 1982). However, direct evidence on this point is lacking.

2.3.3 Effect of Neural Activity on Collateral Sprouting

In the system used by Diamond's group to investigate collateral sprouting in the back skin of adult rats (see Fig. 2.2) it was found that the initial rate of sprouting could be enhanced by activation of the intact innervation with either adequate natural stimulation or electrical stimulation. For example, a single application of 50 pinches throughout the field of the intact nerve immediately after denervation of surrounding skin was sufficient to cause significant expansion of the pinch-sensitive field to occur after only 10 days, compared with about 14 days in control animals. A similar precocious sprouting of A∂ fibres could be induced by electrical stimulation of the intact nerve (10 min at 20 Hz), or by direct electrical stimulation of skin, at a strength suprathreshold for A∂ fibres. After the initial acceleration of sprouting, the rate of expansion reverted to that observed in the absence of stimulation, suggesting that the impulse activity served mainly to reduce the latency of onset, rather than the rate of sprouting. The effect of stimulation was abolished if action potentials in the nerve were blocked with tetrodotoxin (Nixon et al. 1984). More recently, it has also been shown that the initial rate of expansion of the noxious heat-sensitive field of an intact nerve into denervated skin can be accelerated by electrical stimulation of C fibres in the nerve, or by the application of noxious thermal or mechanical stimuli to the normal innervation field. The accelerated sprouting of noxious thermal sensitivity by noxious mechanical stimulation implies that the sprouting C fibres are polymodal. Similarly, noxious heating of the innervated skin caused the precocious expansion of the pinch-sensitive field in the absence of sprouting by A∂ nociceptors (Doucette and Diamond 1987). In contrast to these findings, Greenfield and Devor (1981) claimed that repeated sensory testing by pinching the hindpaw of rats after sciatic nerve section had no noticeable effect on the rate or extent of expansion of the pinch-sensitive field of the saphenous nerve. The reasons for this conflicting observation are unknown, but might be related to the presence of a higher innervation density in the hindpaw than in back skin. Finally, it has been shown that sprouting of the pinch-sensitive field of the

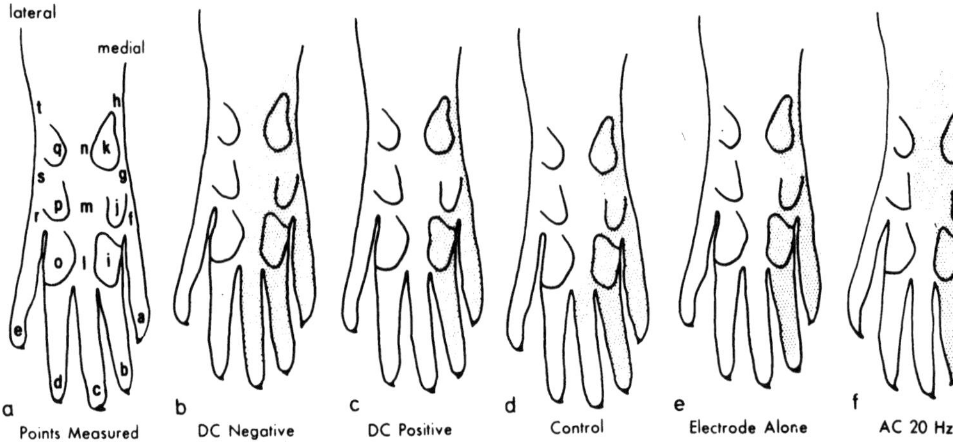

Fig. 2.5 a–f. Effect of applied electrical field on collateral sprouting of high-threshold mechanoreceptive afferents into denervated skin of the adult rat. **a–f** show ventral views of the left hind paw. Areas innervated by high-threshold mechanoreceptive afferents of saphenous nerve (*hatching*) were determined by pinching the skin at the points indicated by small letters in **a** and noting presence or absence of a withdrawal reflex, 3 weeks after section of the sciatic nerve. **b** Effect of applying dc current of 1 μA between the most lateral digit (– ve) and an indifferent electrode on the tail for 6 out of every 7 postoperative days. **c** Same as **b** but toe electrode + ve. **d** Control rat – no electrodes applied. **e** Control with electrodes but no current applied. **f** Same as **b** and **c**, but 20 Hz ac current applied. Note the greater lateral expansion of the saphenous nerve field in **b** compared to the other cases. (From Pomeranz et al. 1984, p. 333)

saphenous nerve after sciatic nerve transection in the adult rat is enhanced by weak dc electric fields if the cathode is placed distal to the growing tips of the nerve fibres (Fig. 2.5). The magnitude of the required field (1 mV/cm) was considered to be much too low to initiate action potentials in the axons (Pomeranz et al. 1984), and the phenomenon was considered to be analogous to the previously demonstrated effects of electric fields on the *in vitro* growth of neurites of amphibian neurons (Hinkle et al. 1981; Patel and Poo 1982).

2.3.4 Collateral Sprouting of Trigeminal Afferents

Collateral sprouting of afferents in the trigeminal region shows some interesting differences from that at lower levels. The lower canine tooth of the cat is innervated mainly by A∂ and C fibres (Holland and Robinson 1983). Most of these travel in the ipsilateral inferior alveolar nerve, while the remainder travel in the lingual nerve (Robinson 1980; Holland and Robinson 1984). Transection of the inferior alveolar nerve reduces the number of axons at the apex of the tooth to 9% of normal 7 days after the lesion. However, electrophysiological experiments (Fig. 2.6) have shown that the anterior teeth, including the canine, become reinnervated within 12–15 weeks by collateral sprouts from fibres of the ipsilateral lingual and mylohyoid

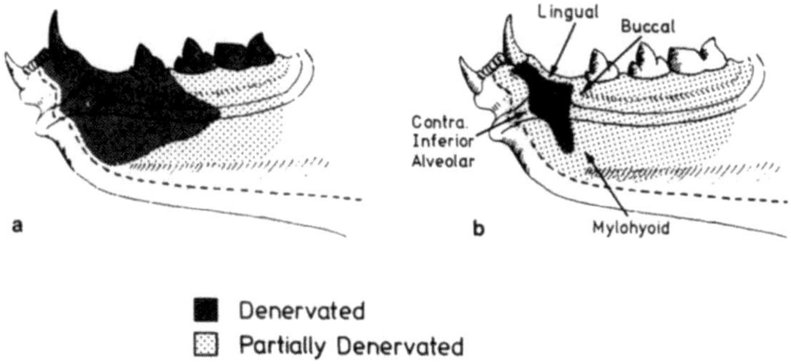

Fig. 2.6 a, b. Collateral sprouting of trigeminal afferents in adult cat. **a** Region denervated immediately after cutting and capping the left inferior alveolar nerve, **b** 15 weeks postoperatively, fibres of the lingual, buccal, mylohyoid and contralateral inferior alveolar nerves have reinnervated much of the denervated region. Innervation was determined by the presence of a jaw-opening reflex in response to electrical stimulation to the canine tooth, mucus membranes, or skin, or by recording electrical activity from the nerves. *Black areas*, totally denervated; *stippled areas*, partially denervated. (From Robinson 1981)

nerves and the contralateral inferior alveolar, lingual and mylohyoid nerves (Fig. 2.6) (Robinson 1981). This situation allows the anatomical and physiological study of an almost pure population of sprouted afferent fibres, in the collaterally reinnervated teeth. Electron-microscopic examination of the innervation of the tooth showed that the total number of axons present was only 33.5% of normal, compared with 80% found after reinnervation by regenerating fibres. There were fewer myelinated fibres than in normal teeth, as well as a highly variable number of unmyelinated fibres. The morphology of the collaterals resembled that of regenerated, rather then normal fibres, inasmuch as the myelinated fibres were smaller and had thinner myelin sheaths, and an unusually large proportion of unmyelinated fibres contained only one axon profile (Holland and Robinson 1987). Surprisingly, whole nerve recordings from sprouted nerves showed that responses to mechanical stimulation of the skin and mucous membranes could be obtained from regions previously denervated by inferior alveolar nerve section. Since it is unlikely that the responses of A∂ fibres would be picked by in this manner (Nixon et al. 1984) it is possible that large low-threshold myelinated afferents may undergo a degree of collateral sprouting in the trigeminal system. Where the cut ends of the inferior alveolar nerve are carefully apposed, reinnervation of the dentine by afferents of the regenerating nerve has been observed at 12 weeks postoperatively (Holland et al. 1987).

2.3.5 Collateral Sprouting and Sensory Recovery in Man

Collateral sprouting appears to contribute to the recovery of cutaneous sensation after nerve injury in man (see Sect. 2.3.1), and this allows a comparison of the events which are known to occur in the skin of animals, with clinical assessments of the recovery of skin sensation in humans (Inbal et al. 1987). Owing to the possibility of confusing abnormal patterns of original innervation with collateral reinnervation, the most convincing evidence for collateral sprouting in man comes only from cases where nerve blocks with local anaesthetics were used to confirm that reinnervated skin which had previously been totally anaesthetic after a nerve injury was now supplied by a different nerve. In other cases, where the site of nerve injury was far removed from the anaesthetic region of skin, some recovery of cutaneous sensibility occurred long before fibres from the regenerating nerve could possibly have reached and invaded the anaesthetic region (Inbal et al. 1987). Collateral reinnervation appears to produce a region that is hypoaesthetic compared with normal skin. However, although the threshold of sensibility is supernormal, the first perceived sensation produced by mechanical stimulation of such a region is described as 'touch' or 'pressure', rather than pain (Inbal et al. 1987). This is very different from the situation where denervated skin is reinnervated by regenerating fibres of the damaged nerve, and where hyperalgesia nearly always precedes the return of specific mechanosensitivity (Sunderland 1978). Since it is widely held that activation of A∂ HTMs should evoke pain, this observation is difficult to reconcile with the observed sprouting of A∂ but not Aβ afferent fibres in animals. Sprouting of low threshold mechanoreceptive C fibres is unlikely to provide an explanation since the vast majority of C fibres in subhuman-primate and human peripheral nerves are also thought to be nociceptors (see review by Willis 1985).

2.3.6 Fate of Collateral Sprouts After Regeneration of Original Nerve

It is generally agreed that, after regeneration of the original innervation, the collateral innervation of skin is displaced. This has been shown both for sprouted high-threshold mechanosensory afferents in adult rats (see Fig. 2.1c) (Devor et al. 1979; Kinnman and Aldskogius 1986) and for low-threshold myelinated afferents in infant rats (Jackson and Diamond 1981; Kinnman and Aldskogius 1986). Collateral sprouts are even displaced by regenerating axons of transplanted foreign nerves, including Aβ fibres, as well as by those of the original nerve. Unlike their collateral sprouts, the regenerating sprouts of low-threshold myelinated afferents are able to cross dermatomal boundaries (Jackson and Diamond 1981). These phenomena may be explained by the competition of nerves for a limited supply of a trophic sprouting factor produced by peripheral tissues (Diamond et al. 1976), and there is a great deal of evidence from other systems, outside the scope of this review, that actively growing or regenerating axons are able to compete successfully for target tissue with neurons whose innervation fields are already extended well beyond their normal limits. Of possibly direct relevance to the mechanism of this competitiveness is the recent demonstration that the collateral sprouting of C and A∂ nociceptive afferents into denervated dorsal skin of the adult rat, demonstrated by Diamond and his co-workers (see Sect. 2.2.2), may be completely inhibited by treatment of the animals with antiserum to NGF. In contrast, such treatment fails to affect the reinnervation of denervated skin by actively regenerating sensory nerve axons, which continue to sprout, not only into skin of their original innervation field, but also into surrounding denervated skin, even after they have been cut off, by dorsal rhizotomy, from any central source of trophic support (Diamond et al. 1987).

The fate of the collateral innervation upon regeneration of the original nerve has not been intensively studied morphologically. Zander and Weddell (1951) claimed that the collateral innervation formed after lesioning the cornea of the rabbit was later withdrawn without any signs of degeneration. However, this opinion has been questioned for lack of convincing evidence, and secondary degeneration of collateral axon sprouts in the rabbit cornea has more recently been demonstrated (Beuerman and Rozsa 1984).

Whether collateral sprouts in skin are withdrawn completely, or merely become non-functional, is not known. One obvious approach to this question would be to cut a nerve with a formerly expanded field and to look for degenerating axons in the region of expanded pinch or noxious thermal sensitivity after reinnervation by the original nerve. In addition, it would be interesting to know whether the rate of collateral reinnervation of denervated skin is greater when the regenerated nerve is cut or crushed a second time, and whether functional sprouting of low-threshold myelinated afferents will occur in the adult in regions where it has previously been induced by nerve transection during infancy.

2.4 Regeneration of Somatic Sensory Afferent Fibres

The third and final major feature of plasticity in the peripheral sensory system, and probably the most important in terms of its effect on central sensory neurons, is regeneration of the peripheral processes of primary afferent fibres after peripheral nerve injuries. If a cutaneous nerve is divided, either by transection or crush, return of normal or near-normal cutaneous sensibility will depend ultimately on the degree to which the damaged axons regenerate and re-establish appropriate lines of communication between cutaneous receptors and the CNS. The degree of functional recovery which occurs after such a lesion will be determined, at least in part, by the number of regenerated axons as well as the extent to which they regain their original functional and somatotopic specificities. Any recovery of function additional to that explicable in terms of this peripheral regeneration must of necessity involve some degree of functional reorganization within the CNS either of the kind discussed elsewhere in this review or as yet unrecognized.

There is a large body of clinical evidence dating back to the turn of the century showing that recovery of somatic sensory function after nerve transection and repair is relatively poor, but that after nerve crush recovery can be quite complete (Walshe 1942; Sunderland 1978). Experimental evidence of this dates from 80 years ago when Osborne and Kilvington (1909) studied reflex responses to nerve stimulation after crush or transection of the sciatic nerve in adult dogs. On the basis of their results they proposed that the good recovery of sensation occurring after nerve crush might be due to reconstitution of a near-original innervation pattern by some mechanism providing guidance of regenerating fibres back to their original innervation fields on the skin.

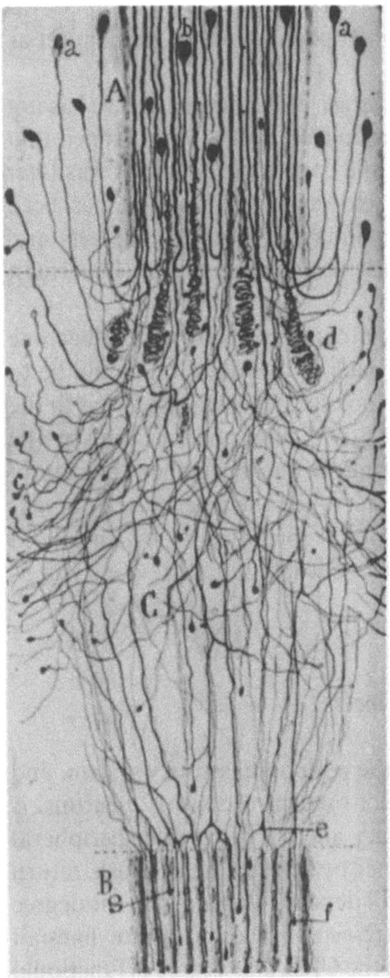

Fig. 2.7. Diagrammatic representation of regeneration of axons and neuroma formation in mammalian nerve from the classical studies of Ramon y Cajal. The drawing identifies the major features of the early stage of neuroma formation and nerve regeneration (Ramon y Cajal 1982) in a case where the stumps of the cut nerve are approximated. *A*, proximal stump; *B*, distal stump. Note the presence of recurrent axon branches (*b*) within the proximal stump, and the multiple branches of axons (*f*) and (*g*) entering Schwann tubes of the distal stump. (From Ramon y Cajal 1982, p. 243)

It is widely believed that the guidance is provided by the endoneurial tubes of Schwann cells in the distal stump of the nerve. The continuity of the endoneurial tubes is not destroyed by simple crush of the nerve, and they could therefore guide the growing axon back to its original end-organs, precisely recreating the original innervation pattern (Sunderland 1978). Following nerve transection the sequence of events is more complex. In his classic studies of nerve regeneration, Ramon y Cajal (1928) observed regenerating axons crossing into the distal stump of a transected nerve where they formed multiple branches, some of which entered the distal Schwann tubes (Fig. 2.7). Having once entered a Schwann tube an individual axon sprout grows along it and only rarely alters course by leaving the tube or doubling back. Therefore, unless one of the sprouts of an axon enters the distal segment of the axon's own former Schwann tube, it will be unable to regain its original cutaneous receptive field. Evidence, described below, indicates that afferent fibres of different submodalities may tend to regain their original submodality after regeneration.

This suggests either that the sprouts show a strong preference for Schwann tubes which previously contained afferents of the same submodality, or that on reaching the skin the regenerating fibres, which have an "identity" based on such properties as their conduction velocity, and the connections they make within the CNS, induce the *de novo* formation of the appropriate kind of receptor. As will be seen below both these notions have received support from time to time. However, though indirect, the most recent evidence favours the first rather than the second of these alternatives. The mechanism by which regenerating axons are attracted to appropriate Schwann tubes in unknown, but presumably depends, as suggested by Ramon y Cajal (1928), on the presence of specific identifying molecules, or trophic factors, possessed by the Schwann tubes. The specificity of this mechanism may be somewhat hierarchical, in that there is some indication that certain rapidly adapting types of afferent fibre will reinnervate a range of rapidly adapting cutaneous receptors. In contrast, regenerating afferent fibres show little or no somatotopic specificity, as might be expected from the specific conduit guidance model of regeneration outlined above.

2.4.1 Numbers of Axons in Nerves Regenerating After Crush or Transection

Obviously, one important factor affecting the nature of cutaneous sensibility after peripheral nerve regeneration must be the degree to which the original complement of functional sensory fibres is re-established. Attempts to study this by counting axons in nerves proximal and distal to crush or transection and repair have produced variable results. One of the earliest studies of this kind dealt with the effect of sciatic nerve transection and resuture at upper thigh level in the rhesus monkey (Davenport et al. 1937, 1939). Myelinated fibres were counted in osmic acid stained sections, and myelinated plus unmyelinated axons were counted in silver stained material, all taken from the nerve within 10 mm of the lesion. In the distal stump 3 months after the lesion the number of fibres which had regenerated and acquired sufficient myelin to stain with osmium was about 59% of the number of myelinated fibres in the contralateral intact nerve, and about 64% of the number in the injured nerve proximal to the lesion. After 6 months there was about 81% restoration of myelinated fibres distal to the lesion when compared with the contralateral intact nerve. For technical reasons the number of unmyelinated axons in the intact nerves and in the proximal part of the lesioned nerves could not be ascertained. However, in the regenerating section of the nerve at both 3 and 6 months after the lesion there was estimated to be about twice as many unmyelinated as myelinated fibres. The authors interpreted this to mean that once a regenerating sprout of a mylinated axon has entered the distal stump there is relatively little delay before it becomes myelinated. In their classical study of nerve regeneration in the rabbit, Gutmann and Sanders (1943) showed that the number and size of myelinated fibres distal to the lesion returned to normal about 150 days after *crushing* the peroneal nerve. In contrast, after nerve *transection and repair*, there was a permanent decrease of around 30–35% in the number of myelinated fibres in the distal segment. Fibres of diameter $> 8\ \mu m$ were more severely affected than smaller fibres, to the extent that they were reduced in number by 65% after 12 months. However, the real loss of smaller myelinated fibres in the distal stump was probably underestimated due to the branching of regenerating large fibres (see below). Similar reductions in myelinated fibre counts distal to nerve transection and repair have been reported for the sural nerve of the rabbit (Shawe 1955) and the ulnar nerve of the cat (Cabaud et al. 1976).

Anatomical studies such as these are, of course, unable to distinguish between a situation where many axons proximal to the lesion each support only one distal process, and one where only a few axons successfully regenerate across the transection site, but by branching support a greater number of distal processes. It is known that axons in regenerating crushed or transected nerves produce many sprouts (Greenman 1913; Ramon y Cajal 1928; Morris et al. 1972; Thomas 1974), suggesting that the latter alternative might be the case. However, until recently opinion was divided on whether a number of these sprouts, or only one, survive after completion of regeneration (Osborne and Kilvington 1909; Seddon 1943; Sunderland and Laverack 1953; Shawe 1955; Fullerton and Gilliat 1965; Roth 1978, 1979). In an attempt to answer this question Horch and Lisney (1981a) used a combination of electrophysiological and anatomical techniques to investigate the regeneration of the sural and posterior femoral cutaneous nerve (PFCN) of the cat. These nerves innervate hairy skin around the heel and on the thigh, respectively. The sural nerve splits into two branches just above the heel, and separate stimulating electrodes were placed on each branch. Single unit recordings were then made from the sural nerve just distal to its emergence from the sciatic nerve. A sample of 128 fibres was studied in normal control nerves, and 504 fibres were examined in nerves which had been thoroughly crushed between the levels of stimulation and recording 1–6 months earlier. Of these, none could be activated from both branches of the sural nerve. In contrast, 1–9 months after transection of the nerve at the same level about 10% of an unstated number of fibres could be excited from both sural nerve branches, and some of these fibres had two separate cutaneous receptive fields. In these experiments no attempt was made to approximate the cut ends of the nerves, and a neuroma formed at the site of transection. Like the sural nerve, the PFCN also divides into two or more branches just before reaching the skin. In 4 cats, 18 months after cutting the PFCN several centimeters proximal to the branch point, stimulation of one branch led to reflected impulses being recorded from the other branch(es). These responses were never seen in normal nerves. By switching stimulating and recording electrodes around, they could be shown to be due to branching of individual axons. As the responses were abolished by cutting the nerve just distal, but not proximal to the neuroma, it was concluded that the branching occurred within the neuroma itself. The number of myelinated fibres successfully regenerating across the crush or transection site in the sural nerve was estimated by recording from single fibres several centimeters proximal to the neuroma, and determining what proportion of these could be activated by electrical stimulation of the nerve 2–3 cm distal to it. Three months after crushing the nerve all fibres recorded proximal to the lesion had regenerated past the level of stimulation. Following transection of the nerve recovery proceeded more slowly, but was still extensive, with about 87% of fibres having regenerated across the site of the lesion by 18 months. Both $A\beta$ and $A\delta$ fibres regenerated equally well. In a second type of experiment, regeneration of PFCN fibres was examined more than 20 months after resection of a 1–2 mm length of the nerve. A recording electrode was placed on the nerve 10–20 mm proximal to the neuroma (or at a similar position in control animals) and a teased filament of dorsal root was stimulated supramaximally such that a limited number of separate antidromically conducted action potentials could be recorded. The nerve was then stimulated distal to the neuroma, and the number of antidromic spikes which were blocked by the resulting orthodromic volley was noted. The procedure was then repeated for all dorsal root filaments producing detectable activity at the recording site. In this way, it was found that after nerve resection about 75% of myelinated fibres had regenerated across the site of the lesion, significantly fewer than had regenerated after simple transection of the sural nerve.

Since in these experiments any branching of regenerated axons was known to occur either at or distal to the neuroma (see above) and the proportion of proximal fibres crossing the neuroma was known, it was possible, by counting the number of axon profiles in the nerve distal to the neuroma, to determine the average number of processes in the distal segment of the nerve supported by each proximal fibre that had sprouted across the neuroma. Proximal to the transection there was no change in the number of myelinated axons, when compared to contralateral control nerves. However, in the nerve distal to the neuroma, there was a 50% surplus of fibres. Thus, of the 87% of fibres which had regenerated into the distal segment of the nerve, each supported, on average, 2.7 branches. Similar results were obtained in 4 animals 20 months after resection of a piece of the PFCN, and in which only 75% of fibres had successfully regenerated (see above).

In these cases the entire original population (determined from transverse sections of the resected piece of nerve) of myelinated fibres was present proximal to the injury, and a 25% excess of axons was present distally. Since the great majority of regenerated fibres possess a single receptive field (see Sects. 2.5.1, 2.5.2) it follows either that only one distal process normally forms a receptive field, or that the multiple branches of an axon must innervate contiguous regions of skin (Horch and Lisney 1981a).

The observed increase in myelinated fibres distal to the lesion in this study is difficult to reconcile with the results of earlier studies, which show a decrease. Since fibres distal to a transection undergo a dramatic reduction of diameter, the most likely explanation is that in the early light-microscopic studies some very small myelinated fibres were confused with degenerating myelin figures and/or not counted at all. Horch and Lisney (1981a) counted axons in the sural nerve in osmium stained, plastic embedded sections under the light microscope, and studied the PFCN with the electron microscope. Significantly, and due probably to the improved resolution that these techniques provided, their data show a reduction of the number of fibres with diameters $> 5\,\mu m$ but a greater increase in the number of fibres of diameter $< 5\,\mu m$ in the nerve distal to transection. Moreover, a recent study in which the electron microscope was used to count axons in the sural nerve of the rat 8 weeks after transection and repair of the sciatic nerve proximal to the emergence of the sural nerve (Jenq and Coggeshall 1984) revealed a 57% increase in the number of myelinated fibres, along with a sharp decrease in their diameters. The total number of axons in the sural nerve fell by about 20%, and this could adequately be accounted for by a large decrease of 43% in the number of unmyelinated fibres.

2.4.2 *Size of Regenerated Axons*

It is widely believed that nerve crush or transection reduces the diameter and conduction velocity of myelinated nerve fibres, both proximal and distal to the lesion (Erlanger and Schoepfle 1946; Sanders and Whitteridge 1946; Sanders 1948; Hodes et al. 1948; Schroder 1972; Devor and Govrin-Lippmann 1979; Horch and Lisney 1981a; Sanders and Zimmermann 1986). Early speculation suggested that distal to the transection the reduction of fibre diameter might be due to shrinkage of the empty endoneurial Schwann tubes down which the axons regenerate (Sanders and Young 1944). Fibres which fail to regenerate through the neuroma are thought to have slower conduction velocities than regenerated fibres. Nevertheless, many of them still conduct at velocities in the Aβ range (Weiss et al. 1945; Burgess and Horch 1973; Davis et al. 1978; Horch and Lisney 1981a). Burgess and Horch (1973) measured the conduction velocities of 397 single myelinated afferent fibres in the sural nerve of the cat which had regenerated after nerve transection. If anything, they found a slight *increase* in the median conduction velocities of both Aβ and A∂ fibres proximal to the lesion, compared with a sample of fibres from normal nerves, but this was not statistically significant. However, of those fibres projecting to upper cervical levels, the median conduction velocity in the dorsal columns was slightly but significantly *reduced*, from 31 to 27 m/s.

The reduction of conduction velocity was particularly severe for fibres in the distal stump. For example, Horch and Lisney (1981a) found that the conduction velocities of axons distal to a transection neuroma in the cat sural nerve were still only 50% of normal 6 months after the lesion, and showed no recovery over the subsequent year. The slowing was uniform across the whole size range of myelinated axons. In contrast, condution velocities had almost returned to normal both proximal and distal to the lesion 6 months after crushing the nerve.

Whether the slowing of conduction is itself of any functional significance for somatosensory perception is doubtful. Thus, the slowing which occurs proximal to the lesion may be less than the progressive slowing which occurs with increasing age in man (Dorfman and Bosley 1979). The greatly reduced distal and dorsal column conduction velocity might affect the timing relationships between inputs in different afferent fibres, especially when nerves are transected a long way from the skin. However, it is not known whether this slowing is causally related to the paraesthesiae and other sensory disturbances commonly observed in man after nerve injury and regeneration (Gilliatt 1982).

2.4.3 *Effect of Denervation on Specialized Cutaneous Mechanoreceptors*

Studies on the effect of denervation of the skin on its complement of specialized receptor structures have shown that these structures do not completely disappear but that they undergo varying degrees of degenerative change. Among the earliest investigations of this kind were the classical studies of Boeke and his colleagues. They first studied a specialized sensory end-organ (the organ of Eimer) in the skin of the mole's snout, and found that this receptor degenerated after transection of the trigeminal nerve (Boeke 1922). Later, Boeke and Dijkstra (1932) reported that trigeminal nerve section caused degeneration of the Grandry corpuscles and the Herbst corpuscles in the skin of the duck's bill (the Grandry corpuscle is a sensitive mechanoreceptor analogous to the mammalian Merkel cell-neurite complex, and the Herbst corpuscle is a Paciniform receptor).

Lee (1936) denervated Pacinian corpuscles in the cat's mesocolon, and observed degeneration of their axons within 2 weeks, but at this stage no apparent effect on the lamellar structure of the corpuscle. Much later, the same receptors were studied in more detail by Schiff and Loewenstein (1972), who stated that the capsule survives after many months of denervation if its blood supply is intact. Electron-microscope studies of cutaneous Pacinian corpuscles dissected from the cat's foot-pad showed that the inner core of lamellar cells phagocytose the degenerating unmyelinated nerve terminal within 4 days of transection of the sciatic nerve in the thigh, but the receptors were not examined beyond the 9th postoperative day (Korthals et al. 1974). In the monkey's hand, axon terminals disappeared from Pacinian corpuscles within 3–4 weeks of median nerve transection. However, corpuscles lacking a nerve terminal were still recognizable 9 months later (Wong and Kanagasuntheram 1971).

Brown and Iggo (1963) showed that when the saphenous nerve of the cat was either crushed or transected the touch domes within the field of the nerve, and their associated Merkel cells, began to atrophy within 4 days. The domes were flattened 20–30 days after nerve transection and their epidemis was histologically similar to the adjacent epithelium. At this stage Merkel cells were rare or absent, but the location of a dome was still recognizable by its tuft of capillaries and dermal papilla. Burgess et al. (1974) extended these observations by examining the gross appearance and histological structure of touch domes on the posterior thigh of the cat supplied by the femoral cutaneous nerve (FCN) at various times after cutting the nerve. In one animal the domes were grossly normal 16 days after nerve transection, but were severely atrophic by 109 days. In a second animal, examined only 35 days after nerve section, none of the domes originally innervated by the cut nerve (these had previously been identified by recording from the nerve and marked by tattooing the skin) could be readily seen by examination of the skin surface under $\times 25$ magnification. Histological examination of touch domes excised at various times after nerve transection essentially confirmed the earlier findings of Brown and Iggo (1963). In one cat the Merkel cells disappeared from denervated domes by 1 week; in another they persisted for 16 days but they had disappeared by 109 days. Electron microscopy was later used to confirm that the Merkel cells disappeared completely from denervated touch domes, and it was found in addition that they were phagocytosed by Schwann cells (English 1977).

Some of the above changes noted in touch domes of the cat have been described in varying degrees following denervation of similar domes in various other species, including opossums and monkeys (Jalowy 1935; Palmer 1965). In contrast, in the

rat, touch domes and their Merkel cells appear to persist for at least 18 months after denervation (Smith 1967; Jackson and Diamond 1981). It has also been reported that the ultrastructure of Merkel cells associated with sinus hairs in the upper lip of the rat does not change up to at least 169 days after denervation by ION transection (Benkenstein 1979).
The effects of denervation on Meissner corpuscles of glabrous skin have been less intensively studied than its effect on other receptors. In the fingertip skin of humans with long-term (2.5–12 months) injury to the median or ulnar nerve, Meissner corpuscles were still recognizable, although smaller than normal (Silver et al. 1964). Similarly, Meissner corpuscles in the fingertips of the rhesus monkey were found to be severely shrunken, but identifiable, 9 months after resection of 10 mm of the median, ulnar and superficial radial nerves at the wrist (Dellon et al. 1975). The fate of denervated low-velocity rapidly adapting receptors in the glabrous skin of other species, for example the Krause end-bulb of the cat, is unknown.

2.4.4 Reinnervation of Cutaneous Receptors by Regenerating Sensory Fibres

Since, with the possible exception of the Merkel cell-neurite complex, specialized cutaneous receptors persist for long periods of time after nerve section, they remain available for reinnervation by regenerating sensory fibres. Despite this, different types of receptors do not become reinnervated with equal success. In their pioneering neurohistological studies Boeke and his colleagues noted that the Grandry corpuscles and the Herbst corpuscles of the duck's bill were reinnervated by regenerating axons after transection of the trigeminal nerve. Following reinnervation, the Merkel-like cells of the Grandry corpuscle, which had partially degenerated, returned to normal size (Boeke and Dijkstra 1932). The ability of specialized mechanoreceptors to become reinnervated by regenerating nerves has since been investigated for several types of mammalian cutaneous receptors.

Reinnervation of Pacinian Corpuscles. As stated above, Pacinian corpuscles tend to retain their capsular structure for long periods after denervation. However, the available evidence indicates that they do not readily become functionally reinnervated. Schiff and Loewenstein (1972) investigated the effect on the functional state of Pacinian corpuscles in the cat's mesocolon of transection and regeneration of the inferior mesenteric nerve (IMN). To encourage regeneration, the two ends of the cut IMN were drawn together into a fibrin clot within a porous plastic tube, and care was taken to preserve the blood supply of the receptors. Despite these precautions, only 6 out of a total of 11 IMN reunions resulted in functionally reinnervated corpuscles, and in these 6 successful cases only 8 out of 56 corpuscles tested 30–40 days after nerve repair produced generator potentials in response to mechanical stimulation. Wong and Kanagasuntheram (1971) studied receptors in the palmar skin of the rhesus monkey with silver staining. Ten months after *crushing* the median nerve at the wrist there appeared to be a few reinnervated Pacinian corpuscles, but these were not evident at similar times after *transection* of the nerve. Finally Krishnamurti et al. (1973) examined the ultrastructure of 5 Pacinian corpuscles in the finger skin of a primate, the slow loris, 75 days after crushing the median nerve in the forearm and found no evidence that the receptors had been reinnervated.

Reinnervation of Meissner Corpuscles. In contrast to Pacinian corpuscles, Meissner corpuscles appear to become readily reinnervated after regeneration of cutaneous nerves in monkeys (Wong and Kanasungtheram 1971; Dellon 1976) and man (Jabaley et al. 1976). For example, in silver stained sections of fingertip skin 6 weeks after median and ulnar nerve transection and repair at the wrist in the rhesus monkey, an occasional Meissner corpuscle was observed to have a thin nerve terminal. By 3 months, two-thirds of the Meissner corpuscles were reinnervated, and by 7–9 months most had a normal appearance (Dellon 1976).

Reinnervation of the Merkel Cell Complex. The recovery of the structure and innervation of touch domes in the cat after nerve crush or transection has been studied in great detail by Horch and his colleagues using both anatomical and electrophysiological techniques (Burgess et al. 1974; Horch 1979, 1981). By visual inspection and mechanical stimulation of the skin while recording from the nerve, they were able to count all touch domes in and around the field of the FCN, and to identify those domes receiving an axon from the nerve. The positions of all visible domes in the region were mapped in relation to permanent tattoo marks which were placed in the skin around the FCN field. In this way it was possible to compare the spatial distribution of domes and their innervation pattern both before and after nerve injury and regeneration. In control experiments it was first established that the dome pattern was normally quite stable. For example, in one cat examined 113 days after a "sham" operation, 122 domes were in the same positions, 10 had disappeared, and 17 had appeared at new locations. Thus the locations of 88% of the domes in the area outlined by tattoo marks coincided with the positions of domes at the time of the sham operation. However, the fact that domes spontaneously disappear from some locations and appear at others suggests that even in normal skin they are not necessarily permanent structures. One animal was examined 689 days after crushing the FCN. There were 50 domes innervated by the FCN originally, and after regeneration this had increased to 65. All these domes were in the original FCN field, and all non-FCN domes were outside this area. The locations of 68% of FCN domes and 69% of non-FCN domes corresponded with dome locations preoperatively. Within the FCN field this degree of correspondence of post- with preoperative dome locations was significantly greater than that (12%) expected if domes had simply formed at random locations. In five cats distribution of FCN and non-FCN domes in and around the FCN field was mapped before and 393–1010 days after transection of the FCN. After regeneration only about 50% of the original number of domes were found in the FCN field in four cats, but in the fifth there was a 52% increase in the number of domes. Two different kinds of dome distribution were found. In two animals, the locations of those FCN and non-FCN domes situated within the original (marked) FCN field showed a significant correspondence with the preoperative locations of FCN domes. In a further three animals, the coincidence of dome location within the original FCN field was poor. In these animals there was a tendency for the regenerated domes to occur in clusters. The fifth cat showed an intermediate type of dome distribution. The coincidence of dome location after FCN regeneration with preoperative locations was statistically significant (at the 1% level) in four out of the five cats.

In these experiments it was assumed that those domes within the original FCN field which were not reinnervated by the FCN nerve had been reinnervated by collateral sprouting from surrounding non-FCN nerves. For example, in one cat there were 84 FCN-innervated domes mapped originally, and 41 domes were present in the FCN field 1010 days after nerve transection. Of these, 24 were apparently not innervated the type I axons of the FCN, and were therefore presumed to have been innervated by collateral sprouting from adjacent non-FCN skin. Since it was later found that collateral sprouting of large diameter myelinated cutaneous afferents did not occur under these conditions it is difficult to explain these results in any way other than to suggest that either a proportion of domes (varying from 16%–58% in four cats) do not regain normal function following their reinnervation, or that the recording technique used was inadequate to detect all type I afferent fibres in the FCN nerve. The latter possibility seems more likely, as it would also explain the presence of a few apparently non-FCN domes in the middle of the FCN field in normal animals, as well as the unexpected atrophy of some of these after FCN transection, as noted by Burgess et al. (1974). Fortunately, these problems do not appear seriously to affect the author's main conclusion, which was that regenerating type I fibres are directed by some mechanism preferentially to sites in the skin formerly occupied by domes, and they suggested that the most likely mechanism might be specific conduit guidance along endoneurial Schwann tubes of the distal stump of the nerve.

Further indirect evidence in favour of the specific conduit guidance theory was later obtained by Horch (1979), using the same system of touch domes and their type I afferent fibres in the cat. As before, the dome population of the FCN (or PFCN) was mapped, and reference tattoos were placed in the skin around the border of the FCN's low-threshold field so that the locations of domes after nerve regeneration could be compared with those prior to nerve injury. In addition, in some experiments, the pattern of innervation of the domes by individual type I afferent fibres in the FCN was carefully mapped. In four cats in which the FCN consisted of two separate fascicles the dome populations supplied by type I fibres in each fascicle were mapped before and about 8 months after transection of the entire nerve. After nerve regeneration, only about one half of type I fibres in each fascicle proximal to the lesion were found to be in the same fascicle distal to the lesion. Due to the somatotopic organization of the distal branches of cutaneous nerves, each fascicle would normally supply afferent fibres to different, and only slightly overlapping, parts of the FCN field. However, after regeneration the number of "correctly" and "incorrectly" innervated domes determined for each fascicle was similar to that expected if type I fibres had regenerated randomly to domes in either part of the FCN field. Normally, a single type I fibre innervates 1–4 domes, and an individual dome is usually innervated by only one afferent fibre (Horch et al. 1974). After regeneration the detailed pattern of innervation depended upon whether the nerve had been crushed or transected. Maps made in seven cats before and 3 months after crushing the FCN showed that almost all (98%) domes reappeared at the

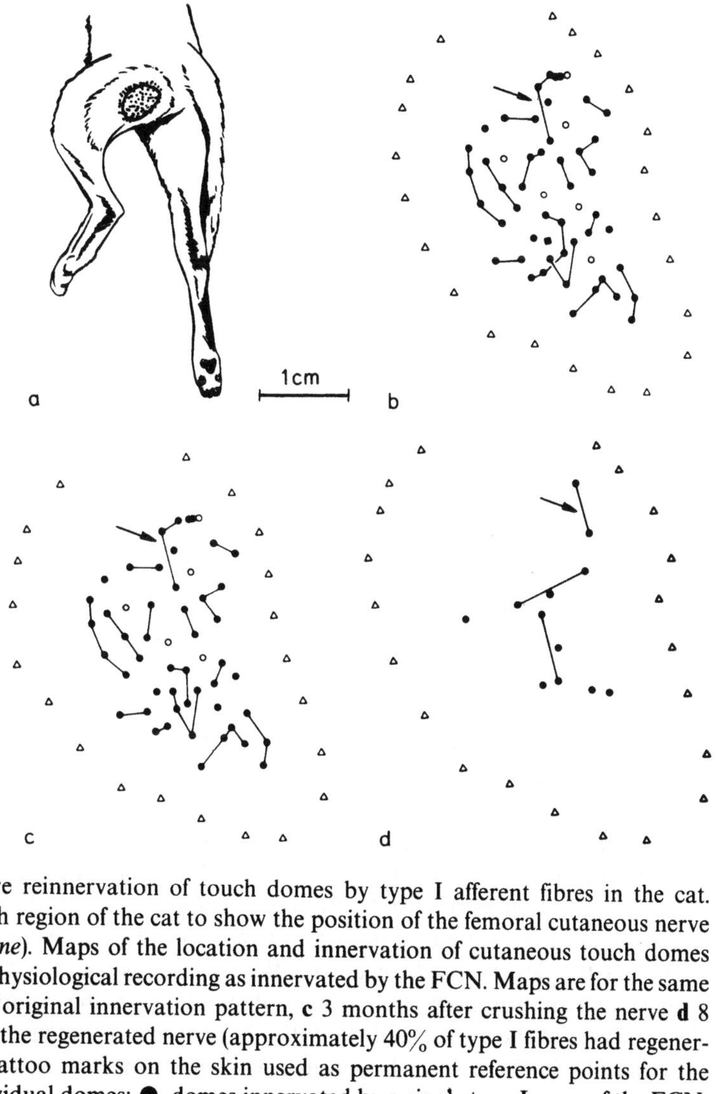

Fig. 2.8 a–d. Selective reinnervation of touch domes by type I afferent fibres in the cat. **a** Posterolateral thigh region of the cat to show the position of the femoral cutaneous nerve (FCN) field (*heavy line*). Maps of the location and innervation of cutaneous touch domes identified by electrophysiological recording as innervated by the FCN. Maps are for the same animal as follows: **b** original innervation pattern, **c** 3 months after crushing the nerve **d** 8 months after cutting the regenerated nerve (approximately 40% of type I fibres had regenerated). *Symbols*: Δ, Tattoo marks on the skin used as permanent reference points for the identification of individual domes; ●, domes innervated by a single type I axon of the FCN; ○, domes innervated by more than one type I fibre of the FCN. Lines join domes innervated by the same type I axon. Note that after nerve crush and regeneration the original pattern of domes innervated by the crushed nerve persists, and sets of domes originally innervated by one axon again become innervated by one axon. After nerve transection and regeneration there is a significant tendency for domes to reappear at the same position, but they are fewer in number and have a different innervation pattern. Only two domes that were originally innervated by the same axon are now innervated by a single axon (*arrow*). (**a** from Burgess et al. 1974, p. 62; **b** to **d** modified from Horch 1979, p. 1443)

locations of original domes, and that the innervation pattern of these domes was identical to that seen originally. Thus where two or more domes had been originally innervated by a single type I fibre, they were also innervated by a single fibre after regeneration (Fig. 2.8b). In 7 of the same 11 cats, selected at random, the FCN was then transected. The remaining cats again served as controls, and final maps of the innervation pattern of domes in the FCN field were made in all 11 animals 7–8 months later. The outcome of nerve regeneration after transection was very different from that after the earlier crush: not only were there now fewer domes but also their innervation pattern had changed. As in earlier work (Burgess et al. 1974) there was a significant tendency for domes to form at the original locations. About 60% of type I fibres in the cut FCN had evidently regenerated back to the skin and innervated domes, but as there were also, on average, only about half as many domes supplied by each type I fibre, the population of domes innervated by the FCN was reduced not by 40%, but by about 65%. There were few instances of two domes being innervated by a single type I fibre of the regenerated FCN, but in most (67%) of these cases the two domes had also been supplied by a single axon prior to FCN transection (Fig. 2.8c). The results of these experiments again strongly suggest that regenerating type I axons are guided back to the skin along specific conduits, which probably correspond to the endoneurial tubes of Schwann cells. The fact that domes tend to re-form at the same locations in the skin, and that multiple domes that were originally supplied by the same fibre become reinnervated by a single fibre after nerve crush indicates that the endoneurial tubes leading to the site of these receptors remain intact, even though the Merkel cells and epithelial specializations disappear after degradation of the afferent fibre. Presumably, there remains in the region previously occupied by a dome some special local property, possibly related to cells of the dermal papilla, which enables new Merkel cells and other dome specializations to redevelop under the trophic influence of the regenerated afferent fibre.

In the case of other specialized cutaneous receptors, which do not disappear after denervation, the endoneurial Schwann tubes would guide appropriate axons back to the partly degenerated receptor. Whether a particular type of receptor is reinnervated or not may then simply depend upon the number of regenerated afferent fibres of the correct submodality in proportion to the number of receptors. For example, in the case of Pacinian corpuscles, where each receptor is normally innervated by a single afferent fibre, there would be relatively few regenerated Pacinian afferents available, and reinnervation would be rare. In the case of Meissner corpuscles, where each receptor is innervated by several afferent fibres, there would be more regenerated fibres available, and reinnervation would be more common (Dellon 1981).

2.5 Modality Specificity of Somatosensory Nerve Regeneration

When myelinated afferent fibres regenerate back to the skin there are several possible ways in which recovery of the peripheral sensory apparatus could occur. First, each regenerating afferent fibre might induce the information of an appropriate type of receptor (e.g. Meissner corpuscle, Pacinian corpuscle, etc.). This possibility has been investigated in experiments involving transplantation of skin grafts or the reinnervation of healed wound sites. However, studies of the innervation of healed wounds in skin have been thoroughly reviewed by Dellon (1981), who concluded that what several workers had assumed to be new receptors were, in fact, preexisting ones physically transposed from adjacent uninjured skin by a process of "wound contracture". Second, afferent fibres might innervate whichever denervated receptor they happen to reach by growing along any available endoneurial tube, then make changes in their central connectivity appropriate to the new receptors. Among the more widely quoted evidence ostensibly showing that regenerating sensory fibres will reinnervate a different class of receptor than those they initially innervated, are the experiments of Boeke and colleagues on skin transplantation. When skin was transplanted from the bill to the leg of the duck, characteristic Grandry corpuscles and Herbst corpuscles developed in the transplanted skin, although these are not normally present in leg skin (Boeke and Dijkstra 1932; Dijkstra 1933). Of course, it could not be determined whether the afferent fibres which had reinnervated and induced the regeneration of these receptors had become fully functional, but in any case they could easily have belonged to an appropriate submodality class (slowly adapting for the Grandry corpuscle and Pacinian for the Herbst corpuscle). The importance of the reinnervated tissue in determining the pattern and type of sensory reinnervation has subsequently been stressed in a range of other studies of the reinnervation of transplanted skin (Adeymo and Wyburn 1957; Fitzgerald et al. 1967). However, the receptor structures formed when skin is transplanted from the foot to the snout of rabbits and guinea-pigs appears to be determined by the new trigeminal innervation of the graft (Kadanoff 1925). In a more definitive experiment, Schiff and Loewenstein (1972) transected the inferior mesenteric nerve (IMN), which carries the afferent fibres from Pacinian corpuscles in the cat's mesocolon, and united the distal cut end with the proximal end of the transected hypogastric nerve (HN), which normally carries visceral afferent fibres from the urinary bladder. Of a total of 8 HN/IMN grafts, all resulted in the reinnervation of Pacinian corpuscles within 45–47 days, and of 64 corpuscles with foreign innervation, 23 were found to produce generator potentials upon mechanical stimulation. The authors attributed the much improved reinnervation, when compared with experiments in which the two ends of the IMN were reunited (see Sect. 2.4.4), to improvement of operative technique (the HN/IMN grafts were done after the IMN repairs). A puzzling feature of these experiments was that the conduction velocities of many of the HN/IMN regenerated Pacinian afferents were much higher than those found after IMN repair. The mean conduction velocity of the normal Pacinian afferents was 28.5 m/s, but after IMN repair it dropped to only 3.8 m/s, whereas after HN/IMN union it was 18.2 m/s. The authors suggested that the Pacinian corpuscles had been reinnervated by regenerating afferent axons which originally supplied other types of receptor in the bladder wall (which has no Pacinian corpuscles), and that before transection some of these axons might have been of larger diameter than those normally innervating the corpuscles. Unfortunately, the functional properties of the several large-diameter (7–10 μm) and 40–70 medium-sized (4–6 μm) myelinated visceral afferent fibres known to run in the HN (Harris 1943) were not studied, and it is possible that those which reinnervated the Pacinian corpuscles had originally innervated a class of rapidly adapting mechanoreceptor. A further complicating factor is that many visceral afferent fibres appear to bifurcate in the inferior mesenteric ganglion, proximal to the level of IMN or HN transection (Harris 1943). Since both nerves originate from the ganglion it is conceivable that Pacinian afferents in the IMN also have collateral branches (perhaps non-functional) in the HN.

Third, afferent fibres may be restricted to growing down only those endoneurial tubes which formerly contained an afferent of the same functional type. Relatively few studies have been carried out to determine the functional properties of regenerated myelinated cutaneous sensory fibres in experiments in which transected or crushed nerves have been allowed to regenerate back into the region of skin they formerly innervated. However, the results of all such experiments strongly suggest that regenerating cutaneous sensory fibres are highly restricted in the type of functional endings which they form in the skin.

2.5.1 Regeneration of Myelinated Afferent Fibres to Hairy Skin

Since it is impossible at present to identify individual afferent fibres and their receptors before and after nerve injury and regeneration, the evidence regarding the specificity of regeneration of sensory afferent fibres must be indirect, and depends upon a comparison of a range of related functional properties of afferent fibres before and after injury and regeneration. These related properties may include (1) the receptive field size and submodality of an afferent, (2) its size (conduction velocity grouping), (3) its characteristic central projections (e.g. whether or not it projects to the dorsal column nuclei), and (4) the presence or absence in it, under certain conditions, of dorsal root reflexes. Additional evidence in favour of specific regeneration has been derived from a comparison of the relative abundances of different functional types of afferent fibres in regenerated nerves with those in normal nerves. This approach was first taken by Burgess and Horch (1973), who examined the functional properties and central projections of a large number of myelinated sensory fibres in the sural nerve of the cat between 300 and 750 days after transecting or crushing the nerve. Single afferent fibres were recorded 20–30 mm proximal to the level of the lesion with microelectrodes, and were located by electrical stimulation of the sciatic nerve. Of 397 fibres tested for projection to C2 after transection and regeneration, the proportion projecting (42.4%) was not significantly different from that for 387 fibres in control nerves (47.6%). Of 365 fibres studied in control sural nerves nearly all could be excited by mechanical stimulation of the skin, and 11 functional fibre types were identified on the basis of their receptor properties, essentially as found in an earlier study by Burgess et al. (1968). Various functional properties of these fibres are shown in Table 2.3 (Burgess and Horch 1973). Only one cat was studied after crushing the nerve, and the functional properties of all of a sample of 49 fibres were indistinguishable from normal. The functional properties of 445 fibres after transection and regeneration are shown in Table 2.4. About 10% of regenerated fibres lacked any discernible receptive field, though these tended to be fewer at later than at earlier times. About 19% of

Table 2.3. Properties of afferent fibres in control nerves (n = 635)

Afferent type	% of sample	Mean peripheral conduction velocity m/s	Percent projecting to C2	Mean dorsal column conduction velocity m/s
Insensitive hair	1.6	68	100	31
G_1 hair	12.8	75	96	34
Intermediate hair	4.1	66	83	35
G_2 hair	8.7	53	78	25
D hair	15.7	21	2	11
SA I	18.8	65	0	—
SA II	10.2	54	100	36
F_1 field	5.1	62	85	31
F_2 field	7.3	50	63	19
Tap	1.9	70	67	30
Nociceptor	13.0	27	0	—
Subcutaneous	0.6	60	33	39

Table 2.4. Properties of afferent fibres in regenerated nerves (n = 445)

Afferent type	% of sample	Mean peripheral conduction velocity m/s	Percent projecting to C2	Mean dorsal column conduction velocity m/s
Insensitive hair	2.7[a]	77[a]	100[d]	30
G_1 hair	3.4[b]	81[d]	93[d]	36
Intermediate hair	6.3[c]	73[d]	86[d]	30
G_2 hair	4.5[d]	61	35[b]	24
D hair	8.5[b]	23[d]	0[d]	—
SA I	10.6[b]	65[d]	0[d]	—
SA II	1.8[b]	60[d]	75[d]	26
F_1 field	6.5[a]	65[d]	76[d]	29
F_2 field	15.3[c]	60[c]	28[b]	22
Tap	3.6[d]	72[d]	81[d]	30
Inexcitable	9.0[c]	37	53	23
Nociceptor	6.3[d]	40[a]	0[d]	—
Subcutaneous	0.2[d]	68[d]	0[d]	—
Field or hair (phasic)	8.5	70	87	35
Field or hair (weak tonic)	5.6	55	40	30
Inexcit. or nocicept.	7.6	37	31	22

Statistical comparison of regenerated and control (Table 2.3) values:
[a] Regenerated significantly larger at 5% level.
[b] Regenerated significantly smaller at 1% level.
[c] Regenerated significantly larger at 1% level.
[d] No significant difference $P > 0.05$, two-tailed test.

myelinated fibres in the control sural nerves were type I afferents. Each typically supplied 2–4 touch domes and none was found to project to upper cervical levels. After regeneration, about 11% of fibres were classified as type I, and again none of these projected to C2. The appearance of the domes, sensitivity of the receptors and adaptation rate of the fibres were all normal. However, unlike normal type I fibres which supply 2–4 domes, the regenerated ones usually supplied only a single dome. Hair follicle afferent fibres in normal sural nerves were classified as G_1, G_2 or intermediate types identified according to: (1) conduction velocity, (2) the rate of hair movement required for excitation, and (3) the ease with which they could be activated by different anatomical types of hair (Burgess et al. 1968). Regenerated afferent fibres classified as G_1 or G_2 according to the effective rate of hair movement were found also to (1) innervate the appropriate types of hair, (2) have the appropriate conduction velocity, and (3) project in the normal ratio to upper cervical levels. The relative abundance of all types of Aß hair follicle afferents was normal at around 26% of the sample. Type II fibres, which normally terminate in Ruffini endings in the hairy skin of the cat (Chambers et al. 1972), were rarely found in the regenerated nerves, although they made up about 10% of fibres in normal nerves. Nevertheless, the few that were found had normal conduction velocities and central projection patterns. Fibres with "field" type receptive fields were more common than normal after regeneration, but the authors suggested that some of these might in fact have been regenerated, but poorly differentiated type I and type II fibres. Of the A∂ fibres, both D hair and mechanical nociceptor afferents were present after regeneration. Although the functional properties and projections of D hair afferent fibres were normal, there was a significant decrease in their numbers. Regenerated A∂ mechanical nociceptor afferent fibres also tended to have normal functional properties, except that they usually had abnormally small receptive fields, in some cases consisting of a single sensitive spot. Finally, while dorsal root reflexes could be evoked by stimulation of the sciatic nerve in most D hair fibres, they were usually absent from A∂ mechanical nociceptor afferents, and this distinguishing feature was also evident after regeneration.

While this study provides a large amount of *indirect* evidence in favour of there being submodality-specific regeneration of afferent fibres, a major interpretive problem is that it is not possible to tell whether a particular afferent fibre has specifically reinnervated the same type of receptor it previously supplied, or whether it has reinnervated any type of receptor, and has then been induced by some trophic influence from the receptor to adopt the appropriate functional properties and pattern of central projection. This problem was recognized by Horch (1976), who attempted to see whether evidence could be found of any rearrangement of the central projections of regenerating afferent fibres. The dorsal funiculus projection and peripheral and central conduction velocities of regenerating myelinated afferent fibres in the sural nerve of the cat were examined at periods of 1, 2, 4, 6, and 12 months following transection of the nerve. Fibres that could the activated by gentle mechanical stimulation of the skin were also classified by submodality. After 1 month only 3% of recorded afferents had low-threshold receptive fields, but by 4 months this had increased to more than 70%. By 6 months the whole sural nerve field had been reinnervated. All times the proportions of sural nerve fibres which projected as far as L3, T13, or C3 remained the same, and similar to controls. When the relationship between conduction velocity and frequency of projection to the three levels was examined, it was found that the slower ($A\partial$) fibres tended not to project: the faster ($A\beta$) fibres projected to each level with a frequency that was largely independent of conduction velocity, with the exception of fibres conducting at around 60 m/s which included the non-projecting type I afferents. These patterns were observed in both control and regenerated nerves. Horch (1976) considered that the lack of any measurable change in the proportion of fibres projecting to different levels during nerve regeneration indicated that there had been no substantial rearrangement of the central projections of regenerating afferent fibres. However, he also noted that a few regenerated fibres that did not reach C3 had abnormally high conduction velocities (> 20 m/s) between the L3 and T13 levels. These could have been fibres which originally projected to the DCN but were in the process of withdrawing this projection. However, since only 2 out of a total of 20 such fibres had a cutaneous receptive field, an alternative, and more likely, explanation is that these were fibres that had atrophied severely at C3 level owing to their inability to regenerate successfully back to the skin (Horch 1976).

2.5.2 Regeneration of Myelinated Afferent Fibres to Glabrous Skin

Electrophysiological studies of the functional properties of regenerated cutaneous myelinated afferent fibres have now been carried out in a number of different species, including man. These studies generally provide further evidence that, while regenerating afferent fibres regain their original functional specificity, they lack somatotopic specificity.

Glabrous Skin of the Hand in Subhuman Primates. Studies of the functional properties of regenerated myelinated afferent fibres in primate glabrous skin after nerve crush or transection have been carried out on the ulnar nerve of the baboon (Dykes and Terzis 1979; Terzis and Dykes 1980). As in cat hairy skin, after nerve crush, recovery of the functional properties of afferent fibres was virtually complete, in this case by 5 months after the lesion (Dykes and Terzis 1979). Surprisingly, the submodality of regenerated afferents was recognizable at a very early stage of recovery, at a time when the receptive field was still small and the threshold abnormally high. At this time rapidly adapting fibres had tuning curves characteristic of their submodality (either low velocity or Pacinian-like) and type I and II fibres could sometimes be distinguished from each other. Cutaneous receptive fields were initially small and irregular, but became larger and more regular in outline with time. After nerve transection with repair, a similar sequence of events occurred, but with a much slower time-course (Terzis and Dykes 1980). At 1 month after transection (just proximal to the wrist) many of the regenerating axons had entered the distal stump, but virtually none had formed a receptive field. However, by 3 months the proportion of fibres lacking a receptive field had fallen to about 50%, which is normal for the ulnar nerve. Unlike receptive fields after nerve crush, many (42%) of the receptive fields were split into two or more separate

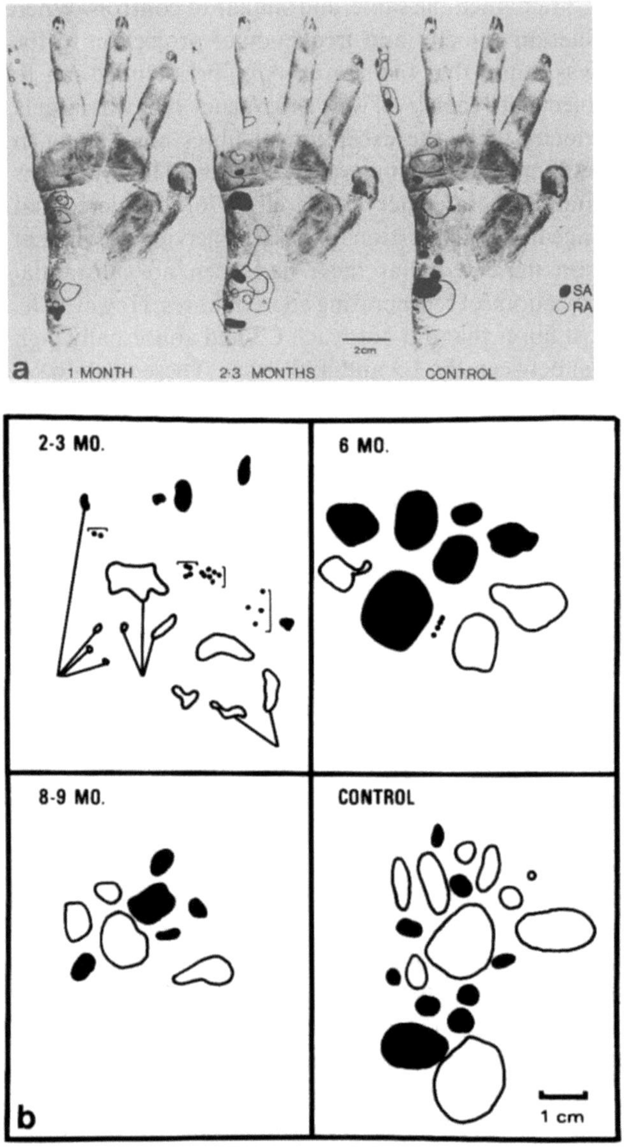

Fig. 2.9 a, b. Reinnervation of the glabrous skin in the hand of the baboon after peripheral nerve lesion. **a** Receptive fields of regenerated afferent fibres determined by single unit recording, at different times after regenerating fibres had demonstrably reached that area of the skin, after crushing all nerves to the hand except the median nerve. Receptive fields became progressively larger and more regular with time. Note that at 2–3 months there are axons with multiple receptive fields (joined by *lines* at left). **b** Typical receptive fields of cutaneous afferent fibres of the ulnar nerve in normal animals (*CONTROL*) and at the different times shown following transection of the nerve with epineurial repair. Receptive fields of regenerated afferents become progressively larger and more regular with time. Axons with multiple receptive fields (*lines* and *square brackets*) were common at 2–3 months, but had disappeared completely by 6 months.

areas (Fig. 2.9a). In all cases each part of the receptive field had the same submodality. The split receptive fields disappeared by 10 months and at intermediate times there was a tendency for the distance between their individual parts to be reduced (Fig. 2.9b). The mechanisms responsible for this coalescence of split receptive fields in the glabrous skin of primates is unknown. Presumably the phenomenon depends upon a continuous sprouting and retraction of axon branches, but it is not known whether this occurs within the skin, or at the site of nerve injury (Terzis and Dykes 1980). Certainly split receptive fields seem to be much more prevalent in primate glabrous skin than in cat hairy skin (Burgess and Horch 1973; Horch and Lisney 1981a), and their gradual disappearance is puzzling in view of the fact that regenerated fibres in the cat continue to support multiple branches in the distal stump long after nerve regeneration (Horch and Lisney 1981a).

Glabrous Skin of the Rat. Recently, similar results have been obtained in experiments where the plantar nerves supplying the glabrous skin of the hindpaw of the rat have been crushed about 10 mm proximal to the heel (Sanders and Zimmermann 1986). Initially, after 1 week, only responses from high-threshold mechanoreceptors could be obtained, and these had receptive fields clustered at the point where the plantar nerves enter the foot. Such high-threshold receptive fields preceded the appearance of low-threshold receptive fields as they spread distally across the paw. Unlike the low-threshold receptive fields of regenerated fibres in baboon skin, those in the rat did not appear to increase in size with time. Low-threshold, low-velocity rapidly adapting and slowly adapting mechanoreceptive afferents were recognizable after only 2 weeks, and Pacinian-like afferents could be found a week later. At this time the proportion of specific, low-threshold afferents increased to control levels, while the proportion of fibres with high-threshold receptive fields fell. It seems likely therefore that the high-threshold responses originated from the outgrowing sprouts of sensory fibres, which had not yet made functional connection with specific receptor structures.

Glabrous Skin of the Human Hand. In humans the technique of microneurography has been used to investigate the functional properties of regenerated sensory afferent fibres in glabrous skin of the hand (Mackel et al. 1983; Hallin et al. 1981). These studies have shown that three types of mechanoreceptors, corresponding to rapidly adapting, type I and type II afferents, are reinnervated. In most respects, these afferents regain their original functional properties, although the receptive field size of type I units is reduced. Reinnervation of Pacinian corpuscles may be very limited, as in baboon glabrous skin (Terzis and Dykes 1980), or may not occur at all. Other long-term abnormalities include a greater than normal proportion of unidentified units innervating deep tissue, a reduced proportion of type II afferents, and an abnormal spatial distribution of receptors, which become more numerous in

◀ **Fig. 2.9 a, b** (*continued*). At intermediate times there was a gradual reduction of the separation between the parts of multiple fields. *SA*, slowly adapting afferents (*filled fields*); *RA* rapidly adapting afferents (*open fields*). (**a** From Dykes and Terzis 1979, p. 1474; **b** from Terzis and Dykes 1980, p. 1218)

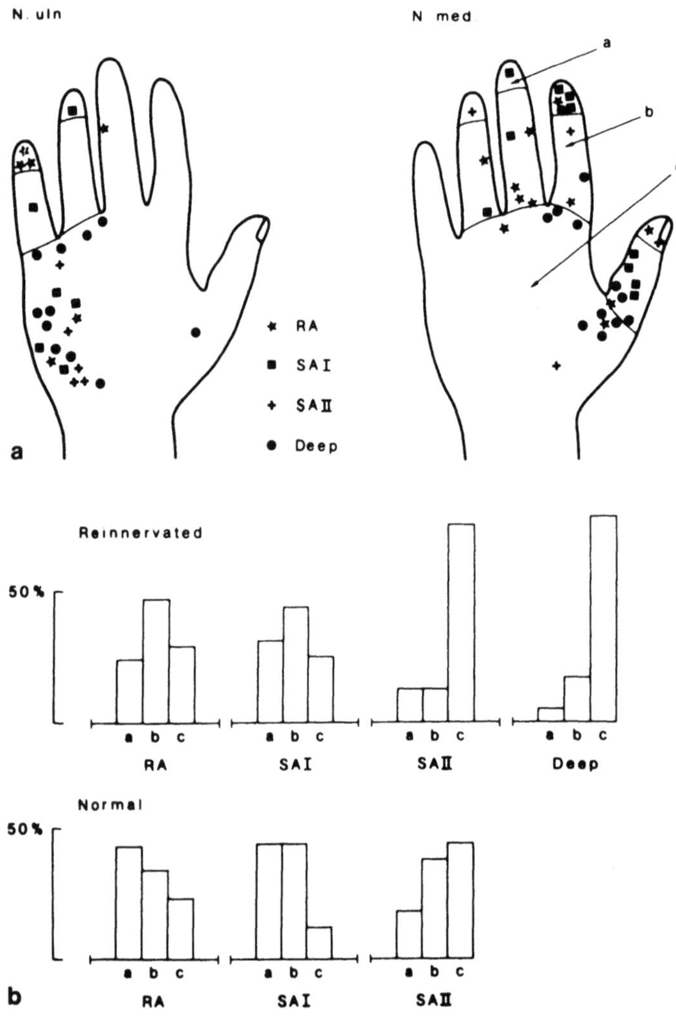

Fig. 2.10 a, b. Reinnervation of mechanoreceptors in human glabrous skin. **a** Location and distribution of the receptive fields of different functional types of mechanoreceptor afferent recorded by microneurography from regenerated ulnar and median nerves in patients who had previously suffered accidental transection and surgical repair of either nerve. The size of receptive fields is not indicated. **b** The distribution of receptive fields of different types of afferent in various parts of the hand in regenerated nerves (*upper*) compared to nerves of normal subjects (*lower*) (adapted from the data of Johansson and Vallbo 1979). Note the increased incidence of all types of receptor in more proximal skin after nerve regeneration. There is also an increase in the proportion of afferents with deep receptive fields. *RA*, rapidly adapting; *SA I*, slowly adapting type I; *SA II*, slowly adapting type II. (From Mackel et al. 1983, p. 59)

proximal skin of the digits and palm, rather than in the tips of the digits (Fig. 2.10). In general, the location and frequency of occurrence of reinnervated receptors could be correlated with varying clinical measures of the degree of recovery of sensory perception, in different regions of the hand.

Although the above studies certainly indicate a strong tendency for specific reinnervation of receptor types, a recent study by Koerber et al. (1989) in the cat has produced important and compelling new evidence that conflicts with this conclusion. As we have already seen, different kinds of cutaneous primary afferent fibres have distinctive regions of termination within the spinal dorsal horn (see Sect. 4.2). Other, non-cutaneous, primary afferent fibres also terminate in different regions of the spinal gray. For example, Ia afferents that innervate muscle spindles terminate in the ventral horn, in addition to the deep dorsal horn, while Ib afferents from Golgi tendon organs terminate only in the deep dorsal horn (for a review, see Fyffe 1984). Koerber et al. (1989) examined the relationship between the exteroceptive properties and conduction velocity of regenerated afferents and certain physiological indicators of their mode of termination within the spinal dorsal horn. In addition, the central termination of a small number of regenerated afferents was determined by intraaxonal staining with HRP. The physiological indicator used was the presence or absence, and characteristic spatial spread, of a cord dorsum potential (CDP) in response to intracellular stimulation of the single afferent either in the dorsal columns, or in the L7 DRG. In an earlier study, a clear and unique relationship between the single-fibre CDP and exteroceptive properties of an afferent had been established for a wide range of afferent fibres (Koerber and Mendell 1988). The regenerated afferents studied were from a mixed nerve, the tibial nerve in the cat, which had been transected and repaired between 6 and 14 months earlier. Of a total of 196 regenerated fibres studied, the adequate stimulus could be determined for 142, and of these, at least 72 (51%) were assessed to have reinnervated inappropriate receptors.

Abnormalities of regeneration included the following. (1) Cutaneous afferents that had reinnervated the wrong type of cutaneous receptor (e.g. a fibre which when stained with HRP, had spinal terminations characteristic of those of an A∂ high threshold mechanoreceptive afferent, but which had a very sensitive rapidly adapting receptive field on a footpad). (2) Muscle afferents that had apparently innervated cutaneous receptors (e.g. a fibre with spinal terminations and conduction velocity characteristic of a muscle Ia afferent that innervated a sensitive type I receptor in the central footpad). (3) Cutaneous afferents that had reinnervated muscle receptors (e.g. a probable original type I afferent that could be activated only by probing a small toe muscle). Koerber et al. (1989) concluded that many types of regenerating afferent fibres in a large mixed nerve can reinnervate inappropriate receptors, and that the outcome of regeneration may depend on the particular nerve chosen for study, and the nature of its peripheral field. While small homogeneous nerves supplying hairy skin, such as the sural nerve or the PFCN (see above) may exhibit apparently specific reinnervation due to the availability of appropriate targets, nonspecific reinnervation may be more usual in larger, mixed nerves. A similar conclusion has recently been derived from studies of the reinnervation of the mystacial vibrissae of rodents following transection of the infraorbital nerve

(Arvidsson and Johansson 1988) (Sect. 3.3.1). These results indicate that, in effect, a primary afferent fibre sprouting into the peripheral stump of a transected nerve may enter any available endoneurial tube, perhaps limited only by diameter, and grow down to reinnervate receptor structures associated with that tube's original axon. This finding has important consequences both for our approach to the surgical repair of nerve injuries (where it highlights the benefit of providing fascicular continuity), and for our understanding of sensory disturbances that follow nerve injury in humans.

2.5.3 Regeneration of Unmyelinated Afferent Fibres

Despite the considerable variability in the degree of recovery of cutaneous sensibility after nerve injury and regeneration in man (Walshe 1942; Sunderland 1978; Dellon 1981) there is general agreement that the ability to perceive strong stimuli as painful recovers earlier than the ability to perceive light tactile stimuli or innocuous warming or cooling (Rivers and Head 1908; Trotter and Davies 1909; Boring 1916; Sharpey-Schafer 1930; Dellon et al. 1972). In many cases of severe nerve injury recovery never progresses past the stage of the return of the sensation of pain. Furthermore the pain sensation may have a particularly unpleasant character (hyperalgesia), and may be evoked by innocuous stimuli (hyperpathia or allodynia). Allodynia, hyperalgesia and often chronic pain (causalgia) may persist for years following the nerve injury (Oester and Davis 1956; Onne 1962; Seddon 1943; Sharpey-Schafer 1928, 1930; Sunderland 1949; White and Selverston 1956). Many previous and authoritative reviews have covered these clinically important aspects of somatosensory plasticity, including several which have appeared during the last decade (Ochoa 1982; Ochoa et al. 1985; Devor 1984a,b; Wall and Devor 1978; Wall 1987).

Functional Properties of Regenerated Cutaneous C Fibres. Although the sensation of pain in man and animals has long been associated with activity in unmyelinated fibres (C fibres) (Clark et al. 1935; Collins et al. 1960; MacKenzie et al. 1975) and lately with C polymodal nociceptors (Willis 1985; Perl 1985), there have been relatively few studies of the regeneration of C fibres. Shea and Perl (1985) examined the receptive properties of 234 isolated regenerated C fibres in the great auricular nerve of the rabbit between 1 and 8 months after transection and repair of the nerve. Within 5 months of injury the proportion of C fibres which could be excited by cutaneous stimuli known to activate C fibres in the normal nerve returned to normal levels. The rate of regeneration of all types of C fibres was the same (1 mm per day) and was similar to that of myelinated fibres. Of the different functional types of C fibre, the lowthreshold mechanoreceptors and specific thermoreceptors had receptive properties like those in normal nerves. Polymodal nociceptors initially had elevated mechanical thresholds, but these returned to normal more than 2 months after injury. About 25% of polymodal nociceptors still had reduced thresholds to heat as long as 5 months after nerve transection, the mean lowering of threshold amounting to 3.8 °C. A very few units showed unusual bursting activity during gentle mechanical stimulation of the skin, and this could be modulated by changes in skin temperature. Still other units were excited by pressure applied to the transection neuroma, but these too were comparatively rare.

In the cat, C fibres responsive to radiant heat have a similarly reduced thermal threshold as well as an increased discharge rate after regeneration (Dickhaus et al. 1976). In agreement with the absence of any lowering of the mechanical thresholds of regenerated polymodal nociceptors in animals, the pain induced by gentle mechanical stimulation after nerve injury in man has been shown to be dependent on conduction in A but not C fibres (Raja et al. 1984; Meyer et al. 1985). However, it is not yet known whether the allodynia results from sensitization of A∂ mechanical nociceptive afferent fibres, from abnormal central processing of activity in low-threshold Aβ fibres, or from the central effects of abnormal activity patterns in Aβ fibres (see Sect. 8.5.1).

Thus, whether the abnormalities of regenerated C fibres are sufficient to explain the painful phenomena which occur in nerve-injured humans, or whether, in addition, some central mechanism must be involved, is uncertain. Since regenerated nerves have normal proportions of nociceptive and low-threshold mechanoreceptive C and A∂ fibres it can be stated with certainty that the abnormalities of cutaneous sensibility are not due to reinnervation of specialized low-threshold cutaneous receptors by "pain fibres", as originally proposed by Stopford (1926). Neither is it likely that the earlier return of noxious cutaneous sensibility is due to the faster regeneration of a distinct system of afferent fibres responsible for the sensibility of noxius heat, cold and pain—the so-called "protopathic" system of Head (Head et al. 1905; Rivers and Head 1908). A more plausible explanation for the early return of noxious sensibility is the collateral sprouting of intact A∂ and C fibres from neighbouring innervated skin; this provides a measure of protective innervation pending regeneration of the damaged nerve.

2.6 Major Conclusions

The response of the peripheral somatosensory system to nerve injury by crush or transection involves three stages, each of which may have an important bearing on the subsequent degree of recovery of cutaneous sensibility. These stages are: (1) death and degeneration of a number of peripherally axotomized sensory cells, (2) the partial reinnervation of denervated regions of skin by collateral sprouting of intact sensory fibers from adjacent skin, and (3) the reinnervation of denervated skin by regeneration of axotomized sensory neurons.

(1) Effects of Axotomy on Neurons of Sensory Ganglia

(a) The proportion of axotomized sensory neurons which die and degenerate following peripheral nerve injury without regeneration in adult mammals is probably quite small, not more than about 5%–10% per 100 postoperative days. Owing to this low rate of attrition, where nerve regeneration occurs within a matter of weeks the total loss of sensory neurons is likely to be insignificant.

However, where regeneration is delayed or prevented, as in some cases of nerve injury, or following amputation of a limb in humans, the loss of neurons may be substantial.

(b) At early times (a few days to a few weeks) after nerve section in rats and mice, some axotomized myelinated afferents transiently develop unusually high spontaneous activity originating at impulse-generating sites in the transection neuroma. After about 30 days, unmyelinated afferents (C fibres) may also develop spontaneous firing. These effects have not been observed following nerve injury in the cat.

(c) The proportion of axotomized sensory neurons which die and degenerate after peripheral nerve injury in immature animals may be substantial. For example, in the rat all axotomized DRG neurons probably die rapidly after peripheral nerve crush or transection during the first 9 days of life. The loss of neurons is probably due to their dependence on a supply of trophic substances (for example NGF) from the periphery—a supply that is cut off by peripheral nerve lesion.

(d) There is no clear evidence to show that either small or large sensory neurons are more likely to die following nerve transection in adult animals. However, in prenatal and neonatal animals, treatment with NGF antiserum preferentially causes the death of the small sensory neurons.

(2) Collateral Sprouting of Intact Sensory Axons into Denervated Skin

(a) When an area of skin is denervated by transection of a peripheral nerve there is rapid partial reinnervation of the denervated region by collateral sprouts of intact sensory axons of neighbouring nerves whose fields normally overlap that of the cut nerve. Sprouting does not proceed indefinitely; there are limits on the extent of collateral reinnervation, but the nature of these limits is poorly understood.

(b) In adult mammals, only small sensory fibres ($A\partial$ and C fibres) are able to *functionally* reinnervate denervated skin by collateral sprouting, although there is some *anatomical* evidence of sprouting by coarse cutaneous afferents. Collateral sprouting may be accelerated by the presence of afferent neural activity in the sprouting sensory axons, and by the presence of a steady electrical field across the denervated skin. There is evidence that in the trigeminal nerves some large diameter myelinated afferent fibres may also be capable of forming functional collateral sprouts.

(c) In neonatal rats, both small ($A\partial$ and C) and large ($A\beta$) diameter afferent fibres are able to sprout into a region of denervated skin. The ability of large diameter myelinated afferents to sprout may be restricted to a brief period between 14 and 19 days after birth.

(d) When the original axons regenerate, they reinnervate the denervated skin, and the collateral innervation is withdrawn. Collateral sprouting in adult rats is inhibited by treatment with NGF antiserum, but reinnervation of skin by regenerating sensory axons is unaffected. It is not known whether the collateral sprouts actually degenerate or simply become non-functional.

(3) Regeneration of Somatosensory Afferent Fibres

(a) After nerve transection and repair, recovery of sensory function is relatively poor, but after a simple crush recovery can be almost complete. The likely reason is that after nerve crush, regenerating axons are able to grow back down their original endoneurial Schwann tubes, which guide them back to their original receptors in the skin. After nerve transection, axons form multiple branches only a proportion of which are able to find their way into an endoneurial tube in the distal stump, and thus regenerate to the skin. The proportion of fibres which achieve such complete regeneration may thus be dependent on the approximation of the proximal and distal stumps.

(b) Most cutaneous low-threshold mechanoreceptors remain in the skin for a prolonged period following denervation, although they may undergo a variable degree of atrophy. In most cases (with the possible exception of the Pacinian corpuscle) they may be reinnervated by regenerating axons, and regain their structural integrity. Receptors thus tend to reappear at the same position in the skin that they occupied originally. There is recent evidence that regenerating large diameter myelinated afferents do not necessarily innervate the same type of peripheral receptor with which they were originally connected. These new findings underscore previous data which show that after nerve section regenerating afferent fibres do *not* make somatotopically appropriate connections with the skin.

(c) In the early stages of regeneration afferent fibres may innervate abnormally small and multiple areas of skin. Later these multiple receptive fields tend to coalesce to form a single receptive field of normal size. However, some axons may maintain multiple branches in the distal stump for long periods after regeneration. Thus some branches may become non-functional.

(d) C fibres regenerate at about the same rate as myelinated afferents. There is indirect evidence that on reaching the skin regenerated C fibres acquire the same submodality that they had originally. Initially, regenerated polymodal nociceptors have elevated mechanical thresholds, but these return to normal after about 2 months. Some regenerated polymodal nociceptors have a reduced threshold to thermal stimulation, and show spontaneous firing. Abnormalities of regenerated C fibres may be partially responsible for the painful syndromes which occur in nerve-injured humans. However it seems likely that central mechanisms are also involved in this particular form of plasticity.

3 Plasticity and the Mystacial Vibrissae of Rodents

3.1 General Account of Pathway

The neural pathway linking the mystacial vibrissae of rats and mice with the face region of the contralateral SI cerebral cortex has been used more extensively for studying the plasticity of neural connections than any other part of the somatosensory system. To the experimenter this system offers three unique advantages. Firstly, each vibrissa is easily detectable in prenatal, neonatal and adult animals (Vincent 1913; Zucker and Welker 1969; Woolsey and Van der Loos 1970; Van der Loos and Woolsey 1973; Van der Loos and Dorfl 1978; Yamakado and Yohro 1979; Andres and Van der Loos 1983, 1985). Secondly, in animals older than one week, the spatial arrangement of the vibrissae is replicated in histochemical and/or cytoarchitectonic features at the levels of the TNC, the VB complex and the cortex (Woolsey and Van der Loos 1970; Killackey et al. 1976; Van der Loos 1976; Belford and Killackey 1979; Killackey and Belford 1979; Ma and Woolsey 1984; Bates and Killackey 1985). Thirdly, these neuroanatomical correlates of the vibrissae are not apparent at birth but develop over the first week of postnatal life (Rice and Van der Loos 1977; Killackey and Belford 1979; Belford and Killackey 1979). These features underscore the importance of the vibrissa system in answering questions that relate to the plasticity of neural circuitry during its establishment. It is generally conceded that such questions are intimately bound to understanding the factors controlling the normal development of the brain rather than those mechanisms which confer upon the mature nervous system the property of plasticity.

Arrangement and Innervation of the Mystacial Vibrissae. The large mystacial vibrissae of rats and mice are arranged in five roughly horizontal rows along the whisker pad. In dorsal to ventral order these rows are labelled A, B, C, D and E with rows A and B containing fewer large vibrissae (mouse and rat 4) than rows C, D and E (mouse 6–7, rat 5–7). The vibrissae in each row are numbered such that number 1 is the most proximal (Woolsey and Van der Loos 1970; Arvidsson 1982; cf. Zucker and Welker 1969). Proximal of these rows is a dorsoventrally orientated row of four large vibrissae known as the *straddlers* (Woolsey and Van der Loos 1970) which are individually named alpha, beta, gamma and delta, in dorsal to ventral order (Fig. 3.1a, b). The whole mystacial region may be moved by four facial muscles – the *extrinsic* muscles. In addition, a striated, follicular muscle forms a sling around the follicle of most mystacial vibrissae. Contraction of these *intrinsic* muscles produces protraction of individual vibrissae (Dorfl 1982, 1985).

Both the intrinsic and extrinsic muscles are innervated by the facial nerve. In contrast, the sensory innervation of the whisker pad is provided by the infraorbital nerve (ION), the somata of the primary afferent fibres being contained in the

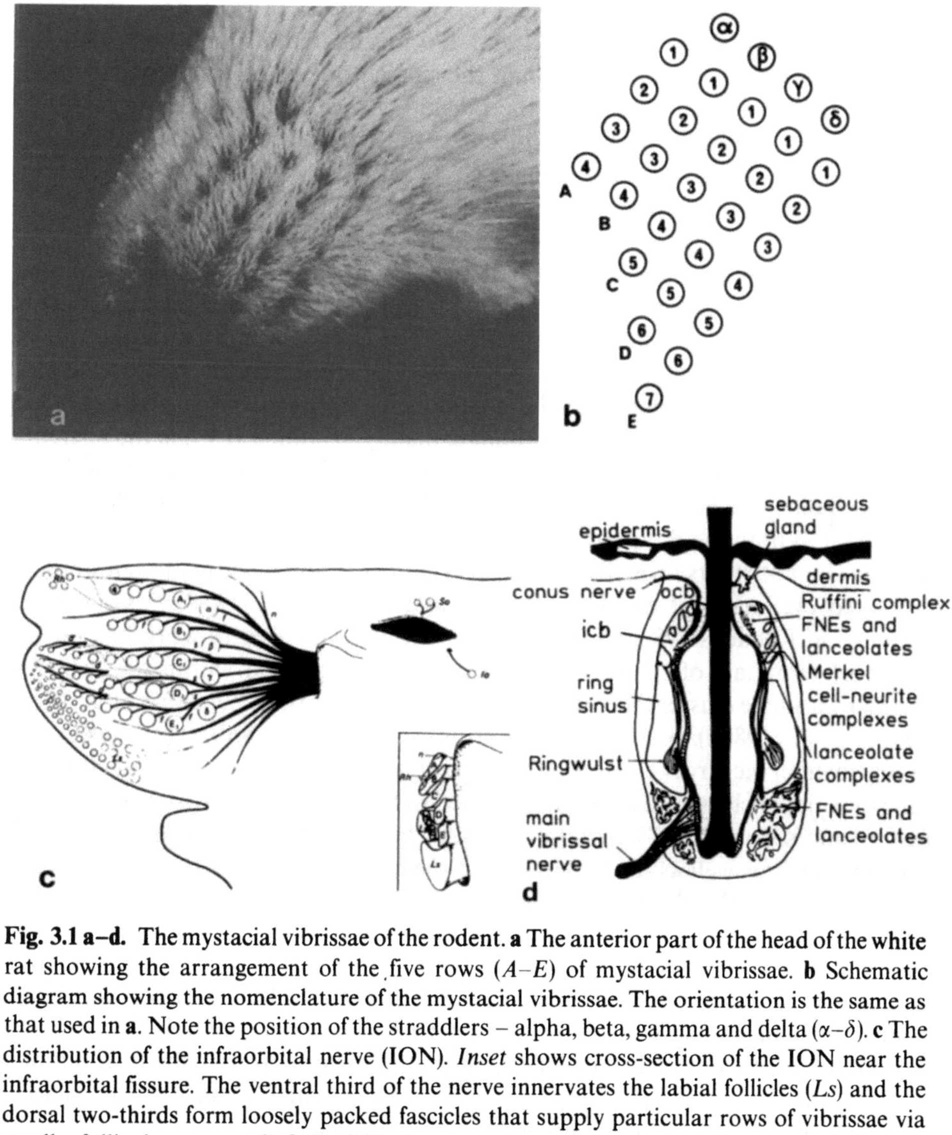

Fig. 3.1 a–d. The mystacial vibrissae of the rodent. **a** The anterior part of the head of the white rat showing the arrangement of the five rows (*A–E*) of mystacial vibrissae. **b** Schematic diagram showing the nomenclature of the mystacial vibrissae. The orientation is the same as that used in **a**. Note the position of the straddlers – alpha, beta, gamma and delta (α–δ). **c** The distribution of the infraorbital nerve (ION). *Inset* shows cross-section of the ION near the infraorbital fissure. The ventral third of the nerve innervates the labial follicles (*Ls*) and the dorsal two-thirds form loosely packed fascicles that supply particular rows of vibrissae via smaller follicular nerves (*f*). **d** The follicular structure and innervation of a mystacial vibrissa. Note the various receptor types and the free nerve endings (*F N E*) associated with the follicle and the two nerves – the small conus nerve and the main vibrissal or follicular nerve. *Other abbreviations*: *Rh*, rhinal follicles; *So*, supraorbital follicles; *Io*, infraorbital follicles; *d* and *s*, deep and superficial nerves; *ocb* and *icb*, outer and inner conical body. (**a, b** From Arvidsson 1982, p. 85; **c** from Dorfl 1985, p. 178; **d** from Renehan and Munger 1986, p. 131)

trigeminal (Gasserian) ganglion (Fig. 3.1c) (Dorfl 1985). In mice each mystacial vibrissa is innervated by between 52 and 170 myelinated afferents such that the larger vibrissae receive more fibres than the smaller ones (Lee and Woolsey 1975; Welker and Van der Loos 1983). In the rat, 150–200 axons have been reported to innervate each vibrissa (Fig. 3.1d) (Vincent 1913; Renehan and Munger 1986) and recordings from single trigeminal ganglion cells have shown that each afferent fibre innervates one and only one vibrissa (Zucker and Welker 1969).

Central Projections of the Vibrissa Afferents and Vibrissa-Related Structures in the TNC. Fibres from the trigeminal ganglion project to the principal sensory nucleus (PrV) and to *each* of the 3 subnuclei (oralis (SpVo), interpolaris (SpVi) and caudalis (SpVc)) of the spinal nucleus of the ipsilateral TNC (Fig. 3.2). In the rat evidence suggests that afferents supplying the common fur between the vibrissae project principally to the SpVc where they terminate mainly in lamina II while the afferents supplying the mystacial vibrissae project to the PrV, SpVo, SpVi and the magnocellular part of the SpVc of the complex (Morgan and Sharp 1986; Sharp et al. 1988). Within the PrV, SpVi and SpVc the terminals of the mystacial vibrissae and sinus hair afferents form discrete *clusters* (Fig. 3.2) which occupy longitudinally orientated cyclinders of neuropil (Belford and Killackey 1979; Arvidsson 1982; Jacquin and Rhoades 1983; Durham and Woolsey 1985). Recent work has shown that most terminals of single vibrissa afferents are confined within the cylinder formed by a single cluster (Hayashi 1980, 1985; Arvidsson 1982; Jacquin et al. 1986; Jacquin et al. 1986b). In transerve sections of the PrV, SpVi and SpVc, the arrangement of the larger clusters reflects the arrangement of the mystacial vibrissae while the smaller clusters are probably related to the smaller more rostral sinus hairs (Belford and Killackey 1979; Arvidsson 1982). The clustering of terminals in the PrV, SpVi and SpVc as well as the thalamus and the cortex (Fig. 3.2) is both a persistent condition and a developmental process which has been termed *segmentation* (Belford and Killackey 1979; Killackey and Belford 1979). Although Ivy and Killackey (1982) and Killackey (1985) later used this term to describe the aggregation of thalamocortical and trigeminothalamic cells into vibrissa-related groups, we shall restrict its use to the clustering of neuron terminals.

In the adult rat, retrogradely labelled trigeminothalamic neurons in the PrV, but not those in the SpVi or SpVc, occur in *aggregates* which replicate the spatial organization of the vibrissae (Erzurumlu et al. 1980; cf. Fukushima and Kerr 1979). Furthermore, in the mouse, but not the rat (Bates and Killackey 1985), there are cytoarchitectonically detectable specializations in the PrV, SpVi and SpVc that again reflect the arrangement of the vibrissae (Ma and Woolsey 1984).

Responses and Structure of Cells in the TNC. In the TNC of barbiturate-anaesthetized rats many cells respond to movement of more than one vibrissa. Nevertheless some degree of somatotopic organization, with respect to the mystacial vibrissae, does appear to exist in the PrV, SpVi and the magnocellular portion of the SpVc, as well as in the SpVo which does not show segmentation of afferent terminals (Fig. 3.3) (Nord 1967, 1968; Shipley 1974; Waite and Cragg 1982; Woolston et al. 1982; Waite 1984). Recently, intracellular staining of cells and

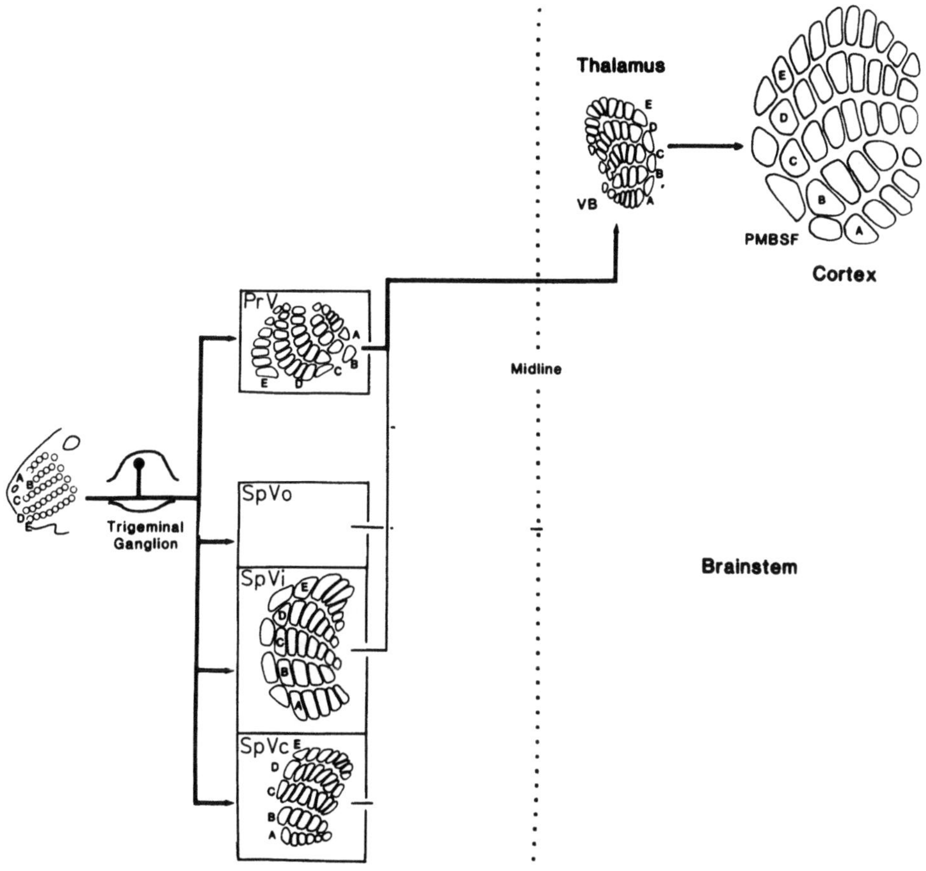

Fig. 3.2. Schematic diagram of the major ascending connections between the mystacial vibrissae and the SI cortex in rodents. Note the isomorphic representation of the vibrissae in macroscopically detectable structures or modules (clusters, barreloids or barrels) in the principal sensory nucleus (*PrV*), subnucleus interpolaris (*SpVi*) and subnucleus caudalis (*SpVc*), of the trigeminal nuclear complex, the ventrobasal (*VB*) complex of the thalamus and the posteromedial barrel subfield (*PMBSF*) of the SI cortex. No isomorphic representation has been described within subnucleus oralis (*SpVo*). (Modified from Durham and Woolsey 1984, p. 429)

primary afferent axons in SpVc and SpVi has shown that in the transverse plane the area contained within the dentritic tree of a single neuron is much larger than that covered by the terminals of an afferent supplying a single vibrissa (compare Fig. 3.4a–d with 3.4e–g) (Jacquin et al. 1986; Jacquin et al. 1986a). However, only in the SpVi (where trigeminothalamic cells were identified and stained) was there a relationship between the size of the dentritic tree and the number of vibrissae exciting a single neuron (Renehan et al. 1986a; Jacquin 1986a). Clearly, as for some cuneothalamic neurons in the cat (Fig. 5.22 and Sect. 5.4.3), the large dendritic trees of many neurons in the SpVi and SpVc, provide ample opportunity for them to receive inputs from primary afferents that supply regions of the body surface that lie

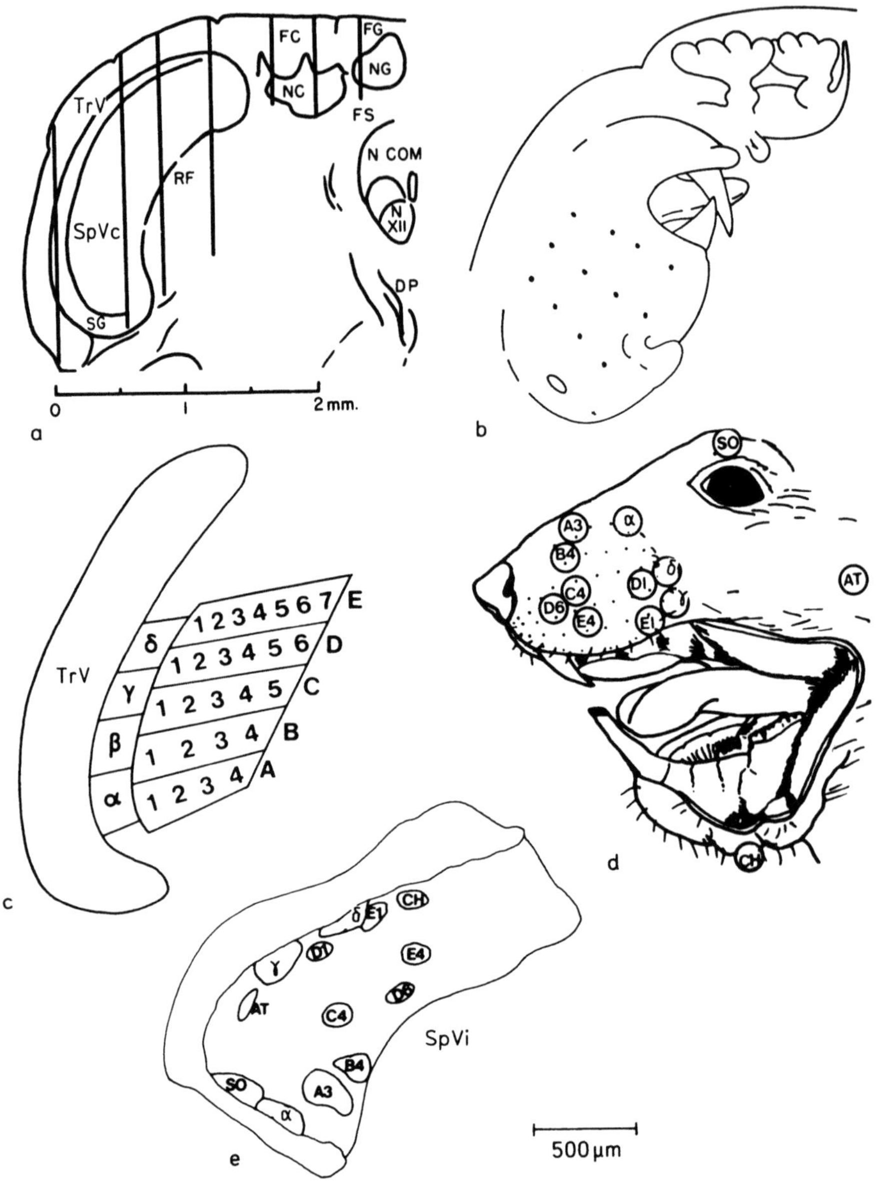

Fig. 3.3 a–e. Organization of primary afferent projections to the spinal nucleus of the trigeminal nuclear complex. **a** Transverse section through the SpVc, 1.0 mm caudal of the obex in the rat. **b** Ratunculus mapped across the SpVc and the dorsal column nuclei (*NC* and *NG* in **a**) at the same level as shown in **a**. Note the inverted representation of the face and representation of the forelimb in nucleus cuneatus (*NC*) and the hindlimb in nucleus gracilis (*NG*). **c** Schematic diagram of the position of the afferent terminals of single vibrissae in the non-laminated trigeminal sensory nuclei; relative to the spinal tract of the trigeminal nerve (*TrV*). **d**, **e** The position of the collateral arborizations **e** of 13 intra-axonally-stained, single vibrissa afferents within the SpVi. Ten afferents supplied mystacial vibrissae and 3 (*SO*, *AT* and *SH*) supplied large vibrassae elsewhere on the face **d**. Compare the position of the arborizations of the mystacial vibrissa afferents with the schema shown in **c**.

well outside their receptive fields as defined in anaesthetized animals. This is presumably a source of ineffective afferent input that may permit alterations of somatotopic organization in adult animals.

Vibrissa-Related Structures in the Ventral Posterior Medial Nucleus. In the ventral posterior medial nucleus (VPM – referred to by some authors as VBm) of the mouse thalamus curved, rod-shaped domains of about 70 μm in diameter which are poor in perikarya compared with the surrounding lattice have been described and termed *barreloids* (Fig. 3.5a–c) (Van der Loos 1976; Woolsey et al. 1979). Barreloids have not been described in adults rats. However, the injection of HRP into the parietal cortex reveals in 3–9, but not 15-day-old rats, curvilinear arrays of labelled thalamocortical neurons in VPM which resemble the barreloids seen in mice (Fig. 3.5f) (Wise and Jones 1978; Ivy and Killackey 1982; Jones 1986). Furthermore, in the VPM of rats histochemical localization of the mitochondrial enzyme succinic dehydrogenase (SDH) has revealed that the terminals of contralateral trigeminothalamic neurons are segmented into curved cylindrical columns of neuropil which resemble the groups of retrogradely labelled thalamocortical neurons (compare Fig. 3.5d with Fig. 3.5f) (Belford and Killackey 1979; Akers and Killackey 1979; Jones 1986). Like the clusters of vibrissa afferent terminals in the TNC of mice and rats, the barreloids in mice and the barreloid-like aggregations of thalamocortical cells in young rats, the terminal clusters of trigeminothalamic neurons in both mice and rats are arranged in a pattern resembling the arrangement of the mystacial vibrissae (Figs. 3.2 and 3.5a–c). In both mice and rats the optimal plane for viewing the isomorphic representation of the vibrissae in the VPM is close to the horizontal (Figs. 3.5a–c and 3.17b). Finally, the anterograde transport studies of Peschanski (1984) have suggested that in the rat the PrV, SpVi and SpVc all have overlapping projections to the VPM.

Responses of Cells in the Ventral Posterior Medial Nuclues. Electrophysiological studies of the dorsomedial part of the VB complex in anaesthetized mice and rats have revealed a somatotopic representation of the mystacial vibrissae of the contralateral face (Verley and Onnen 1981; Waite 1973a). In rats under urethane or light barbiturate anaesthesia, most thalamic neurons responded to one and only one vibrissa (Waite 1973b). A few neurons responded to between 4 and 6 vibrissae or to guard hair deflection or skin indentation. Neurons responding to the common fur of the contralateral vibrissal pad have been recorded in a small rostral region of the VB complex (Verley and Onnen 1981). Although no electrophysiological evidence has been found for a thalamic representation of the vibrissae or fur of the ipsilateral face, Fukushima and Kerr (1979) and Peschanski (1984) have reported a small ipsilateral projection from the PrV and SpVc, respectively (but see Lund and Webster 1967b; Smith 1973).

◀ **Fig. 3.3 a–e** (*continued*). *Other abbreviations*: *FC*, fasciculus cuneatus; *FG*, fasciculus gracilis; *SG*, substantia gelatinosa; *N COM*, nucleus commissuralis; *N XII*, nucleus of the 12th cranial nerve; *DP*, decussation of the pyramids. (**a**, **b** From Nord 1967, pp. 350, 352; **c** modifies from Arvidsson 1982, p. 89; **d** and **e** modified from Jacquin et al. 1986e, p. 278)

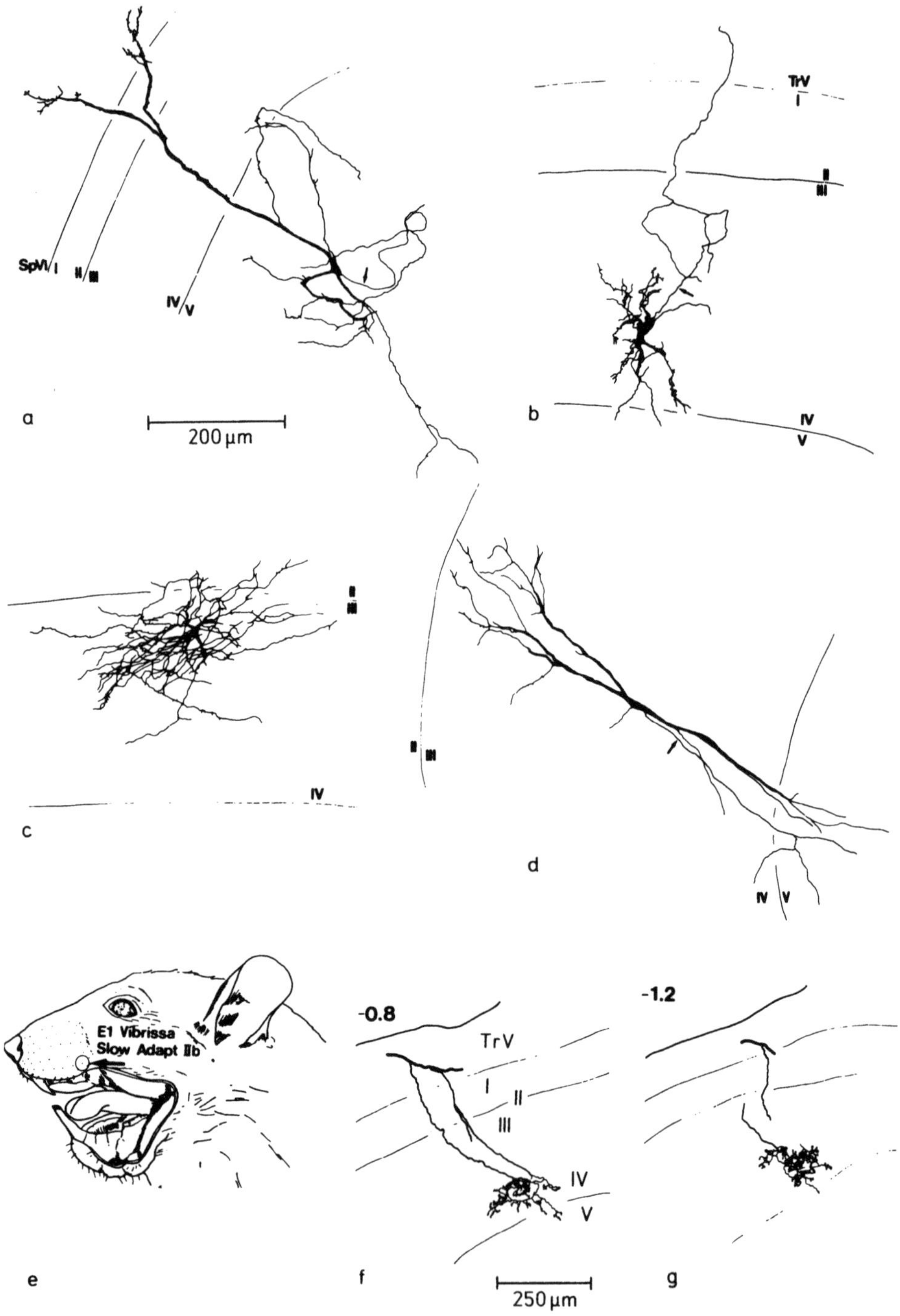

Fig. 3.4 a–g. The morphology of physiologically typed cells and vibrissa afferent arbors in the magnocellular part of subnucleus caudalis of the trigeminal nuclear complex. **a–d** intracellularly stained cells in laminae III, IV and V of subnucleus caudalis that were activated by movement of the following facial hairs:

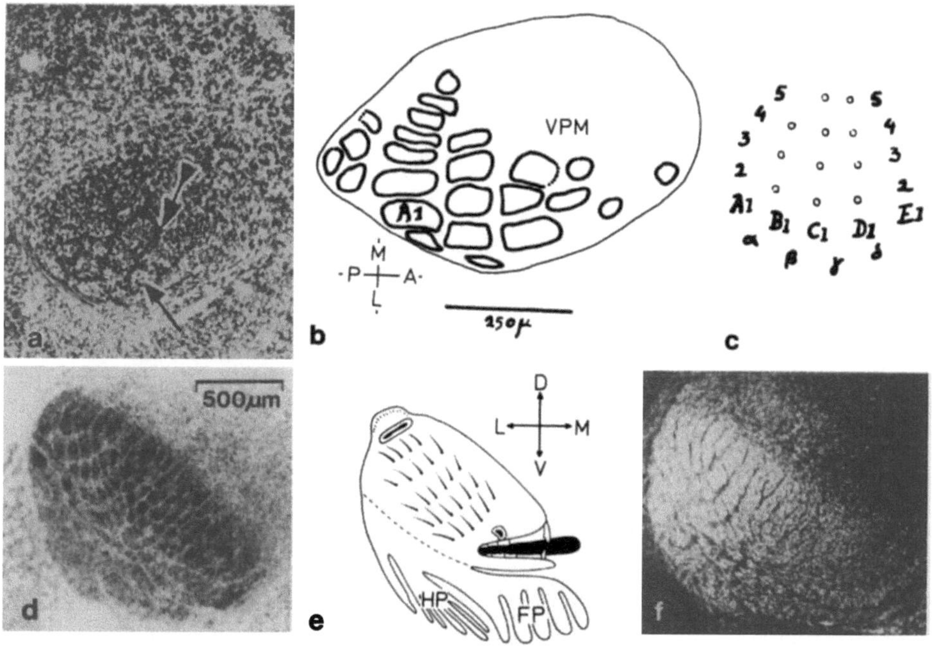

Fig. 3.5 a–f. Anatomical organization of the vibrissa representation in the VB complex of the thalamus. **a** Approximately horizontal, Nissl-stained section through the barreloids of the right VPM of the mouse thalamus. The *plain arrow* points to a cell-sparse core and the *winged arrow* to the cell-dense wall. **b** Reconstruction of the barreloids from two adjacent sections showing the position of the barreloid that corresponds to the A1 vibrissa. **c** Plan of all the barreloids showing their arrangement and nomenclature according to the mystacial vibrissae. **d–f** Organization in the coronal plane of the VB complex of the rat. **e** The somatotopic organization of the head and body in the coronal plane of the VB complex of the adult rat. **d** The segmentation of the terminals of trigeminothalamic neurons in the face representation of a rat on P 7 (SDH staining). **f** The distribution of retrogradely, HRP-labelled, thalamocortical cells in a rat on P 8 (darkfield illumination). Note the similarity in the pattern of the trigeminothalamic neuron terminals **d** and the cell bodies of thalamocortical neurons **f** within the representation of the mystacial vibrissae **e**. *Other abbreviations*: *FP*, forepaw; *HP*, hind paw; *D*, dorsal; *V*, ventral; *L*, lateral; *M*, medial; *P*, posterior, *A*, anterior. (**a–c** From Van der Loos 1976, pp 2, 4; **d–f** modified from Ivy and Killackey 1982, p. 1)

◀ **Fig. 3.4 a–g** (*continued*). **a** A single vibrissa in the D row; **b** the gamma straddler and the C1 and D1 vibrissae; **c** the gamma and delta straddlers, the C1, D1, D2, E1, E2, E3 and E4 vibrissae and guard hairs between these vibrissae; **d** the beta straddler, the A1, A2, A3, A4, B1, B2 and B3 vibrissae and the guard hairs between rows A and B. **e–g** the central arborizations, within subnucleus caudalis, of a single intra-axonally-stained afferent that was activated by movements of the E1 vibrissae. Note that the dendritic trees of cells shown in **a** and **d** are far more extensive than the arborizations of this single afferent (**f** and **g**) or the territory within the vibrissae representation, occupied by the terminations of all the afferents supplying a single vibrissa (see Fig. 3.3c, e). Anaesthetic used, pentobarbitone 60 mg/kg I.P. (**a–d** From Renehan et al. 1986a, p. 1192; **e–g** from Jacquin et al. 1986d, p. 1162)

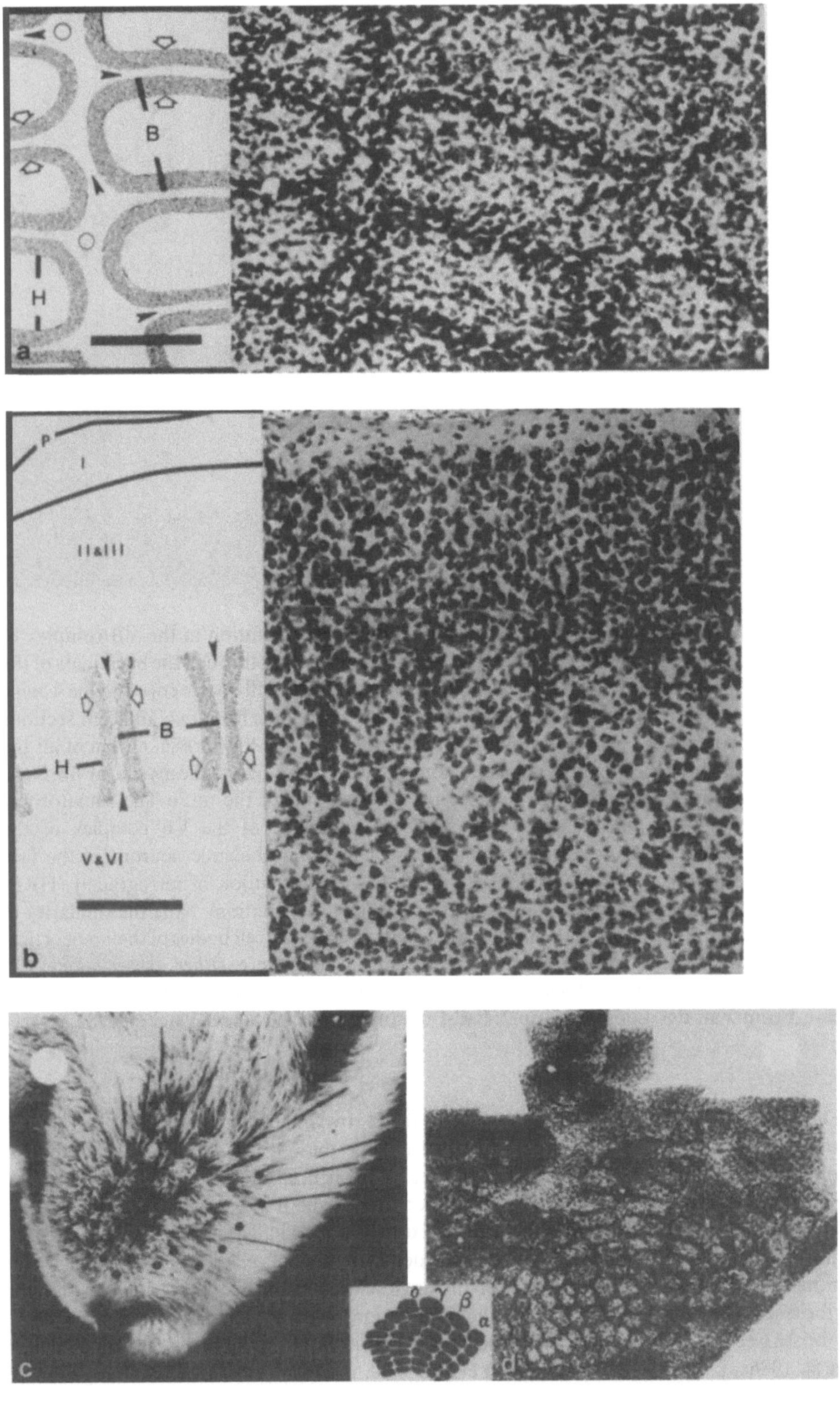
a
B
H
b
P
I
II&III
B
H
V&VI
c
d
δ γ β α

Vibrissa-Related Structures in the SI Cortex. In layer IV of the face representation in SI cortex discrete, cytoarchitectonically identifiable structures called barrels have been described in both rats and mice (Fig. 3.6a). In sections perpendicular to the cortical surface these structures look like the sides of tiny barrels that rest on layer V and support layer III (Fig. 3.6b). Viewed in the tangential plane the largest barrels are arranged in a pattern which resembles the arrangement of the mystacial vibrissae and the area of cortex containing them is called the posteromedial barrel subfield (PMBSF) (Figs. 3.2 and 3.6c, d). In this plane the PMBSF barrels of mice are elliptical and about 280 μm along the major axis whereas those of rats are circular and measure about 400 μm in diameter. In both rats and mice the barrels consist of cell-dense walls which are separated from the walls of adjacent barrels by regions called septa that contain few cells (Fig. 3.6a). In mice the barrel centres are called hollows and contain few cells while in rats the barrels are filled with small neurons (Woolsey and Van der Loos 1970; Welker and Woolsey 1974; Pasternak and Woolsey 1975). In mice there are from 1035 to 2624 neurons in the 6 barrels which are topographically related to the 6 vibrissae of row C – the number of neurons in each barrel being related to the number of myelinated afferents (mean of 69–162) supplying the appropriate vibrissa (Pasternak and Woolsey 1975; Lee and Woolsey 1975). In mice Golgi studies have shown that in layer IV of the PMBSF, 85% of cells lie in the barrel wall or hollow and have dendrites that are confined to that barrel. The remaining 15% of cells extend dendrites into 2 or more adjacent barrels. Of these two-thirds (10%) are located in the septa while one-third (5%) are located in the hollows (Woolsey et al. 1975). In mice, neurons located in the barrel walls have dendrites which project into the hollow of the barrel (Fig. 3.7a–d) (Woolsey et al. 1975; Killackey and Leshin 1975; Steffen 1976; Steffen and Van der Loos 1980). Morphological studies of the dendritic trees of the barrel neurons of rats have revealed a similar organization (Simons and Woolsey 1984).

Anatomical studies utilizing anterograde degeneration produced by lesions in the VB complex (Petrovicky and Druga 1972; Killackey 1973; Caviness et al. 1976; Cragg 1975; Donaldson et al. 1975; Killackey and Leshin 1975; Killackey et al. 1976; Waite 1977), SDH histochemistry (Killackey and Belford 1979; Pidoux et al. 1980) and cytochrome oxidase (CO) histochemistry (Wong-Riley and Welt 1980; Land and Simons 1985) have shown that the terminals of many thalamocortical fibres are arranged into clusters each of which is centred on a single barrel within the PMBSF (Fig. 3.2). In both rats and mice these clusters are located in layer IV and the

◀ **Fig. 3.6 a–d.** The morphology and organization of the barrels in the posteriomedial barrel subfield (PMBSF) of the SI cortex of the mouse. **a** Tangential, methylene blue-stained section through layer IV of the PMBSF and a drawing of part of that section. **b** Coronal section through the PMBSF and a drawing of part of that section showing the curvature of the walls which gives these aggregrations of cells the appearance of barrels. *P* indicates the pia on the surface of the cortex. In **a** and **b** the *open arrows* indicate the barrel walls, the *filled arrows* indicate the cell-sparse septa separating the walls of adjacent barrels and *H* shows the hollow of one barrel. *B* indicates, in both **a** and **b**, the minor axis of a barrel. **c, d** show the mystacial vibrissae of a mouse and, at the *top*, the isomorphic arrangement of the large barrels that make up the PMBSF (see *inset*). (From Woolsey and Van der Loos 1970.) *Scale-bar* **a** and **b**-100 μm

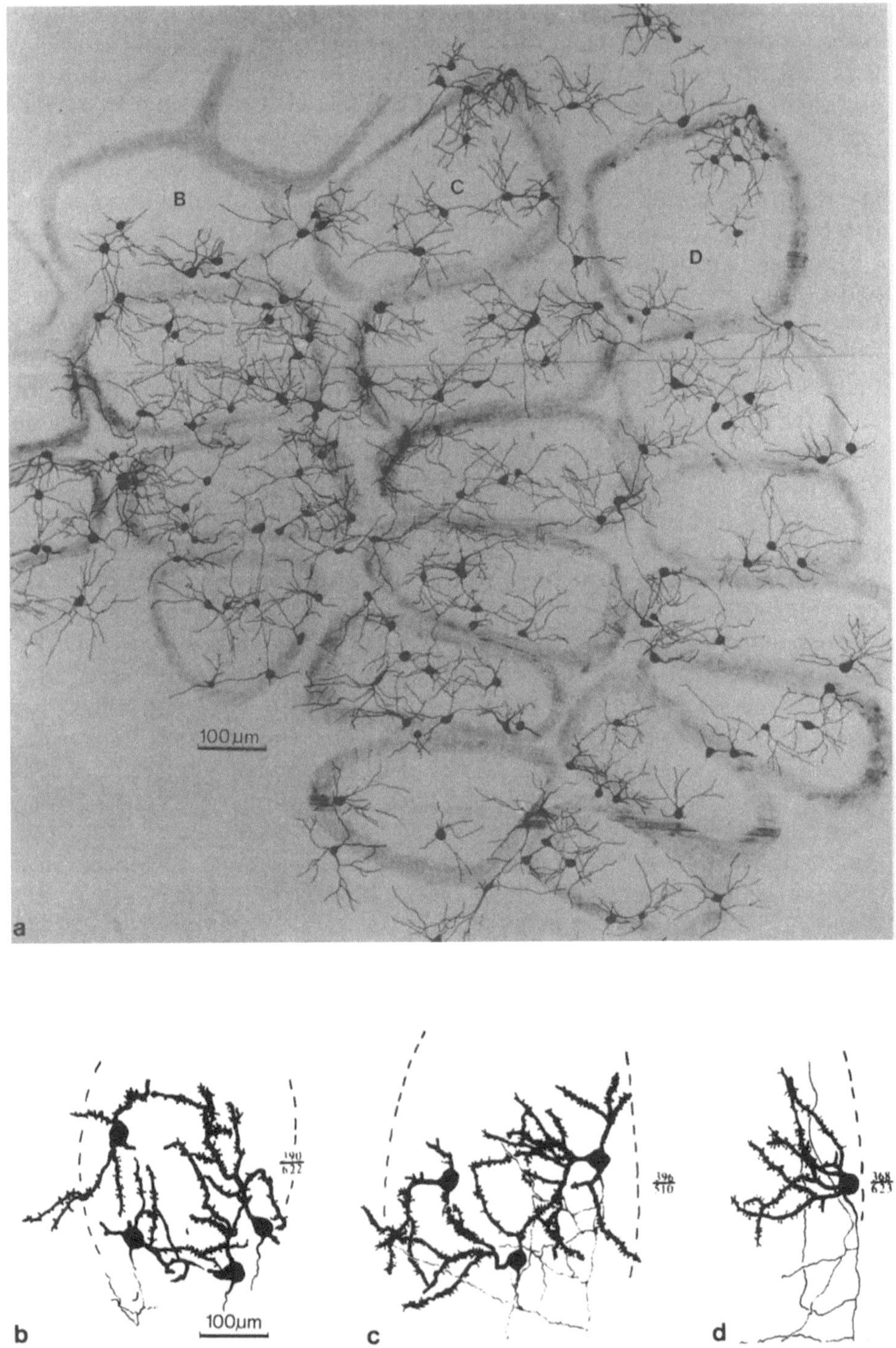

Fig. 3.7 a–d. The dendritic architecture of layer IV neurons with respect to the barrels in the PMBSF of the mouse. **a** Camera lucida drawing of cells in a Golgi-Cox/Nissl-stained, tangential section through the barrel rows *B*, *C* and *D*.

ventral portion of layer III where their arrangement is isomorphic to that of the mystacial vibrissae. All workers agree that lesions within the VPM result in degenerating terminals within the walls and hollow, or centre, of cortical barrels and some also report degeneration in the septa between barrels (Donaldson et al. 1975; Cragg 1975; Caviness et al. 1976) – a region which has been shown to receive projections from the medial portion of the posterior group of the thalamus (Chmielowska et al. 1989). In addition, diffuse degeneration in layers I, II and in patches in layer VI underlying individual barrels has been reported (Petrovicky and Druga 1972; Caviness et al. 1976; Cragg 1975; Donaldson et al. 1975; Jones and Leavitt 1974) but Wise and Jones (1978) have concluded that, in that rat, the VB complex projects to lower layer III and layers IV and VI and that thalamocortical projections to layer I probably arise from the intralaminar nuclei. Attempts to demonstrate unequivocally that the cells in a single barreloid project exclusively to a single barrel have so far been inconclusive (Donaldson et al. 1975; Land et al. 1986).

Responses of Cells in SI Cortex. In rats and mice vibrissae in the same position on the ipsilateral and contralateral mystacial pad activate neurons within concentrically arranged areas of the PMBSF although at a particular locus in the PMBSF the number of ipsilateral vibrissae represented is greater than the number of contralateral vibrissae (Pidoux and Verley 1978). This means that in the experiments of Pidoux and Verley (1978) a single ipsilateral vibrissa was represented over a larger area of cortex than its contralateral homologue. Evidence suggests that the ipsilateral input arrives via the corpus callosum from neurons in the PMBSF of the contralateral hemisphere (Pidoux and Verley 1978; White and De Amicis 1977; Olavarria et al. 1984). As in the thalamus, the common fur of the contralateral mystacial pad (together with the smaller sinus hairs) is represented in a separate area that abuts the representation of the mystacial vibrissae (Pidoux and Verley 1978; Nussbaumer and Van der Loos 1985).

While there are a number of studies of the cortical representation of the contralateral vibrissae in rats and mice there are discrepancies in the reported number of vibrissae (i.e. the size of the receptive field) which are effective in firing a single cortical neuron. Single unit and multiunit studies in mice anaesthetized with methoxyflurane/nitrous oxide (Nussbaumer and Van der Loos 1985) and in barbiturate-anaesthetized rats (Welker 1971, 1976) have shown that neurons in the PMBSF are excited by a single vibrissa. In contrast, Axelrad et al. (1976) found, in barbiturate-anaesthetized mice and rats, that many cortical cells responded to more than one vibrissa. In mice and rats allowed to recover from pentobarbitone anaesthesia 58% and 85% of single layer IV neurons, respectively, responded exclusively to a single vibrissa. Across all layers 51% of cells in mice and 55% in rats respond to only one vibrissa. In layer IV recording loci in the septa yielded responses to larger numbers of vibrissae (Simons

◀ **Fig. 3.7 a–d** (*continued*). **b–d** Camera lucida drawings of cells in Golgi-Colonnier-stained, coronal sections through single barrels, the sides of which are indicated by *stippled lines*. *Fractions* indicate the level in the cortex; the *numerator* is the distance in μm below the surface and the *denominator* is the distance from the white matter. (**a** From Steffen and Van der Loos 1980, p. 423; **b–d** from Steffen 1976, p. 58)

1978; Simons and Woolsey 1979). Recently, a quantitative study in rats under light barbiturate and halothane anaesthesia has shown that in layer IV single cells respond to movements of 1–4 vibrissae. In the supragranular layers cells often responded to any vibrissa throughout a row or an arc while in the intragranular layer cells responded to any vibrissa within a large group (Chapin 1986). Finally, with urethane anaesthesia no cells in 7-day-old rats responded to a single vibrissa (Armstrong-James 1975) although in the adult rat under urethane 38% of cells in layer IV, 69% of cells in lalyer V and 38% in all layers respond to more than one vibrissa (Ito 1981; Princince and Donoghue 1986). The finding that a higher percentage of cells in layer IV than in other layers respond to a single whisker is in keeping with the anatomical findings that the terminations of VB neurons projecting to the PMBSF are mostly confined to within the barrels (see above). When the cortical activity produced by the stimulation of a single vibrissa has been assessed using the 2-deoxyglucose (2-DG) method in very young (Fig. 3.21) or adult rats, uptake appears to be confined to vertical columns of tissue that are centred on the appropriate barrel. These columns extend throughout all layers and spread tangentially slightly beyond the territory of the appropriate barrel (Hand 1982; Kossut and Hand 1984a, b; Levin et al. 1988). Contrary to expectations based on the electrophysiological observations of Chapin (1988) and others, the tangential spread of 2-DG uptake in the supra- and infragranular layers does not include the representations of whole rows or arcs of vibrissae. What tangential spread is observed could easily be accounted for by short interneuronal pathways between neighbouring columns or barrels or even the extension of the dendrites of cells within one column or barrel into a neighbouring column or barrel (Fig. 3.7) (Simons and Woolsey 1984).

The existence of cells responding to inputs from two or more contralateral vibrissae has led Simons (1978) and Simons and Woolsey (1979) to refer to the *principal* whisker as that vibrissa which elicited the most vigorous response in a cortical neuron. Where cortical recording loci have been accurately marked it has been shown that with very few exceptions, the principal or single vibrissa to which a given neuron, or group of neurons, in layer IV responds, is the vibrissa that is topographically matched to the barrel containing the neuron, or group of neurons (Welker 1971, 1976; Simons 1978; Simons and Woolsey 1979; Nussbaumer and Van der Loos 1985). It may thus be concluded that while the PMBSF contains a relatively precise somatotopic map of the contralateral vibrissae, the receptive fields of barrel neurons are often composed of more than one vibrissa.

3.2 Normal Development of the Vibrissae and Their Neural Connections to the Cerebral Cortex

Follicle Formation, Trigeminal Ganglion Cells and the TNC. In the following text the day of conception will be referred to as gestation day 0 (G 0) and the day of birth as postnatal day 0 (P 0). The gestation period of rats is 22 days and that of mice 19 days. In mice at G 10 nerve fibres from the trigeminal ganglion invade the whisker pad without making contact with the epidermis (Tello 1923; Andres et al. 1981;

Table 3.1. Temporal sequence of development in the vibrissa system of the rat and the mouse

Developmental process	Mouse	Rat
Sensory neuron and skin		
Trigeminal ganglion neurons born	G 10–13	G 11–13
Primary afferents invade whisker pad	G 10	G 13
Primary afferents invade the TNC	G 10	G 13–14
Vibrissa pattern formed on skin	G 12–15	G 14–18
Primary afferent terminals cluster in TNC	?	P 0
Trigeminal nuclear complex (TNC)		
TNC neurons born	?	G 12–15
Trigeminothalamic axons in VB complex	P 0	G 20–P 0
Trigeminothalamic cells aggregate in TNC	?	P 0–2[a]
Trigeminothalamic terminals cluster in VB complex	P 3–6	P 1–4
VB complex of thalamus		
VB complex neurons born	?	G 16
Thalamocortical axons in SI cortex	G 15–17	G ?–P 0
Thalamocortical cells aggregate to form barreloids	P 3	P 2–8[b]
Thalamocortical terminals cluster in PMBSF of SI	P 2	P 3–6
PMBSF of SI cortex		
Neurons of layer IV of SI cortex born	G 14–16	G 18
Layer IV cells aggregrate to form barrels	P 3–6	P 3–6

Gestation periods: Rats 22 days; Mice 19 days.
Abbreviations: TNC, trigeminal nuclear complex; G, gestation day; G 0, first 24 h after fertilization; P, postnatal day; P 0, the first 24 h after birth. Where a range is shown it represents the time taken for the completion of the developmental process (see exception, footnote [b]); ?, time unknown.
[a] Principal sensory nucleus only.
[b] Ephemeral feature, disappears by P 15.

Andres and Van der Loos 1985). At the same time they also make contact with the alar plate of the brain stem (Davis et al. 1981). The vibrissal follicle pattern is laid down between G 12 and 16 (Fig. 3.8c) (Yamakado and Yohro 1979; Van Exan and Hardy 1980; Andres and Van der Loos 1983) and has been shown to develop even in noninnervated embryonic skin that has been placed in culture (Fig. 3.8b) (Andres and Van der Loos 1982, 1983). At G 10 the rudiments of the vibrissae first appear as domes on two longitudinal ridges on the muzzle, the lateral nasal and the maxillary prominences, that are separated by the nasolacrimal groove (Fig. 3.8a). Rows A and B develop on the lateral nasal prominence and rows C, D and E develop on the maxillary prominence (Yamakado and Yohro 1979). Within the TNC the latest cell division takes place between Gs 10 and 13 (Taber Pierce 1970) but selective stimulation of the vibrissae in row C does not produce 2-DG labelling until Ps 2–4 (Melzer et al. 1986).

Current knowledge regarding the sequence of developmental events in the vibrissa system is summarized in Table 3.1. In rats follicle formation is detectable at G 14 and

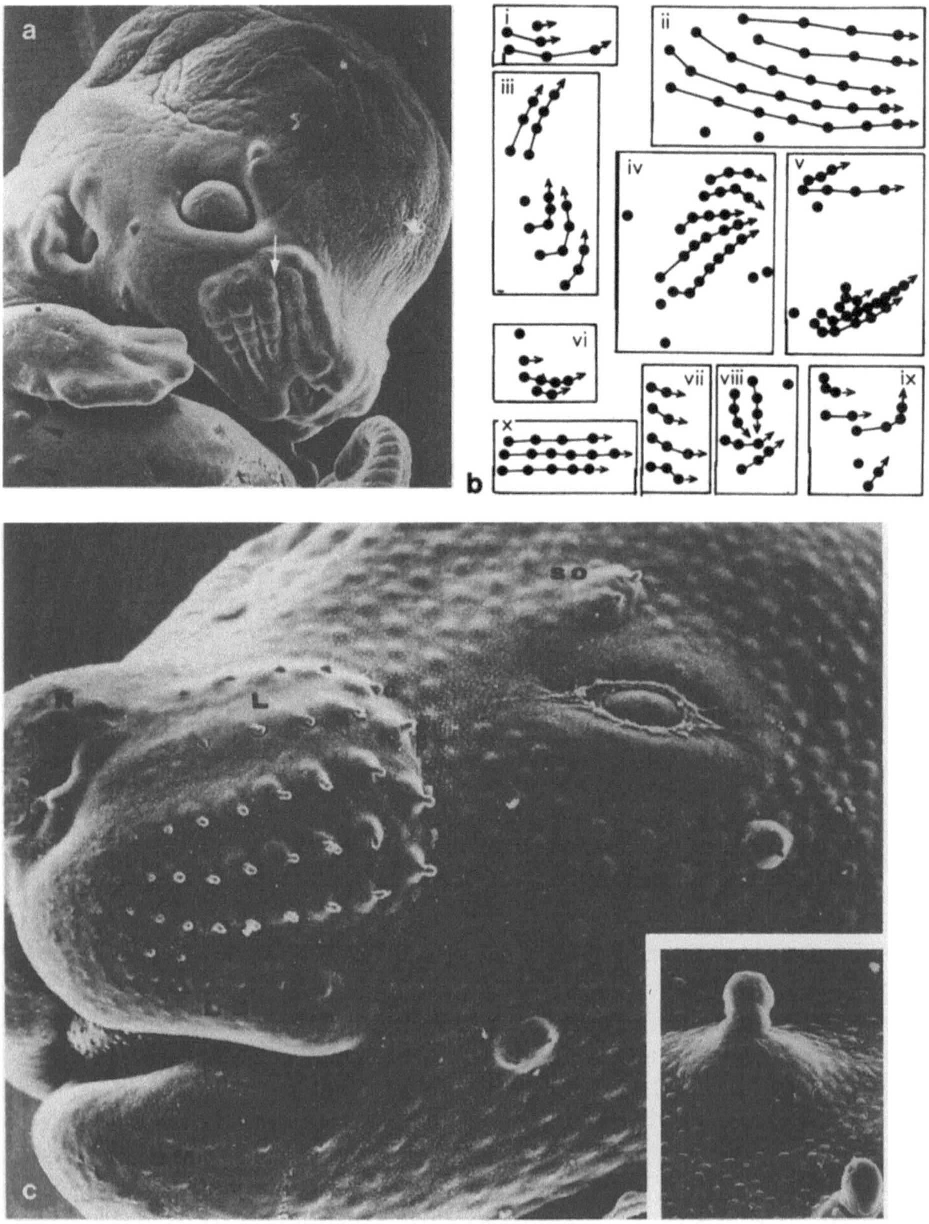

Fig. 3.8 a–c. The *in vivo* and *in vitro* development of the mystacial vibrissae on the skin. **a** Scanning electron micrograph of the head of a mouse embryo on G 11. The rudiments (domes) of the mystacial vibrissae are already developed and lie in the five ridges that correspond to vibrissae rows A to E. The *white arrow* indicates the nasolacrimal groove that separates the lateral nasal rudiments (rows A and B) from the maxillary rudiments (rows C, D and E). **b** A computer-aided reconstruction of the position of the mystacial vibrissae on the snout of a 12-day-old (**bi**) and a 13-day-old (**bii**) embryo and on skin explanted (before innervation, see Table 3.1) from a 9-day-old embryo and cultured, for 6–7 days, on the chorioallantoic membrane of the chicken egg (**biii** to **bx**).

some primary afferent fibres reach the periphery before this time (Erzurumlu and Killackey 1982; Killackey 1985). Most large neurons of the trigeminal ganglion are born between G 11 and 13 with the last evidence of the birth of small neurons being on G 15 (Altman and Bayer 1982; Forbes and Welt 1981). By G 13 the three divisions of the trigeminal ganglion are apparent and the trigeminal afferents have formed broad topographical relations with the face and with the brainstem (Fig. 3.9a). Fasciculation of the nerve into bundles which are related to rows of vibrissae occurs between Gs 15 and 18 (Fig. 3.9b) (Erzurumlu and Killackey 1982; Killackey 1985) and it has been suggested that Merkel cells in the embryonic whisker pad are foci that are attractive to the peripheral terminals of developing primary afferents (Killackey and Belford 1980). Within the TNC itself neurons are born between Gs 12 and 15 (Altman and Bayer 1982; Nornes and Morita 1979; see also Clark 1979). At birth segmentation of primary afferent terminals is just detectable in the PrV, SpVi and SpVc (Belford and Killackey 1979; Erzurumlu and Killackey 1983) while the ephemeral aggregation of trigeminothalamic cells into vibrissa-related groups occurs over Ps 0–2 (Erzurumlu et al. 1980; Killackey 1985). Segmentation has not been reported in the mouse TNC but in the hamster (gestation period 16 days) segmentation in the TNC is evident on but not before P 5 – a developmental stage equivalent to P 0 in the rat (Jacquin and Rhoades 1987).

The VB Complex. In the rat, neurons in the VB complex are born abruptly on G 16 (McAllister and Das 1977) and by G 20 (2 days before paturition) trigeminothalamic fibres have reached the caudal and ventral portions of the VB complex (Killackey 1985). At birth the terminals of trigeminothalamic neurons and neurons of the dorsal column nuclei have reached the rostral and medial portions of the VB complex (Fig. 3.5). Although there is already a high level of SDH activity in the dorsomedial portion of the complex (Belford and Killackey 1979), the density of fibres in this region increases further over the following 4 days (Killackey 1985). At P 1 the terminals of trigeminothalamic fibres have clustered into rows which relate to the rows of vibrissae and from P 2–4 further segmentation occurs until each cluster may be clearly related to a single vibrissa (Belford and Killackey 1979; Belford et al. 1984; Killackey 1985). In the first two weeks of life changes occur in the distribution of thalamocortical cells across the VB complex. Thus at Ps 0 and 1 they are uniformly distributed but between Ps 2 and 8 they are aggregated into discrete curvilinear arrays which, like the barreloids of the mice, reflect (in coronal sections) the arrangement of the vibrissae. By P 15 these aggregations are no longer detectable (Ivy and Killackey 1982). By comparison, in the mouse at P 0 a high level of CO

◀ **Fig. 3.8 a–c** (*continued*). In **biii** to **bx** the *arrows* indicate what are considered to be rows of vibrissae that have developed on the cultured, non-innervated, skin. In three cases (**biii, biv** and **bv**), five rows are discernible. **c** Scanning electron micrograph of the head of a mouse embryo on G 13. Vibrissae are now growing from the top of the rudiments (see *inset*) and rudiments of the primary pelage hair (small domes) now appear to cover most of the facial skin. The lateral nasal (*L*) and maxillary (*M*) divisions of the whisker pad and the supraorbital vibrissa (*SO*) are indicated. (**a, c** Modified from Yamakado and Yohro 1979, p. 165; **b** modified from Andres and Van der Loos 1983, p. 311)

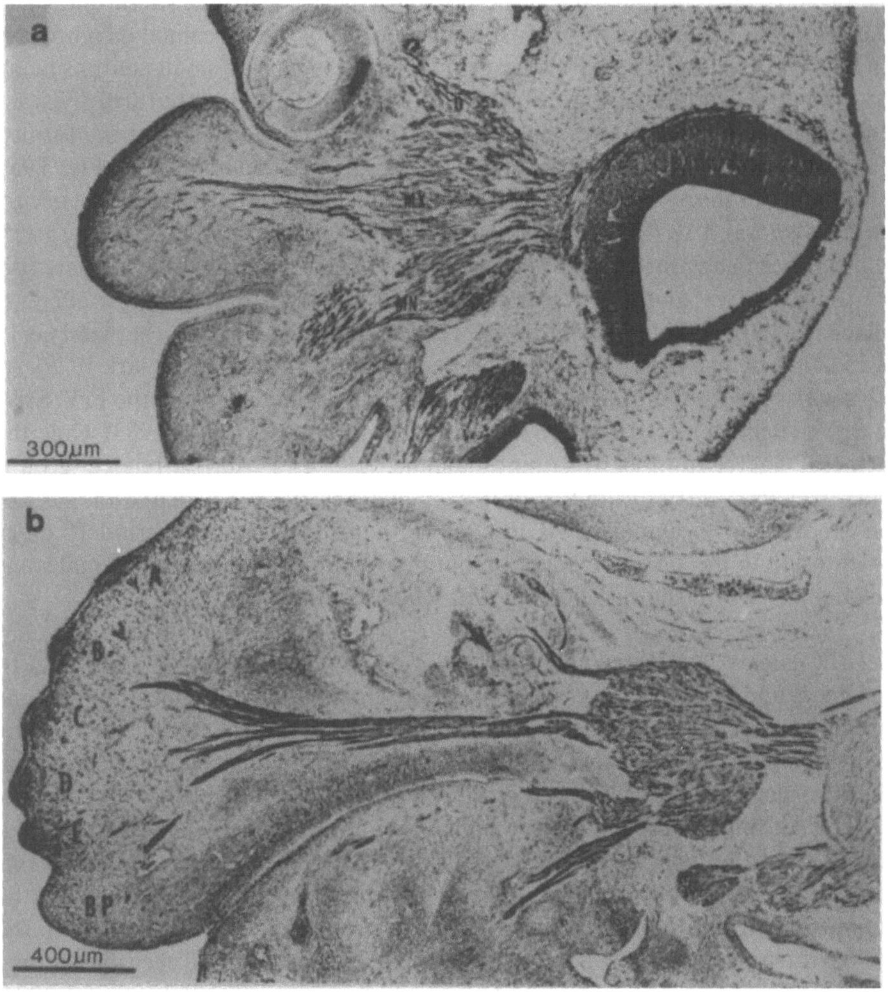

Fig. 3.9 a, b. Development of the central and peripheral connections of the trigeminal ganglion of rat. **a** Parasagittal section through the head of an embryo on G 13 showing the three distinct divisions of the trigeminal ganglion (*O*, ophthalmic; *MX*, maxillary; *MN*, mandibular) and the linear arrangement of their respective processes. Note that the fasciculated axons of the maxillary nerve have arrived at the maxillary process and that they are close to the epidermis. **b** A photomontage of two successive parasagittal sections through the head of an embryo on G 15. The fascicles of axons from the maxillary nerve fan out toward the five rows of vibrissae (labelled *A* to *E*) and the buccal pad (*BP*). The ophthalmic nerve has reached the supraorbital vibrissae (*arrow*). (Modified from Erzurumulu and Killackey 1983, pp. 368, 369)

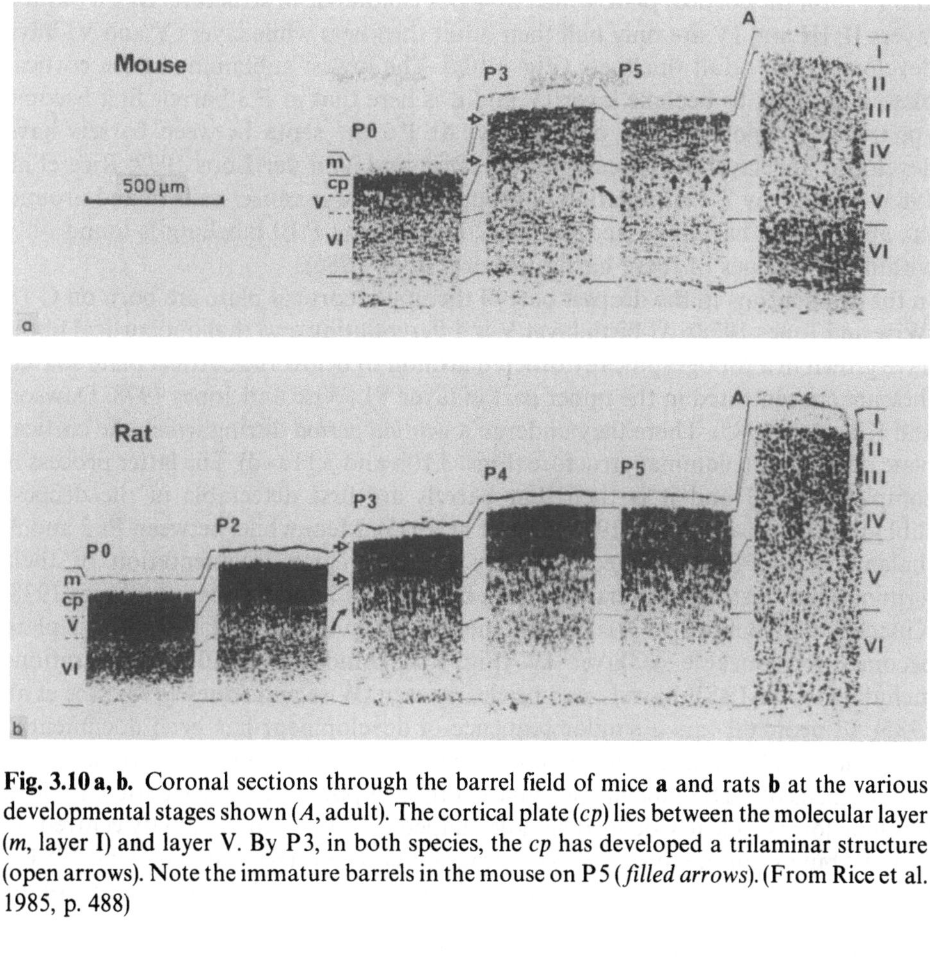

Fig. 3.10 a, b. Coronal sections through the barrel field of mice **a** and rats **b** at the various developmental stages shown (*A*, adult). The cortical plate (*cp*) lies between the molecular layer (*m*, layer I) and layer V. By P3, in both species, the *cp* has developed a trilaminar structure (open arrows). Note the immature barrels in the mouse on P5 (*filled arrows*). (From Rice et al. 1985, p. 488)

activity is apparent in VPM suggesting the presence of trigeminothalamic terminals. However, neither the barreloids nor segmentation of trigeminothalamic terminals are detectable until P 3 (Woolsey et al. 1979; Yamakado 1985). Furthermore, although by P 4 stimulation of the vibrissae causes cells in the VB complex to accumulate 2-DG (Melzer et al. 1986), septal delineation of individual barreloids is not established until P 6 and maturation of synaptic connections in the barreloid neuropil is incomplete until P 13 (Yamakado 1985).

The SI Cortex. The development of the cytoarchitecture in the cortex, and of the projection patterns of thalamocortical afferents, has been intensely studied. In the mouse rough topographical ordering of thalamocortical projections is present at Gs 15–17 (Crandall and Caviness 1984). However, not until P 2 are presumptive clusters of thalamocortical axon terminals detectable in what will become layer IV (Waite 1977). Neurons of layers VI and V are born on Gs 10–13 and by P 0 layer V is differentiating from the cortical plate (Fig. 3.10a). Cells of layer IV are born on G 14 after which cell birth shifts from layer IV to layer II by G 18. Together layers II, III and IV form the cortical plate which by P 3 is trilaminar in structure. By Ps 5 and 6 layers II, III and IV are only half their adult thickness while layers V and VI have developed their adult thickness (Fig. 3.10a). The lowest sublamina of the cortical plate is destined to become layer IV and it is here that at P 3 barrels first become apparent as regions of low cell density. At P 6 the septa between barrels have developed and layer IV is recognizable (Rice and Van der Loos 1977; Rice et al. 1985). Finally, by P 7 stimulation of groups of vibrissae causes cells in and around the appropriate barrels to accumulate 2-DG and by P 10 labelling is found only within the confines of these barrels (Melzer et al. 1986).

In the rat, neurons in the deepest part of the upper cortical plate are born on G 18 (Wise and Jones 1978). At birth layer V is differentiating and thalamocortical fibres have grown in a topographically ordered fashion to below the cortical plate so that they are concentrated in the upper part of layer VI (Wise and Jones 1978; Dawson and Killackey 1985). There they undergo a *waiting period* during which the cortical plate develops a trilaminar structure (Figs. 3.10b and 3.11a–d). The latter process is complete by P 3 and it is then that barrels are first detectable in the deepest sublamina (Wise and Jones 1978; Rice et al. 1985). Meanwhile, between Ps 2 and 3, thalamocortical fibres project diffusely to this region, segmentation of their terminals into distinct clusters occurring between Ps 3 and 6 (Wise and Jones 1978; Killackey and Belford 1979). On P 4 the deepest sublamina of the cortical plate becomes recognizable as layer IV (Fig. 3.10b) and the cellular aggregrations including the PMBSF barrels, can be clearly seen (Wise and Jones 1978; Rice et al. 1985). In neonatal rats a similar sequence of development has been documented using the 2-DG technique and selective stimulation of the C3 vibrissa. Between P 4 and 6 a focus of 2-DG labelling is apparent in the upper third of the cortex. By P 12 labelling forms a clearly defined spindle-shaped coloum of high activity centred on layer IV but extending to supra- and infragranular layers (Kossut and Hand 1984a). At all ages the tangential distribution of 2-DG activity exceeds the boundaries of the appropriate barrel showing that the intracortial influence of one vibrissa extends into adjacent barrels (Fig. 3.21) (Hand 1982; Kossut and Hand 1984b).

General Conclusions on Development. The chronology of the events which establish connectivity between the mystacial vibrissae and the cerebral cortex of mice and rats is summarized in Table 3.1. Near complete data are only available for the rat but what are available for the mouse suggest that it follows a similar developmental sequence. The following general statements may be made concerning the development of connections in this part of the rat somatosensory system. First, the birth of neurons at all four levels (trigeminal ganglion, TNC, thalamus and cortex) occurs within an 8-day period (Gs 10–18) with cells at the higher levels being born after cells at the lower levels. Towards the end of this period the follicles of the mystacial vibrissae differentiate in a manner that requires no innervation and yet establishes the pattern of receptive structures which will eventually be manifested at all three levels of the somatosensory system. Second, shortly after birth the axons of sensory neurons, trigeminothalamic neurons and thalamocortical neurons are found in those regions of the periphery and/or central nervous system that they will occupy in the mature animal. These cell groups thus appear to establish independently the

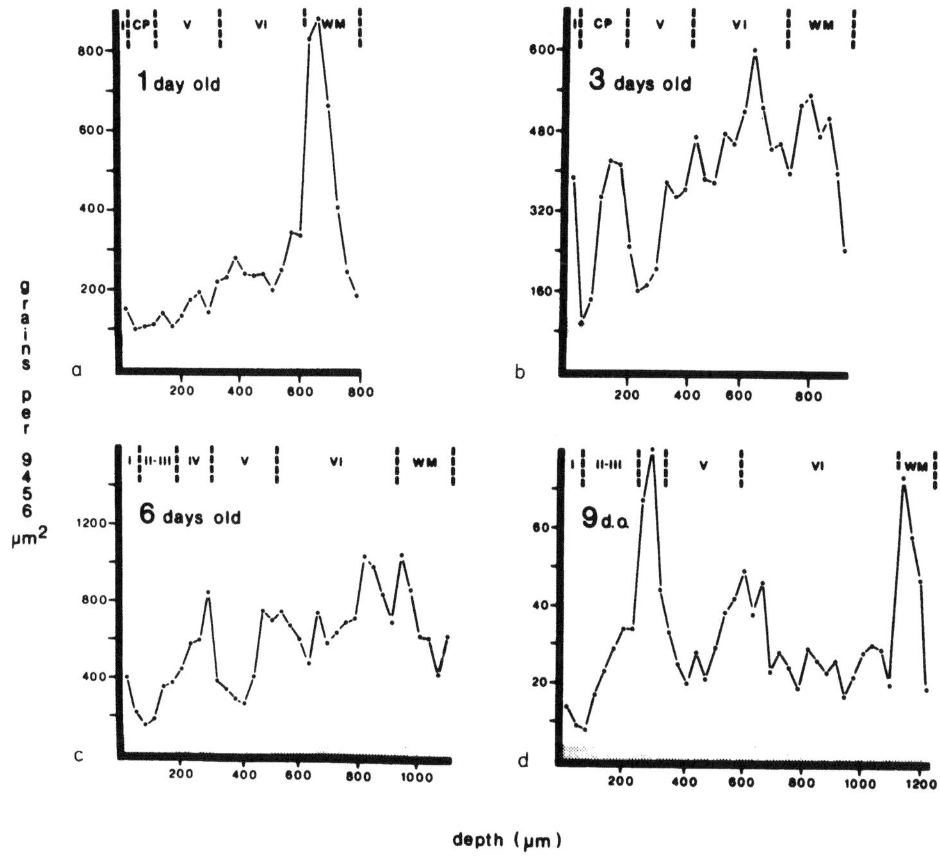

Fig. 3.11 a–d. The development of connections from the VB complex to the SI cortex in rats: grain density counts throughout the depth of the SI cortex in animals of various ages following injection of tracer into the VB complex. **a** In one-day-old rats the terminals of thalamocortical neurons are concentrated in the subcortical white matter (*WM*). Until the development of the trilaminar cortical plate (P 3, see Fig. 3.10b) terminals undergo a waiting period in upper layer VI. **b** In three-day-old rats there is a high concentration of terminals in the ventral part of the cortical plate (*CP*). **c** In six-day-old rats this part of the *CP* is recognizable as layer IV and can be seen to contain a high concentration of thalamocortical terminals. **d** The distribution of labelled terminals in the nine-day-old rat is indistinguishable from that in the adult. (From Wise and Jones 1978, p. 199)

broad topographical relationships between the various components of the system prior to the formation of those precise neuronal connections which associate them with individual follicles and/or neuronal aggregrates at the next higher or lower level of the neuraxis. Third, at all levels of the central nervous system most or all of the terminals of those afferent neurons which relate to a particular follicle become organized into a discrete cluster (segmentation). Fourth, shortly after the onset of segmentation, neurons in the PrV, in the VPM of the VB complex and in the PMBSF of the parietal cortex become arranged into groups (the barreloids and barrels) each of which is centred on a cluster of afferent terminals. The emergent picture is one of two stages: (1) The birth of neurons and the growth of their axons to the general target region, and (2) the development of precise connectivity between a small receptive surface and a single group of cortical neurons. As will be seen below it is this second process that seems to be controlled by a message from the skin itself.

3.3 Effects of Lesions and Manipulations in Prenatal, Neonatal and Developing Animals

The original observation that the pattern of mystacial vibrissae in mice was reflected in the pattern of the largest barrels in the PMBSF led rapidly to attempts to alter the arrangement of the latter by surgical removal of groups of vibrissae. Providing this procedure was carried out before P 3 it consistently resulted in adult mice which lacked the barrels that were spatially related to the lost vibrissae (Fig. 3.17) (Van der Loos and Woolsey 1973). Since these early experiments removal of vibrissae from neonatal animals has been widely used in many studies of plasticity within the TNC, the VB complex and the PMBSF of the SI cortex. In addition, the effects of hypo- or hyperstimulation of vibrissae, damaging the ION, producing supernumerary vibrissae and reducing the size or altering the structure of the cerebral cortex have all been tested. In the following text it is therefore proposed to document the effects which have been seen following each type of manipulation and so to establish, for this system, a set of conclusions concerning the factors controlling the development of neural connections and the effects of injuries on the nervous system of the neonatal and the adult animal.

3.3.1 Damage of the Infraorbital Nerve

The ION of rats supplies all the mystacial vibrissae, the skin of the whisker pad, the hairs and skin of the upper lip and the skin at the side of the nose (Waite and Cragg 1982). The principal objective of sectioning this nerve has been to test whether cells in those regions of the TNC and the PMBSF that normally receive input from the mystacial vibrissae become responsive to skin normally supplied by the mandibular and ophthalmic-maxillary branches of the trigeminal nerve. In relation to the mechanism(s) underlying any observed changes it has also been important to establish (1) the fate of trigeminal afferent fibres following section of the ION in neonatal and adult animals, (2) the effects on the peripheral projections, and on the central axonal arborizations, of surviving ION afferents and on other trigeminal afferents, and (3) the effects on the dendritic trees of cells in the TNC.

Survival of ION Afferents. The ION of the neonatal rat contains an average of about 42400 axons of which 42051 ($n=3$, SD$=2083$) are unmyelinated and 168 ($n=3$, SD$=47$) are myelinated (Renehan and Rhoades 1984). The ION of normal adult rats contains about 9000 fewer axons (total = approximately 33000) but, presumably as a result of postnatal myelination, there is a marked increase in the proportion of myelinated axons (mean no. of myelinated axons $=19740$, SD$=2054$, $n=4$, cf.

mean no. unmyelinated axons = 13 319, SD = 1159, $n = 4$; Jacquin et al. 1984). In agreement with this work, Waite (1984) had reported mean numbers of myelinated axons in the ION of adult rats between 18 211 and 18 547.

Jacquin et al. (1984) have reported that there are 33 000 axons in the ION of adults rats. This number is surprisingly high, as three separate reports place the mean total number of cells in the *whole trigeminal ganglion* at 30 233 (Aldskogius and Arvidsson 1978), 49 350 (Gregg and Dixon 1973) and 52 372 (Forbes and Welt 1982). Furthermore, Aldskogius and Arvidsson (1978) point out that 40% of the trigeminal ganglion cells project into the mandibular nerve and that many of the remainder do not project into the ION. Thus it would seem that taken most liberally (< 60% of the largest total of 52 372 ganglion cells) there are many more axons in the ION than there are cells in the entire trigeminal ganglion which project exclusively to this nerve (see Sect. 2.2.4).

After sectioning the ION at P 0 and thereafter at weekly intervals up to an age of 60 days, Waite (1984) found only about 2000 (11%) myelinated axons are found in the proximal stump. In a different study the anterograde transport of HRP along the ION in 3- to 45-day-old, neonatally-lesioned rats resulted in labelling of 4392 (SD = 1316, $n = 4$) ganglion cells – a number which was similar to that obtained (4244) if the ION was sectioned on P 0 and at weekly intervals, thereby preventing the regeneration of peripheral connections (Chiaia et al. 1986; Rhoades et al. 1987b). On the basis that in the adult the ION contains a total of 33 000 axons (Jacquin et al. 1984) the figure of 4300 would suggest a 13% survival of those ganglion cells that project into the ION but recently it has been shown that in neonatally nerve sectioned, adult rats not all axons in the regenerated ION arise from cells that supplied the ION in the neonate (Rhoades et al. 1987c). Certainly, a high rate of afferent mortality is suggested by the observation that sectioning the ION at birth abolishes the high level of SDH activity in the TNC that is normally associated with terminals of vibrissa afferents (Fig. 3.12a) (Killackey and Shinder 1981).

Further evidence for the poor survival of ION afferents following neonatal and repeated section of the ION is the report by Waite (1984) that these procedures cause an 18–29% reduction in the cross-sectional area of all the trigeminal nuclei of the adult rat (Figs. 3.12b and 3.13b). Anterograde transport of HRP from the ION in these animals and in individuals sectioned only at birth, produces heavy labelling in laminae I and II of the SpVc and the rostral cervical dorsal horn. Only very sparse labelling was observed in the PrV, SpVo, SpVi and the magnocellular part of the SpVc – regions of the TNC which normally contain dense somatotopically arranged clusters of vibrissa afferent terminals (Arvidsson 1982; Jacquin and Rhoades 1983; Chiaia et al. 1986). These results suggest that when the ION in the neonate is cut, the ensuing mortality of sensory cells is selective against those afferents which normally supply large myelinated fibres to the vibrissae rather than the unmyelinated afferents or the afferents supplying the fur, both of which project to laminae I and II of the SpVc and the cervical cord (Aldskogius and Arvidsson 1978; Morgan and Sharp 1986; Dubner et al. 1978; Waite 1984).

In comparison, repeated transection of the ION in adult rats has been reported to produce no decrease in the number of myelinated fibres in the proximal stump or in the cross-sectional area of the TNC (Waite 1984). However, Aldskogius and

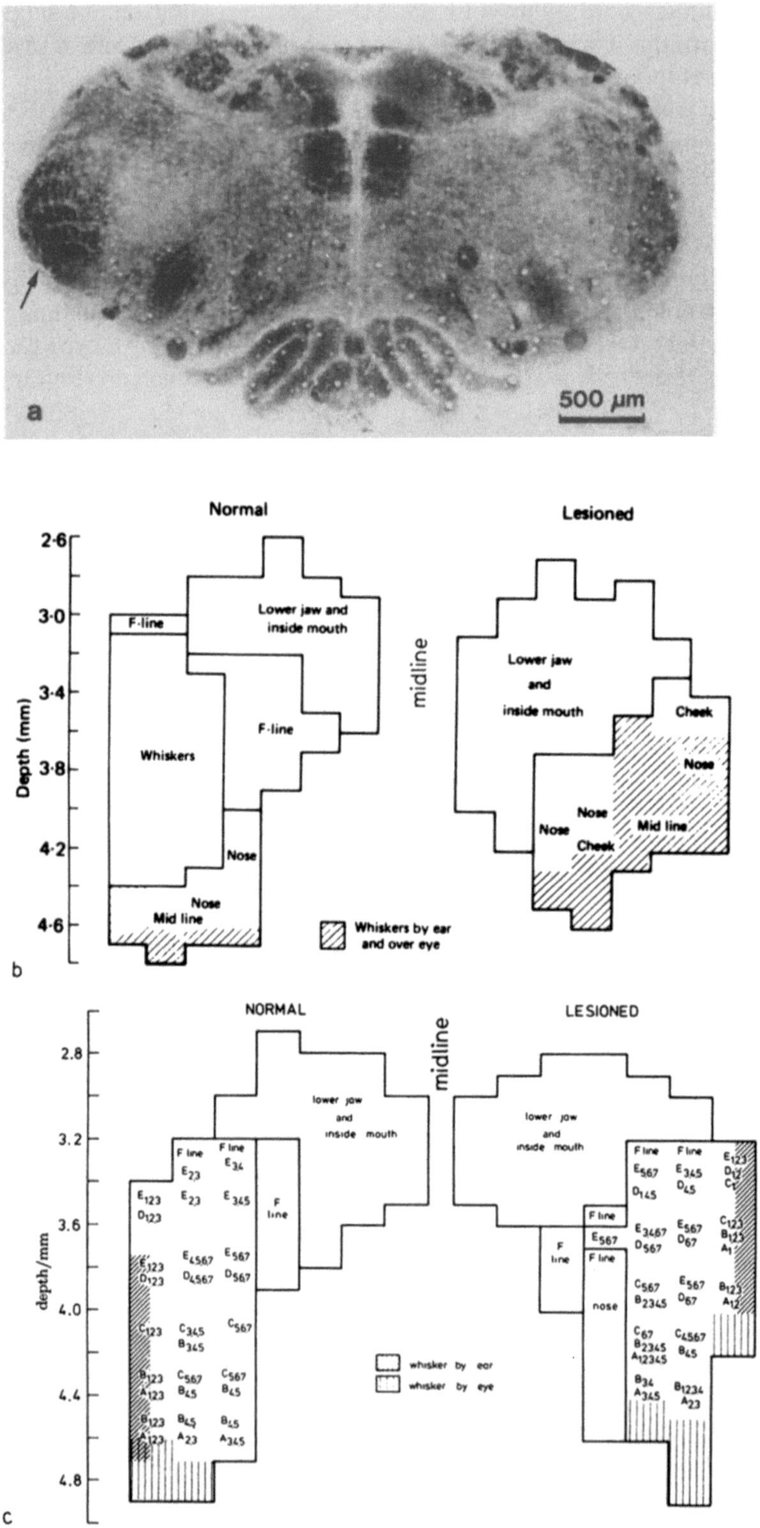
a
500 μm
Normal
Lesioned
Depth (mm)
2·6
3·0
3·4
3·8
4·2
4·6
F-line
Lower jaw and inside mouth
Whiskers
F-line
Nose
Nose
Mid line
midline
Lower jaw and inside mouth
Cheek
Nose
Nose
Nose
Cheek
Mid line
Whiskers by ear and over eye
b
NORMAL
LESIONED
midline
depth/mm
2.8
3.2
3.6
4.0
4.4
4.8
lower jaw and inside mouth
F line
nose
whisker by ear
whisker by eye
c

Arvidsson (1978) reported a 10%–17% decrease in the number of cells (mean = 14% or approximately 4200 cells) in the trigeminal ganglion 60–70 days after sectioning the ION in the adult and there exists some evidence that peripheral nerve section produces transganglionic degeneration of trigeminal afferents that is selective against the largest fibres (Arvidsson and Grant 1979).

Reinnervation of the Periphery. Transganglionic labelling of single vibrissa nerves following transection and regeneration of the ION in the adult rat has been used to study the specificity of peripheral regeneration. Unlike the normal animal vibrissa nerves of the regenerated ION gave rise to a broad and diffuse set of central terminations (Arvidsson and Johansson 1988). If regeneration is permitted following neonatal section of the ION the vibrissae are found to be reinnervated by about 39% of the number of myelinated fibres seen innervating the vibrissae in normal adults. In contrast, neonatal crush results in an increase to 45% of the normal number of vibrissa afferents (Waite and Cragg 1982). The original identity of the sensory axons that reinnervate the periphery remains problematic. Following neonatal section many afferents in the regenerated ION have unusual receptive fields and some completely lack a detectable receptive field (Renehan et al. 1986; Jacquin, et al. 1987c). In preliminary studies five single primary afferents that had reinnervated the vibrissae were intra-axonally stained revealing that only three had central arborizations resembling the vibrissa afferents in normal animals (Jacquin et al. 1986a). The other two had widespread "topographically inappropriate" arborizations (Renehan et al. 1986b). The vibrissae afferent terminals are known to be clustered by P 0 (Table 3.1; Belford and Killackey 1980; Erzurumlu and Killackey 1983; Jacquin and Rhoades 1987) and so it is tempting to assume that the morphology of the terminal arborizations can be used to establish the identity of single afferents which have regenerated peripheral connections after neonatal section of the ION. This question of identity of regenerated sensory axons has formed the basis of recent experiments that are reviewed in the following paragraphs.

◀ **Fig. 3.12 a–c.** The effects of damaging the infraorbital nerve (ION) in the neonatal rat, on SpVi of the trigeminal nuclear complex. **a** Transverse section through the SpVi that has been processed with SDH histochemistry. The maxillary branch of the trigeminal nerve was severed unilaterally in this animal on the day of birth and the animal was killed on P 5. The *arrow* indicates the clusters of SDH-labelled terminals of vibrissal afferents (segmentation) on the unoperated side. Note the absence of segmentation and the reduced SDH staining on the deafferented side. **b, c** The somatotopic organization of the SpVi in the transverse plane of 60- to 110-day-old rats following neonatal (P 0) and weekly section of the ION **b** and neonatal (P 0) crush and subsequent regeneration of the ION **c**. In **b** note the complete absence of a representation of the whiskers and the F-line, the invasion of the area normally containing these representations (see *left panel*) by inputs from skin elsewhere on the face and the reduction of the cross-sectional area of the SpVi (mean reduction = 18%). In **c** note the restoration of relatively normal somatotopy on the lesioned side and the absence of any reduction in the cross-sectional area of the SpVi. Somatotopic maps in **b** and **c** were made under pentobarbitone anaesthesia (60 mg/kg, I.P.). (**a** Modified from Killackey and Shinder 1981, p. 122; **b** modified from Waite 1984, p. 433; Waite and Cragg 1982, p. 202)

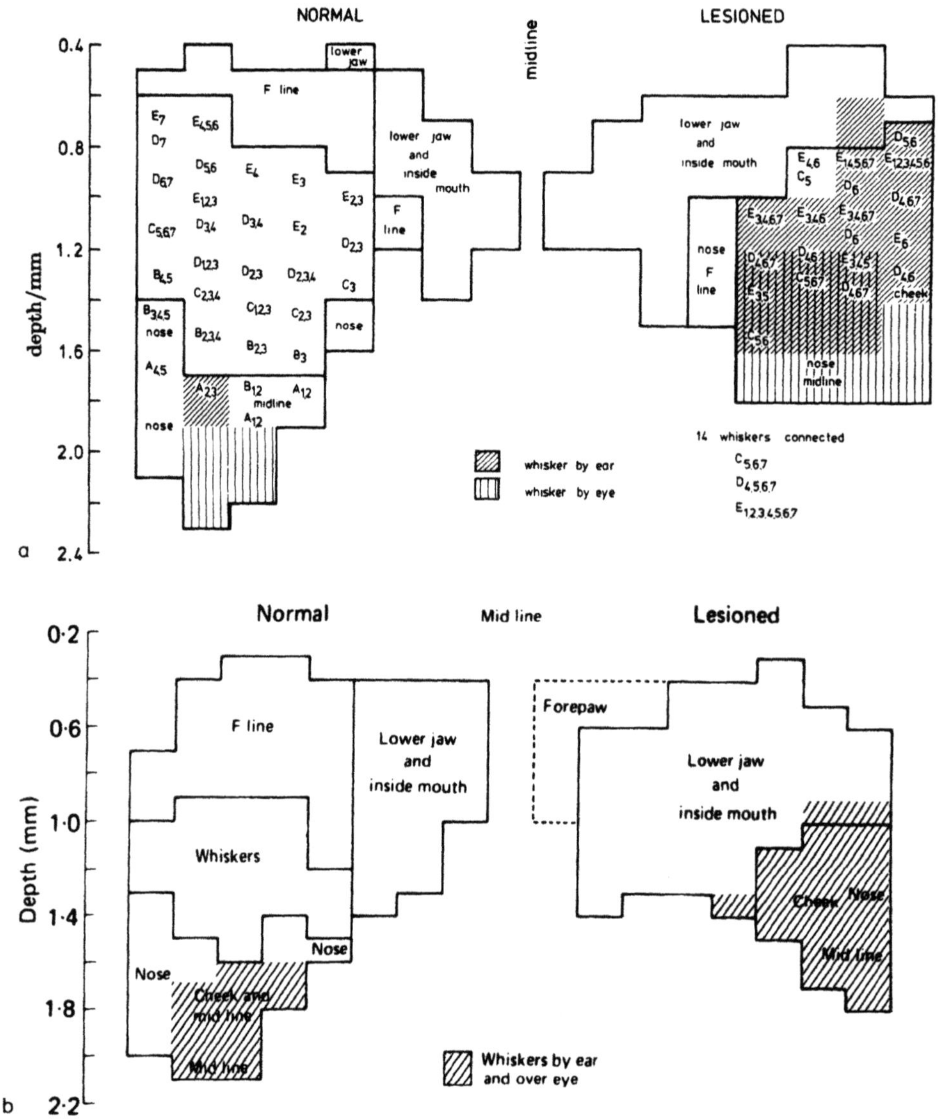

Fig. 3.13 a, b. The effects of damaging ION in the neonatal rat, on the somatotopic organization of the SpVc of the trigeminal nuclear complex. The somatotopic organization of the SpVc in the transverse plane of 60- to 110-day-old rats following neonatal (P 0) and subsequent regeneration of the ION **a** and neonatal (P 0) and weekly section of the ION **b**. In **a** note that somatotopic organization is disrupted and that only 14 of the vibrissae were found to be represented. These same 14 vibrissae were found to be represented in the SI cortex of the same animal (see Fig. 3.15e). Also note in **a** the massive invasion of the whisker representation by whiskers near the eye and the ear (see *hatching*). In **b** note the complete absence of a representation of the whiskers and the F-line, the invasion of the mean area normally containing these representations (see *left panel*) by inputs from skin elsewhere on the face and the reduction of the cross-sectional area of the SpVc (mean reduction = 29%). Somatotopic maps were made under pentobarbitone anaesthesia (60 mg/kg, I.P.). (**a** Modified from Waite and Cragg 1982, p. 209; **b** modified from Waite 1984, p. 433)

Identity of Regenerated ION Afferents. In a series of three experiments Rhoades et al. (1987b) used anterograde tracing methods in an attempt to establish the peripheral projections of trigeminal primary afferents in adult rats, before and after neonatal section and subsequent regeneration of the ION. In the first experiment the injection of HRP into the ION of neonatally nerve-sectioned, adult (60-day-old) rats "proximal to the point of neonatal transection" labelled 6 to 15 (mean 10.9) axons in the mandibular sensory nerves. Such labelling was never seen in normal rats. In the second experiment these workers injected diamidino yellow (DY) into the ION of 25 neonatally nerve-sectioned, adult rats and true blue (TB) into the "peripheral fields" of the ophthalmic (8 operated rats) or mandibular (7 operated rats) or both trigeminal divisions (10 operated rats). In 4 unoperated rats a maximum of only six cells/ganglion were labelled with both dyes. In contrast, in neonatal nerve-sectioned rats 216.9 (SD = 98.1) cells were double-labelled following injection of TB into both mandibular and ophthalmic "peripheral fields" while an average of 87.3 or 147.3 cells were double-labelled after injections into the mandibular or ophthalmic peripheral fields, respectively. All labelled cells were located in the mediolateral region of the ophthalmic-maxillary part of the ganglion.

Rhoades et al. (1987d) concluded from these data that neonatal section of the ION induces some non-ION afferents to send a second axon down the regenerated ION. The distribution of the diameters of single- and double-labelled cells was very similar to the distribution of the diameters of all trigeminal ganglion cells (Forbes and Welt 1982). In somatic afferents, soma diameter is related to conduction velocity and, in turn, conduction velocity is loosely related to modality. On this basis it would seem that all physiological types of primary afferents may participate in the regenerating ION.

In the third experiment of Rhoades et al. (1987d), a two-stage, double-labelling technique was employed to identify within single animals the nerve supplied at birth by some trigeminal cells within the trigeminal ganglion. Three protocols were used: (1) In 14 neonatal rats injections of TB were made into non-ION territory supplied by other divisions of the trigeminal nerve and the ION was transected. On reaching adulthood (60 days old) the regenerated ION was injected with DY. Two to 3 days later an average of 264.3 (SD = 87.2) double-labelled cells were found in the ophthalmic-maxillary part of the ganglion. These results were interpreted as suggesting that trigeminal afferents that would normally have innervated only non-ION territory were induced, by neonatal section of the ION, to sprout a second axon into the regenerated ION; (2) In seven neonatal rats TB was injected into the whisker pad – a major part of the peripheral field supplied by the ION. Six hours later the ION was transected. On reaching adulthood injections of DY were made into non-ION territory supplied by other divisions of the trigeminal nerve. Two to 3 days later an average of 92.6 (SD = 27.2) double-labelled cells were found in the trigeminal ganglion suggesting that neonatal section of the ION caused some ION afferents to innervate non-ION territory; (3) In five neonatal rats TB was injected into the ION skin and the ION sectioned 6 h later. On reaching adulthood DY was injected into the ION and 2–3 days later 434.6 (SD = 118.4) cells were found to be double-labelled. A comparison of the results of (3) with data from identical experiments on normal animals (ION intact) leads to the conclusion that 75% of the ION afferents that survive neonatal transection of the ION regenerate axons back into the ION.

The experiments of Rhoades et al. (1987d) suggest that section of a peripheral nerve at the developmental stage of a newborn rat may (a) induce afferents in that nerve to project into other nerves, and (b) induce afferents in surrounding intact nerves to sprout a second axon into the regenerating nerve. These data may thus account for the unusual central terminations of trigeminal primary afferents which are activated by movements of the vibrissae in adult rats that have undergone neonatal section of the ION (see above; Renehan et al. 1986b). However, of more general importance is the proposal of Rhoades et al. (1987d) that the data could account for the changes in the central representation of the tactile input from the body surface in experiments that involve neonatal nerve section in rats and all other mammals – changes that have been considered to manifest plasticity of neural circuitry in the central nervous system. Certainly it would seem that the branching of afferents into foreign nerves could explain why, after regeneration of the ION following neonatal section, there exists a quite abnormal representation of the periphery at central sites where normally there is a clear somatotopic representation of the mystacial vibrissae (Figs. 3.13a and 3.15e) (Waite and Cragg 1982; Waite 1984).

Fortunately, there are two qualifying points which must be addressed before entertaining too seriously the despairing proposal that all central effects of neonatal nerve section are indeed mainfestations of aberrant regeneration of the peripheral axons of sensory neurons. First, because of pronounced differences in the gestation period of different mammals, the exact developmental stage of a neonatal animal is species-dependent such that raccoons, cats and primates are relatively mature at birth when compared with hamsters, mice and rats. Neonatal nerve section might thus elicit less profound peripheral changes in the former group. Second, in five intact rats only an average of 2443 (SD = 985.4) cells were labelled after injection of dye into the whisker pad at birth and into the ION 60 days later (see Table 3.2 of Rhoades et al. 1987d). When this figure is compared with even a low estimate of the number of trigeminal cells supplying axons to the ION (say 20000; see above, it must be concluded that only a small (12.5%) portion of ION afferents are labelled. It is thus possible that the increase in the number of double-labelled cells in neonatally nerve-sectioned, adult rats might represent an effect of nerve section on the uptake and transport of the dyes and that in normal animals many trigeminal afferents actually do send axons into several trigeminal nerves. In fact, as concluded above, the ION of normal adult animals does contain many more axons than there are cells which project exclusively to this nerve – an observation that suggests that in normal rats either some cells send multiple axons down the ION or that many cells with an axon in a different nerve (e.g. the mandibular) actually project an axon into the ION.

Somatotopic Representation of the Mystacial Vibrissae. The effects of ION damage in the newborn rat, on the somatotopic representation of the vibrissae within the mature central nervous system, have been evaluated in three types of experiments: (1) ION crush followed by regeneration, (2) ION transection followed by regeneration, and (3) ION transection at weekly intervals in order to prevent regeneration.

Table 3.2. The effects of cutting and crushing the infraorbital nerve on those regions of the trigeminal nuclear complex and SI cortex where infraorbital nerve afferents are normally represented

Procedure	% of vibrissae represented	Size of ION representation	Condition of somatotopy	Text figure
Trigeminal nuclei				
Neonate:				
(1) ION crush & regen.	100	Unchanged	Normal	3.12c, 3.14a
(2) ION cut & regen.	33–35	Reduced	Disrupted	3.13a
(3) ION cut & recut	0	Reduced[a]	Reorganized	3.12b, 3.13b, and 3.14b
Adult:				
(3) ION cut & recut	0	Unchanged	No representation	3.14c
Cortex				
Neonate:				
(1) ION crush & regen.	100	Unchanged	Normal	3.15d
(2) ION cut & regen.	33–35	Reduced[b]	Disrupted	3.15e
(3) ION cut & recut	0	Reduced[c]	Reorganized	3.15c
Adult:				
(3) ION cut & recut	0	Reduced[d]	No representation	3.15b

For normal representation in SI cortex, see Fig. 3.15a. Anaesthetic used in these studies: sodium pentobarbitone (60 mg/kg i.p.). Data from Waite and Cragg (1982) and Waite (1984).
[a] Size of trigeminal nuclei reduced.
[b] Responses to vibrissae found within an area that is only about 36% of the size of the normal vibrissa representation.
[c] Massive invasion of the central ION territory (most of which is normally the vibrissa representation) by inputs from the digits, lower jaw, cheek and whiskers over the eye and the ear. Cells in the remaining 12% of the ION representation lack receptive fields.
[d] Some invasion of the central ION territory (most of which is normally the vibrissa representation) by inputs from the digits, lower jaw and whiskers over the eye and the ear. Cells in the remaining 75% of the ION representation lack receptive fields.

Procedure (3) has also been applied to adult animals. The results of these procedures are summarized in Table 3.2 and Figs. 3.12–3.15.
Following neonatal crush and regeneration of the ION there was no reduction in the size of the representation of the ION at either the level of the TNC (compare Figs. 3.12c and 3.13a) or the SI cortex (compare Fig. 3.15a with Fig. 3.15d). At each level of the central nervous system each vibrissa was represented and somatotopic organization was normal. In contrast, neonatal section and regeneration of the ION resulted in a considerable reduction in the size of the representation of the ION at trigeminal and cortical levels and only 33%–50% of the vibrissae were represented (Fig. 3.13a, compare Fig. 3.15a with Fig. 3.15e). The representation of these vibrissae did not follow an orderly somatotopic plan and many cells in the TNC and

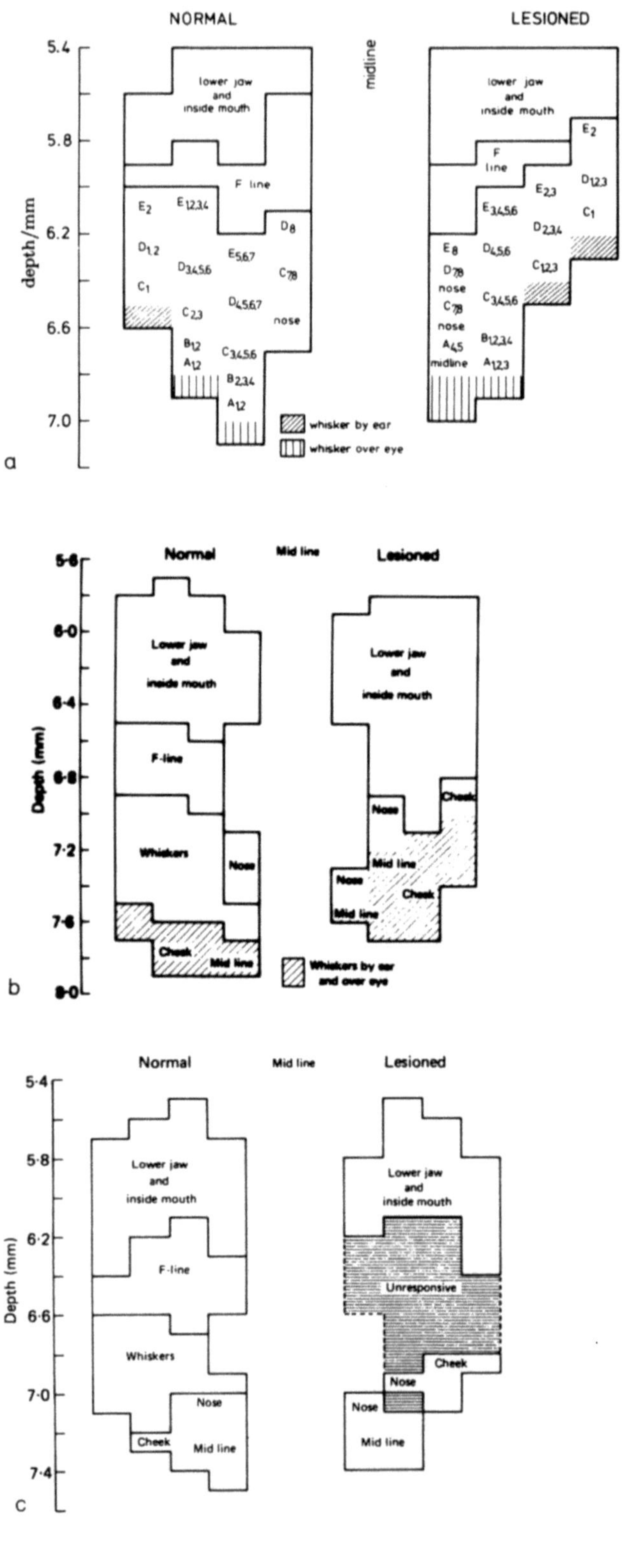
NORMAL
LESIONED
midline
depth/mm
lower jaw and inside mouth
F line
nose
whisker by ear
whisker over eye
a
Normal
Mid line
Lesioned
Depth (mm)
Lower jaw and inside mouth
F-line
Whiskers
Nose
Cheek
Mid line
Whiskers by ear and over eye
b
Normal
Mid line
Lesioned
Depth (mm)
Lower jaw and inside mouth
F-line
Whiskers
Nose
Cheek
Mid line
Unresponsive
c

the cortex had unusual receptive fields (Waite and Cragg 1982; Stennett et al. 1986). Finally, neonatal and weekly section of the ION resulted in a radically altered somatotopic map within that region of the TNC and SI cortex normally devoted to the representation of ION afferents. Within this region the skin *surrounding* the peripheral territory of the ION was represented in a *topographically ordered* manner (Figs. 3.12b, 3.13b and 3.14b, compare Fig. 3.15a with Fig. 3.15c). In the cortex only those cells within a circumscribed region that constituted 12% of the normal cortical representation of the ION, lacked receptive fields (Fig. 3.15c; Table 3.2). In contrast, if the ION is sectioned at weekly intervals in adult rats, about 75% of the normal ION representation in both the TNC and the cortex remains unresponsive to peripheral stimulation (Figs. 3.14c and 3.15b). However, it is most interesting to note that in the cortex of these adult-lesioned animals, there is an approximately 500 μm encroachment into the normal ION representation of inputs from the digits, the lower jaw and the whiskers over the eye and ear (Waite 1984) (compare Fig. 3.15a with Fig. 3.15b; Sect. 3.4).

Mechanisms Underlying Somatotopic Alterations. The explanation for the changes in somatotopy following neonatal nerve section and subsequent regeneration of the ION, seems to be adequately served by available data concerning the connections of primary afferents both to the periphery and the TNC, the cortex merely reflecting those changes visible at the trigeminal level (Waite and Cragg 1982; Waite 1984). However, it must be emphasized that the normal clustering of trigeminothalamic terminals in the VB complex (Fig. 3.16) and of the thalamocortical terminals in the PMBSF of the cortex does *not* occur following neonatal transection of the ION (Killackey and Shinder 1981; Waite and Cragg 1982). Nevertheless, recent evidence suggests that after this operation at least some of the surviving ION afferents send axons into intact trigeminal nerves (see above; Rhoades et al. 1987d). Also many afferents with regenerated axons in the ION have unusual and large receptive fields that may include many vibrissae or a multiplicity of receptor types. In addition, some trigeminal ganglion cells are found with receptive fields in the ION territory but no axon in the ION, while others are found with mandibular receptive fields and an axon in the ION (Renehan 1986b; Jacquin et al. 1987). These aberrations are

◀ **Fig. 3.14 a–c.** The effects of damaging the ION in the neonatal *and* the adult rat, on the somatotopic organization of the principal sensory nucleus (PrV) of the trigeminal nuclear complex. The somatotopic organization of the PrV in the transverse plane of 60- to 120-day-old rats following neonatal (P 0) crush and regeneration of the ION **a**, neonatal (P 0) and weekly section of the ION **b** and section of the ION in a 60-day-old rat followed by weekly section of 60 days **c**. In **a** note the restoration of relatively normal somatotopy on the lesioned side and the absence of any reduction in the cross-sectional area of the PrV. In **b** note the complete absence of a representation of the whiskers and the F-line, the invasion of the area normally containing these representations (see *left panel*) by inputs from skin elsewhere on the face and the reduction of the cross-sectional area of the PrV (mean reduction = 21%). In **c** note the complete absence of cells responsive to peripheral stimulation within those parts of the PrV that on the normal, unoperated side contain the representation of the whiskers and the F-line. The somatotopic maps were made under pentobarbitone anaesthesia (60 mg/kg, I.P.). (**a** Modified from Waite and Cragg 1982, p. 201; **b**, **c** modified from Waite 1984, p. 432)

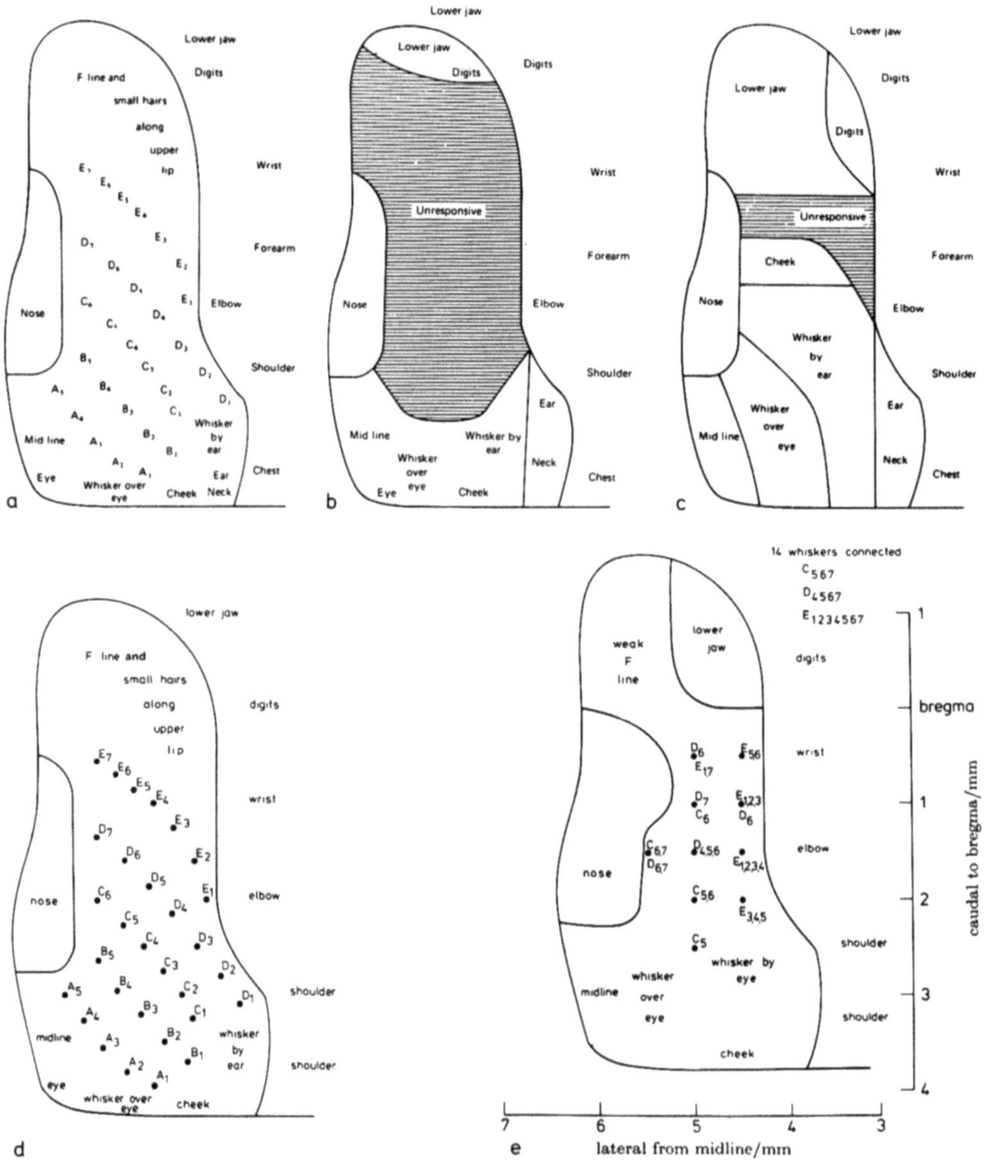

Fig. 3.15 a–e. The effects of damaging the ION in the neonatal *and* the adult rat, on the somatotopic organization of the SI cortex. **a** The somatotopic organization of a normal adult rat. **b** The somatotopic organization in a 120-day-old rat following weekly section of the ION from the age of 60 days. Note the large region in which cells were unresponsive to peripheral stimulation and the encroachment into the normal representation of the ION (F-line and the whiskers; see **a**) of inputs from the digits, lower jaw and the whiskers over the eye and the ear. **c** The somatotopic organization in a 60-day-old rat following neonatal and weekly section of the ION. Note the massive invasion of the ION representation by inputs carried by intact nerves and the movement of the representations of some of these inputs (e.g. the skin of the cheek).

apparent in both the receptive fields of cells in the TNC (SpVc and SpVi) which are often larger and show an abnormally high degree of receptor convergence, and the somatotopic organization of these nuclei, which is disorganized (Fig. 3.13a) (Waite and Cragg 1982; Stennett et al. 1986). The possible contribution of structural changes in the dendritic trees of these cells or the terminals of temporarily disused, surviving ION afferents remain to be properly established (Stennett et al. 1986; Woerner et al. 1986). There is, however, clear evidence that, as well as abolishing the segmentation of afferent terminals in the PrV, SpVi and SpVc (Fig. 3.12a), neonatal ION section abolishes the aggregrations of trigeminothalamic cells in the PrV normally detectable in young rats (Killackey and Shinder 1981; Bates et al. 1982). In contrast to nerve section, crush of the ION in the neonate results in the restoration of a clear somatotopic representation, within the TNC, of the skin supplied by the ION (Figs. 3.12c, 3.14a and compare Fig. 3.15a with Fig. 3.15d) (Waite and Cragg 1982). Although there is not yet any evidence that, after crush of the ION, single afferents regenerate back to precisely the receptors or peripheral positions they innervated prior to crush (see Sect. 2.3), given that regeneration has occurred, such a presumption is considerably simpler than one which involves the massive and yet orderly alteration of central pathways which would be necessary to restore somatotopy to a set of central neurons receiving a somatotopically scrambled set of input lines. The shorter time-course of the restoration of peripheral connections following crush of the ION (2–3 days), compared with that following nerve section (7 days), perhaps reflects the relative facility with which sensory axons might grow along a channel (the nerve distal to the crush) that is still structurally connected to the nerve central to the damage (Waite and Cragg 1982). The presumption that specific reinnervation occurs following neonatal nerve crush is thus reminiscent of the developmental sequence in which the general topography between fascicles of the developing ION and parts of the whisker pad clearly precedes a precisely ordered chain of connections between a single vibrissa and a group of neurons in the TNC (Fig. 3.9b; Table 3.1) (Erzurumlu and Killackey 1983; Killackey 1985). Finally, it is surprising that after neonatal crush of the ION there is such a large degree of restoration of normal somatotopic organization since neonatal crush of the sciatic nerve appears to cause the death of the entire complement of damaged sensory neurons (Sect. 2.2.1). If it were not for the short recovery period (2–3 days) one might think that all the axons in the regenerated ION might have come from sensory cells that normally do not have an axon in this nerve.

◀ **Fig. 3.15a–e** (*continued*). **d** The somatotopic organization in an adult rat following neonatal nerve crush and regeneration. Note the relatively normal somatotopic organization (cf. **a**). **e** The somatotopic organization in an adult rat following neonatal section and subsequent regeneration of the ION. Note that the normal somatotopic organization is disrupted and that only 14 of the vibrissae were found to be represented. These same 14 vibrissae were found to be represented at the trigeminal level in the same animal (see Fig. 3.13a). The somatotopic maps were made under pentobarbitone anaesthesia (60 mg/kg, I.P.). (**a–c** From Waite 1984, pp. 430, 431; **d, e** from Waite and Cragg 1982, pp. 198, 200)

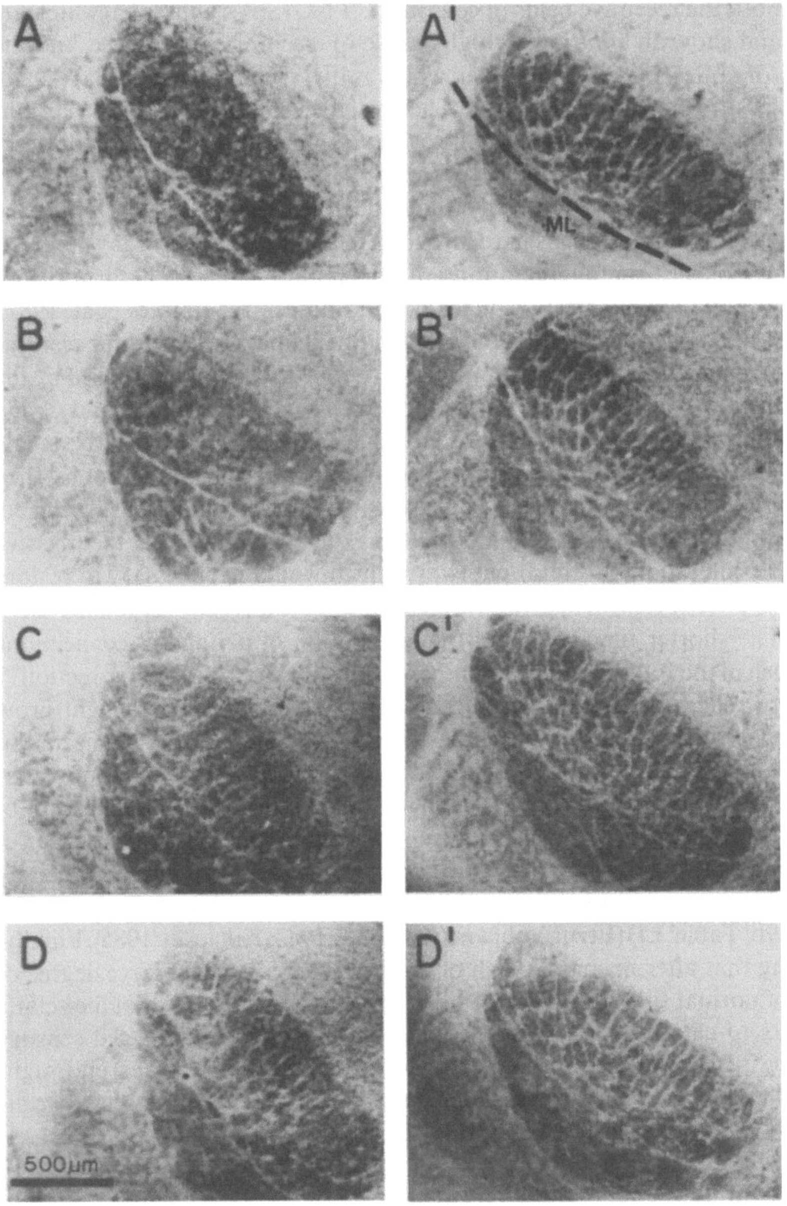

Fig. 3.16 A–D. The effects on the segmentation of trigeminothalamic terminals in the VB complex of 5-day-old rats, of sectioning the trigeminal nerves on P 0 (**A**), P 1 (**B**), P 2 (**C**) and P 3 (**D**). **A**, **B**, **C** and **D** show coronal sections through the VB complex contralateral to the denervated side while **A**′, **B**′, **C**′ and **D**′ show reversed photomicrographs of the normal VB complex contralateral to the unoperated side of the same individuals. The darkly stained areas indicate the presence of SDH in the terminals of the trigeminothalamic fibres and the terminals of medial lemniscal fibres (*ML*, **A**′). Note that at these times, nerve section does not alter the density of SDH activity, or, presumably, the density of terminals, but that the earlier that nerve section is performed the greater is the disruption of the segmentation of the trigeminothalamic terminals. Note also that the boundary between the medial lemniscal and the trigeminal projections (*dashed line* in **A**′) is maintained in all cases. (Modified from Killackey and Shinder 1981, p. 124)

Neonatal and repeated section of the ION results in an orderly representation of skin innervated by the intact trigeminal nerves within the territory normally devoted to the representation of ION afferents (Figs. 3.12b, 3.13b, 3.14b and 3.15c). There are four simple ways this could occur. First, some mandibular and ophthalmic-maxillary primary afferents in the neonate might have ineffective projections to those parts of the trigeminal nuclei in which cells are responsive to ION afferents – ineffective projections that become effective following permanent section of the ION and the degeneration of many of its afferent fibres. Second, the central portions of mandibular or maxillary afferents are induced to sprout into the ION territory following permanent, neonatal section of the ION. Third, the dendrites of cells in the ION territory reach out to the mandibular or ophthalmic-maxillary afferents. Fourth, there is a reorganization of interneuronal pathways between spared afferents and the deprived cells in the TNC.

Accepting that anterograde transganglionic labelling permits the visualization of all projections of a nerve (see Snow and Wilson 1989), the first of the above possibilities seems to be excluded by experiments in which the overlap between the central projection of the ION and the mandibular branch of the trigeminal nerve was studied in the neonatal hamster (gestation period = 16 days) and the neonatal rat (gestation period = 22 days). Despite the earlier developmental stage of the newborn hamster, the detectable overlap in the central projections of these nerves was, in both species, minimal (Jacquin and Rhoades 1985). The second possibility is equally unacceptable for, even together with cauterization of all the vibrissae to prevent regeneration, neonatal section of the ION did not produce in the adult rat a detectable invasion of the central ION territory by terminals of mandibular afferents or surviving ophthalmic-maxillary fibres (Jacquin and Rhoades 1985). There is, however, some evidence of such sprouting in identical experiments carried out in the hamster (Jacquin and Rhoades et al. 1983; Jacquin and Rhoades 1987). Finally, although in the SpVc reorganization is complete by 21 days after neonatal section of the ION, suggesting that the growth of neuronal processes may underlie these changes (Waite 1984), there is, as yet, no data relating to the third and fourth possibilities. It would seem that experiments aimed at achieving finer-grain somatotopic maps and incorporating intracellular and intra-axonal staining in the same experiment will hold the key to this mystery which is clearly bound up with conditions that prevail exclusively during development. A possible explanation for the encroachment of non-ION inputs into the ION representation following ION section in adult animals is considered elsewhere (Sect. 3.4).

3.3.2 Lesions to One or More Vibrissae

Most studies of the effects of the subtraction or addition of vibrissae at particular points on the whisker pad have utilized such gross microscopical features as the aggregation of cells or the clusters of afferent terminations which, at trigeminal, thalamic or cortical levels, reflect the peripheral arrangement of the vibrissae (Sect. 3.1). The strategy adopted has been to alter the number of vibrissae in the perinatal period and by studying the resultant abnormalities to ascertain something

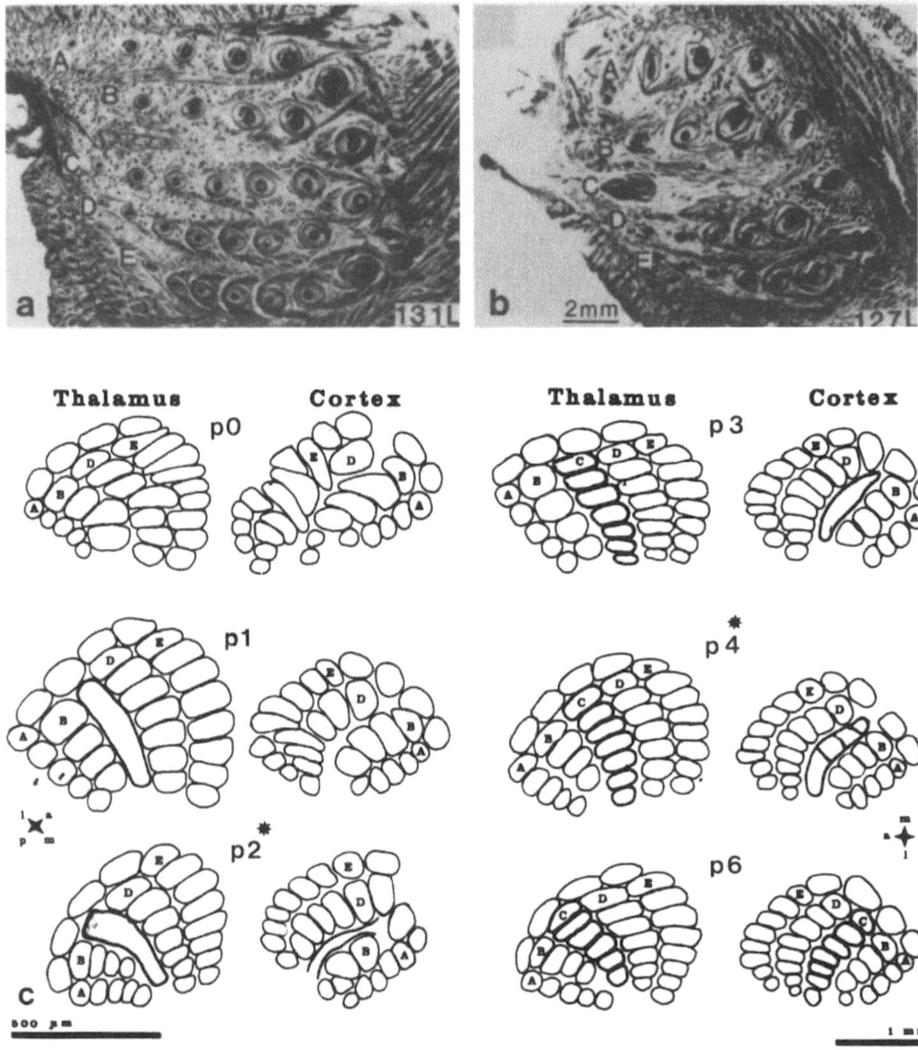

Fig. 3.17 a–c. The effects of neonatal lesion of the vibrissae of row C on the thalamic barreloids and cortical barrels in the adult mouse. Tangential section through the whisker pad of a normal adult mouse (**a**) and an adult mouse following cautery of the row C vibrissae on the day of birth (**b**). **c** Drawings of the thalamic barreloids and the cortical barrels of the PMBSF in 60-day-old mice following cautery of the vibrissae of row C on P 0, P 1, P 2, P 3, P 4 and P6. Structures that are related to the row C vibrissae are drawn in *heavy lines.* Note that the thalamic barreloids are insensitive to vibrissa lesions performed on or after P 3 but that the cortical barrels are sensitive to lesions performed up to and on P4. The *scales* and *coordinates* for the thalamic sections are shown on the *left* and those for the cortex sections are shown on the *right*. The *asterisks* (*) indicate cases where the drawings of the thalamus and the cortex were made from the same side of the same animals. Compare with Fig. 3.19. (**a, b** Modified from Durham and Woolsey 1984, p. 433; **c** modified from Woolsey et al. 1979, p. 373)

of the nature of the factors controlling normal development. The most commonly used procedure is the removal of a row of vibrissae by electrocautery. The effects, on the mouse whisker pad, of performing this operation neonatally are shown in Fig. 3.17a, b.

Vibrissa Lesion and Death of Primary Afferent Neurons. In normal mice the 29 vibrissae of the whisker pad are innervated by an average of 2788 myelinated axons (Welker and Van der Loos 1983). Unilateral electrocoagulation of all the mystacial vibrissae at birth results in a 36.5% cell loss in the ophthalmic-maxillary part of the trigeminal ganglion (Savy et al. 1981), an 11%–38% (2823 to 7007 fibres) decrease in the number of myelinated axons in the trigeminal nerve root (Waite and Cragg 1979; Savy et al. 1986) and a 33% decrease in the volume of the SpVi (Hamori et al. 1987) of adult mice. In these studies the normal intact side acted as a control. The mortality of trigeminal ganglion neurons, following this procedure, is less if cautery is performed later in development. Thus Waite and Cragg (1979) found that lesioning before P 5 produced a 38% (7007) decrease in myelinated axons by P 19 whereas lesion on Ps 7, 8 and 9 produced a 20% decrease and lesion in one 60-day-old adult produced only a 4.4% decrease by P 119. Qualitatively, the age-related effects of this procedure on the number of afferent fibres in the ION resemble the effects of sectioning the ION (see below and Sect. 3.3.1).

Clustering of Terminal in the TNC. In the newborn rat the terminals of trigeminal sensory neurons are already organized into clusters that are arranged in a pattern that is isomorphic to the arrangement of the vibrissae (Fig. 3.2; Table 3.1) (Belford and Killackey 1979). Electrocautery of a row of vibrissae on P 0 elicits, in the PrV, SpVi and SpVc, a loss of terminal clustering within the region between those clusters that are related to the vibrissae of the two neighbouring rows. After 24 h there appears to be some disruption of the patches of high SDH activity in the deafferented region and by 72 h only a single amorphous band of relatively high SDH activity is apparent (Belford and Killackey 1979). By P 7 and 8 only a homogeneous region of low SDH activity remains (see A in Fig. 3.18a) (Belford and Killackey 1980) that presumably reflects a massive transganglionic degeneration of primary afferent terminals (see Waite and Cragg 1979; Hamori et al. 1987).
Electrocautery of the vibrissae of row C on Ps 0, 1, 2, 3 and 4, as well as, in the rat, P 5, has been used to assess the time-course of sensitivity of segmentation in the TNC of the 7- to 8-day-old rat and the 15- to 100-day-old mouse, to the removal of vibrissae (Belford and Killackey 1980; Durham and Woolsey 1984). In the rat lesions at Ps 0, 1 and 2 resulted in decreased SDH activity so that there was no apparent segmentation in the regions of the PrV, SpVi and SpVc in which row C is normally represented. Following lesions on Ps 3 and 4 some patches of high SDH activity remained and after lesions on P 5 the patches were more distinct resembling the normal segmentation in the representations of intact rows of vibrissae (Fig. 3.18a) (Belford and Killackey 1980). In 15- to 100-day-old mice the trigeminal representation showed low SDH activity after vibrissa damage at all ages tested (Fig. 3.19a) (Durham and Woolsey 1984). That this reflects the degeneration of primary afferents is supported by the reports of Waite and Cragg (1979) and Savy et al. (1986) that

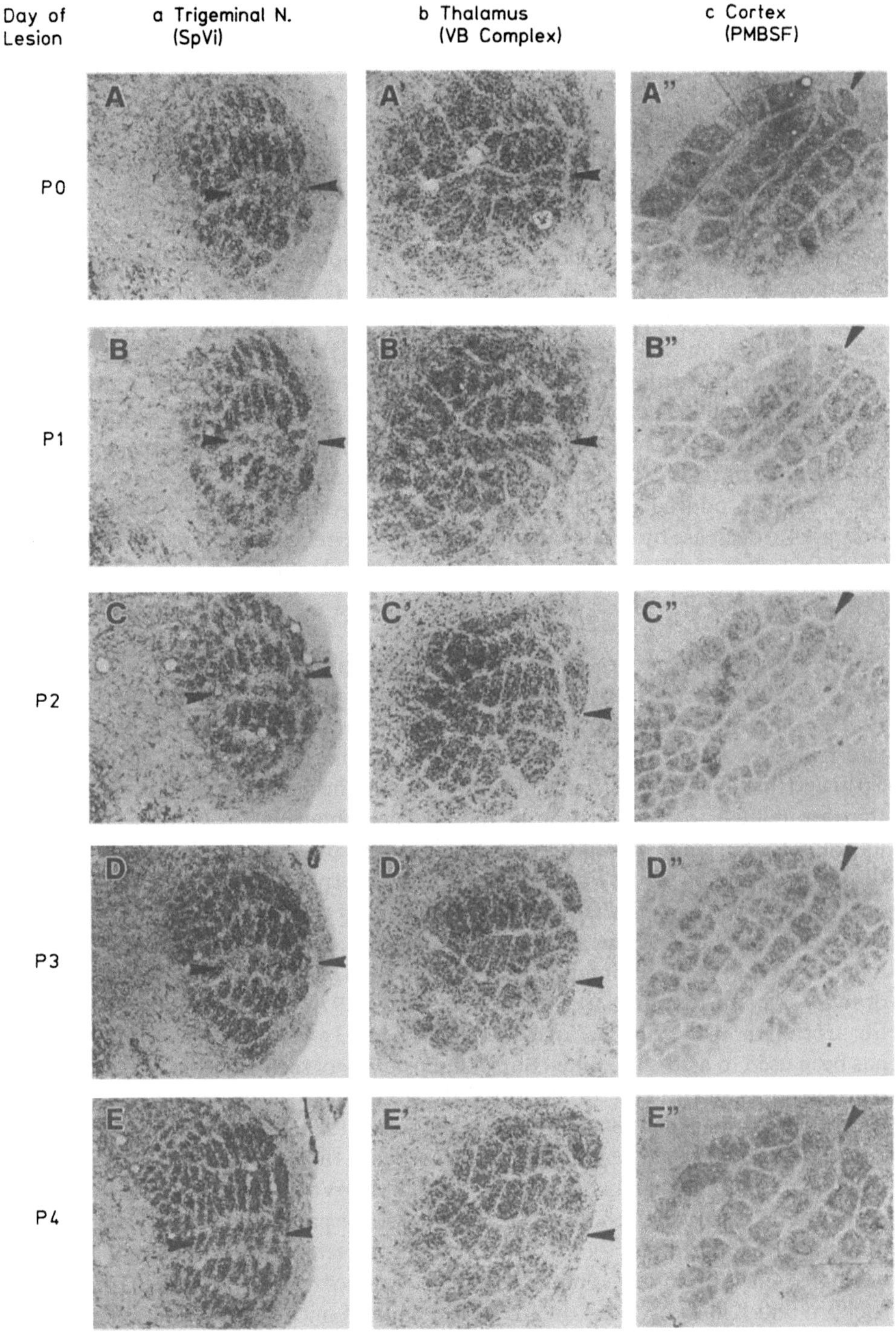

Fig. 3.18 a–c. Legend see p. 95

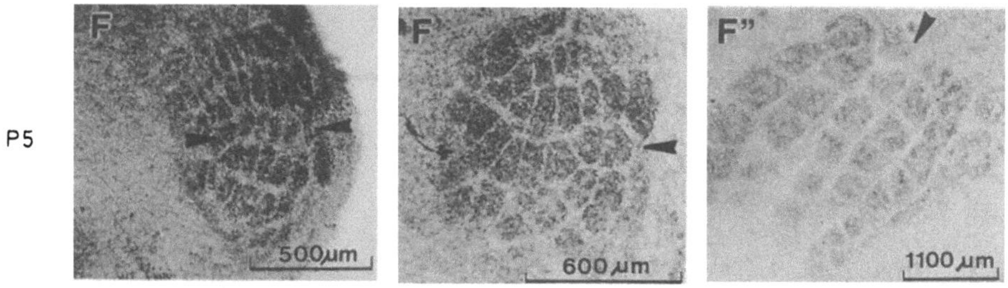

Fig. 3.18 a–c. The effects on the isomorphic clusters of terminals in the TNC, the thalamus and the cortex of 8-day-old rats of cauterizing the vibrissae of row *C* on the first 6 days (P0-P5) after birth. *Darkly stained areas* indicate the presence of vibrissa afferent (**a**), trigeminothalamic (**b**) and thalamocortical (**c**) terminals containing SDH. **a** Transverse sections through the SpVi following cautery on Ps 0, 1, 2, 3, 4 and 5. Note the loss of SDH activity in the row *C* representation (*arrow*) following lesions on Ps 0, 1 and 2 but the progressive increase in activity in this region when lesions were placed at times approaching the age when rats were sacrificed. (P 8). See text for further discussion. **b**, **c** Horizontal sections through the VB complex (**b**) and tangential sections through the PMBSF (**c**) following cautery on Ps 0, 1, 2, 3, 4 and 5. In both the VB complex (**b**) and the PMBSF (**c**), note the fused band of high SDH activity in the row *C* representation (*arrow*) in animals lesioned before P 2 but the relatively normal segmentation in this region in animals lesioned after P 2. (Modified from Belford and Killackey 1980, p. 340)

neonatal cautery of all vibrissae produces, in the maxillary nerve of adult mice, a loss of at least as many myelinated fibres (2823–7007) as are found innervating the vibrissae of adults (2785 fibres; Welker and Van der Loos 1983). Thus despite the fact that the gestation period of rats exceeds that of mice by 3 days it seems likely that the segmentation observed on Ps 7 and 8 in the rat following cautery on Ps 5 and 6 represents the organization of the terminals of vibrissa afferents which have not yet undergone transganglionic degeneration. It would not be surprising if some changes in the TNC were delayed until after the central processes of dying afferents began to degenerate. Thus in order to assess whether late (P 5 and 6) cautery still evokes central changes it would be interesting to know more of the morphology of both cells in the TNC and the central arborizations of surviving trigeminal afferents, in rats older than 10 days (Belford and Killackey 1980).

In neither the TNC of rats nor that of mice does neonatal cautery of a row of vibrissae elicit an increase in the size of the terminal clusters of adjacent intact rows (Belford and Killackey 1980; Belford and Killackey 1979; Durham and Woolsey 1984). These observations suggest that there is no invasion of the territory in which the lesioned row is normally represented. There are, however, changes in the distribution of retrogradely labelled trigeminothalamic cells in the PrV of the rat. Cautery of one row at P 0 results in a loss of those aggregations of trigeminothalamic cells in the appropriate region (Killackey and Belford 1980). It thus seems that some postnatal cell migration and/or cell survival in the TNC is dependent on the presence of clusters of terminals of intact vibrissa-specific afferents.

Waite (1984) has reported an 18% reduction of the cross-sectional area of the SpVi following neonatal and repeated ION section in the rat. More recently, Hamori

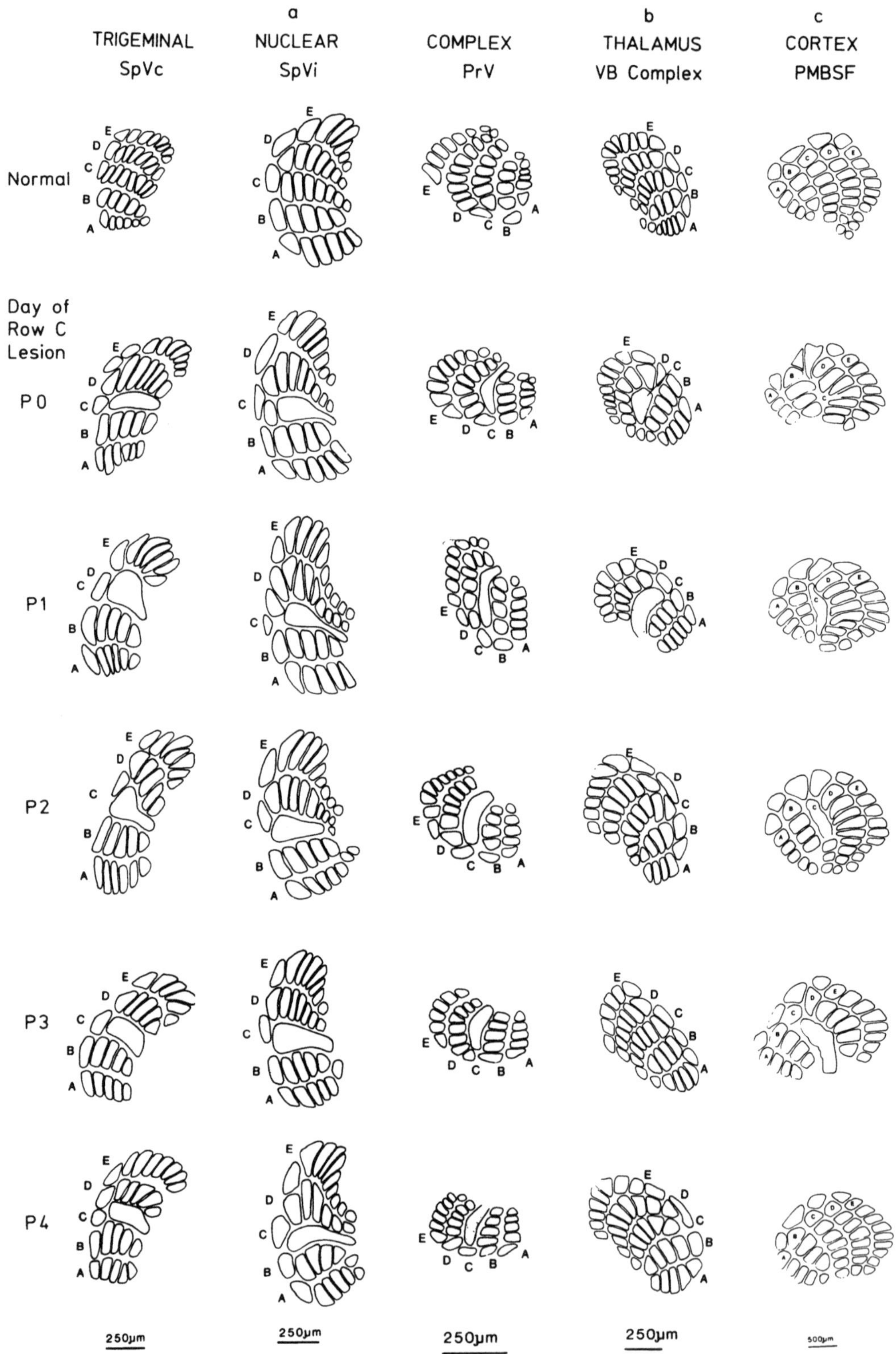
a
b
c
TRIGEMINAL
NUCLEAR
COMPLEX
THALAMUS
CORTEX
SpVc
SpVi
PrV
VB Complex
PMBSF
Normal
Day of
Row C
Lesion
P0
P1
P2
P3
P4
250µm
250µm
250µm
250µm
500µm

et al. (1987) have reported, in the SpVi of mice, a 33% decrease in the volume, a 20% increase in the density of cells and pronounced degeneration of terminals, following neonatal cautery of all vibrissae. These results have raised the possibility that transsynaptic degeneration may occur at the thalamic and cortical levels of the vibrissa system following neonatal damage to the vibrissa afferents. This, however, does not seem to be the case as Hamori et al. (1987) were unable to detect any degenerating cells in TNC or degenerating terminals in the VB complex of mice subjected to neonatal cautery of their vibrissae. Furthermore, cautery of the vibrissae (or section of the ION) in neonatal rats or mice does not reduce the level of SDH activity in the dorsomedial VB complex below that of normal animals or that of the adjacent projection zone of the DCN (Figs. 3.16, 3.18b and 3.19b) (Belford and Killackey 1979; Belford and Killackey 1980; Durham and Woolsey 1984). Nor does this procedure influence the overall SDH activity level in the PMBSF of the cortex of either species (Killackey and Belford 1979; Durham and Woolsey 1984). These observations suggest that the changes evoked at the thalamic and cortical levels by neonatal cautery of the vibrissae probably result from alterations in the organization of connections rather than a gross reduction in the population of any of the contributing elements.

Clustering of Terminals in the VB Complex and the SI Cortex. In both rats and mice, cautery of a single row of vibrissae on P 0 alters the pattern of segmentation in both the VB complex and the PMBSF such that within the region between the representations of adjacent rows there develops a single band of high SDH activity instead of individual vibrissa-related patches (Figs. 3.18b, c and 3.19b, c) (Killackey and Belford 1979; Belford and Killackey 1979). Although there is a postnatal increase in the density of trigeminothalamic terminals in the dorsomedial portion of the VB complex of the rat (Killackey 1985), high levels of SDH activity in the vibrissa representation are already evident at birth and the first signs of segmentation subsequently occur over Ps 1 and 2 (Table 3.1) (Belford and Killackey 1979). In contrast, the PMBSF of rats shows a uniformly low level of SDH activity on P 0 (Killackey and Belford 1979), presumably because thalamocortical terminals do not invade the region destined to become layer IV until P 2–3 (Wise and Jones 1978) – ages when there is a detectable increase in the SDH activity in the PMBSF

◀ **Fig. 3.19 a–c.** The effects on the isomorphic clusters of terminals in the TNC **a**, the thalamus **b** and the cortex **c** of 16- to 100-day-old mice of cauterizing the vibrissae of row *C* on the first 5 days after birth. **a** Segmentation within the SpVc, SpVi and the PrV in the normal animal and in animals subjected to cautery of the row *C* vibrissae on Ps 0, 1, 2, 3 and 4. The region outlined within the normal representation of row *C* contained only *low* levels of SDH activity suggesting that many of the afferents supplying row *C* had undergone transganglionic degeneration by the time of sacrifice (between P 16 and 100). **b** Shows that only before P 2 does cautery of the row *C* vibrissae produce a fused band of *high* SDH activity in the VB complex. **c** shows that only before P4 does cautery of the row *C* vibrissae produce a fused band of *high* SDH activity in the PMBSF of the SI cortex. Compare with the effects on cytoarchitecture shown in Fig. 3.17. (Modified from Durham and Woolsey 1984, pp. 436, 440, 441)

(Fig. 3.10b; Table 3.1). Thus in both the cortex and the thalamus segmentation appears to result from the clustering of recently arrived terminals rather than the penetration of the target tissue (VPM or layer IV in the SI cortex) by terminals that are already grouped into vibrissa-specific clusters. At both levels early continuity of afferent fibres with single vibrissae is a necessary condition for normal development of these features.

Row-Related Clustering of Terminals. Cautery of all the vibrissae, crush of the ION on P 0 or cutting the ION on P 1–2 results in five bands of high SDH activity in the PMBSF of 5- to 60-day-old rats (Waite and Cragg 1982; Killackey and Shinder 1981; Killackey and Belford 1979). Each of these bands runs along the representation of a single row of vibrissae suggesting that the clustering of the terminals of thalamocortical neurons which receive input from the vibrissae of a single row is more resistant to early postnatal damage of primary afferents than is the clustering of terminals from thalamocortical neurons subserving single vibrissae within a row. Indeed in the rat it requires ION section on P 0 to fully disrupt the row-related bands of thalamocortical terminals (Killackey and Shinder 1981; Waite and Cragg 1982). It is interesting to note that although section of the ION on P 0 in the mouse results in the development of five "barrel-like" strips (see below) in place of the barrel rows A to E (Andres and Van der Loos 1985), lesion of all vibrissae on P 0 results in a continuous sheet of high SDH activity in the PMBSF (Pidoux et al. 1980).

To assess the tendency to cluster of the terminals of trigeminothalamic or thalamocortical neurons which represent adjacent vibrissae in neighbouring rows, Killackey and Belford (1980) cauterized an arc of vibrissae (a line of vibrissae which crosses the rows) at P 0. By Ps 4 and 8 the region of the PMBSF in which the damaged vibrissae are normally represented contained only 2–4 small clusters of terminals (Fig. 3.20a, c). A fused arc of high SDH activity like that resulting from cautery of all vibrissae of a single row, was never observed. Similarly, a small lesion placed in the PMBSF of neonatal mice disrupts the arcs, but not the rows, of the terminal clusters of thalamocortical neurons (Fig. 3.20e, d; Sect. 3.3.6) (Ito and Seo 1983). The factors underlying this apparent intra-row affinity (or inter-row repulsion) between the terminals of the appropriate central projection neurons are unknown but possibly reflect the association of trigeminal primary afferents with the five branches of the ION that each supply one of the five rows of the mystacial vibrissae and the five row-related ridges on the muzzle at a developmental stage when individual vibrissae have not yet formed (Figs. 3.1c and 3.8a; Table 3.1) (Killackey and Belford 1980).

Effects on Cytoarchitectonics in the VB Complex and the SI Cortex. The effects of early vibrissa removal were first noted in cytoarchitectonic studies of the PMBSF of mice at Ps 12–60 (Van der Loos and Woolsey 1973; Jeanmonod et al. 1981). While there is some variation between individual animals the general effect of cauterizing row C on P 0 is to prevent any sign of barrel development in the region between the barrels of rows B and D (Jeanmonod et al. 1977). In addition, there is an enlargement of the barrels of rows B and D such that they encroach upon the territory normally occupied by row C barrels (Fig. 3.17a–c) (Van der Loos and Woolsey 1973; Steffen

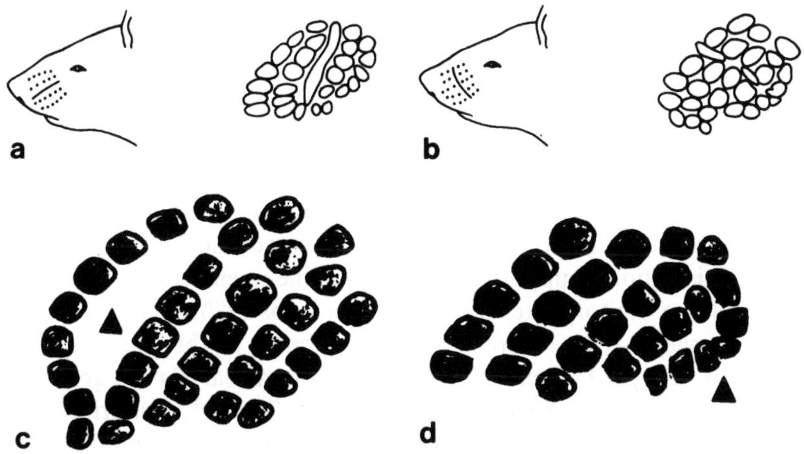

Fig. 3.20 a–d. Inter and intra-row affinity of the clusters of thalamocortical terminals as shown by the effects of neonatal cautery of vibrissae and microlesions in the PMBSF of the neonatal cortex. The relative effects of neonatal (P 0) cautery of a row **a** or an arc **b** of vibrissae on the segmentation of terminals in the PMBSF of 4- to 8-day-old rats. Note that cautery of row C leads to a fused band of terminals **a** while cautery of an arc **b** leads to several isolated and disorientated clusters. **c, d** The effects of lesions placed in the presumptive PMBSF of rats on P 0. Note that when the lesion happened to be centrally positioned in the PMBSF (**c**) the rows of terminal clusters developed around the lesion such that the continuity of single rows was maintained. In this case the area containing the barrels was normal. If, however, the lesion fell near an edge of the PMBSF then the isomorphic pattern of terminal clusters was normal except that it was confined to a smaller area of cortex. See text, Sect. 3.3.6 for further explanation. (**a, b** From Killackey and Bedford 1980, p. 205; **c** and **d** from Ito and Seo 1983, p. 601)

and Van der Loos 1980; Jeanmonod et al. 1981; Woolsey and Wann 1976; Woolsey et al. 1979). Similarly, neonatal lesion of the vibrissae of row C has been shown to produce an enlargement of those terminal clusters of thalamocortical neurons that are isomorphic to the vibrissae of rows B and D, along with a shrinkage in the area of those terminations that are centred on the normal representation of the row C vibrissae. In place of the barrels of row C there develops, over the same time-course as the barrels of intact vibrissae, an elongated zone of low cell density surrounded by a cell-dense band ("barrel-like" territory) and occasionally traversed by indistinct lines of high cell density (Jeanmonod et al. 1977; Jeanmonod et al. 1981, Woolsey and Wann 1976). Lesion on Ps 1 and 2 more often result in an elongated zone of uniform cell density ('barrel-less" territory) which may have a poorly defined cell-dense boundary. The enlargement of the barrels of rows B and D is greatest in these mice (see Table 3.1 in Jeanmonod et al. 1981) despite the fact that there is no change in the area of the PMBSF following cautery at any of the postnatal times studied (Woolsey and Wann 1976). Lesion on P 3 results in poorly defined barrels while lesion on Ps 4, 5 and 6 results in sharply defined barrels which are sometimes smaller than normal (Jeanmonod et al. 1981). Lesion on P 7 does not disrupt the normal appearance of cortical barrels. Similarly, lesion on or after P 4 results in comparatively normal vibrissa-specific clustering of the terminals of thalamocortical neurons which represent row C (Durham and Woolsey 1984).

Although, in the mouse, the axons of thalamic neurons are present in the SI cortex on Gs 15–17 (Crandell and Caviness 1984) clusters of thalamocortical axon terminals are not observed in the region which will become layer IV of the PMBSF until P 2 (Waite 1977). Furthermore, barrels are not detectable until P 3 (Rice and Van der Loos 1977). It thus seems likely that at Ps 2–3 cortical cells aggregrate around the vibrissa-specific clusters of thalamocortical terminals (Killackey et al. 1976). This notion is supported by a study of the orientation of cortical cell dendrites in barrel-like territory (Steffen and Van der Loos 1980) of 60-day-old mice in which the vibrissae of row C were cauterized at birth. In this study it was shown that dendrites of layer IV cells in the wall of normal barrels are generally orientated towards the centre of the hollow (Fig. 3.7a–d). Similarly, dendrites in the cell-dense wall of the barrel-like territory project dendrites preferentially into the central region indicating that the developing dendrites of layer IV cells are attracted to the clusters of terminals of thalamocortical neurons. Since in both the rat and the mouse neonatal cautery of a row of vibrissae causes the terminals of thalamocortical neurons representing that row to form a band (Killackey et al. 1976; Durham and Woolsey 1984), it would seem that the aggregation of cortical cells around the terminals of thalamocortical neurons would indeed form a barrel-like territory. Ironically the same conclusion was reached by Harris and Woolsey (1979) who showed that in barrel-less territory more layer IV cells projected dendrites into the barrels of rows B and D than were found to do so within the barrels of row C in normal animals. In this case, the argument was made that the terminals of those thalamocortical neurons which respond to intact vibrissae are more attractive to the dendrites of layer IV cortical neurons than the terminals of thalamocortical neurons representing afferents of the damaged vibrissae. One could propose a third theory that in individual animals one or the other of the above theories is correct with the result that one observes either the barrel-like or the barrel-less configurations.

Changes in Cortical Activity. The surgical removal of all but the C3 vibrissa on P 2 results in a tangential expansion of 2-DG labelling of the C3-evoked uptake of 2-DG in and around the contralateral cortical representation of the C3 vibrissa, in 6- to 12-day-old rats. The magnitude of this expansion is equal to 35%–140% of the area of metabolic activity evoked in the opposite hemisphere by stimulation of C3 on the opposite (intact) whisker pad (Hand 1982; Kossut and Hand 1984b). This expansion is accompanied by a 40% increase in the size of the C3 barrel although, as in the intact animal, the 2-DG labelling exceeds the boundaries of the C3 barrel (Fig. 3.21). These findings suggest that the intracortical spread of activity in the tangential plane is, like other features of the system, influenced by the state of the connections with the periphery (Hand 1982; Kossut and Hand 1984b).

The effects of cauterizing a row of vibrissae on Ps 0, 1, 2 and 3 on segmentation in the VB complex and the PMBSF of 7- to 12-day-old rats, have been tested. Lesion on Ps 0 or 1 produces in both the VB complex and the PMBSF a fused band of high SDH activity while after a lesion on P 3 relatively normal segmentation occurs across the area of representation of the lesioned row (Fig. 3.18b, c) (Belford and

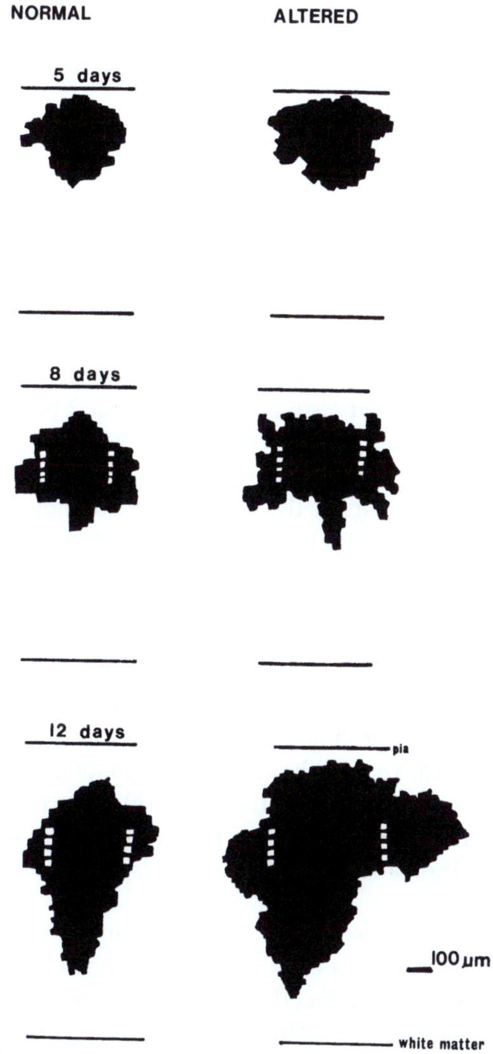

Fig. 3.21. Effects on metabolic activity produced in the SI cortex by stimulation of the C3 vibrissa on Ps 5, 8 and 12, after surgically removing all vibrissae *except* C3 on P 2. Figure shows isodensity maps of the 2-DG autoradiograms produced in coronal sections through the C3 barrel, the borders of which are shown by the *dashed vertical lines*. Note that even on P 5 neonatal removal of the other vibrissae produces an increase in the tangential spread of evoked metabolic activity over that seen in normal animals and that this difference increases with longer survival times and is accompanied by an increase in the width of the C3 barrel in neonatally lesioned animals. (From Kossut and Hand 1984b)

Killackey 1980). In a similar study conducted on mice, Durham and Woolsey (1984) found a greater difference between the periods of sensitivity of clusters in the VB complex and the PMBSF to vibrissa lesion, than has been reported in the rat. Thus lesion of row C after P 2 did not disrupt segmentation in the VB complex while segmentation in the PMBSF only became insensitive to vibrissa lesion on P 4

(Fig. 3.19b, c). While in both the rat and the mouse it seems that there exists a critical period over which normal connections to the periphery are necessary for correct development of functionally related central nervous pathways, there is no good explanation as to why this period is the same for the thalamus and the cortex in rats yet quite different for these regions in the mouse (Durham and Woolsey 1984).

Time-Course of Effects. In the mouse, the relative time-course of sensitivity to peripheral lesion of the thalamic barreloids and the cortical barrels has also been examined. Lesion of the row C vibrissae on P 0 completely abolished the row C barreloids in the mouse thalamus but similar lesions on Ps 1 and 2 produced between the barreloids of rows B and D a band of low cell density surrounded by a thin boundary of high cell density (Fig. 3.17c) (Woolsey et al. 1979). As mentioned above, similar lesions on Ps 0, 1 and 2 produced a shrunken band of terminals in place of the clusters which in the VB complex and the PMBSF normally represent the single vibrissae of row C (Fig. 3.19b, c). Cautery of the vibrissae on or after P 3 does not disrupt the thalamic barreloids or the clustering of trigeminothalamic terminals of adult mice (Woolsey et al. 1979). However, in the same animals the cortical barrels and the segmentation of thalamocortical terminals were disrupted if lesions were performed on or before P 4 and P 3, respectively (Figs. 3.17c and 3.19c). Woolsey et al. (1979) suggested that the cortical barrels became insensitive to vibrissa lesion after P 5 whereas Jeanmonod et al. (1981) reported that the barrels were stable after P 4. The important points, however, are that (a) in the mouse the thalamic barreloids and the clustering of trigeminothalamic terminals both become resistant to peripheral damage about 2 days before the cortical barrels and the thalamocortical terminals, and (b) in the thalamus or the cortex both the clustering of afferent terminals and the aggregrations of postsynaptic cells become insensitive to vibrissa lesions at about the same time.

Effects on Somatotopic Organization and Receptive Fields. Some controversy exists over the effects on somatotopic organization of removing all the vibrissae. Thus Waite and Taylor (1978) report that dewhiskering rats at P 0 produces in 15- to 26-week-old individuals total invasion of the PMBSF by other inputs from the face surrounding the peripheral territory of the ION. The abnormal inputs are represented in a somatotopically ordered manner within the PMBSF. In contrast, Pidoux et al. (1979, 1980), Verley and Onnen (1981) and Farkas-Bargeton et al. (1986) have shown that following neonatal whisker removal cells in the VPM and in the PMBSF of mice and the PMBSF of rats respond to the common fur of the whisker pad – a receptor sheet which normally is represented in small areas that abut the thalamic and cortical representations of the vibrissae. Similarly, Killackey et al. (1978) have reported that neonatal removal of a single row of vibrissae results in cells in the appropriate region of the cortex becoming responsive to the fur on the skin underlying the damaged row. Unfortunately these authors did not identify histologically the cortical laminae containing their recording sites. Furthermore, it has been suggested that the relatively high cauterizing currents used by Waite and Taylor (1978) produced extensive damage to all afferents innervating the whisker pad including those supplying the common fur (see Pidoux et al. 1980; Simons, et al.

1984). Indeed the electrocautery carried out in mice by Waite and Cragg (1979) resulted in the death of more than twice the number of myelinated afferents that innervate all the mystacial vibrissae (Savy et al. 1986). Further support for this suggestion derives from the fact that the somatotopic reorganization observed by Waite and Taylor (1978) in the cortex resembles that reported by Waite (1984) following neonatal and repeated section of the ION. Finally, Simons et al. (1984) have also suggested that the reports of responses from the common fur in denervated vibrissa territory reflects reinnervation of non-vibrissal structures in the whisker pad by regenerating vibrissa afferents (Sect. 3.3.1; Rhoades et al. 1987).

In their own study, Simons et al. (1984) made recordings in the PMBSF of unanaesthetized adult mice and rats following neonatal cautery of the row C vibrissae. They were careful to localize cortical recording sites and to assess histologically the innervation of the whisker pad. Their results showed that layer IV cells between the enlarged barrels of row B and D were either unresponsive to any peripheral stimuli or were activated by stimulation of the row C and scar tissue or the vibrissae in rows B and D. These responses in fact resemble those seen in layer IV neurons within the PMBSF of normal animals except that cells in the row C region have been deprived of their normal input (Simons 1978; Simons and Woolsey 1979; see Sect. 3.1). Thus while Waite and Taylor (1978), Killackey et al. (1978), Pidoux et al. (1979), Pidoux et al. (1980), Verley and Onnen (1981) and Farkas-Bargeton et al. (1986) agree that the reorganization they observed most likely reflects changes in the central circuitry at the level of the TNC, Simons et al. (1984) have emphasized the close relationship between peripheral damage and cortical representation of the vibrissae, marvelling at the stability with which the normal central representation of the intact elements of the system develops. Certainly, the fact that the comparatively traumatic procedure of neonatal thalamotomy does not result in the invasion of the deafferented laminae or cortical area by the terminals of commissural neurons (Wise and Jones 1978) underscores the inherent stability of mechanisms controlling connectivity between higher centres during development.

3.3.3 The Effects of Supernumerary Vibrissae

The observation of Andres and Van der Loos (1983) that the non-innervated whisker pad from 9- to 10-day-old mouse embryos will, if placed in culture, develop the usual vibrissa pattern suggests that it is the organization of the skin which is reflected in the central nervous system and not vice versa. Nevertheless, despite many attempts to study the effects of removing vibrissae on brain development, there have been, until recently, very few attempts to *increase* the number of vibrissae. The first indication of the effects of such a procedure were the reports of Van der Loos and Dorfl (1978) and Yamakado and Yohro (1979) in which it was shown that in mice with an extra vibrissa there was an extra barrel in the appropriate region of the PMBSF (Fig. 3.22a, b). Similar results have since been reported in a number of strains of mice lacking a particular vibrissa or with supernumerary vibrissae (Van der Loos and Welker 1985). However, not all extra vibrissae are accompanied by an extra barrel and Welker and Van der Loos (1983) have reported that those

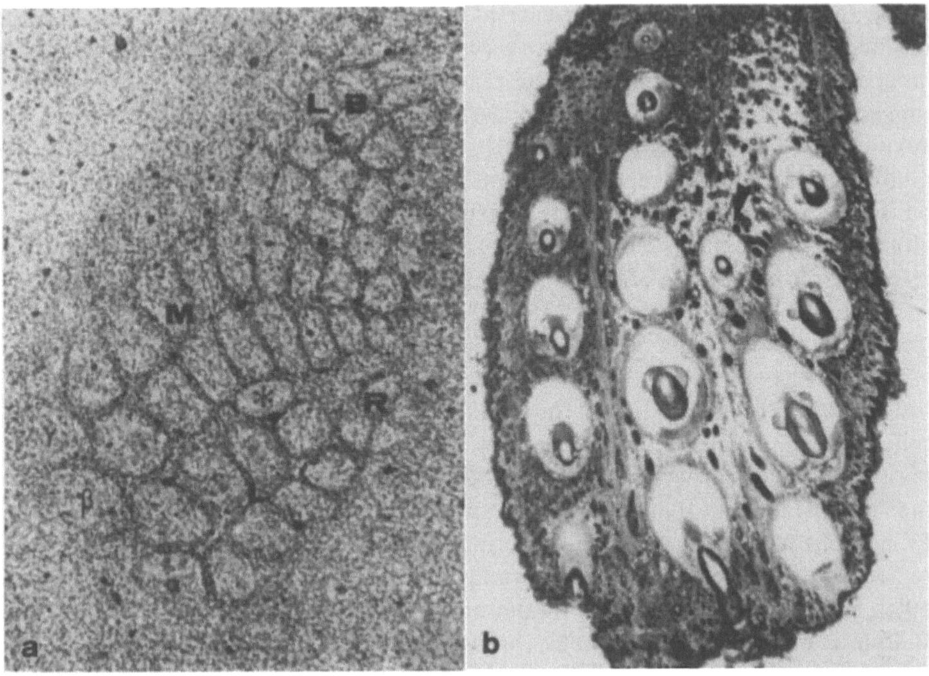

Fig. 3.22 a, b. Effects of a supernumerary mystacial vibrissa on the PMBSF of the SI cortex in the mouse. Photomicrographs of tangential sections through the PMBSF and the contralateral whisker pad of a BALB/c (nu/+) mouse. In **b** note the supernumerary vibrissa (*arrowhead*) and, in **a** the corresponding supernumerary barrel (*) that lies between the lateral nasal (*L*) and the maxillary (*M*) barrels. The rhinal (*R*) and the labial barrels (*LB*) are normally formed and the positions of the barrels of the alpha beta and gamma straddlers are shown. (From Yamakado and Yohro 1979, p. 171)

supernumerary vibrissae that are innervated by less than 40 myelinated fibres are not represented by a cortical barrel. The establishment of a threshold level of peripheral innervation density which must be reached in order for a vibrissa to be represented in the cytoarchitectonics of the cortex is itself highly suggestive of a causal relationship between the structure of the periphery and the central nervous system organization. In a valiant attempt to establish this firmly, Andres and Van der Loos (1985) attempted to alter the arrangements of vibrissae in embryonic (Gs 16–18) and newborn (P 0) mice. Operations performed included transplanting an additional whisker pad, rotating a whisker pad, removing and replacing a whisker pad and removing a whisker pad. The arrangement of barrels was then examined between Ps 12 and 303. In all cases the arrangement of barrels was disorganized but difficulties in associating individual barrels with individual vibrissae prevented more than the general conclusion that prenatal modification of the *already innervated* periphery (Table 3.1) does influence structure in the central nervous system.

3.3.4 The Effects of Lesioning Unmyelinated Afferents

Capsaicin, the active ingredient in paprika, has been shown to kill somatic C-fibres if given to neonatal animals and to transiently deplete the adult C-fibres of neuropeptides. Its actions on circuitry in the dorsal horn of the spinal cord have therefore been the subject of a considerable number of investigations which are reviewed in Chap. 4. The effects of administering capsaicin to neonates (Ps 2 and 4), on the cortical representation of the vibrissae in adults, or its application to the ION of adults has been investigated (Wall et al. 1982; Nussbaumer and Wall 1985). In each study 20 μm tungsten microelectrodes were used to record multiunit responses in the cortex of mice anaesthetized with methoxyflurane in O_2/N_2O. Recording sites were localized histologically using microlesions. The results of 18 penetrations in five neonatally treated mice compared with 9 penetrations in two control mice, suggested an increase from 1.0 to 3.6 in the mean number of vibrissae represented at multiunit recording sites (MRSs) of treated animals. Unfortunately, the very small number of penetrations per animal (3.6) and the small number of animals surveyed does not encourage acceptance of these conclusions. Furthermore, the relationship to barrel cytoarchitecture of what are presumably the same 9 control recording sites is not given in either report. There is therefore no acceptable evidence for the conclusion of Wall et al. (1982) that the neonatal destruction of C-fibres by capsaicin increases the size of the receptive fields of neurons in the PMBSF of adult mice.

The effects of application of capsaicin to the ION of adults has been measured by comparing a total of 101 MRSs in experimental mice with 68 MRSs in control mice. Responses to more than one vibrissa could be recorded at 81% of the 48 MRSs in the barrel hollows whereas in control mice only 15% of the 27 MRSs yielded responses to more than one vibrissa (Nussbaumer and Wall 1985). Considering that rather large-tipped microelectrodes were used to make multiunit recordings there remains some question as to whether receptive field sizes in this study reflect those of single cells and, if so, whether these cells were actually positioned within layer IV. As pointed out in Sect. 3.1, cells in the supra- and infragranular layers are often excited by a number of vibrissae (Chapin 1986).

These studies are interesting in that they centre on the question of how unmyelinated fibres might regulate the inflow of tactile information to the SI cortex. Given the potential for interactions between myelinated and unmyelinated afferent input both within the spinal cord and the thalamus (Chaps. 4 and 6) one might expect capsaicin poisoning of peripheral nerves to have some effect on the cortical representation of tactile input. However, it is difficult to see how a useful interpretation could emerge from this work considering the complex, yet poorly understood, functional implications of such a lesion.

3.3.5 Hyper- and Hypostimulation of Vibrissa Afferents

Plucking the vibrissae from the whisker pad of adult mice does not damage the follicles and presumably does little damage to the terminals of afferent fibres. The

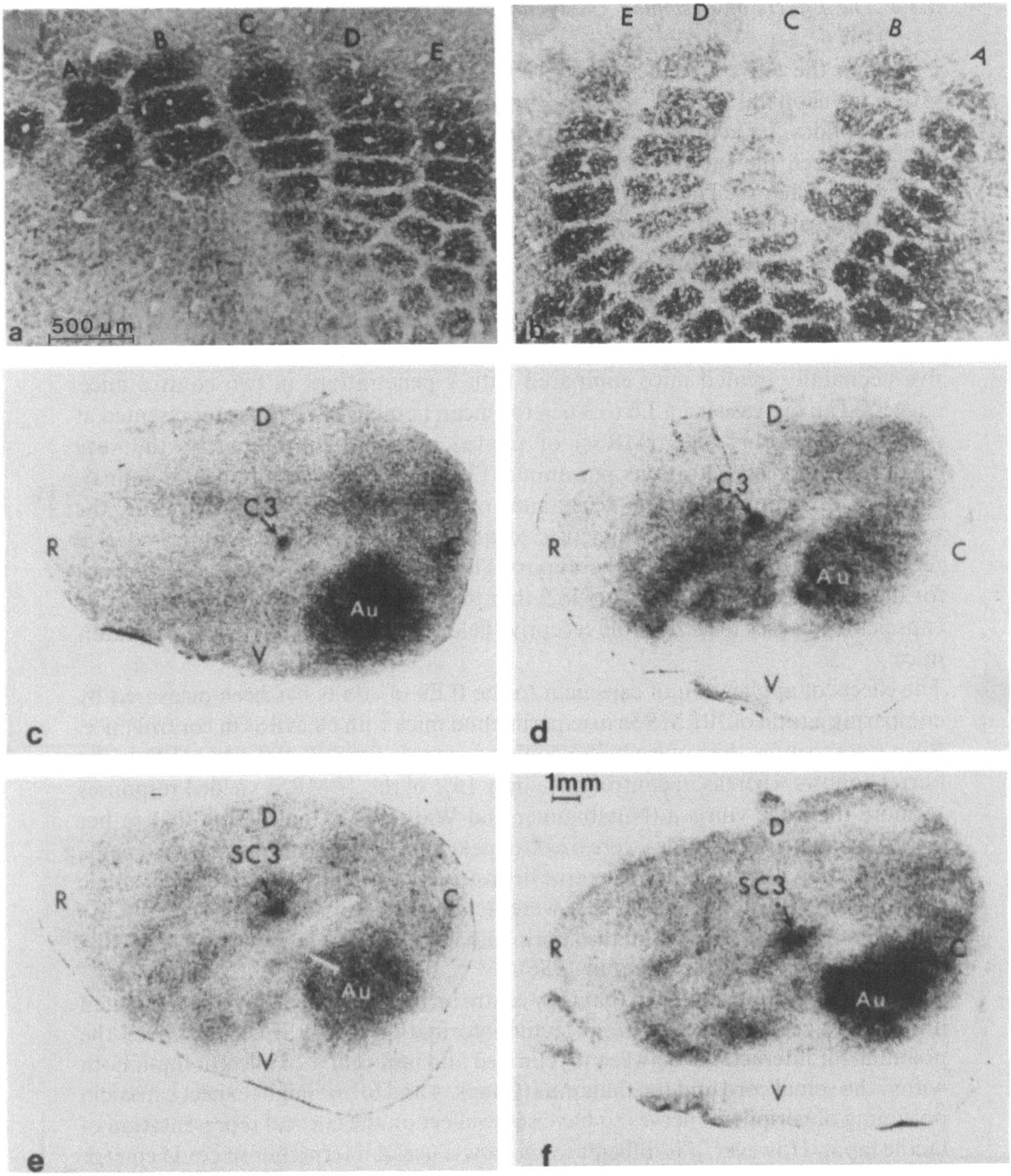

Fig. 3.23 a–f. Effects of deprivation of input from vibrissa afferents in neonatal and adult rodents, on neural activity within the SI cortex, **a**, **b** Cytochrome oxidase (CO) activity in the PMBSF of a normal adult mouse (**a**) and an adult mouse in which the row C vibrissae were repeatedly plucked for 80 days prior to sacrifice (**b**). Notice the decreased CO activity in the row C representation (**b**) following repeated dewhiskering in the adult. **c–f** Autoradiograms of tangential sections through layer IV in the region of the PMBSF of the rat cerebral cortex, showing the distribution of 2-DG uptake elicited by stimulation of the C3 vibrissa.

major functional effect of this procedure is thus thought to be a decrease in the sensory activation of the afferents supplying the hairless follicles and, in fact, a decrease in the uptake of 2-DG by cells in the appropriate cortical barrels has been reported (Durham and Woolsey 1978). Similarly, repeated clipping or plucking of some vibrissae results in a decrease in the level of the mitochondrial enzyme cytochrome oxidase (CO) in the appropriate barrels of adult rats (Fig. 3.23a, b) (Wong-Riley and Welt 1980; Land and Simons 1985). There is, however, no evidence for cytoarchitectural changes in the cortical barrels of mice even if vibrissae are clipped repeatedly from birth (Verley and Axelrad 1977b). These results suggest only that when no stimulation is applied to a group of vibrissae, there is a measurable decrease in neural activity in that area of cortex in which they are normally represented. More interesting are the findings of Hand (1982) that regular *clipping* of all but the C3 vibrissa for 90 days beginning on P0 or P89 (Fig. 3.23d, f) results in an expansion of the C3-evoked uptake of 2-DG around the C3 barrel without, again, any change in barrel size (compare neonatal *lesion*, Fig. 3.21, Sect. 3.3.2).

A probable corollary of the observations of Hand (1982) and Kossut and Hand (1984b) on early sensory deprivation (Figs. 3.21 c, e and 3.23) is the recent demonstration that repeated trimming, from P0, of the row C vibrissae results in altered physiological properties in the cortical neurons of adult rats (Fig. 3.24) (Simons and Land 1987). In particular, neurons in deprived barrels showed reduced angular tuning, increased responsiveness and abnormal temporal patterns of stimulus-evoked discharges, an increased spontaneous activity and enlarged receptive fields. Thus such deprived neurons responded equally well to movements of several adjacent vibrissae, as they do to movements of the appropriate (or principal) vibrissa, whether or not these vibrissae (like the appropriate vibrissa) had been subjected to trimming. In contrast, neurons in non-deprived barrels were responsive to the appropriate (or principal) whisker (Simons 1978) and only poorly responsive even to intact neighbouring vibrissae of the same row (Fig. 3.24). The mechanisms underlying these changes are unknown but the authors suggest that sensory deprivation might reduce both the effectiveness of the focal excitatory actions of thalamocortical neurons as well as the intracortical inhibitory mechanisms which they say depress the spontaneous activity of cortical neurons and

◄ **Fig. 3.23a–f** (*continued*). **c–e** The effects of sensory deprivation from the day of birth: autoradiograms of the left (**c**) and right (**e**) cortex of a 90-day-old rat in which all the vibrissae on the left whisker pad, *except* C3, were repeatedly clipped from P0. In the deprived right cortex (**e**), note that the tangential spread of spared C3-evoked uptake of 2-DG (SC3) exceeds the spread of C3-evoked uptake of 2-DG on the left or control side (**c**). **d, f** The effects of sensory deprivation in the adult: autoradiograms of the left (**d**) and right (**f**) cortex of a 180-day-old rat in which all the vibrissae on the left whisker pad, *except* C3, were repeatedly clipped from P90. In the deprived right cortex (**f**) note that the labelling around the focus of the C3 representation is denser than on the left or control side (**d**). *R*, *C*, *D* and *V* refer to the rostral, caudal, dorsal and ventral portions of the tangential sections. *Au* designates the auditory cortex which normally labels densely. (**a, b** Modified from Wong and Riley 1980, p. 2334; **c–f** from Hand 1982, p. 58)

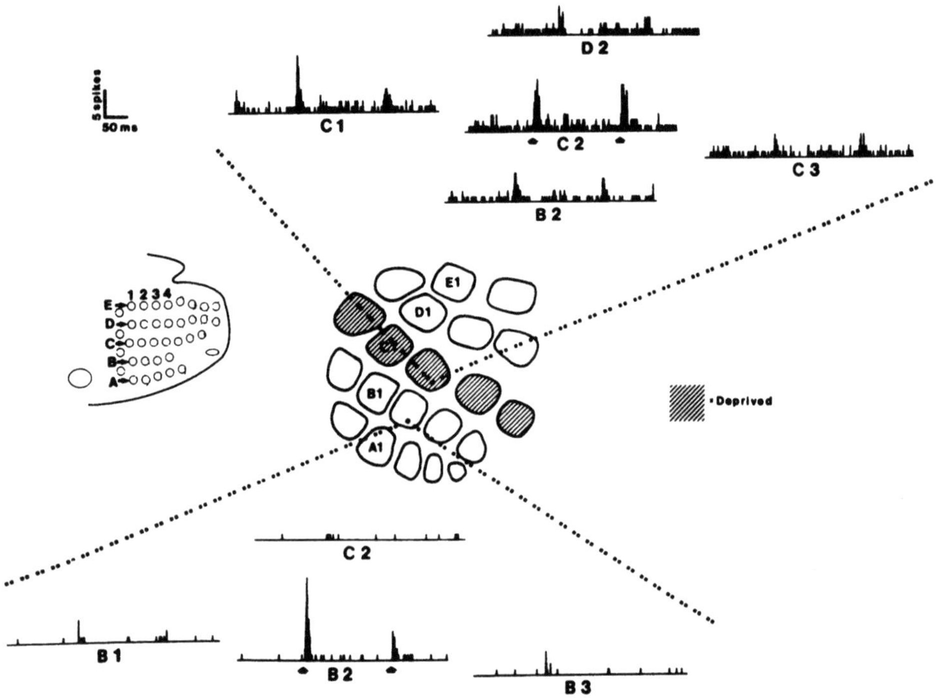

Fig. 3.24. Effects of sensory deprivation during postnatal development on the responses to vibrissa stimulation of single neurons in the PMBSF of the rat. Comparison of the responses elicited by deflection of the principal whisker and adjacent whiskers for neurons recorded in a deprived barrel and one recorded in a homologous part of a non-deprived barrel in the same animal. The left row C whiskers have been trimmed from birth until P 54 and the recordings were obtained 83 days later. The deprived barrels are shown (*hatching*) on a reconstruction of a CO labelled, tangential section through the PMBSF of the right (deprived) cortex. The upper set of peristimulus time histograms (PSTs) show the responses of a neuron in the deprived C2 barrel to mechanically controlled deflections of the regrown C2 vibrissa and of four adjacent vibrissae (C1, D2, B2 and C3). The lower set of PSTs show responses of a neuron in a non-deprived barrel (B2) to deflection of the B2 vibrissa and three adjacent vibrissae (B1, C2 and B3). Note that the responses of the neuron in the deprived C2 barrel to stimulation of the C2 vibrissa are similar in magnitude to the responses elicited by stimulation of both other ablated vibrissa (C1 and C3) of the same row, and unablated vibrissae of neighbouring rows. (From Simons and Land 1987)

shape their responses to maintained whisker displacements (Simons and Land 1987).

While there are no other data relating specifically to these suggestions, evidence from elsewhere indicates that sensory deprivation may have a very clear effect at the level of the TNC. In these preliminary studies the vibrissae on one side of the rat's face were clipped at P 0 and every 2 days thereafter for 60 days. On the deprived side the terminal arborizations of nine intra-axonally-stained, vibrissa afferents in the SpVi were compared with the arborizations of six intra-axonally-stained, vibrissa afferents from the intact side and six from normal rats. The arborizations of the deprived afferents were significantly wider (mean = 272.4 μm compared with 187.7 μm) and bore significantly fewer boutons per collateral (mean = 60.0 as compared with 124.3) than those from non-deprived vibrissae (Woerner et al. 1986). Despite the fact that deprivation was more extensive in this study compared with that inflicted by Simons and Land (1987), it remains quite likely that morphological changes in the terminal arborizations of many primary afferents might account for the altered cortical responses following early-postnatal, chronic, sensory deprivation.

3.3.6 Cortical Alterations

Although many studies have dealt with the effects of peripheral lesions few have looked at the effects of damage to the central nervous system. In the mutant mouse, known as *Reeler*, the cortex is thinner, lacks layer I and has a poorly defined boundary between layers IV and V. Furthermore, the medium-sized pyramidal cells normally present in layer III are found deep to layer IV. Despite these abnormalities Reelers do have cortical barrels and these are of normal size when viewed in the tangential plane (Caviness et al. 1976; Cragg 1975). In normal mice the thalamocortical terminals occupy a layer of cortex about 150 μm thick which contains the layer IV granule cells and the medium-sized pyramidal cells of lower layer III. In contrast, the thalamocortical terminals of Reelers extend radially through the cortex for 400 μm and so are located not only in layer IV but also at deeper levels which contain the abnormally placed, medium-sized pyramidal cells. Consequently, Caviness et al. (1976) have suggested that the radial extent of thalamocortical projections is determined by the distribution of cell types in the cortex while the position of the terminal clusters in the tangential plane determines the position of the barrels. The independence of cortical lamination from thalamic innervation has been supported by the subsequent findings of Wise and Jones (1978) that, following thalamotomy on P 0, the cortical laminae in the PMBSF develop normally except for the absence of barrels in layer IV. However, in microencephalic rats lacking cortical layers II, III and IV, Yurkewicz et al. (1984) have shown that thalamocortical axon terminals are located within two depth ranges in a manner resembling the bistratified organization seen in normal rats (Sect. 7.5.5). It therefore remains possible that in the absence of a well-defined inner granule-cell layer, the approximate laminar distribution of thalamocortical axon terminals is determined by properties intrinsic to the thalamic neurons.

There are only two studies of the effects of surgically produced cortical microlesions within the vibrissa representation. In the first of these, small (300 μm diameter) electrolytic lesions were created in the presumptive barrel field of rats on P 0 and the pattern of clusters of thalamocortical terminals examined using CO histochemistry in the adult (Fig. 3.20c, d) (Ito and Seo 1983). While this procedure did not alter the number of terminal clusters detectable in adult rats, it did elicit and altered spatial organization of these clusters (and so presumably the cortical barrels) around the lesion. Two altered patterns of segmentation were seen. First, in some animals all the vibrissa-related clusters of thalamocortical terminals were displaced *en masse* to one side of the lesion (Fig. 3.20d). In these cases the area in the tangential plane occupied by either the individual clusters or the entire set of clusters was significantly less than in normal animals. This suggests that only a limited amount of cortex is allocated, or specified, for barrel formation and that if that portion is reduced (e.g. by lesion) then vibrissa-related inputs must be distributed to whatever is left. This suggestion is supported by observations that in adult rats, rendered microencephalic by exposure to the cytotoxic agent methylazoxymethanol acetate at G 15–17, the barrels are smaller and less cell-dense than in normal animals (Welker 1973). In contrast, however, Yurkewicz et al. (1984) have found that this agent applied on G 14 or 15 resulted in an absence of barrels in the microencephalic adult progeny (Sect. 7.5.5). It would be interesting to know whether clustering of thalamocortical terminals occurs within the vibrissa region of these animals, which were reported to lack entirely cortical layers II, III and IV. Second, Ito and Seo (1983) found that in some neonatally lesioned rats, the lesion was positioned between the thalamocortical clusters of rows C and D or rows D and E but never did they observe a case of a lesion occurring between arcs of clusters (Fig. 3.20c, Sect. 3.3.2). This suggests that when the cortical damage cannot be totally avoided by those ingrowing thalamocortical fibres that represent *en masse* the total complement of mystacial vibrissae, then these fibres will simply distribute themselves around the cortical damage in a way that least disrupts both their vibrissa-specific and row-specific allegiances.

In the second study, Seo and Ito (1987) made microlesions in the PBMSF region on Ps 0, Ps 5, Ps 10, Ps 14 and Ps 20 and 90 days after birth used CO and SDH histochemistry to study the distribution of thalamocortical terminals. They reported that lesions after P 10 resulted in disruption of the vibrissae-related pattern of thalamocortical clusters. In normal animals the barrels become visible between P 3–6 (Table 3.1). Thus the lesions placed between P 5 and P 10 presumably damaged not just one or more clusters of thalamocortical terminals but also the associated aggregations of granule cells. The subsequent restoration of the correct number of terminal clusters around the lesions suggests that between P 5 and P 10 not only do damaged thalamocortical axons regrow into intact nearby cortex but that such space can be made available presumably by some movement of undamaged terminal clusters in the tangential plane of the cortex. Unfortunately, Seo and Ito (1987) did not further study the effects of the microlesions made at P 5 and P 10 on the barrels themselves and so it is not clear whether a redistribution of granule cells accompanies the reorganization of thalamocortical terminals.

Taken together, studies of the role of the cortex in formation of the barrels suggests that at birth, within both the depth and breadth of the cortex, certain cell groups in particular laminae are endowed with a greater capacity than others to attract ingrowing thalamocortical terminals (see Sect. 7.5.5). As discussed above (Sects. 3.2 and 3.3.2), it is only after the vibrissa-specific clusters of thalamocortical terminals have penetrated layer IV that these cortical cells aggregrate into barrels and send their dendrites preferentially into the intrabarrel neuropil. The role, if any, of prenatally formed, topographically ordered thalamocortical projections in the specification of particular fields within the SI cortex (e.g. vibrissa versus limb cortex) remains to be determined. In relation to this question Yurkewicz et al. (1984) have observed that in microencephalic rats the receptive fields of cortical neurons in SI are of normal size, the embryonic thalamocortical projections are topographically ordered but the peripheral parts of the body representation are missing (Fig. 7.28). These features point to a reduction in the area of cortex that is preordained to receive thalamocortical projections from the VB complex and a concomitant loss of those thalamocortical fibres which, as a result of their position in the topographically ordered embryonic projection, find themselves in inappropriate cortex. These authors emphasize the similarities of the effects of neonatal cautery of vibrissae which results not just in the loss of the associated barrels but an expansion of the barrels of neighbouring vibrissae into what appears to be available SI cortex (see Sect. 3.3.2).

3.4 Plasticity in the Vibrissa System of Adult Animals

3.4.1 The SI Cortex

In the rat subtle changes occur in the structure of barrels between P20 and P34 (Rice 1985) and 2-DG labelling is not adult-like until P21 (Kossut and Hand 1984a). Similarly, in the mouse 2-DG labelling and the intrabarrel synaptology are not adult-like until P10 and P13, respectively (Yamakado 1985; Melzer et al. 1986). However, mice do not reach sexual maturity until about 4–7 weeks while in rats 7–14 weeks are generally required (Green 1966; Baker et al. 1979). If we consider sexual maturity as indicative of adulthood then there have been only three attempts to demonstrate plasticity in the vibrissa system of adult rodents In the first study Hand (1982) and Levin et al. (1988) have shown that 60–90 days after *removal or denervation* of all but the C3 vibrissa from adult rats there is a 30%–39% expansion in the 2-DG labelling in layer IV of the C3 representation (Fig. 3.23d, f), a 31% increase in the area of the column of 2-DG labelling that extends throughout the depth of the cortex, and a 14% increase in the size of the C3 barrel. *Clipping* all but the C3 vibrissa from adult rats for 90 days, produced a 32% expansion of 2-DG labelling but no change in the size of the C3 barrel (Hand 1982). A number of explanations are discussed by Hand (1982) but the most persuasive is that in both cases the subthreshold excitatory influence of the C3 vibrissa afferents on cells around the C3 barrel is raised to a threshold level. The increase in the size of the C3 barrel following lesion of the other vibrissae suggests further that even in the adult there remains some capacity for a spatial redistribution of cells in layer IV of the cortex and possibly also the terminals of thalamocortical neurons. Furthermore, these results suggest that this potential is only expressed following chronic damage to specific groups of primary afferent axons. The role of afferent mortality in this process in unknown (Sects. 3.3.1, 2).

Recently, Levin et al. (1988) have studied the interactions of the norepinephrine-containing projections from the locus coeruleus to the cortical barrels, with the deafferentation-induced changes in the C3-evoked 2-DG uptake. The stereotaxic injection of 6-hydroxydopamine into the locus coeruleus was used to produce unilateral depletion of cortical norepinephrine. In rats with intact vibrissae this lesion resulted in an increase in the area of the C3-evoked uptake of 2-DG both within layer IV and throughout the depth of the cortex. In contrast, when lesion of the locus coeruleus was combined with chronic denervation of all vibrissae except the C3, stimulation of C3 resulted in 2-DG uptake that was of similar areal magnitude to that seen in normal (unlesioned) animals with intact vibrissae.

The authors conclude from these interesting experiments that norepinephrine may be effecting long-term structural changes both within the PMBSF and at subcortical sites. However, some attention to the pharmacological effects of norepinephrine in the SI cortex suggests that structural alterations are not necessarily indicated by these data. Waterhouse and Woodward (1980) have shown that norepinephrine enhances the excitatory responses of many cortical cells as well as suppressing their ongoing background activity. In addition it augments postexcitatory inhibition or, as it is referred to elsewhere in this review, in-field inhibition (Figs. 6.7 and 8.5; Sects. 4.2.5, 5.2.3, 6.3, 7.3.1 and 8.5.1). Norepinephrine thus has properties that would permit it to focus, both spatially and temporally, cortical responses to stimulation of a point on the skin surface. As all norepinephrine in the mammalian cortex is derived from the locus coeruleus, destruction of this nucleus might thus be expected to result in transcortical expansion of the area of cortex activated by stimulation of a single vibrissa. That norepinephrine from the locus coerulus is a necessary requirement of denervation-induced expansion of the representation of intact inputs is more puzzling. Waterhouse, Moises and Woodward (1980) have shown that norepinephrine enhances the inhibitory effects of iontophoretically applied GABA on cortical neurons and the excitatory effects of iontophoretically applied ACh. As discussed elsewhere, ACh also exaggerates, sometimes for long periods, the excitatory responses of cortical neurons to natural stimulation or to glutamate (see Sect. 7.3.3) – an observation that has led Metherate et al. (1987) to propose that an enhancement of its effects are a condition of denervation-induced expansion of intact somatosensory inputs. Perhaps this enhancement requires the presence of norepinephrine (and so the locus coeruleus) or at least is offset by the removal of this substance and the consequential decreased effectiveness of ACh. Clearly, modulation of somatosensory input to the SI cortex may be achieved by several centres outside the somatosensory axis without the need to postulate structural changes (Sect. 8.5).

The second example of adult plasticity is that of Waite (1984) who mapped the SI cortex of 120-day-old rats, 60 days after sectioning and resectioning the ION (Fig. 315a, b). It seems likely that even in the adult this operation causes death of many trigeminal ganglion cells although it does not produce a decrease in the size of the trigeminal nuclei (Sect. 3.3.1). Furthermore, trigeminal neurons within the projection zone of the ION do not acquire new receptive fields. In contrast, at the cortical level neurons within 25% of the area of the normal ION representation become responsive to the digits, lower jaw and the whiskers over the eye and the ear (Table 3.2). The reorganized region lies at either end of the ION representation, within about 500 μm of the regions where these novel inputs are normally represented. Cells in the central 75% of the ION territory remain unresponsive (Waite 1984). This distance is characteristic of transcortical shifts in somatotopic organization shown to result from the transection of limb nerves in adult rats and primates (see Sects. 7.2.8 and 7.5.2).

The encroachment of the digit inputs into the ION territory is especially interesting as it constitutes a case in which inputs from spinal afferents have established, in an adult animal, an influence over cortical territory normally allocated to trigeminal afferents. The only other report of a nerve-section-induced invasion of a cortical representation of trigeminal inputs by inputs from the spinal cord is that of Merzenich et al. (1983b) in the adult monkey (Sect. 7.2.8). The changes in functional connectivity which underlie these alterations in cortical somatotopy are unknown but evidence in the cat suggests that they do not involve the establishment of projections from primary afferents supplying receptors on the body surface to cells within the upper cervical spinal segments and the SpVc (Kerr 1972, 1976). It is, however, possible that they reflect altered efficacy of somatotopically inappropriate thalamocortical neurons such as have been described in the SI cortex of the cat (Snow et al. 1988).

The third example of plasticity in the whisker-to-barrel pathway of the adult rodent involves mature mice in which the effects of vibrissa removal on the GABA levels in the PMBSF was assessed by the immunocytochemical labelling of glutamate acid decarboxylase (GAD) (Welker et al. 1989). Within 3 days of vibrissa removal there was a reduction of GAD within the appropriate barrels. This reduction was maximal at 2 to 4 weeks but at longer survival times (up to 7 months) the level of immunoreactivity had returned to normal.

The multiple representation of vibrissae at the cortical, thalamic and trigeminal levels, the projection of the dendrites of many barrel neurons to adjacent barrels and the large size of the dendritic trees of the SpVi neurons compared with the spread in the transverse plane of the terminal arbors of single trigeminal, primary afferents (Sect. 3.1) all suggest that within the vibrissa system there is a substrate, albeit functionally weak, for much convergence and divergence. Given the prolonged effects of vibrissa lesion on GABA levels within the corresponding barrels (Welker et al. 1989) it is not difficult to imagine how, even in the mature animal, chronic alteration of afferent input might elicit an equally prolonged disinhibition of previously subthreshold inputs from surviving afferents. Further studies of the morphology of identified afferent axons and postsynaptic cells may do much to establish the *mechanisms* underlying those examples of plasticity that have already been documented (see Arvidsson and Johansson 1988). In addition, however, a most productive new line of exploration would appear to be the study of the responsiveness of cells in barrels deprived of input from their principal vibrissa, to the principal vibrissae of neighbouring barrels of the *same row* (Simon and Woolsey 1979; Princince and Donaghue 1986).

3.4.2 The Ventral Posterior Medial Nucleus

Afferents supplying the mystacial vibrissae project to all four subdivisions of the TNC. Three of these subdivisions, the PrV, SpVi and SpVc have overlapping projections to the VPM of the thalamus (Peschanski 1984) (Fig. 3.2; Sect. 3.1). In anaesthetized rats trigeminothalamic cells in the SpVi typically respond to several vibrissae (Jacquin et al. 1986a) whereas in the PrV, Rhoades et al. (1987a) have suggested, from rather little evidence, that responses to a single vibrissa are more common. In anaesthetized rats there is good evidence that most cells in the VPM respond to only one vibrissa (Waite 1973b; Rhoades 1987a). On the basis of the similarity in the receptive field sizes of PrV cells and VPM cells, Rhoades et al. (1987a) suggested that cells in the VPM might derive their effective input primarily from the PrV and that the widefield inputs from the SpVi might be ineffective. In order to test whether destruction of the PrV might reveal the presumably subthreshold effects of inputs from the SpVi, they injected kainic acid into the PrV in adult rats. Recordings were made in animals anaesthetized with a combination of xylazine and ketamine followed by urethane. In normal animals prepared in the same manner, 85% of VPM cells responded to vibrissae stimulation. From 0 to 10 h after kainic acid injection most cells in the VPM remained unresponsive to cutaneous stimuli. Over 6 days the number of responsive cells increased until 47% were found to respond to movement of the vibrissae. After 30–90 days the number of cells responding to the vibrissae increased to 59%. In all animals sustaining chronic lesions of the PrV single VPM cells were activated by an average of about six vibrissae suggesting that in these animals the efficacy of widefield trigeminothalamic inputs from the SpVi had increased to the extent that they now drove VPM cells and so determined the receptive characteristics of those cells. This interpretation was supported by the observations that the injection of kainic acid into the SpVi in four chronically lesioned animals abolished responses in many VPM cells while in normal animals chronic or acute destruction of the SpVi did not alter the responses of vibrissa-sensitive cells in the VPM.

Rhoades et al. (1987a) concluded from these experiments that (a) no immediate expression of new inputs to the VPM occurred after PrV lesions, and (b) that chronic lesions to the PrV resulted in long-term changes during which VPM neurons developed inputs that are characteristic of SpVi cells. This experiment is very interesting because it testifies to the interaction of thalamic inputs in the regulation of somatic receptive fields and because in the VB complex of the rat there are very few inhibitory interneurons (Chap. 6). Many neurons developed new receptive fields over the 6 days that followed lesion of the PrV. On the basis of the measured time-course of synaptogenesis in the VPL of the rat following lesion of the DCN (Fig. 6.10; Sect. 6.5) (Wells and Tripp 1987), it seems unlikely that these early changes in functional connectivity could result from anatomical changes in the VPM. Perhaps the receptive fields of VPM neurons are regulated with a very slow time-course by trigeminal influences on the reticular nucleus of the thalamus which is known to exert a GABA ergic control over the VB complex (Chap. 6) (Jones 1986; Salt 1989).

Finally, it is important to note from the above study the interaction between trigeminal inputs, that could (by virtue of their small receptive fields – typically one vibrissa) be regarded as classically *lemniscal* and those inputs with large receptive fields that we might refer to as *widefield* in character (see Sect. 6.3). The dominance of lemniscal inputs over widefield inputs is a conspicuous characteristic at both cortical and thalamic levels of the somatosensory system and is one which is discussed in some detail in Chap. 6.

3.5 Major Conclusions

The vibrissa system has been widely used principally for three reasons. First, the accessibility of the receptor sheet to pre- and neonatal manipulations. Second, the one-to-one relationship between histologically detectable but multiunit *modules* (barrels, clusters, barreloids) and a single receptor – the mystacial vibrissa. Third, owing to the fact that these modules develop after birth, their dependence on the receptor innervation has been testable. Simple studies of the developmental sequence along the pathway that links the vibrissa with the SI cortex (Table 3.1) have suggested a causative role for certain factors in the developmental process. To this knowledge experimental manipulations of the system has added both confirmatory and new knowledge relating to these controlling factors. More recently detailed studies of the system in mature animals has negated some of the early (and theoretically unnecessary) assumptions as to the degree of one-to-one representation between a receptor and a module within the central nervous system. The interpretation of some observations is complicated by the acceptance of multiunit recordings in situations in which cells near the electrode tip are known to have intact inputs from nearby the denervated areas of skin. Nevertheless, many of the studies mentioned above are establishing a firm data base which will permit the use of the vibrissa system in experiments designed to understand that less malleable structure, the mature central nervous system.

The following statements may be made from work on the vibrissa system:

(1) *General Topography.* The general topography of connections at all levels of the neuraxis are determined prenatally around the time of differentiation of the target tissue (follicles, TNC, VB complex and cortex). The role (if any) of these projections in specifying the target tissue, or vice versa, is unknown.
(2) *Peripheral to Central Sequencing along the Neuraxis.* The formation of precise synaptic connections proceeds in a peripheral to central sequence which spans parturition.
(3) *Necessity of the Periphery.* For development to proceed in the normal manner intact afferent connections with the periphery are a necessary condition.
(4) *Deterministic Influence of the Periphery.* While both the precise connections between central modules at different levels of the neuraxis, and the number of modules at any one level are dictated by the position, number and innervation density of receptive structures on the skin, the spatial pattern of these receptive structures develops independently of the nervous system.
(5) *Influence of Target Tissue.* Some unknown factor in that part of the SI cortex destined to contain the representation of the vibrissae specifies it as that cortex which can receive and maintain topographically organized terminals from those thalamocortical neurons receiving input from the vibrissae. However, it is unclear whether, within the specified region, the final location of ingrowing axon terminals it determined by the distribution of layer IV granule cells or factors intrinsic to the thalamocortical neurons.
(6) *Cluster Formation and the Periphery.* Axon terminals at all levels show an ordered affinity that is strongest between axons representing principally the

same receptive structure (or vibrissa) and next strongest for axons representing receptive structures which have a close developmental and structural association (e.g. the vibrissae of one row).

(7) *Afferent Terminals and Barrel or Barreloid Formation.* Postsynaptic cells aggregate around, and project dendrites into clusters of afferent axon terminals.

(8) *Plasticity in the Adult.* Many relatively weak connections exist that do not support the original hypothesis of a one-to-one relationship between receptor (or vibrissa) and centrally located neural modules at various levels of the mature neuroaxis. Altered efficacy of such connections could result in functional, if not structural, plasticity in the mature central nervous system. This mechanism possibly underlies the two instances in which plasticity in the vibrissa system of the adult has been reported. Norepinephrine released from the cortical projections of the locus coeruleus appears to influence in some complex manner the changes of functional connectivity that result from partial deafferentation of the whisker pad.

4 Plasticity and the Spinal Dorsal Horn (with Notes on Homologous Regions of the Trigeminal Nuclei)

4.1 Experimental Strategies for Demonstration of Plasticity in the Dorsal Horn of the Spinal Cord and Trigeminal Nuclei

Although there is a vast literature on the anatomy and physiology of the dorsal horn, much of this is controversial. As much as anything, this reflects the rapid growth of our knowledge of the functioning of this neuronal system over the past two decades. As in other parts of the somatosensory system, many attempts to demonstrate plasticity in the dorsal horn have predated the acquisition of important data concerning the functional and structural organization of that region in normal animals. Thus, we have briefly reviewed relevant aspects of this knowledge and brought these to bear, where possible, upon the many and diverse studies that relate to plasticity.

The dorsal horn of the spinal cord has several advantages for studies of plasticity. Like the DCN (Chap. 5) and the TNC (Chap. 3), the dorsal horn receives input from primary afferent fibres, and its neurons have cutaneous receptive fields that are topographically organized. The distribution and termination of different types of primary afferent fibres in the dorsal horn is now well known for myelinated afferents, and difficulties in defining the projections of unmyelinated afferents are now yielding to advances in techniques of intra-axonal staining. The functional properties, morphology and somatotopic organization of dorsal horn neurons have been extensively studied, and for some projection neurons, such as the feline spinocervical tract (SCT) cell, the relationship between primary afferent connections and receptive field synthesis is now better understood than it is for any other central somatosensory neuron. Historically, studies of somatosensory plasticity in the nervous system of adult mammals could be said to have begun in the dorsal horn with the now familiar studies by Liu and Chambers and their colleagues in the 1950s of long-distance axon sprouting in the spinal cord following extensive dorsal rhizotomies or corticospinal tract lesions (see Sect. 4.5.5). It is therefore not surprising that, in the modern era of neurophysiology, the dorsal horn has been an area where there have been many attempts to define and to understand plasticity of receptive fields of somatosensory neurons.

Studies of plasticity in the dorsal horn fall into several overlapping categories. Perhaps the most publicized and controversial experiments are those in which the somatotopic organization of dorsal horn neurons has been examined following crush or transection of peripheral nerves or following dorsal rhizotomy or section of descending fibres (by spinal hemisection). In some cases this work has led to a concern for whether structural changes such as axonal sprouting might underlie alterations in somatotopy. However, the evidence for structural changes is equivocal, and some investigators have turned to the examination of alternative explanations for changes of somatotopy. The best known of these are experiments purporting to show that dorsal horn neurons have a special class of normally ineffective inputs originating from somatotopically inappropriate projections of primary afferent fibres. It has been claimed that the existence of such inputs may be demonstrated by electrical stimulation of afferents that supply skin outside the receptive fields of the spinal neuron.

In a different series of experiments, claims have been made that activity in, or the selective blockade, or removal, of unmyelinated afferents (C fibres) can cause long-term biochemical and physiological changes in the dorsal horn, as well as long-lasting changes in the tactile receptive fields of neurons in the dorsal horn and spinal trigeminal complex. These claims have led to various hypotheses of both neurophysiological and vegetative roles for C fibres in the control of somatotopy. However, some of these studies are controversial and therefore require re-appraisal. Finally, an attempt has been made to relate changes in somatotopy of dorsal horn neurons following peripheral nerve injury to the presence of somatotopically inappropriate collaterals of cutaneous afferent fibres in the dorsal horn. These collaterals are simple in structure and normally have few or no synaptic boutons, and thus they are presumed to be non-functional in the neurophysiological sense (Meyers and Snow 1984; Snow and Wilson 1989, 1990; Wilson and Snow 1990a, b). Local proliferation and synaptogenesis by these collaterals in regions where cells have lost their major cutaneous input might lead to the appearance of new receptive fields (Snow and Wilson 1989, 1990).

4.2 Overview of Dorsal Horn Organization

4.2.1 Laminar Cytoarchitectonic Organization

Cytoarchitectonics. It is appropriate to begin a discussion of plasticity in the dorsal horn with a summary of Rexed's (1952) cytoarchitectonic scheme of the grey matter of the cat's spinal cord. The laminar organization devised for the cat by Rexed (1952), on the basis of the shape, size and distribution of cell bodies, has been applied successfully to many other mammalian species, and will be referred to repeatedly in this Chapter. Since the cytoarchitectonic organization of the dorsal horn has been described many times in recent years (for reviews see Willis and Coggeshall 1978; Brown 1981; Scheibel 1984; Wall 1984), only the briefest of accounts will be given here.

Lamina I was described by Rexed (1952) as a thin veil of grey substance, forming the dorsalmost part of the grey matter. Dorsally, lamina I is bordered by the dorsal columns, and its ventral border is characterized by the presence of bundles of longitudinally running fibres. The cell bodies in lamina I are primarily horizontally arranged, and vary in size from 5 μm × 8 μm to about 25 μm × 30 μm. Most functional investigations of lamina I have dealt mainly with the large marginal cells of Waldeyer which, as Rexed (1952) himself noted, make up only a small proportion of all lamina I neurons. Laterally, lamina I bends around the apex of the dorsal horn and extends downwards over the lateral side of the horn.

Lamina II is much thicker than lamina I. It curves ventrally around the lateral edge of the dorsal horn, and has been equated by Rexed (1952) to the substantia gelatinosa of Rolandi. It consists of tightly packed small neurons about 5–10 μm in diameter which are arranged in a primarily radial fashion.
Lamina III forms a broad band across the dorsal horn. Its neurons are slightly larger (about 8 μm × 12 μm on average) and more loosely packed than those of lamina II.
Lamina IV is a thick layer extending across the whole width of the dorsal horn. Neurons in this lamina vary greatly in size from around 7 μm × 10 μm to 35 μm × 45 μm, the majority being about 10 μm × 15 μm. The large cells are relatively uncommon, but are so prominent that they give the impression that lamina IV is a lamina of large neurons.
Lamina V occupies the neck of the dorsal horn, and it is the most dorsal lamina not to show any ventral curvature at its lateral edge. Lamina V is characterized by the presence of large bundles of fibres, which give the lateral region of the lamina a reticulated appearance. The cell density is lower than it is in laminae I–IV, and the cells vary in size from 8–10 μm to 30–45 μm.
Lamina VI is a thick layer in the cervical and lumbar enlargements, but is much reduced or even absent in other regions of the cord. The neurons of lamina VI are smaller and more regularly arranged than those in lamina V. The cells vary from 8–35 μm in size, with most being around 12–15 μm.
Golgi studies have shown that a characteristic pattern exists for the general morphology and orientation of dendritic trees of dorsal horn neurons in each lamina, and this pattern varies greatly between the different laminae (Fig. 4.1) (Scheibel and Scheibel 1968b). In the cat, the dendrites of lamina I cells run horizontally within the lamina, forming disc-like dendritic domains. However in primates (Beal 1979), including man (Schoenen 1982), ventral dendrites of some lamina I marginal neurons have been seen to enter lamina II. In the cat, Scheibel and Scheibel (1968b) concluded that lamina II the neurons are most often bipolar in structure, with quasi-two-dimensional dendritic domains extending rostrocaudally in a parasagittal plane (Fig. 4.1), but other workers have suggested a wider spread of dendritic fields of lamina II neurons (Ramony Cajal 1928; Mannen 1975). The dendritic branches of these neurons are well endowed with large spines. Morphologically, lamina III is dominated by the dentrites of large lamina IV neurons. Lamina III neurons themselves have the same two-dimensional dendritic structure as lamina II neurons. Neurons of lamina IV were described as having dendrites radiating in medial, lateral and dorsal directions from the soma. The dendritic domain of these cells thus resembles a cone with the cell body at the apex (Fig. 4.1). The dendrites of neurons in laminae V and VI are bushy and tend to become more extensive the greater the depth of the cell body in the dorsal horn. The dendritic orientation in these deeper laminae in predominantly mediolateral, with little rostrocaudal spread (Fig. 4.1) (Scheibel and Scheibel 1968b; Matsushita 1969; Mannen 1975; Proshansky and Egger 1977).

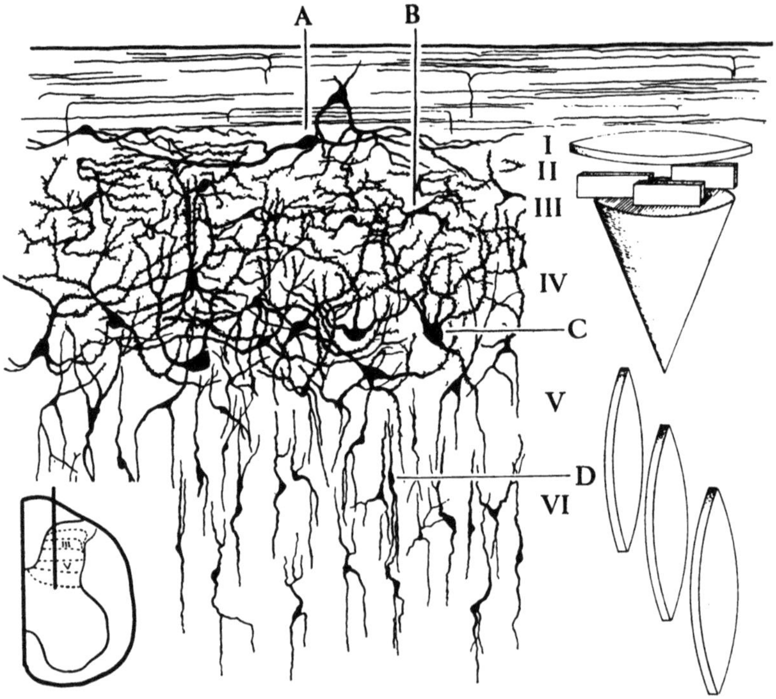

Fig. 4.1. Laminar cytological and dendritic architecture of the mammalian dorsal horn. Schematic parasagittal section through the dorsal horn of the cat showing the characteristic pattern of dendritic organization of cells in different laminae, as seen in Golgi studies. *A*, marginal cell of Waldeyer; *B*, small neurons of laminae II and III; *C*, large lamina IV neuron. The geometric figures on the right correspond approximately to the most common shape of the dendritic domains of dorsal horn neurons in different laminae. (From Scheibel and Scheibel 1968b, p. 38)

4.2.2 Laminar Organization of the Termination of Primary Afferent Fibres

The differential projection pattern of primary afferent fibres to the laminae of the dorsal horn has recently been reviewed by Maxwell and Rethelyi (1987) (Fig. 4.2). Low-intensity cutaneous mechanical stimuli (brush, touch) activate a variety of Aß, A∂ and C afferent fibres. Noxious mechanical stimuli (pinch, prick) activate two classes of nociceptive afferents, the A∂ and C high-threshold mechanoreceptors. Aß afferents form terminal arborizations from the inner part of lamina II to lamina V (reviewed by Brown 1981). A∂ afferents supplying low-threshold receptors terminate in lamina III, while those originating from cutaneous high-threshold mechanoreceptors terminate in lamina I. By following the lateral border of lamina I the latter fibres also reach lamina V, where they form additional terminal arborizations (Light and Perl 1979). Cutaneous C fibres terminate mainly in lamina I and II (Ralston and Ralston 1979; Sugiura et al. 1986). Unmyelinated (C) afferents of visceral origin terminate in laminae I, II and V (Sugiura et al. 1988).

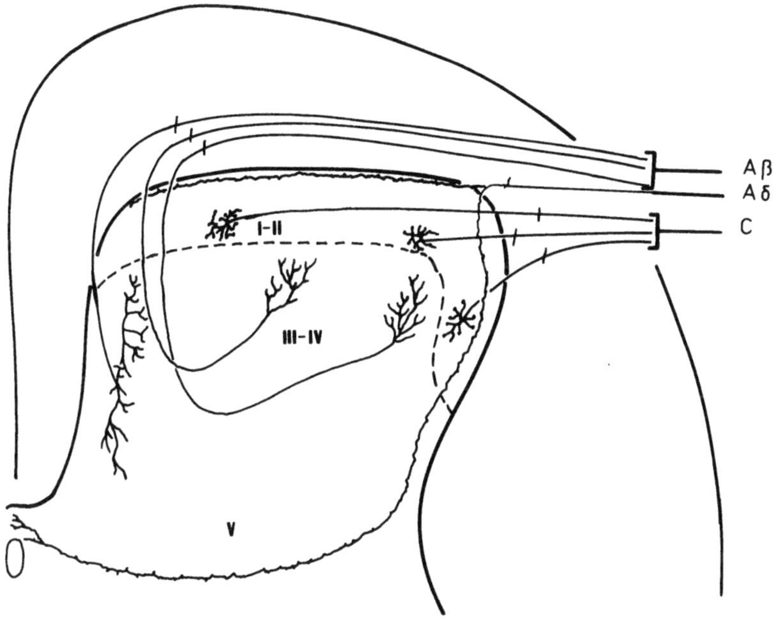

Fig. 4.2. Highly schematic representation of the termination sites of the collaterals of cutaneous afferent axons in the dorsal horn. Note the different laminar distribution of different classes of fibres (*Aβ*, *Aδ* and *C*). (Redrawn from Maxwell and Rethelyi 1987, p. 117)

4.2.3 Microanatomical Organization of Low-Threshold Cutaneous Afferents

Spinal Dorsal Horn. In the last decade, the technique of intra-axonal staining of physiologically identified primary afferent fibres has led to a fairly detailed knowledge of the relationship between the different types of mechanoreceptor innervated, the pattern of projection of their collaterals to the dorsal horn, and the microanatomy of their central terminals. This work has been extensively reviewed in recent years (Brown 1981; Fyffe 1984; Fitzgerald 1984; Maxwell and Rethelyi 1987).

An outstanding feature of all Aß afferents innervating sensitive cutaneous mechanoreceptors is that the mediolateral position of their arborizations in the dorsal horn is maintained throughout the entire rostrocaudal extent of the axon's terminal field. Most synaptic boutons of Aß hair follicle afferents are found in lamina III, but some occur in the inner part of lamina II and the dorsal part of lamina IV. Afferents from other sensitive cutaneous mechanoreceptors issue collaterals which terminate more widely in the dorsoventral axis of the dorsal horn, from the inner part of lamina II to the dorsal part of lamina VI. There are, in addition, differences in the longitudinal and mediolateral distribution of terminals from different Aß mechanoreceptive afferents. A single hair follicle afferent distributes its terminals as a continuous thin (150–400 μm wide) sheet of arborizations occupying the whole depth of lamina III, and stretching for up to several millimeters in the rostrocaudal axis (Fig. 4.3a) (Brown et al. 1977b). Single primary afferent fibres supplying rapidly

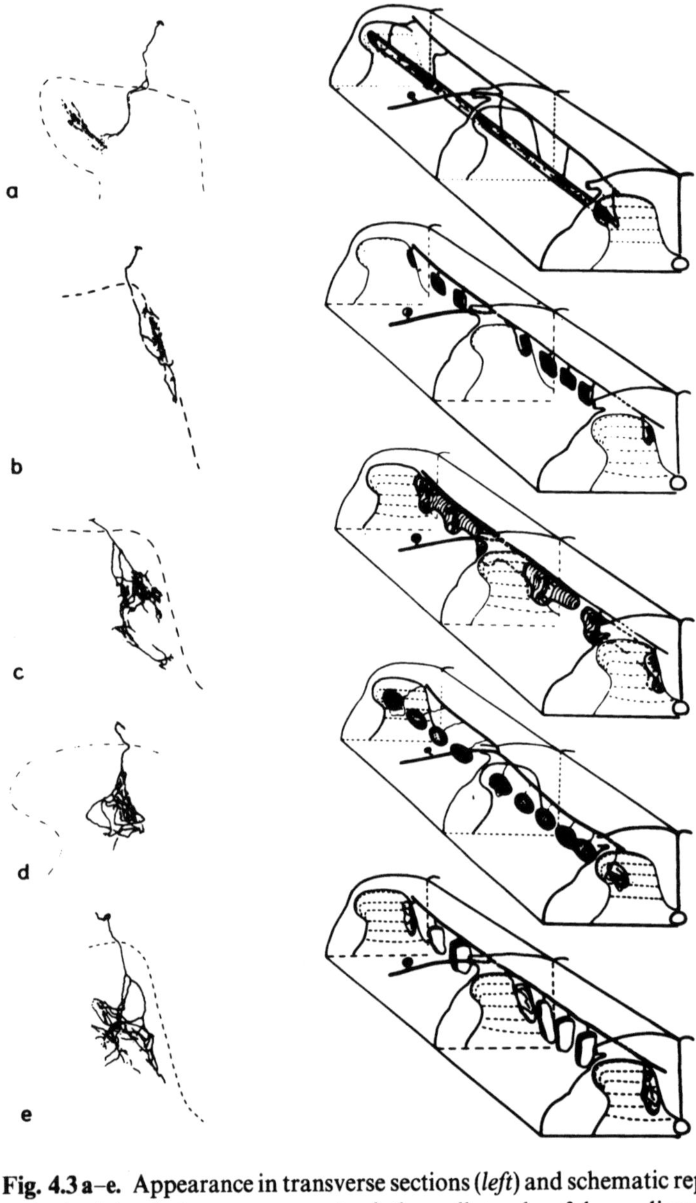

Fig. 4.3 a–e. Appearance in transverse sections (*left*) and schematic representations, drawn to scale (*right*), of the arrangement of the collaterals of large-diameter myelinated axons innervating different types of sensitive cutaneous mechanoreceptors. **a** Hair follicle afferent fibre shown entering the spinal cord and giving off collaterals which terminate in a parasagittally aligned sheet of arborizations in laminae III and IV, **b** a fibre from rapidly adapting receptors in glabrous skin, **c** collaterals of a typical Pacinian afferent from glabrous skin, supplying terminals to laminae III–VI, **d** collaterals of a slowly adapting type I axon, **e** a slowly adapting type II afferent. Note the considerable rostrocaudal extent, and limited mediolateral extent of the arborizations of all these afferent fibres. (From Brown et al. 1981b, p. 476)

adapting mechanoreceptors in the glabrous skin of the pads distribute their collateral arborizations to the most medial part of lamina III. These arborizations superficially resemble the arborizations of hair follicle afferents, but differ in that they do not form continuous longitudinal sheets (Fig. 4.3b). The arborizations of adjacent collaterals occupy 400–600 μm in the rostrocaudal axis, with gaps of 100–700 μm between them. Some of these collaterals also supply boutons to lamina IV (Brown et al. 1980a). Thinly myelinated (A∂) axons supplying low-threshold hair follicle mechanoreceptors (D-hair afferents) distribute their collaterals to the dorsal horn in a similar way to Aß hair follicle afferents (Light and Perl 1979).

The collateral morphology of a typical type I afferent is essentially the same whether the receptors are situated in hairy or glabrous skin. The arborizations of single collaterals are contained within laminae III, IV and V and are ellipsoid in shape, measuring 250–500 μm transversely at their widest point and 250–650 μm dorsoventrally. In the rostrocaudal axis the terminal arborizations stretch for 100–700 μm, and there are gaps of 100–500 μm between the arborizations of adjacent collaterals (Fig. 4.3d) (Brown et al. 1978).

The terminal arborizations of type II afferents also possess a distinctive morphology. Collaterals terminate in a dense arborization extending through lamina III and IV into lamina V. Transversely, the arborizations are wider than most others, extending 500–600 μm, and thus occupying as much as one-third of the total width of the dorsal horn. The rostrocaudal extent of the collateral arborizations is limited to a mere 100–300 μm, and consequently the arborizations of adjacent type II axon collaterals almost never overlap (Fig. 4.3e) (Brown et al. 1981).

Terminal arborizations of collaterals of axons from the rapidly adapting Pacinian corpuscles in the subcutaneous tissue of the foot and toe pads are located in laminae III–VI of the medial third of the dorsal horn. The overall depth of these arborizations is thus much greater than that of hair follicle afferents. In the transverse plane the arborizations are, like those of other rapidly adapting afferents, quite narrow at around 170–350 μm. The dorsal termination region extends about 400–750 μm in the longitudinal axis, and there is usually overlap between the terminals of adjacent collaterals. In contrast, the ventral arborizations are much shorter and rarely show any overlap (Fig. 4.3c) (Brown et al. 1980a).

Recently, it has become possible to visualize the central terminals of single functionally identified C fibre afferents by means of intracellular injection of plant lectins into the cell body in the DRG (Sugiura et al. 1986). The terminal arborizations of the collaterals of C fibres are highly circumscribed, and largely restricted to the superficial laminae of the dorsal horn (laminae I and II). However, a single lumbosacral polymodal nociceptor that was injected gave rise to a few additional small arborizations in laminae III and IV. The rostrocaudal distribution of this axon was traced for over 4.6 mm, while the mediolateral extent of its arborizations was only 200 μm. The extent and density of the arborizations varied remarkably between different collaterals, in a similar way to the variation found in the collateral arborizations of Aß afferents in the dorsal horn of the cat and the rat (see Sect. 4.3.3) (Meyers and Snow 1984; Woolf 1987). The relationship of dendritic morphology and somatotopy of dorsal horn neurons to the somatotopic organization of primary afferent projections is discussed below (Sect. 4.3.4).

Spinal Trigeminal Nuclei. The collaterals of large myelinated primary afferents in the TNC of the cat form terminal arborizations in the outer part of the spinal trigeminal nucleus when they are situated rostral to the obex, in lamina V when in the rostral part of the SpVc, and in laminae III and IV when in the caudal part (spinal dorsal horn region) of the SpVc (Hayashi 1985b). Unlike the situation in the lumbosacral dorsal horn, major morphological differences between the collateral arborizations of different functional types of Aß mechanoreceptive afferents (vibrissa, G_1 hair and type I) were not observed in this study. The collaterals of A∂ (D-hair) afferents were similar to those of Aß afferents, except that their terminal arborizations in the caudal part of the SpVc extended into the inner part of lamina II, as well as laminae III and IV (Hayashi 1985a, b).

4.2.4 Relation of Functional Properties to Lamination of the Dorsal Horn

Neurons in the dorsal horn (laminae I–VI of Rexed) have been shown repeatedly to be excited either monosynaptically or polysynaptically by cutaneous, muscle or visceral afferent fibres (for reviews see Willis and Coggeshall 1978; Brown 1981; Price 1984; Cervero 1986). Only a brief outline of the large and growing body of knowledge concerning the convergent properties of neurons in the dorsal horn will be given here. For present purposes, it is sufficient to emphasize that, generally speaking, the functional properties of dorsal horn neurons have been found to reflect the laminar pattern of the terminations of the different types of primary afferent fibres described in Sects. 4.1.2 and 4.1.3.

Lamina I. Lamina I neurons respond to a variety of cutaneous sensory modalities, but they are mainly concerned with high-threshold inputs from the skin. In the cat dorsal horn, Christensen and Perl (1970) described three classes of response to cutaneous stimulation in different large cells of lamina I: (1) noxious mechanical, (2) noxious heat/noxious mechanical, and (3) innocuous cooling, with weaker responses to noxious heat/noxious mechanical. These responses were later confirmed by intracellular staining as being typical of the large marginal neurons (Light et al. 1979). Most lamina I spinothalamic tract neurons in the chloralose-anaesthetized cat, the chloralose- and barbiturate-anaesthetized rhesus monkey, and the unanesthetized decerebrate macacque are also reported to respond to noxious pinch, noxious heat or cold or innocuous cooling (Price and Mayer 1974; Willis et al. 1974; Craig and Kniffki 1985). In lamina I of the lumbar cord of the decerebrate-spinal rat a wider range of responses to mechanical stimuli has been reported. For example, in one study 5% of cells responded only to light mechanical stimuli (hair movement or light touch), 50% responded only to firm pressure or pinch, and the remainder responded to light tactile stimulation but showed an increasing response to stronger stimuli. Of the last, wide-dynamic-range group, about one half could be driven by brushing hairs. Some of the neurons with high-threshold mechanoreceptive fields also responded to noxious heating of their receptive field (McMahon and Wall 1983a). Both nociceptor-specific and wide-dynamic-range neurons have also been found in the region of lamina I in decerebrate rats (Menetrey et al. 1977) and in chloralose-anaesthetized cats (Cervero et al. 1976).

Laminae II and III. In recent years several groups have described the receptive fields of the small interneurons in lamina II. The early descriptions varied considerably between different laboratories (reviewed by Cervero and Iggo 1980). Three general patterns of responses have been found:
(1) nociceptive specific neurons, responding to noxious mechanical stimuli and noxious heat, (2) wide-dynamic-range mechanoreceptive neurons, and (3) low-threshold mechanoreceptive neurons. A range of unusual properties which have not been described for cells of other laminae have been reported, including (1) cells with variable rates of ongoing discharge which could be more easily inhibited than excited by peripheral stimulation ("inverse" cells, Cervero et al. 1979), (2) cells giving long after-discharges to cutaneous stimulation, (3) cells whose responses to slowly repeated mechanical stimulation showed marked habituation (Fitzgerald 1981), and (4) cells whose receptive fields fluctuate in size (Sect. 4.5.2). In the cat intracellular staining with HRP has shown that within lamina II there is a differential distribution of different functional classes of neuron. Thus, nociceptor-specific and wide-dynamic-range neurons are more commonly found among cells with dendrites that ramify in the superficial part of the lamina (lamina II_o) and low-threshold mechanoreceptive neurons have dendrites that ramify in the deeper part (lamina II_i).

Laminae IV, V and VI. The large neurons of laminae IV–VI were among the earliest to have their functional properties described. According to the classification scheme developed by Wall (1967) and based on the laminar divisions of Rexed (1952), neurons of lamina IV in the cat are excited only by light tactile stimulation (light touch, hair movement), and have no inhibitory region either around or alongside their excitatory receptive field. Neurons of lamina V, in addition to responding to light tactile stimuli, also respond to strong pressure or pinch, and sometimes to visceral inputs (Pomeranz et al. 1968). They have larger receptive fields than cells of lamina IV, and in addition have inhibitory surrounds (Hillman and Wall 1969). In lamina VI, neurons have similar properties to those in lamina V but they had even larger receptive fields and responded also to muscle and joint receptors. Wall (1967) was at pains to point out that these functionally defined laminae did not necessarily correspond with the cytoarchitonically defined laminae of Rexed (1952), but represented zones where cells with the above properties were most concentrated. While there may be a rough gradient of functional properties through laminae IV–VI of the kind suggested by Wall (1967), more recent studies of identified dorsal horn neurons (e.g. SCT neurons, PSDC neurons) have shown that cells contained in the same anatomical lamina can have widely different functional properties (Brown 1981). In addition to the increase in receptive field size with depth, Wall (1967) also reported that latency of responses to electrical stimulation increased with the depth of the recording site. These observations led Wall (1967) to suggest that large myelinated afferents might excite neurons in lamina IV directly, and these in turn excite cells in lamina V, which then excite cells in lamina VI. This proposal of a cascade of activation through the deep laminae of the dorsal horn in the cat has *not* been confirmed by subsequent workers who have been unable, using statistical methods, to demonstrate that the depth of the recording site is related to either an increase in the size of receptive fields or an increase in central response latency following peripheral stimulation (Brown et al. 1975).

4.2.5 Inhibitory Receptive Fields

Like neurons of the somatosensory cortex, thalamus and DCN, many dorsal horn neurons have inhibitory, as well as excitatory components to their receptive fields. The inverse type of lamina II neuron, mentioned above, with predominantly inhibitory responses from the skin is, of course, one example, but more detailed studies of inhibitory inputs have been made for the large dorsal horn neurons of the deeper laminae. In some cases the inhibitory field may be coextensive with the excitatory field or in other cases may extend beyond it, forming an inhibitory surround. Such an arrangement has been described for the typical wide-dynamic-range neurons of lamina V in the decerebrate cat. On the basis of extracellular recordings from these cells, Hillman and Wall (1969) suggested that their receptive fields consisted of three distinct regions: (1) a small central region from which low-threshold fibres excite the cell, (2) a larger area, which includes the low-threshold region, and from which high-threshold fibres are excitatory and low-threshold fibres are inhibitory, and (3) a much larger area, which includes regions (1) and (2), and from which low-threshold fibres are purely inhibitory. Inhibition was thus evoked from all regions within the excitatory receptive field. While typical lamina IV neurons were described as lacking inhibitory fields (Wall 1967), recent intracellular and extracellular studies of SCT cells, which are found mainly in lamina IV, have shown that there are regions of skin, both within and outside the region from which impulses can be evoked, from which both excitatory postsynaptic potentials (EPSPs) and inhibitory postsynaptic potentials (IPSPs) may be generated by light tactile stimuli. Surprisingly, in these studies action potentials in single synaptically coupled Aß hair follicle afferents were never observed to evoke IPSPs, but it is possible that A∂ (D-hair) hair follicle afferents or type I afferents provide this low-threshold in-field inhibitory input to SCT cells. (Sects. 5.2.3, 6.3, 7.3.1 and 8.5.1) (Brown et al. 1987; Noble and Short 1989).

4.3 Somatotopic Organization of the Dorsal Horn

4.3.1 Dorsal Horn Neurons

That the receptive fields of neurons in the lumbosacral dorsal horn are somatotopically organized has been recognized since Wall's (1960) pioneering microelectrode studies of the cat dorsal horn. Wall (1960) noted that cells responding to light tactile stimulation of the skin were present in a well defined horizontal lamina in the dorsal part of the dorsal horn. Within this functionally defined lamina, responses from the distal part of the leg were found medially, while responses from the proximal part of the limb were found laterally. In the rostral part of the lamina (the L4 segment) cells responded to the anterior part of the leg, and in the caudal region (S2) responses were evoked from the posterior part of the leg and the perineal region. A similar conclusion was reached by Bryan et al. (1973) who mapped the receptive fields of dorsal horn neurons in laminae IV and V of Rexed whose axons ran in the dorsolateral funiculus. More recently, several studies in both the cat and the rat have shown the precision of the somatotopic organization of the dorsal horn, not only for the low-threshold cutaneous receptive fields of neurons in laminae IV and V, but also for neurons in more superficial laminae responding to inputs from high-threshold mechanoreceptive afferents (Sect. 4.3.2). With the advent of trans-ganglionic tracing techniques and intracellular staining the question of how the somatotopic organization of dorsal horn neurons relates to that of the primary afferent input has also been addressed in some detail.

The first extensive map of the somatotopic organization of the lumbosacral dorsal horn of the cat was published by Brown and Fuchs (1975), who recorded from single unidentified dorsal horn neurons in unanaesthetized, decerebrate preparations. Most of the recording sites included in their study were verified histologically as being in lamina IV or V. Within the horizontal lamina of mechanoreceptive cells in each spinal segment, the representation of the hindlimb skin was found to correspond well with the dermatomes of the corresponding dorsal roots (Fig. 4.4). The somatotopic gradient in the mediolateral axis of the dorsal horn was shown to have two components, one of which ran essentially proximo-distally along the limb, while the other ran from the dorsal surface to the ventral surface. Brown and Fuchs (1975) noted that in the dorsal horn there were no segmental discontinuities or multiple representations of single limb areas and they pointed out that the area of dorsal horn devoted to the representation of the foot and toes is relatively large in relation to the representation of the much more extensive proximal skin regions.

Further details of the representation of the distal part of the limb were provided in studies of the somatotopy of lumbosacral SCT cells in chloralose-anaesthetized cats. These showed that the toes are represented in a rostrocaudal sequence in the middle of the L6–L7 dorsal horn from the most medial toe (toe 2) rostrally to the most lateral toe (toe 5) caudally (Fig. 4.5) (Brown et al. 1980b; Light and Durkovic 1984; Wilson et al. 1986). The somatotopic organization of other dorsal horn neurons in the same 300–500 μm thick functional lamina in which most SCT cells are found appears to be basically the same as that of the SCT cells (Wilson et al. 1986; Kunze et al. 1987). In the far lateral region of the L6–L7 dorsal horn in the cat is a narrow strip of tissue where neurons have receptive fields on the most proximal skin of the thigh. Due to the curvature of the laminar boundaries in the lateral part of the dorsal horn, cells having receptive fields on the leg are often encountered dorsal to cells with receptive fields on the thigh in single vertical microelectrode tracks (Wilson et al. 1986)–a fact which may well have led to some serious misinterpretations of somatotopy following partial deafferentation of the hindlimb.

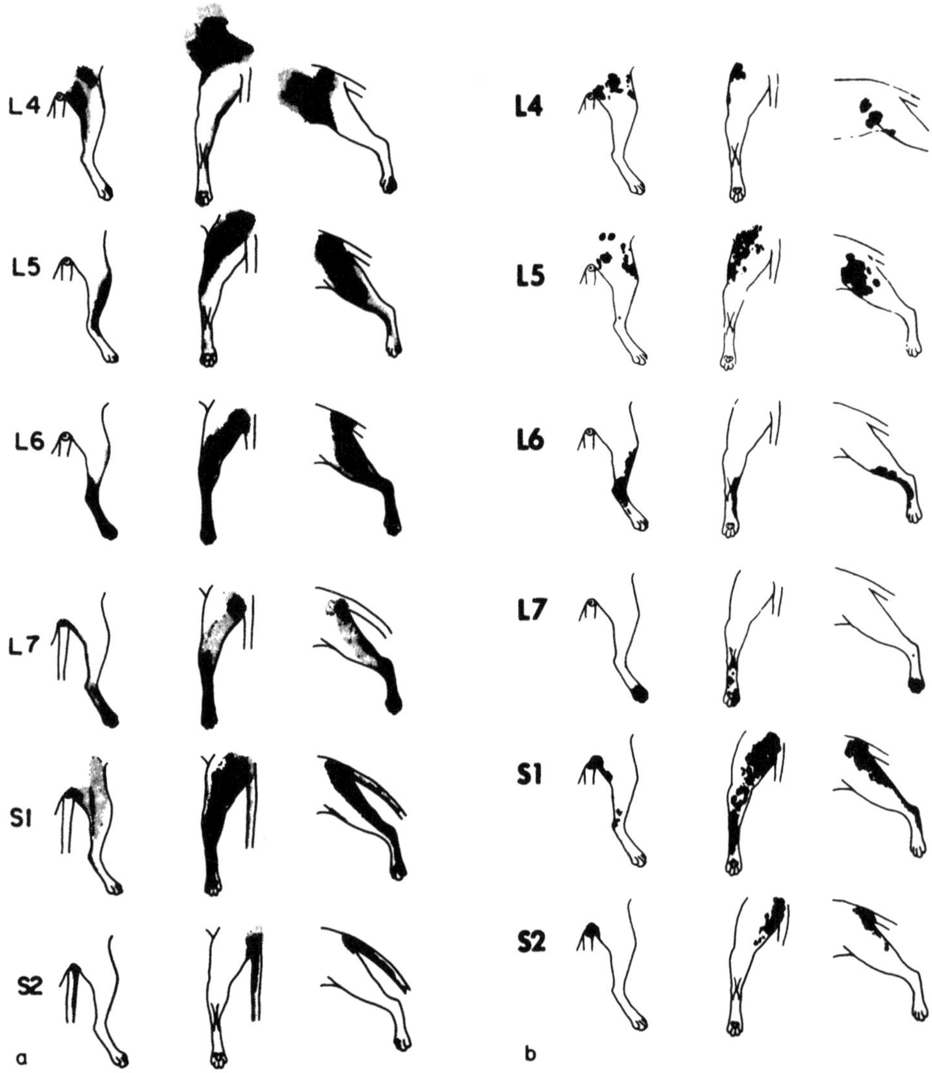

Fig. 4.4 a, b. Comparison of dorsal horn and dorsal root dermatomes of the lumbosacral spinal cord of the cat. **a** Dorsal horn dermatomes determined by extracellular recording from dorsal horn neurons in the appropriate segments of decerebrate spinal animals. In each figurine between 30 and 50 single-unit receptive fields are superimposed, **b** dermatomes of dorsal roots L4–S2 determined by recording from large numbers of dorsal root filaments. Relative density of the *stippling* reflects the relative number of units recorded. (**a** From Brown and Fuchs 1975, p. 4; **b** from Brown and Koerber 1978, p. 262)

4.3.2 Somatotopy and Lamination

The Cat. The nociceptive specific neurons of lamina I have small receptive fields with well-defined edges, and it has thus proven possible to obtain clear somatotopic maps of this lamina (Cervero et al. 1976; Hylden et al. 1987). In the lumbosacral

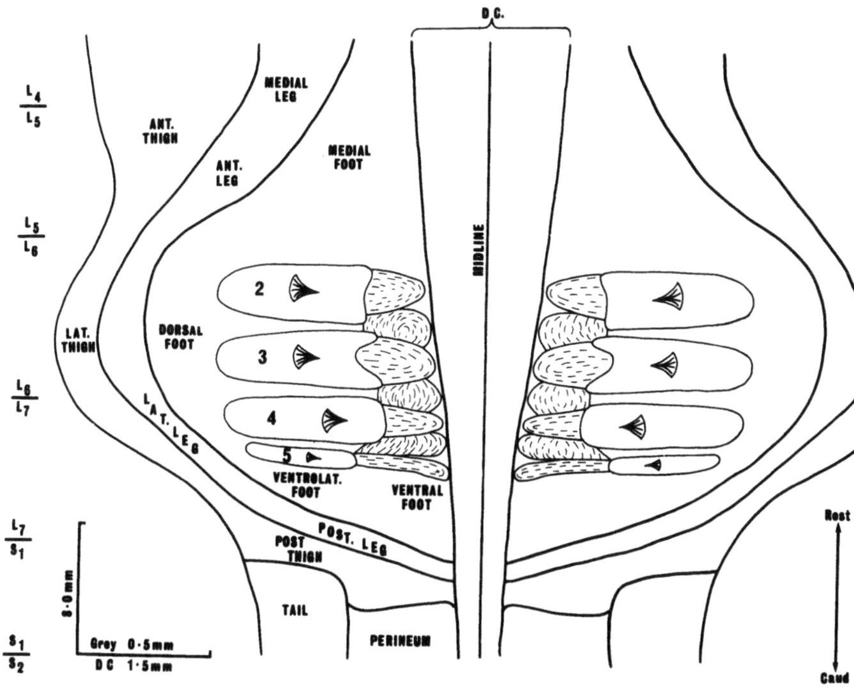

Fig. 4.5. Schema of the somatotopic organization of the low-threshold tactile receptive fields of lumbosacral dorsal horn neurons in the chloralose-anaesthetized cat. Map was obtained from the receptive fields of antidromically identified SCT cells (located in laminae III–V) or other neurons in the same depth range as SCT cells. Note the rostrocaudal sequence of the representation of the toes in the medial part of the L6 and L7 segments. *ANT*, anterior; *POST*, posterior; *LAT*, lateral; *VENTROLAT*, ventrolateral; *DC*, dorsal columns; *Rost*, rostral; *Caud*, caudal. (From Wilson et al. 1986, p. 614)

dorsal horn of the cat, the somatotopic organization of lamina I neurons superficially resembles that of SCT cells and other dorsal horn neurons of laminae IV and V. This is illustrated in Fig. 4.17a. Unfortunately, the receptive fields of neurons from several animals are combined in this figure, and, owing to inter-animal variability, details of the representation of the toes are obscured. However, Fig. 4.17a does show that the most common distal receptive fields in lamina I covered one or more toes or toe pads. In contrast, in laminae IV–V receptive fields covering more than one toe or toe pad, are relatively rare (Wilson et al. 1986), although such fields have been observed among postsynaptic dorsal column (PSDC) neurons (Brown et al. 1983).

Dorsal horn neurons with approximately concentric receptive fields are located in rectangular parasagittally oriented slabs that occupy the same mediolateral position along a fairly extensive rostrocaudal length of the dorsal horn (Fig. 4.6) (Light and Durkovic 1984), and encompassing all laminae of the dorsal horn (I–VI inclusive). The somatotopic organization of the receptive fields of lumbosacral dorsal horn neurons is thus the same whether they respond to light touch, to pressure or to more intense mechanical stimulation, such as pinch (Light and Durkovic 1984).

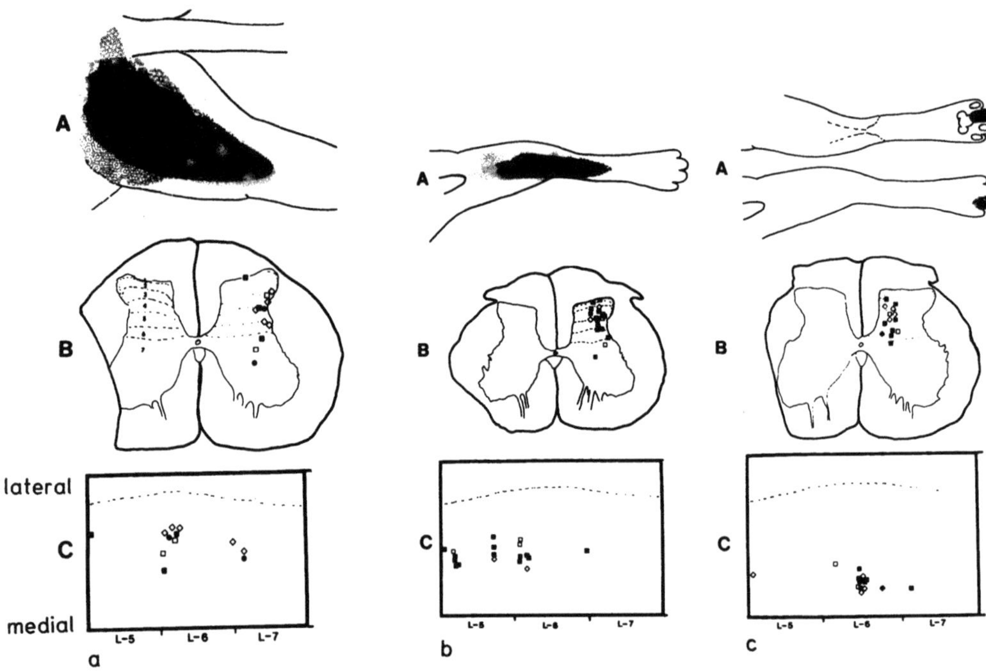

Fig. 4.6 a–c. Relation of somatotopic organization of neurons located in different laminae of the spinal cord in the cat. Receptive fields (*stippling* on parts *A*) and location in spinal cord (transverse (*B*) and horizontal (*C*) views) of neurons in the lower lumbar cord of the unanaesthetized, decerebrate spinal cat. **a** Neurons activated by cutaneous stimulation of the thigh, **b** neurons activated by cutaneous stimulation of the ankle and shin, **c** neurons activated by cutaneous stimulation of toe 3. Symbols indicate effective stimuli: *open diamonds*, light touch only; *open squares*, touch and pressure; *filled diamonds*, touch and pinch; *filled squares*, touch, pressure and pinch; *filled circles*, pinch only. (From Light and Durkovic 1984, pp. 453 and 454)

The Rat. Somatotopic maps of the hindlimb in both superficial (laminae I–III) and deep (laminae IV–V) laminae of the dorsal horn of the urethane-anaesthetized rat (Woolf and Fitzgerald 1986) resemble those obtained from lamina I and laminae IV and V of the cat. Thus in the rat, the medial area of the dorsal horn is occupied by cells with receptive fields on the plantar surface of the toes, while more laterally located cells have receptive fields whose centres are shifted successively more proximally and dorsally. Distal receptive fields in the rat usually extend over a number of toes (Fig. 4.7), unlike those in the cat which are normally confined to a single digit. When pairs of neurons were recorded at superficial and deep levels in the same microelectrode track, Woolf and Fitzgerald (1986) found that while 80% of cell pairs showed some overlap of their receptive fields there was little correspondence between the location of the centres and size of these fields. This difference may be indicative of a greater dependence on convergence over polysynaptic pathways for the synthesis of receptive fields of deep dorsal horn neurons in the rat than in the cat (Sect. 4.3.4), and may have an important bearing on the mechanisms underlying plasticity in the dorsal horn of the two species.

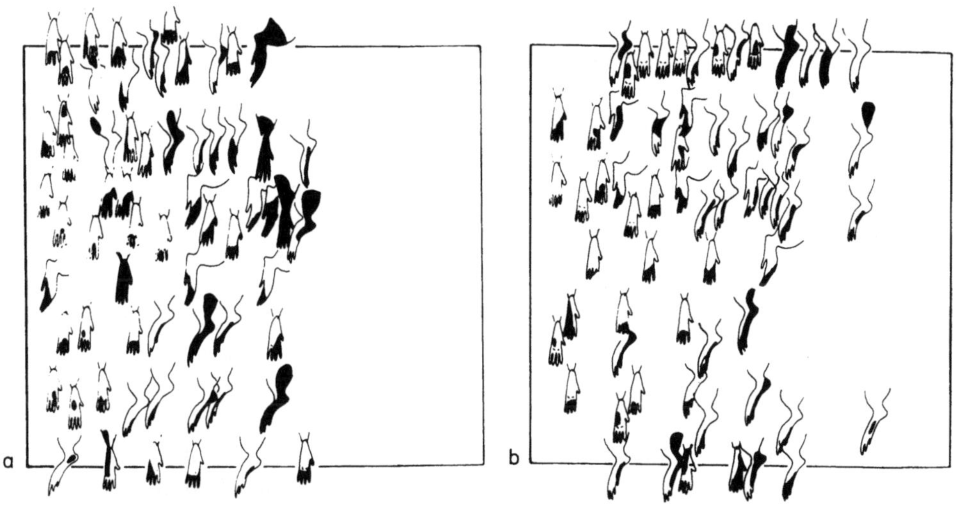

Fig. 4.7 a, b. Somatotopic organization of the hindpaw representation in superficial and deep lumbar dorsal horn of the rat. Schematic view of the 4th lumbar segment in two horizontal planes showing the somatotopic organization of the receptive fields of units recorded in the superficial laminae of the dorsal horn (**a**) and in the deep dorsal horn (**b**). The medial edge of the dorsal horn is on the *left*, the lateral on the *right*. The *top* of the diagram represents the L3/4 boundary and the *bottom* the L4/5 boundary. (From Woolf and Fitzgerald 1986, p. 528)

4.3.3 Relation of Primary Afferent Projections to Dorsal Horn Somatotopy

A thorough knowledge of the extent to which the somatotopic organization and functional properties of dorsal horn neurons reflect the projection patterns and connectivity of primary afferent systems is a prerequisite for any explanation of plasticity, as manifested by changes of receptive field size, location or modality. Tapper et al. (1973) noted that many lamina IV and V neurons send their axons into the dorsolateral funiculus, but not all primary afferent systems which connect with these cells are necessarily represented in each of the ascending pathways within the dorsolateral funiculus. In decerebrate, low-spinal cats Tapper et al. (1973) recorded from neurons at many different locations throughout the SI dorsal horn. For each cell, measurements were made of the latency, approximate central delay and following frequency to electrical stimulation of the posterior femoral cutaneous nerve. In addition, the receptive field was mapped with tactile stimuli which were designed to separately activate four independent afferent systems, namely type I fibres, D-hair afferents, and two kinds of Aß hair follicle afferents. Responses to the four fibre systems appeared to be uniformly distributed throughout the population of dorsal horn cells, and the relative abundance of each was related to the abundance of that kind of afferent in cutaneous nerves. Tapper et al. (1973) concluded that there was nothing about the distribution of the four afferent fibre systems which might explain the segregation of their representation in ascending systems. It therefore appeared that central synaptic mechanisms might be responsible for generating heterogeneous output lines from a fairly homogeneous monosynaptic afferent input. The possibility was thus raised that the *somatotopic organization* of the dorsal horn might also be controlled by intrinsic mechanisms that were largely independent of the distribution of the arborizations of primary afferent fibres.

Presynaptic Somatotopy Hypothesis. Further studies in the cat, utilizing measurements of central delay following mechanical stimulation of single touch domes (Brown et al. 1973), indicated that dorsal horn neurons responding to single impulses evoked in single type I fibres had no significant region of their receptive field that was *not linked monosynaptically* to primary afferent fibres. In other words, no extensive region of the light tactile receptive fields of these dorsal horn neurons appeared to be *entirely dependent* upon polysynaptic inputs. Generalizing to all low-threshold tactile afferents and to all dorsal horn cells responding to these afferents Brown and Fuchs (1975) suggested that all areas of the dorsal horn in which a given area of skin is represented are supplied by at least some afferents from that region of skin. Taken together with their observations on the somatotopic organization and size and shape of receptive fields of dorsal horn neurons, this suggestion led them to propose the working hypothesis that there is a precise somatotopic organization of the presynaptic input to the superficial dorsal horn which mirrors the somatotopic organization of dorsal horn neurons: a corollary of this so-called *presynaptic somatotopy hypothesis* (Brown and Culberson 1981) is that no somatotopically inappropriate afferent projections exist. In order to test such a hypothesis it is of course necessary to determine the anatomical distribution of somatotopically identified primary afferent fibres both within and outside the dorsal horn region in which the receptive field of the afferent is represented in the receptive fields of central cells. Attempts to accomplish this have been made at three different levels of precision, that of the dorsal roots, of peripheral nerves, and of single identified afferent fibres.

Dorsal Roots. Brown and Koerber (1978) determined the extent of hindlimb tactile dermatomes by mapping the receptive fields of large numbers of sensory neurons in the dorsal root ganglion. The extent of the cutaneous receptive fields of dorsal horn neurons within a single spinal segment were somewhat larger and differently organized than the corresponding dermatomes determined by this means (Brown and Fuchs 1975). The L4 segment contained the representation of skin of the L3 and L5 dermatomes which was not included in the L4 dermatome. The somatotopic map in the L5 dorsal horn lacked large regions of abdomen and thigh skin in the L5 dermatome, and included skin of the L6 dermatome around the ankle and anterior leg which was absent from the L5 dermatome. The L6 dorsal horn had substantial representation of the L7 and S1 dermatomes, and the L7 segment contained the representation of the skin of the L6, S1 and S2 dermatomes. Finally, the S1 segment appeared to contain the representation of skin in the S2, L7 and L6 dermatomes. These deviations from simple segmental dermatomal projections to the dorsal horn are thought to be due to a process of fibre sorting, leading to the formation of somatotopically appropriate afferent projections to all regions of the dorsal horn. The anatomical projections of different lumbosacral dorsal roots to laminae III and

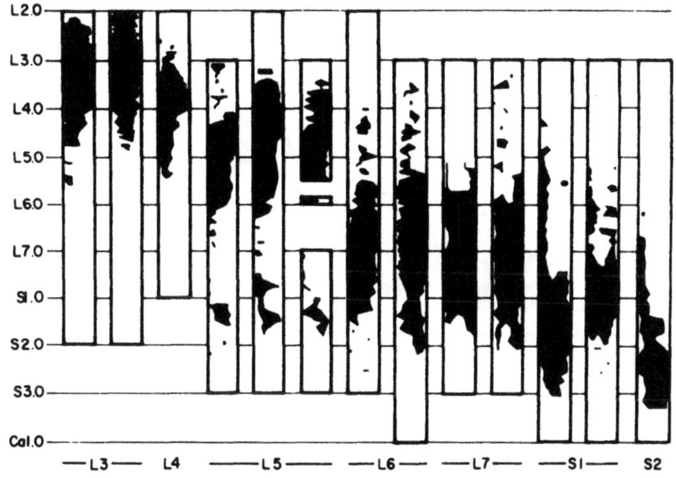

Fig. 4.8. Projections of dorsal root afferents to lumbosacral dorsal horn of the cat. Pattern of axon degeneration shown in dorsal views of the left dorsal horn in 13 cats following transection of the single dorsal root shown at the *bottom* of the figure. Segmental level X.0 rostral end of segment X. (From Brown et al. 1981, p. 47)

IV of the dorsal horn, determined by degeneration techniques, also corresponded well with the somatotopic organization of dorsal horn neurons (Brown and Culberson 1981). Adjacent dorsal roots, which always had overlapping dermatomes, also had overlapping central projections to the dorsal horn (Fig. 4.8), and again at this level of analysis there was no evidence of somatotopically inappropriate projections. Thus, even though the S1 dorsal root projected as far rostrally as the lateral border of the L4 dorsal horn, this projection was not considered somatotopically inappropriate, because skin of the S1 dermatome was often found to be represented there (Fig. 4.4). In fact, the projections of roots with proximal dermatomes (L4, L5, S1, S2) were found to be displaced laterally in the L6 and L7 segments to allow for the greatly enlarged representation of distal skin in the medial dorsal horn (Fig. 4.8). This may be taken as further evidence that the somatotopy of the dorsal horn in the cat is based primarily on the projection pattern of primary afferent fibres.

Peripheral Nerves. In order to obtain a more precise picture of the relationship between the representaion of skin supplied by a given set of afferents and the distribution of their central projections, transganglionic transport of HRP has been used to map the central projection fields of peripheral nerves. In the cat, Koerber and Brown (1982) examined the projection of a number of hindlimb cutaneous nerves to laminae III and IV of the lumbosacral dorsal horn, and compared the map they obtained (Fig. 4.9) with published somatotopic maps of lamina IV dorsal horn neurons. In general, as predicted by the presynaptic somatotopy hypothesis, the degree of overlap or separation between the central projection zones of different cutaneous nerves was found to be related to the overlap or separation of their

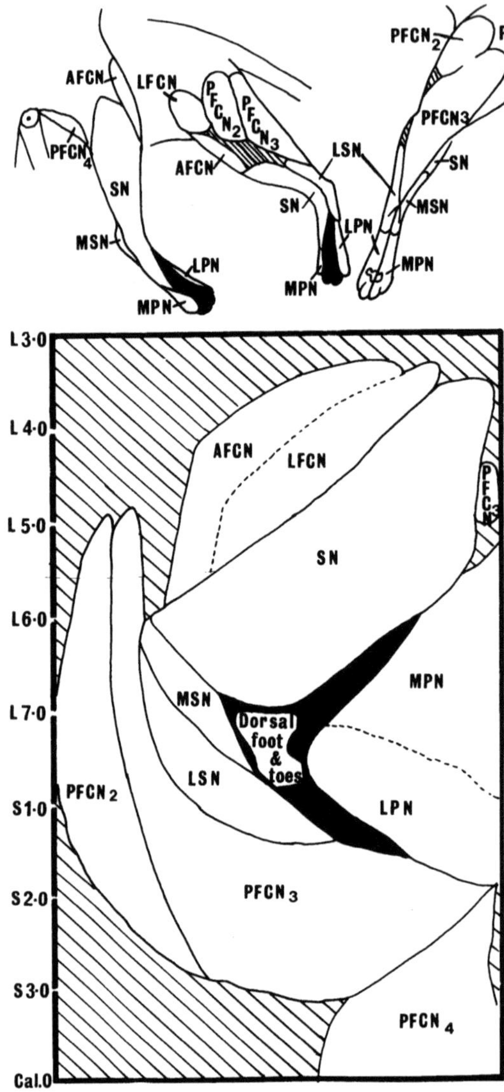

Fig. 4.9. Somatotopic organization of the projection of afferents in cutaneous nerves to the lumbosacral dorsal horn of the cat. Pattern of labelling in a dorsal view of the dorsal horn following application of HRP to the corresponding peripheral nerve. *Abbreviations*: *AFCN*, anterior femoral cutaneous nerve; *LFCN*, lateral femoral cutaneous nerve; *SN*, sural nerve; *MSN*, medial sural nerve; *LSN*, lateral sural nerve; *MPN*, medial plantar nerve; *LPN*, lateral plantar nerve; $PFCN_{2-4}$, branches of posterior femoral cutaneous nerve. The normal peripheral innervation field of each nerve is shown in the cat hindlimb, at *top* of figure. (Redrawn from Koerber and Brown 1982, p. 486)

cutaneous innervation fields. Moreover, the pattern of projection of the nerves closely resembled the somatotopic organization of the dorsal horn. This was particularly true of the mediolateral axis, where it was found that only nerves supplying the most distal areas (the lateral and medial plantar nerves) projected to the medial part of the L6 and L7 dorsal horn. In the rostrocaudal axis there were some signs of possible somatotopically inappropriate projections from branches of the PFCN, which was found to project much further caudally than might be expected from the somatotopy of dorsal horn neurons. For the most part, however, the projections which might have been somatotopically inappropriate were to ventral lamina IV or sometimes lamina V, rather than lamina III, and owing to a lack of detail concerning the laminar and somatotopic organization of the more caudal segments it could not be stated with certainty that any of these more caudal projections of the PFCN were, in fact, somatotopically inappropriate.

A similar study of the projection of forelimb nerves to laminae II and III of the cervical dorsal horn (Nyberg and Blomqvist 1985) also lends support to the presynaptic somatotopy hypothesis of P.B. Brown and co-workers. Again, in the mediolateral axis of the cord, nerves with separate peripheral innervation fields had separate projection fields in the dorsal horn while nerves with adjoining but non-overlapping innervation fields had adjoining but non-overlapping central projections. However, along the rostrocaudal axis of the dorsal horn there was evidence of overlap between the projection zones of nerves having quite separate peripheral innervation fields. Of special interest was the finding of a rostrocaudal overlap of about 1000 μm between the projections of the ventral digital nerves supplying the medial aspect of one digit and the lateral aspect of the neighbouring digit (Fig. 4.10). In the absence of further details of somatotopic organization of dorsal horn neurons in the cervical enlargement, it is impossible to know whether the region of overlap indicates the presence of somatotopically inappropriate projections, or whether it contains neurons with receptive fields spanning the innervation zones of both nerves (see Snow and Wilson, 1989).

In physiological experiments, large-diameter myelinated afferents entering the cat's spinal cord over rostral lumbar dorsal roots have been shown to project as far caudally as the S1 segment. This was demonstrated by recording single afferent spikes in filaments of a dorsal root, and antidromically activating the afferents by electrical stimulation through a microelectrode in the distant segment (Wall and Werman 1976). Using this method all filaments of the L2, L3 and L4 dorsal roots were found to contain some afferents that projected as far as S1. That these long-ranging afferents are actually present in substantial numbers was indicated by the fact that, even in the most rostral of these roots, around 16% of afferents sampled could be antidromically activated from the S1 segment. In most cases the axons could be activated from the grey matter of the dorsal horn as well as from the dorsal columns. For a few axons in the L2 dorsal root the cutaneous receptive field was mapped and found to be in the dermatome of the L2 root. Somatotopic maps of the L6–S1 dorsal horn do not indicate any representation of the L2 dermatome at these segmental levels (Sect. 4.3.1) and it therefore seems likely that at least some of the long-ranging afferents were projecting to somatotopically inappropriate regions of the dorsal horn. It is interesting to note that the single L2 afferent collateral whose

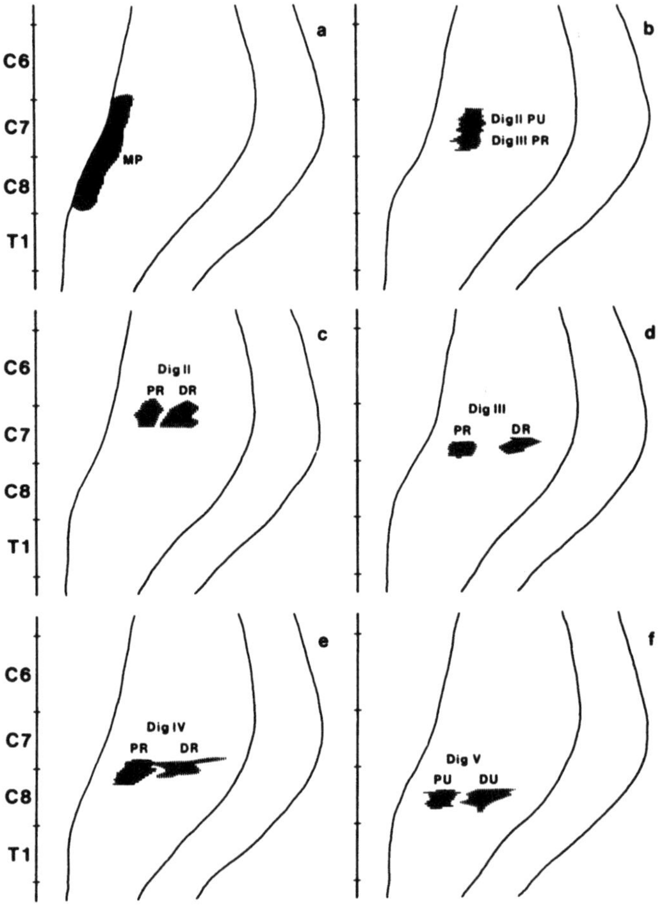

Fig. 4.10 a–f. Projection of different cutaneous nerves (**a**–**f**) of the forepaw to the cervical dorsal horn of the cat. Pattern of terminal labelling in laminae II and III shown in a dorsal view of the right dorsal horn following application of HRP to single digital nerves. Where two nerves are shown in the same figure, one was labelled on the *right* side and one on the *left*, and the two projections then superimposed. Note in **b** the overlap in the rostrocaudal axis between the projections of adjacent plantar digital nerves from digits 2 and 3. Note also the lack of mediolateral overlap between the dorsal and ventral digital nerves of a single digit. *Abbreviations*: *MP*, nerve of metacarpal pad; *PU*, plantar ulnar digital nerve; *PR*, plantar radial digital nerve; *DU*, dorsal ulnar digital nerve; *DR*, dorsal radial digital nerve; *Dig II–V*, digits 2–5. (From Nyberg and Blomqvist 1985, p. 34)

terminal trajectory in S1 was illustrated by Wall and Werman (1976) appeared to occupy a compact region of dorsal horn. While the significance of this observation may be limited due to the relatively low resolution afforded by the described methodology (stimulus points were separated by 200 μm mediolaterally, by 100 μm dorsoventrally and threshold-stimulating currents were often much higher than the 1–2 μA expected for focal microstimulation of axon terminals, Curtis 1979; Snow et al. 1988), such a profile corresponds with the simple morphology of single Aß afferent collaterals in identified somatotopically inappropriate regions of the dorsal horn (Meyers and Snow 1984) (Sect. 4.3.3).

The projections of a variety of hindlimb nerves in the rat have been studied by two groups of workers using transganglionic tracers. Swett and Woolf (1985) applied WGA-HRP to the cut ends of the posterior cutaneous nerve of the thigh, the lateral sural, sural, saphenous, superficial peroneal, and tibial nerves. Terminal labelling was found to be concentrated in laminae I and II of the dorsal horn, indicating that small afferents were labelled preferentially (Sect. 4.2.2). Consequently, the somatotopic maps constructed by these workers were limited to these superficial laminae. The sural, lateral sural, superficial peroneal, and tibial nerves (all subdivisions of the sciatic nerve) each projected to a separate and distinct zone of the superficial dorsal horn. Collectively, they formed a "U"-shaped zone of terminal labelling extending from the L2 to the L5 segment, with the open end of the "U" facing rostrally. The central part of the "U" was occupied by the projection of the saphenous nerve, which extended from L2 caudally to the L3/4 boundary. Molander and Grant (1986) used free HRP to study the projections of the tibial, medial plantar, lateral plantar, common peroneal, saphenous, sural, and lateral femoral cutaneous or obturator nerves. However, while these workers found a broadly similar somatotopic organization, their results differed in two important respects from those of Swett and Woolf (1985). First, the labelling in lamina I, unlike that in lamina II, did not appear to be somatotopically organized. Second, there was a variable amount of overlap between the lamina II projections of nerves innervating non-overlapping cutaneous territories. The reasons for these disparities are not clear, but the difference in overlap may be related simply to differing interpretations of what constitutes a boundary between labelled and unlabelled tissue. Such considerations may become especially important when using this method to detect the presence of sprouting by primary afferent fibres (Sect. 4.5.5). Unfortunately, neither of these studies contributes much to our knowledge of the somatotopic organization of peripheral nerve projections to the deeper laminae of the dorsal horn.

Correlative physiological studies of dorsal horn somatotopy and anatomical transganglionic tracer studies of hindlimb peripheral nerve projections to laminae III and IV in the rat lumbosacral cord have provided some support for the presynaptic somatotopy hypothesis. In these experiments (Woolf and Fitzgerald 1986) transganglionic HRP labelling of the terminals of coarse myelinated afferents in laminae III and IV was obtained by the strategy of using unfixed tissue together with a long survival time. The majority of dorsal horn neurons with receptive fields on a given skin area were found to lie within the region of dorsal horn containing afferent projections from the nerve supplying that area of skin. However, both in the superficial (laminae I and II) and deep (laminae III and IV) dorsal horn, a significant number of neurons had receptive fields on skin supplied by peripheral nerves that projected only to other regions of the dorsal horn. Since the neuronal receptive fields and nerve projection zones were mapped in different animals, it is necessary to be cognisant of such problems as inter-animal variation of somatotopy, variability of staining, size of the cord and accuracy of determining the location of recorded neurons. These possible sources of error tend to be played down by the authors because some neurons were apparently located as far as 1.3 mm from the region of

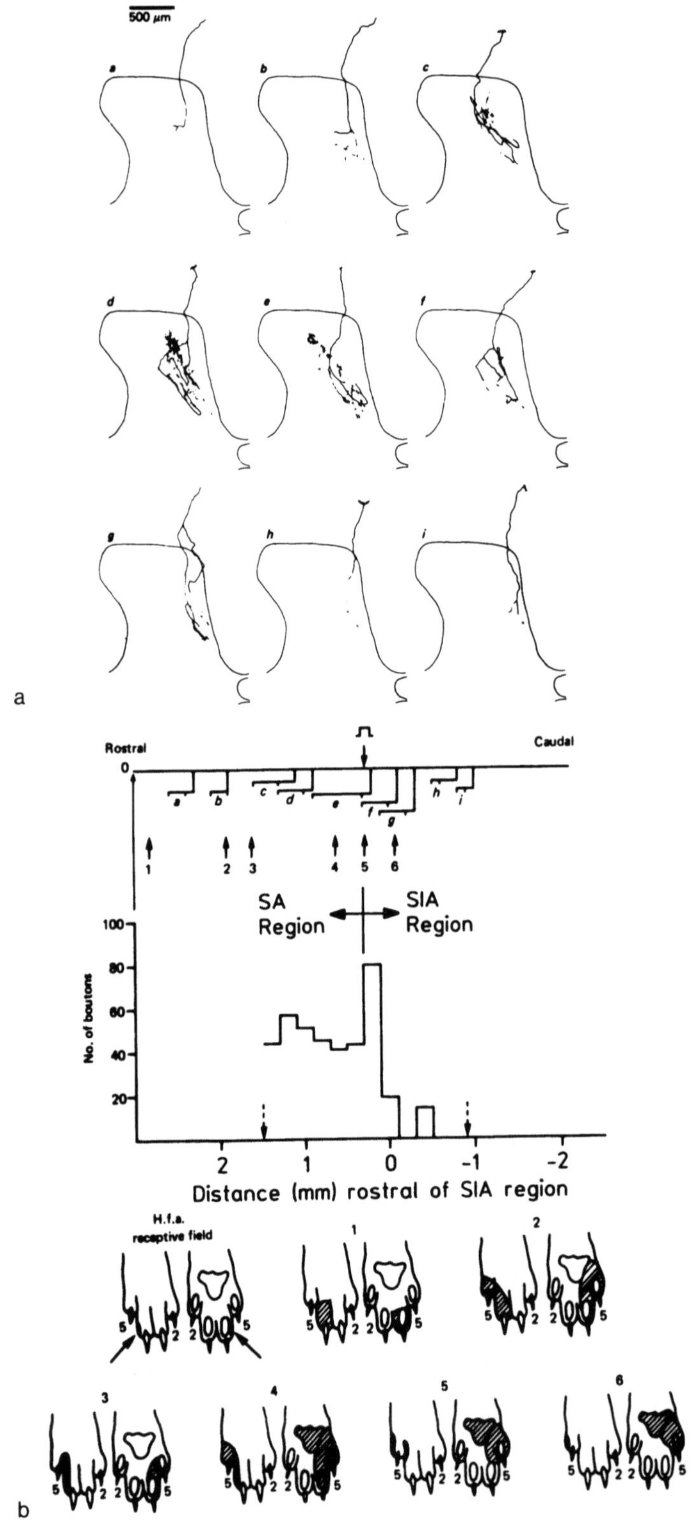
500 μm
a
b
c
d
e
f
g
h
i
a
Rostral
Caudal
0
1
2
3
4
5
6
SA Region
SIA Region
100
80
60
40
20
No. of boutons
2
1
0
-1
-2
Distance (mm) rostral of SIA region
H.f.a. receptive field
5
2
b

appropriate nerve projection. However, the data for units with receptive fields that included part of the peripheral innervation field of the tibial nerve field indicate that most of the cells with such *anomalous* receptive fields were within about 500 μm of the projection zone of the tibial nerve (see Fig. 10 in Woolf and Fitzgerald, 1986). In summary, it may be said that studies utilizing degeneration and transganglionic tracers have generally indicated that, in both the lumbar and cervical enlargements of the cat spinal cord, there is a tight correspondence between the distribution of afferent projections from peripheral nerves innervating different areas of skin and the anticipated somatotopic organization of dorsal horn neurons. It is possible that the physiological methods used by Wall and Werman (1976) are capable of detecting somatotopically inappropriate projections that are not demonstrable by these traditional anatomical methods. In the rat, the situation appears to be different and it is possible that there are relatively large mismatches between the projections of peripheral nerves and the observed somatotopic organization of dorsal horn neurons. However, the situation in the rat requires closer examination.

Single Identified Primary Afferents. To minimize errors in comparing dorsal horn somatotopy and primary afferent projections, it is essential to perform both determinations in one and the same animal. This is particularly true when looking for somatotopic mismatches along the rostrocaudal axis, since in this direction in the cat the somatotopic representation of the whole hindlimb skin can vary in position by up to 1.5 segments (Wilson et al. 1986). The projections of individual identified, Aß, hair follicle afferents to somatotopically identified regions of the lumbosacral dorsal horn have been determined in the cat by intra-axonal injection of HRP (Meyers and Snow 1984). Afferents with receptive fields on the toes were selected for study, because the region containing the representation of a single toe has easily definable rostral and caudal boundaries (Fig. 4.5). After injecting an axon, the receptive fields of dorsal horn neurons were mapped in the region con-

◀ **Fig. 4.11 a, b.** Somatotopically inappropriate projections of Aß hair follicle afferents in the lumbosacral dorsal horn of the cat. **a** Reconstructions in the transverse plane of the arborizations of 9 stained collaterals of a single hair follicle afferent fibre injected intra-axonally with HRP. The collaterals are shown in rostral to caudal sequence (*a–i*). Collaterals *a–e* project to somatotopically appropriate regions whereas collaterals *f–i* project to somatotopically inappropriate regions. **b** Rostrocaudal distribution of the arborizations of the same collaterals (*a–i*) shown in **a**, together with a count of the boutons up to 1.2 mm rostral and caudal to the injection site (*down arrow*) and the composite receptive fields of dorsal horn neurons in six mediolateral rows (1–6, *up arrows*) of three adjacent microelectrode tracks (150 μm separation) through the region of dorsal horn containing the stained afferent. Note that in rows 5 and 6 the composite receptive fields of the dorsal horn neurons do not encompass the receptive field of the afferent. Note also the small number of boutons in those parts of the somatotopically inappropriate region more than 500 μm caudal of the somatotopically appropriate region. The axon and the abscissa of the histogram are drawn to the same scale. *SA* region, somatotopically appropriate region, where receptive fields of dorsal horn neurons overlap that of the afferent fibre; *SIA* region, somatotopically inappropriate region, where receptive fields of dorsal horn neurons do not encompass that of the afferent fibre. (Modified from Meyers and Snow 1984, pp. 66 and 69) taining the injection site. The hair follicle afferent collaterals previously described by Brown et al. (1977b) were selected for study because they provide the major source of low-threshold tactile afferent input to the antidromically identifiable cells of the SCT.

The major collaterals of single Aß hair follicle afferents had previously been shown to be complex, and to form unbroken rostrocaudally oriented sheets of terminal arbors in a fixed mediolateral plane of the dorsal horn (Sect. 4.2.3). Meyers and Snow (1984) noted that in addition to these major collaterals each injected axon also gave rise to simpler collaterals that branched less profusely than the major collaterals and bore relatively few or no synaptic boutons (Fig. 4.11a). Furthermore the arborizations of the simple collaterals did not overlap in the rostrocaudal axis. Of all collaterals, 42% were of this simpler form and recent ultrastructural examination has confirmed that they lack synaptic specializations on their terminal branches (Wilson and Snow, 1990a). When the somatotopy of the dorsal horn was studied it was found that the simple collaterals lay in somatotopically inappropriate regions of the dorsal horn that were rostrally or caudally displaced from the somatotopically appropriate region for the afferent. Collaterals that terminated within 500–600 μm of the somatotopically appropriate region had complex terminations and bore many synaptic boutons (Fig. 4.11b). It is interesting to note that this distance is approximately equal to half the rostrocaudal extent of the dendritic trees of medially placed lamina IV cells (Brown et al. 1977a, 1980c). The arborizations in this region could thus preferentially make contact with the dendrites of dorsal horn neurons within the somatotopically appropriate area.

A study of the morphology of the central arborizations of cutaneous Aß afferent fibres in the rat showed that both complex and simple collateral arborizations are also present in hair follicle afferents, rapidly adapting afferents innervating glabrous skin, and type I afferents in the dorsal horn in that species (Woolf 1987). The simple collaterals were usually either situated toward the rostral end of the main region of terminal arborizations or issued from a descending branch of the main axon. In the absence of detailed somatotopic maps of the rat dorsal horn the relationship of the simple collaterals to dorsal horn somatotopy can not be estimated. However, since the band of complex collateral arborizations of a single axon with a glabrous skin receptive field on the tips of two toes occupied about 2 mm in the rostrocaudal axis, or about the length of an entire L5 or L6 segment, it is likely that at least some of the simple collaterals would have been in somatotopically inappropriate regions. In a more recent study Shortland et al. (1989) have examined the collateral morphology of a number of intraaxonally stained primary afferent fibres in the rat. They concluded that collaterals could be divided into three basic types: (1) *complex* collaterals, with many branches and bearing synaptic boutons, (2) *simple* collaterals, with few branches but bearing boutons, and (3) *blind-ending* collaterals, with severely restricted branching pattern and no boutons. In line with the combined electrophysiological and anatomical studies in the cat (Meyers and Snow 1984), many of the simple and blind-ending collaterals of afferents with receptive fields in the cutaneous territory of a given peripheral nerve overlapped in the rostrocaudal axis with the complex collaterals of afferents with receptive fields in the territory of a different nerve. Finally Sugiura et al. (1986) have published pictures of single, intracellularly stained cutaneous C fibre afferents in the guinea-pig that give rise to simple collaterals that are rostrocaudally displaced from a focus of complex collaterals arising from the same axon. The displacement of these simple collaterals from the complex collaterals is such as to strongly suggest that, like Aß afferents, C fibre afferents may also give rise to somatotopically inappropriate afferent terminals rostral and caudal to the somatotopically appropriate region. Just as the somatotopically inappropriate terminations of Aß fibres have been implicated in lesion-induced rostrocaudal shifts in somatotopic organization (Figs. 4.19 and 8.2; Sect. 4.4.3) (Snow and Wilson 1989, 1990; Wilson and Snow 1990b). The existence of similar terminations in C fibres might have important consequences for plasticity in relation to pain states induced by peripheral nerve lesions or limb amputation.

An important feature of the somatotopically inappropriate collaterals of Aß hair follicle afferents in the cat and the "simple" collaterals of primary afferents in the rat is that they maintain the same mediolateral position in the dorsal horn as the main band of collateral arborizations (Fig. 4.11a). Therefore without sprouting into new territory they cannot provide a basis for the mediolaterally directed changes of somatotopy of low-threshold receptive fields in the dorsal horn which have been reported by some workers (Sect. 4.4.3) (Devor and Wall 1981a,b). Hypotheses concerning the possible functional role of these collaterals in normal and nerve-lesioned animals are discussed more fully below (Fig. 8.2a) (Sects. 4.5.5, 8.3 and 8.4) (see also Snow and Wilson, 1989, 1990).

4.3.4 Relation Between Dorsal Horn Cell Dendritic Morphology and Receptive Field

Since the advent of the technique of intracellular staining of physiologically identified neurons with HRP (Snow et al. 1976), several attempts have been made to find correlations between the dendritic envelope of single dorsal horn neurons and the extent, position and modality of their receptive fields. For some of these neurons, such as the SCT cell of the cat, the highly ordered spatial arrangement of the central terminal arborizations of primary afferent fibres that is dependent on the location of their receptive fields is maintained in the functional connectivity between low-threshold afferents and the cell (Brown et al. 1980; Brown and Noble 1982). Ritz and Greenspan (1985) stained lamina V dorsal horn neurons in the sacrococcygeal cord of the decerebrate cat. The average dendritic extent of these neurons was approximately the same, irrespective of whether they were wide-dynamic-range or nociceptor-specific neurons. However, the wide-dynamic-range neurons showed considerably more variability in the spread of their dendrites than the nociceptor-specific cells. Compared with these two types, lamina V neurons responding only to innocuous mechanical stimulation tended to have smaller dendritic spreads, especially in the rostrocaudal dimension. A significant correlation was found between the mediolateral spread of the dendrites and the size of the low-threshold receptive field. However, there was no correlation between the size of the low-threshold receptive field and the rostrocaudal dendritic spread, or between the size of the high-threshold field and either the mediolateral or rostrocaudal dendritic spread. The correlation between the extent of the low-threshold receptive field and the mediolateral dendritic spread is exactly what would be predicted from the presynaptic somatotopy hypothesis, because the somatotopic gradient of the primary afferent input to the dorsal horn is much steeper mediolaterally than rostrocaudally. In other cases any correlation between the dendritic extent of a cell and its functional properties would appear to be more tenuous. Of a small sample of cat lamina III interneurons that were injected with HRP, some responded to light touch, and some to strong pressure. However, although the cells had a wide range of dendritic morphology there was no clear correlation between structure and response modality (Maxwell et al. 1983).

In a recent study in the decerebrate rat, Woolf and King (1987) were unable to find any correlation between the size of either the low- or high-threshold receptive fields of wide-dynamic-range lamina V neurons and the mediolateral spread of their dendrites (Fig. 4.12). This suggests that, in the rat, monosynaptic input from primary afferent fibres may be less important in determining somatotopy than polysynaptic inputs. Egger et al. (1986) injected HRP into cells in the medial dorsal horn of the chloralose-anaesthetized cat with receptive fields on the hindlimb plantar cushion. For a small sample of lamina III neurons, there was a significant correlation between the mediolateral dendritic spread and the size of the receptive field, but no such correlation was found for a sample of lamina IV neurons. However, these cells, situated as they are in the most medial part of the dorsal horn adjacent to the white matter, might be unrepresentative of other lamina IV neurons.

4.4 Effect of Lesions on Somatotopic Organization

4.4.1 Dorsal Rhizotomy

Reorganization of Somatotopy. The first report to suggest that dorsal horn neurons in the spinal cord of adult mammals could, under certain circumstances, lose their receptive field and start responding to stimulation of a different area of skin was that of Basbaum and Wall (1976). These workers recorded from dorsal horn neurons at the L4/L5 junction in two groups of adult cats. In the first (acute) group, recordings were made initially with all spinal roots intact, then immediately after unilateral transection of all dorsal roots caudal to L3, with the exception of S1. In the second (chronic) group the same roots were cut chronically for periods between 1 day and 9 months before the recordings. Prior to dorsal rhizotomy, the most distal receptive fields encountered in the L4/L5 boundary region extended along the rostral border of the lower leg. None extended onto the foot, and no cells were found with receptive fields wholly situated on the foot. Immediately after dorsal rhizotomy cells responsive to cutaneous stimulation could not be found in this region of the dorsal horn in about half the cases, and in the remaining animals one or two cells responded only to heavy pinch on the leg. The long latency of the response ($>$ 15 ms) following electrical stimulation of this receptive field indicated that the cells were coupled to their receptive fields via polysynaptic pathways. One week or more after dorsal root section, cells were again found in laminae IV and V which responded to light tactile

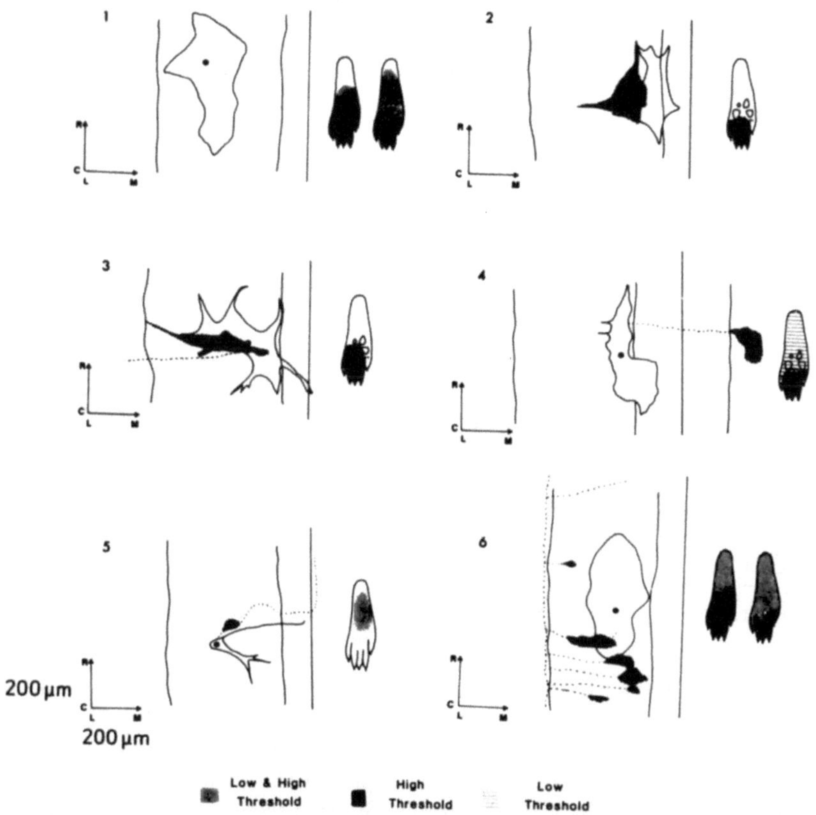

Fig. 4.12. Dendritic spread and cutaneous receptive fields of rat lamina V neurons. Plan views of the dorsal horn showing the maximum extent, in transverse sections, of the dendritic trees of HRP-injected lamina V dorsal horn neurons (1–6), together with their cutaneous receptive fields, in unanaesthetized decerebrate spinal rats. For each diagram the *vertical solid line* on the *left* represents the lateral border of the dorsal horn, the next *vertical line* to the *right* represents the medial border of the dorsal horn, and the *straight line* on the *extreme right*, the midline of the cord. The *dot* represents the cell body, the *continuous line* connects the maximal mediolateral spread of the dendrites in transverse sections, the *dotted line* represents the axon, and the *shaded regions* the distribution of the axon collaterals. The scale applies to all 6 cells. *R*, rostral; *C*, caudal; *L*, lateral; *M*, medial. Note the lack of correspondence between the mediolateral spread of the dendrites, and the size of the receptive field. (Modified from Woolf and King 1987, p. 472)

stimulation of the skin (light brushing, touch or pressure), and after 1 month or more these were found in all microelectrode tracks made at the L4/L5 junction. The newly developed receptive fields were all either within the S1 dermatome, or they were situated much more rostrally on the abdomen and thorax (Fig. 4.13). Some cells had two receptive fields, usually one in the S1 dermatome, and a second on the abdomen or thorax. Electrical stimulation of the skin, sural nerve, or S1 dorsal roots activated a few cells at latencies that were compatible with monosynaptic activation. Basbaum and Wall (1976) interpreted their findings as indicating that new functional connections had formed between intact primary afferent fibres and partially deafferented dorsal horn neurons.

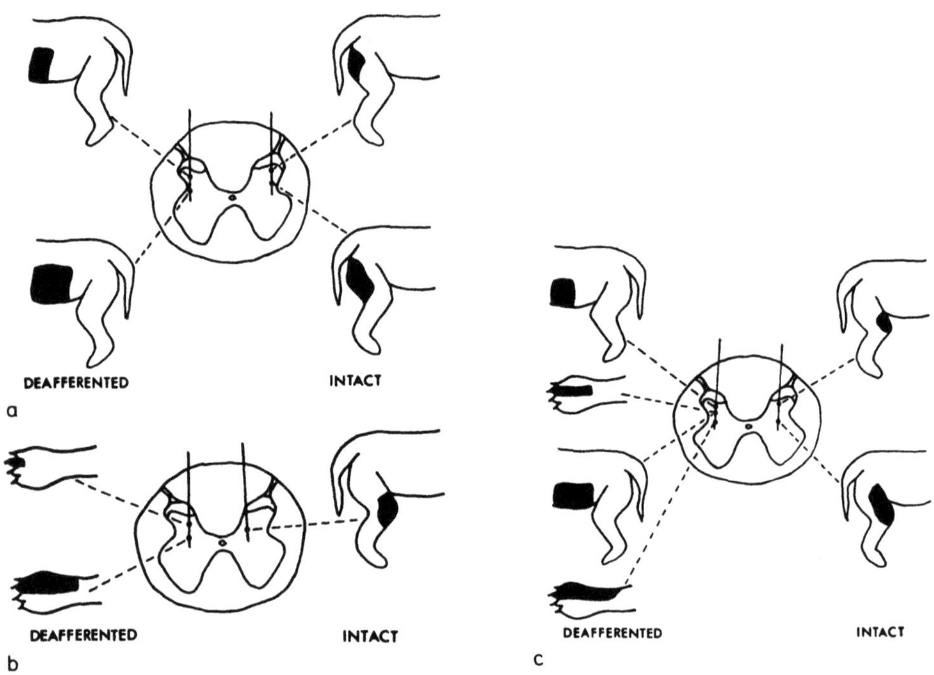

Fig. 4.13 a–c. Receptive fields and locations of dorsal horn neurons in normal and chronically dorsal rhizotomized (dorsal roots caudal of L3 transected, SI spared) segments of the dorsal horn in the cat. In each figurine the *filled region* shows the brush/light touch receptive field of the cell. **a** Examples of receptive fields and location of cells on rhizotomized and intact side of L4 dorsal horn 6 weeks after root section. Cells with abdominal receptive fields were found in the lateral region of the partially deafferented dorsal horn where cells on the intact side had receptive fields on the leg. **b** Examples of receptive fields and location of cells on deafferented and intact side at the junction of the L4 and L5 segments 6 weeks after root section. Cells with distal foot receptive fields were found in the medial part of the partially deafferented dorsal horn where cells on the intact side had receptive fields on the leg. **c** Examples of receptive fields and location of cells on rhizotomized and intact side at the junction of the L4 and L5 segments 32 weeks after root section. Cells with abdominal, distal foot and combined double receptive fields were found midway across the dorsal horn, where cells on the intact side had receptive fields on the upper leg. (From Basbaum and Wall 1976, pp. 194 and 195)

The question of whether the apparent reorganization of receptive fields was due to the formation or strengthing of monosynaptic connections from primary afferent fibres to dorsal horn neurons was subsequently re-investigated using intracellular recording in T13-spinalized cats which were either decerebrated or anaesthetized with Dial (Mendell et al. 1978). In normal cats neurons in the L7 dorsal horn have a natural receptive field situated on the leg, foot or toes. However, in these low spinal preparations about 30% of these cells, particularly those with receptive fields on the upper leg, were found to respond also to electrical stimulation of skin on the flank, which was shown by recording from dorsal root filaments to excite afferents in only the L3 and L4 dorsal roots. Of the cells responding to the flank stimulus, less than

5% had minimal latencies less than 5 ms and followed stimulation at 10 Hz with a fixed latency. Thus there was little likelihood that many of these responses were evoked monosynaptically from primary afferents. Intracellular recording revealed that in about 50% of cells with a hindlimb receptive field, flank stimulation evoked a mixture of EPSPs and IPSPs. Even at a slow stimulus repetition rate of only 1 Hz the EPSPs had variable latencies, indicating that, like the spike responses, they were unlikely to be monosynaptically evoked.

In other cats, dorsal roots L5–S2 were sectioned and intracellular recordings were again made from cells in the L7 segment. The dorsal rhizotomy eliminated the natural receptive fields of these cells, but their responses to electrical stimulation of the flank were similar to those in cats with intact dorsal roots. However, between 38 and 51 days later, many cells throughout the L7 dorsal horn had acquired a natural receptive field on the flank. One-fifth of cells penetrated with microelectrodes had such a receptive field, and the remainder responded with an EPSP to electrical stimulation of the flank. Four times as many cells as normal showed a short latency (< 5 ms) EPSP. Moreover, these EPSPs were now of constant amplitude and latency, and followed repetitive stimulation at up to 100 Hz or more, suggesting they could have been evoked monosynaptically. Mendell et al. (1978) claimed their results showed that a few cells in the L7 dorsal horn normally receive weak polysynaptic excitatory inputs from afferents entering the cord over the L3 or L4 dorsal roots. By about 1 month after the lesion many more cells in the L7 segment acquire excitatory inputs from the nearest rostral intact root, and some of these inputs are now via monosynaptic connections from primary afferent fibres.

In a later study Sedivec et al. (1983) examined the receptive fields of SCT cells in the L6 dorsal horn of the chloralose-anaesthetized cat approximately 6 weeks after transection of dorsal roots L5–S2, and again found evidence of reorganized receptive fields. Although about one-third had no receptive field, most cells responded to skin outside the L6 dermatome. Receptive fields were generally larger than those in the intact cord, and were situated either proximally on the hindlimb or on the abdomen. These abnormal types of fields were not observed when the L7 root, which contributes substantially to the L6 segment, remained intact. Sedivec et al. (1983) suggested that complete removal of primary afferent input to a spinal segment was necessary before cells would develop responsiveness to input from rostral intact dorsal roots.

In experiments where the overall somatotopic organization of the dorsal horn has been examined, or where recordings have been made from a sample of cells, unbiased with respect to the presence or absence of either a peripheral input or spontaneous firing, no evidence has been found for somatotopic reorganization following transection of dorsal roots. In intact cats and in those with all lumbosacral dorsal roots except L6 transected Pubols and Goldberger (1980) mapped the somatotopic organization of the L6 dorsal horn under methoxyflurane anaesthesia. Single units or unit clusters were isolated in extracellular recordings while gross mechanical stimuli were applied to the hindlimb region, and the receptive fields of any responsive units were then mapped using light tactile stimulation of the skin. Immediately following the lesion there was found to be a loss of responsiveness throughout the entire mediolateral extent of the dorsal horn, especially in the lateral part, which normally receives only a slight projection from the L6 dorsal root (Goldberger and Murray 1982).

Within 8 weeks of dorsal rhizotomy, responsiveness appeared to return to the lateral region of the dorsal horn. Skin on the proximal thigh and back (outside the L6 dermatome) was not represented, but the percentage of units with distal receptive fields increased. Pubols and Goldberger (1980) suggested that, following the lesion, cells with large receptive fields covering the upper thigh and extending down to the foot, and thus having convergent input from several dermatomes, initially became *unresponsive to the spared portion of their input*. Thus in the acutely lesioned animals, many cells along the lateral border of the dorsal horn lacked a receptive field. With time, responsiveness to the spared-root input returned, causing lateral cells in the chronic preparations to display receptive fields in the proximal part of the L6 dermatome. The authors further suggested that this recovery might be mediated through *local sprouting* of intact primary afferent fibres projecting to the lateral part of the L6 dorsal horn, as proposed by Goldberger and Murray (1978, 1982) (Sect. 4.5.5).

In a later study Pubols and Brenowitz (1982) compared the somatotopic organization of the receptive fields of lamina III and IV dorsal horn neurons on both sides of the L6 segment of cats with two types of chronic (10–16 week) lesions, either a unilateral lumbosacral dorsal rhizotomy sparing L6, or a unilateral L6 dorsal rhizotomy. The somatotopic organization was similar on both sides of the cord in both groups of animals, with the exception that representation of the proximal part of the hindlimb was absent in the animals with only the L6 root intact. In the cats with L6 dorsal root section the representation of hindlimb skin within the L6 segment was complete, but there appeared to be a minor, although consistent, mediolateral compression of the entire somatotopic map. Pubols and Brenowitz (1982) concluded that the L6 dorsal root alone was capable of maintaining a nearly normal somatotopic map in the L6 segment, and that the relatively minor changes in somatotopic organization observed after chronic partial deafferentation of the L6 segment by dorsal rhizotomy could be attributed simply to the removal of some afferent inputs. These results would be expected on the basis of the large overlap between the hindlimb dermatomes and between the central projections of their afferents to the dorsal horn (Sect. 4.3.3).

Brown et al. (1983b) investigated the effects of chronic lumbosacral dorsal rhizotomy on the somatotopic organization of the L6 and L7 dorsal horn in the cat in two types of experiment. In the first, dorsal roots L6 and L7 were transected, and in the second all lumbosacral roots *except* L6 were cut. In both cases survival times varied between 27 and 75 days. Intracellular recordings were made under chloralose-anaesthesia from SCT cells in the L6 and L7 segments ipsilateral to the dorsal rhizotomy. The receptive fields of the cells were mapped, and some cells were stained by intracellular injection of HRP, so that their exact position within the dorsal horn could be determined. Of 20 SCT cells stained with HRP, only 7 had low-threshold receptive fields, and of 25 neurons recorded extracellularly, only 7 had apparently normal receptive fields. Other cells either had no receptive field at all, or responded only to tapping the limb. Of the few neurons with localized receptive

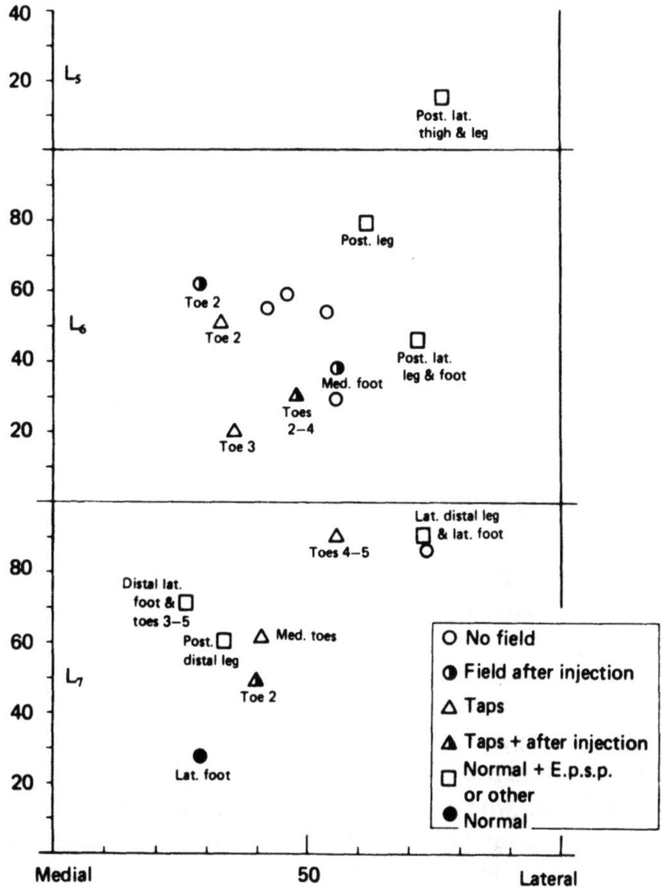

Fig. 4.14. Absence of somatotopically inappropriate receptive fields of SCT cells stained by intracellular injection of HRP in chronically dorsal rhizotomized segments of the cat dorsal horn. Plan view of the dorsal horn showing segments L6 and L7 and the caudal part of L5. The L6 and L7 dorsal roots were chronically sectioned. The rostrocaudal extents of the segments have been normalized and intracellularly stained SCT neurons placed in their appropriate mediolateral and rostrocaudal positions. Note that intracellular current injection revealed a receptive field in some neurons that initially lacked one. All neurons which lacked a receptive field initially were situated in a diagonal band running rostrocaudally and mediolaterally in L6 and rostral and lateral L7. At the caudal edge of the band were neurons responsive to tapping the limb. The only neuron with a somatotopically inappropriate receptive field was the cell in mid L7 responding to tapping toe 2 after current injection. (From Brown et al. 1983a, p. 596)

fields that were injected with HRP, in each case the position of the neuron was found to be appropriate to the location of its receptive field in relation to the normal somatotopy of the dorsal horn (Fig. 4.14). Brown et al. (1983b) also made recordings from a large number of unidentified dorsal horn neurons during the course of these experiments, and found that like the SCT cells, none of these showed any signs of inappropriate inputs. Brown et al. (1983b) concluded that after dorsal rhizotomy, dorsal horn neurons lose all or part of their afferent input and within 75 days show no sign of having developed new or inappropriate afferent connections.

Clearly, it is far from proven that chronic dorsal root section elicits extensive somatotopic reorganization in the dorsal horn. Nevertheless it is *theoretically* useful to consider the mechanisms that might underlie such changes. To explain their results, Basbaum and Wall (1976) favoured the formation of new monosynaptic connections, rather than the opening-up of polysynaptic pathways. However, because of the difficulty of demonstrating unequivocal monosynaptic activation, they were unable to provide rigorous substantiating evidence for this idea. Such an explanation, of course, implies the existence of primary afferent projections from the S1 dorsal root and from some roots rostral of L4, to the L4–L5 dorsal horn in the chronically lesioned animals. These projections could either pre-exist in the normal animal, or develop as a result of sprouting in response to the deafferentation produced by the lesion. The latter option was clearly in the minds of Basbaum and Wall (1976) at the time of writing, for they suggested that the long response latency of cells with anomalous receptive fields might be due to the fact that "the (primary afferent) axons may be unusually fine in diameter or have immature properties". They further pointed out that intact afferents are present in the dorsal columns within 1 mm of the newly connected cells, implying that sprouting might occur with ease over this relatively short distance.

Recent anatomical studies employing the Fink-Heimer technique have shown that terminals of S1 afferents are normally found in laminae III and IV of the dorsal horn as far rostrally as the L4 segment in the cat (Sect. 4.3.3) (Brown and Culberson 1981; Culberson and Brown 1984). These terminals are located at the medial edge of the L4 dorsal horn, where cells with new receptive fields on the foot and toes were apparently mainly found in the chronically rhizotomized animals of Basbaum and Wall (1976). Similarly, although rostral lumbar roots have a more restricted rostrocaudal distribution than caudal lumbar roots, terminals of L3 afferents are normally present in laminae III and IV at the L4/L5 segmental level (Culberson and Brown 1984). Thus long-range sprouting need not be involved in the proposed formation of new primary afferent connections (Sect. 4.3.3). The activation of a few neurons from sural nerve stimulation at short latency (Basbaum and Wall, 1976) is more difficult to explain, since orthograde transport of HRP has not revealed any terminal projections of sural nerve afferents rostral to the L6 segment (Koerber and Brown 1982). While this does not rule out completely the normal projection of a few sural afferents to the L4–L5 junction the implication is that if monosynaptic activation from sural afferents to L4–L5 dorsal horn neurons really exists in these rhizotomized animals, it may only do so as a result of sprouting.

The results quoted in relation to extracellular recording in the chronically dorsal rhizotomized animals in the study of Mendell et al. (1978) are somewhat anecdotal and details of cell positions and receptive fields, which might have lent support to the claims of Basbaum and Wall (1976), are not given. From the data presented and current knowledge of dorsal horn somatotopy it is possible to suggest that no change whatsoever actually occurred. There are, for instance, normally cells in the lateral parts of the L7 dorsal horn with large receptive fields extending onto the flank (Devor and Wall 1967b) and a few of these were in fact found by Mendell et al. (1978)

in intact cats. To assist with intracellular penetration, the authors used a search stimulus applied to the dorsal columns at L5. Thus, in all three types of preparation (intact, acutely rhizotomized and chronically rhizotomized) they only penetrated cells which were excited from fibres running in the dorsal columns at L5. This procedure must introduce a bias in the chronically lesioned animals towards sampling cells having relatively strong excitatory inputs from afferents entering the cord through the L3 root, and in the absence of any data concerning the mediolateral position of all penetrated neurons it is impossible to be sure that lateral cells with somatotopically appropriate projections from flank afferents were not overrepresented in the intracellular sample. However, from their analysis of the properties of EPSPs, Mendell et al. (1978) concluded that there were many more cells in the chronically rhizotomized cord than in the intact (or acutely rhizotomized) cord having a strong, monosynaptic, excitatory input from long-ranging afferents, and this could best be explained by supposing that the dorsal rhizotomy induced either the formation of new connections, or the unmasking of pre-existing but nonfunctional connections. A third possibility, that the change of somatotopy in these experiments might be due to a strengthening of *weak* monosynaptic inputs, was rejected by Mendell et al. (1978) because such inputs were not seen in intracellular records from dorsal horn neurons in intact cats or in cats with acute dorsal root sections. However, before any firm conclusion may be reached on this issue, it will be necessary to reexamine the effect of chronic dorsal rhizotomy on the excitatory inputs to a sample of dorsal horn neurons, such as SCT cells, distributed randomly across the dorsal horn. As discussed above, extracellular recording from such cells in segments partially deafferented by dorsal root section has given no indication of somatotopic reorganization. Moreover, there is little or no anatomical evidence to show either that L3 and L4 afferents normally send projections to the medial part of the L7 dorsal horn in the cat (Brown and Culberson 1981) or that significant intersegmental sprouting of intact primary afferent fibres occurs following dorsal rhizotomy (Sect. 4.5.5). Superficially, however the findings of Pubols and Goldberger (1980) would seem to indicate that chronic dorsal rhizotomy cap produce strengthening of some connections from intact primary afferent fibres to partially deafferented dorsal horn neurons, and to this extent they are consistent with those of Basbaum and Wall (1976) and Mendell et al. (1978).

The use of a subset of dorsal horn neurons, such as SCT cells, by Brown et al. (1983b) to investigate the effects of chronic dorsal rhizotomy on dorsal horn somatotopy has been criticized on the basis that the response of these cells to the lesion might differ from that of many other dorsal horn neurons. However, this approach has the great advantage of providing a sample of neurons that have not been selected on the basis of some property which might itself be affected by the lesion, such as the presence of a receptive field, of spontaneous activity, or of excitation from intact afferents in the dorsal columns. It has been suggested that the SCT may be an especially *hard-wired* system not capable of showing effects of long-ranging afferents (Brown et al. 1983b). Certainly, the small excitatory subliminal fringes of SCT cells (Brown et al. 1987) are not sufficiently extensive for them to be a likely basis of the somatotopic reorganization reported by Basbaum and Wall (1976) and by Mendell et al. (1978). However, the essence of any such reorganization seems to be that it involves the development of new monosynaptic excitatory inputs (whether by sprouting or by some unknown mechanism of functional derepression), rather than the strengthening of subliminal mono- or polysynaptic inputs. It should also be noted that in the experiments of Brown et al. (1983b) unidentified dorsal horn neurons, as well as SCT neurons, both failed to show any sign of somatotopically inappropriate inputs following chronic dorsal rhizotomy.

Change of Cutaneous Modality. Other evidence of changes of connectivity between afferent fibres and dorsal horn neurons has emerged from a detailed examination of their responses to different classes of afferents following dorsal rhizotomy. Scattered reports of increases in the mechanical threshold for activation of dorsal horn neurons appear in various studies involving partial deafferentation of the dorsal horn by dorsal root section (Basbaum and Wall 1976; Loeser and Ward 1967). More recently, however, changes in the convergence of afferents serving different cutaneous modalities onto partially deafferented dorsal horn neurons have been reported from two different laboratories (Pubols and Brenowitz 1982; Sedivec et al. 1983).

Although chronic dorsal rhizotomy failed to produce convincing somatotopic reorganization in their experiments, Pubols and Brenowitz (1982) did notice an increase in the proportion of neurons with high-threshold components in their receptive fields. Further investigation showed that in the L6 spared-root animals, the increase of neurons with high-threshold fields was not real, but was due instead to a relative loss of neurons with low-threshold receptive fields. However, in animals with only the L6 root transected there was no sign of a consistent loss of low-threshold fields in the L6 dorsal horn. Thus in the rhizotomized L6 segment there appeared to be an absolute increase in the number of cells responding to high-intensity mechanical stimulation of the skin. Sedivec et al. (1983) examined this effect in SCT cells of the L6 dorsal horn in chloralose-anaesthetized cats with either acute or chronic (56–156 days) unilateral L5–S2 dorsal rhizotomy, sparing L7. As a result of the chronic lesion the proportion of SCT cells responding only to light tactile stimuli was roughly halved, while the proportion responding to noxious inputs increased. In the acutely rhizotomized segment there was an increased proportion of SCT cells responding only to light tactile stimuli (Fig. 4.15a). Thus, compared to the acute animals, the increase in the proportion of cells responding to high-threshold inputs in the chronically rhizotomized segment was even more marked. In addition to the change of responsiveness to strong pressure, it was noted that the proportion of neurons responsive to moderate pressure, cooling or noxious heat changed in parallel with the changes in the proportion of high-threshold fields. The altered proportions of high- and low-threshold responses found in the acutely rhizotomized state persisted for a period of about 45 days (Fig. 4.15b) suggesting that this represented a new steady-state which was later terminated by rather rapid events occurring in the dorsal horn after a delay of the order of 6–7 weeks.

The mechanisms underlying these changes are unknown, but it seems reasonable to suppose that they are related to alterations in the transmission of inputs from low- and high-threshold receptors via afferent fibres of different diameter. The responses

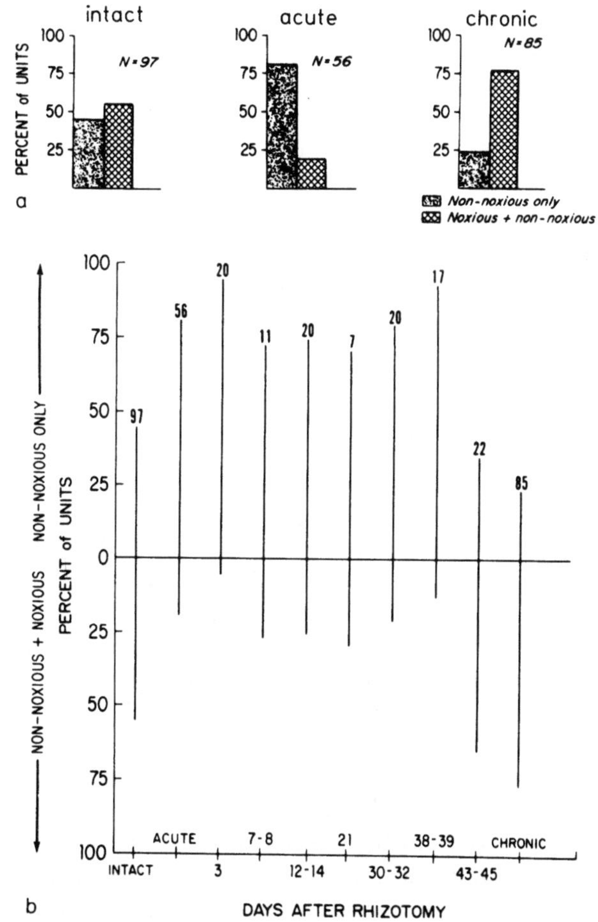

Fig. 4.15 a, b. Increased proportion of cells responsive to noxious stimulation in segments of dorsal horn chronically deafferented by dorsal root section. **a** Proportion of SCT cells responding to non-noxious input only and to noxious + non-noxious input in intact and acutely and chronically rhizotomized animals, respectively. Note change in proportion of SCT cells responding to noxious inputs in the chronic animals, compared with the acutely rhizotomized animals. *N*, number of SCT cells. **b** Time-course of alteration of response modalities of SCT cells after partial dorsal rhizotomy. At all intervals up to 39 days, most cells responded to non-noxious input only. After this time, most SCT cells also responded to noxious input. The number of SCT cells sampled during each time interval is shown above each line. (Modified from Sedivec et al. 1983, pp. 1513 and 1516)

of SCT cells to pressure, pinch, noxious heat, and cold are probably mediated by A∂ and C fibres, whereas responses to hair movement are mainly due to activation of Aß fibres (reviewed by Brown 1981). Following acute section of dorsal roots there might be a tendency for inputs to the affected segment from the small fibres to be selectively blocked, owing to their smaller rostrocaudal distribution in comparison with that of Aß fibres. There is some support for this suggestion from studies of the projections of single small-diameter (Light and Perl 1979; Rethelyi et al. 1982; Sugiura et al. 1986) and large-diameter (reviewed by Fyffe 1984) afferent fibres. However, most small-fibre input to lamina IV neurons is probably relayed through the substantia gelatinosa (Fitzgerald and Wall 1980; Light and Kavookjian 1987), and this explanation therefore requires support from studies of the effect of dorsal rhizotomy on the receptive fields of interneurons in the superficial layers of the dorsal horn. Owing to its late onset, it is conceivable that the increase of responsiveness to small-fibre inputs might be the result of sprouting (Sect. 4.5.5). The selectively enhanced response to small-fibre inputs might be due to these afferents having a greater ability than Aß fibres to undergo sprouting centrally.

4.4.2 Chronic Spinal Lesions

The evidence for sprouting of primary afferent projections to the lumbosacral dorsal horn in cats with chronic spinal hemisection in a more rostral segment or with other chronic lesions of descending projections to the dorsal horn (Murray and Goldberger 1974) (Sect. 4.5.5) led Brenowitz and Pubols (1981) to search for functional signs of the increased primary afferent projection. In cats with acute (5–7 h), subacute (2–5 days) or chronic (88–174 days) spinal hemisection at T13 that spares the dorsal columns, the low-threshold tactile receptive fields of dorsal horn neurons in the L7 segment were mapped under methoxyflurane anaesthesia. Following acute or subacute hemisection, there was no obvious change of somatotopic organization or receptive field size of dorsal horn neurons, though, as shown by their increased spontaneous activity, there was clearly an increase in their excitability (Sect. 4.5.1). In the chronically hemisected animals, although the general somatotopic organization was not disrupted, there was a statistically significant increase in the size of proximal receptive fields of cells in the far lateral part of the dorsal horn on the side ipsilateral to the lesion. This expansion of receptive fields occurred in the absence of any further increase in spontaneous activity, which was in fact lower than that in the acutely hemisected animals. However, it is surprising that even the largest receptive fields found by Brenowitz and Pubols (1981) were no larger than those described in intact cats by Devor and Wall (1976b).

4.4.3 Peripheral Nerve Transection or Crush

The question of whether somatotopic maps in the adult mammalian dorsal horn change as a result of peripheral nerve section is, if anything, more controversial than the effects of dorsal rhizotomy or cord hemisection discussed above.

Peripheral Nerve Transection. The first reported investigation of the somatotopic organization of the dorsal horn following peripheral nerve transection in adult mammals appeared in a short paper by Devor and Wall (1978); this was followed by a more detailed description of their experiments (Devor and Wall 1981a). In these studies, the distal part of the hindlimb, below the knee, was denervated in adult cats by transection and ligation of the sciatic and saphenous nerves at the level of the mid-thigh. As expected, this procedure produced anaesthesia of the entire foot, which persisted for at least 105 days, the longest period of behavioural assessment. Extracellular recordings were made from single neurons in the L6/L7 dorsal horn of unanaesthetized decerebrate preparations. In animals with intact peripheral nerves typical microelectrode tracks yielded an average of 5.5 well-isolated units having a receptive field somewhere on the hindlimb, lower back or perineum. With few exceptions, units having a receptive field which included skin proximal to the ankle were located in the lateral part of the dorsal horn. The few that were encountered more medially were always located in deeper laminae (depth > 1800 μm, corresponding to laminae VI–VII). In tracks made more than 200 μm medial of the dorsal root entry zone cells with a receptive field encompassing skin proximal to the ankle were rarely encountered at depths less than 1800 μm, and no units were encountered that had a receptive field wholly located proximal to the ankle. Immediately after ipsilateral nerve transection cells with receptive fields on the leg and thigh were still found in the dorsolateral part of the dorsal horn, but in tracks passing more than 200 μm medial of the root entry zone very few cells having a receptive field were found, and these were excited by pressure on the leg or thigh. Devor and Wall (1981a) concluded that in the intact cat, nearly all neurons in the medial part of the L6/L7 dorsal horn (and at depths less than 1800 μm) have receptive fields on the foot or toes, and that acute transection of the sciatic and saphenous nerves deprives almost all the cells in this region of their receptive fields. This conclusion is in complete accord with normal dorsal horn somatotopy, and with the known cutaneous distribution of the two nerves (Koerber and Brown 1982).

In animals surviving 28–105 days, the picture was very different from that in the acute cats. In tracks passing more than 200 μm medial of the root entry zone an average of 2.0 cells with clear receptive fields were isolated per track (compared with less than 0.05 cells per track in the acute animals), in addition to many more whose receptive fields could not be accurately identified (presumably the latter cells responded to tapping the limb). Of the cells with localized receptive fields, 25% responded to hair movement, 43% to light touch but not hair movement, and the remaining 32% only to firm pressure or pinch. Devor and Wall (1981a) stated that, in their size and location, the receptive fields of these cells resembled the large fields of cells in the lateral dorsal horn of intact cats (Devor and Wall 1976a). Devor and Wall (1981a) concluded that, while many cells in the region of the deprived toe and foot representation still lacked a peripheral receptive field, large numbers of them had acquired a new receptive field on proximal skin.

These experiments were later repeated in the L4/L5 dorsal horn of the adult rat (Devor and Wall 1981b) with essentially the same results, except that the time-course of the changes in the medial dorsal horn was found to be much shorter than it was in the cat. Thus, in the barbiturate-anaesthetized rat the reorganization of somatotopy was first detected 4–5 days after nerve transection, and became maximal after only 1 week. It is clearly stated by the authors that in rats surviving 6 days or more, 60% of the cells responded to gentle stimulation (hair movement or light touch), while the remainder responded only to stronger stimuli. It is not clear whether the 60% of cells responding to light stimuli also responded to more intense stimulation (i.e. had a wide-dynamic range), but there is no doubt that this study claims to have shown the appearance of low-threshold mechanoreceptive fields on proximal skin in the medial dorsal horn after chronic peripheral nerve transection.

The appearance of new proximal receptive fields among cells in the toe and foot representation of the medial dorsal horn was later reported both for the adult rat by Wall and coworkers (Fitzgerald et al. 1985) and independently in the adult cat (Lisney 1982, 1983). Fitzgerald et al. (1985) followed the protocol of Devor and Wall (1981b) in experiments designed to investigate a possible role for NGF in the reorganization of receptive fields as well as other central changes consequent to peripheral nerve injury in the rat (Sect. 4.5.4). In these experiments an average of only 6 cells with receptive fields per animal was encountered in 5 microelectrode tracks passing through the medial dorsal horn immediately after ipsilateral sciatic and saphenous nerve transection, while 10–12 days after nerve section, a mean of 24 cells with receptive fields was found in each animal. The receptive fields were all situated on the proximal part of the leg, the buttocks or the perineum. However, in contrast to the earlier report (Devor and Wall 1981b), which emphasized the normal sensitivity to gentle stimulation of a majority of these cells (see above), Fitzgerald et al. (1985) reported that while light touch and hair movement evoked weak responses, "clearer activity" was evoked by pinching the skin. Presumably this rather vague statement means that most of the new receptive fields had higher thresholds than normal, but no data are provided on this important point.

The experiments of Lisney (1983) were designed to follow any changes of dorsal horn somatotopy which might occur after nerve transection and regeneration. Lisney (1983) found that 1 month after simple sciatic and saphenous nerve transection *without ligation*, the somatotopic organization of the lumbosacral dorsal horn was changed in a manner similar to that described by Devor and Wall (1978, 1981a) after ligation of the cut nerves. It was concluded that the suppression of regeneration of the cut axons by ligation of the proximal stump was not a factor in producing the change of somatotopy. Lisney (1983) also pointed out the apparent lack of somatotopic organization in the reorganized region. Owing to regeneration of the transected nerves, neurons in the medial part of the dorsal horn in cats examined 9 months after nerve transection had regained receptive field on the toes and foot. However, these receptive fields were not like normal distal foot receptive fields, being larger, fragmented, and lacking somatotopic organization. These features could perhaps be predicted from the lack of somatotopic specificity and the multiple receptive fields of regenerated axons (Sect. 2.5). However, they also serve to show that, even within relatively small regions of the somatotopic map, the ability of the receptive fields of drosal horn neurons to change after peripheral nerve injury is insufficient to bring about a restoration of the normal somatotopic organization.

Various attempts in other laboratories to replicate the findings of Devor and Wall (1981a) and Lisney (1983) in the cat, have met with limited success. Brown et al. (1984) recorded from SCT cells and other neurons having an axon in the dorsolateral fasciculus, PSDC neurons, and other dorsal horn neurons that were activated orthodromically by dorsal column stimulation. To control for any possible long-term depressant effects of the repetitive cervical cord stimulus required for antidromic activation, in some animals unidentified dorsal horn

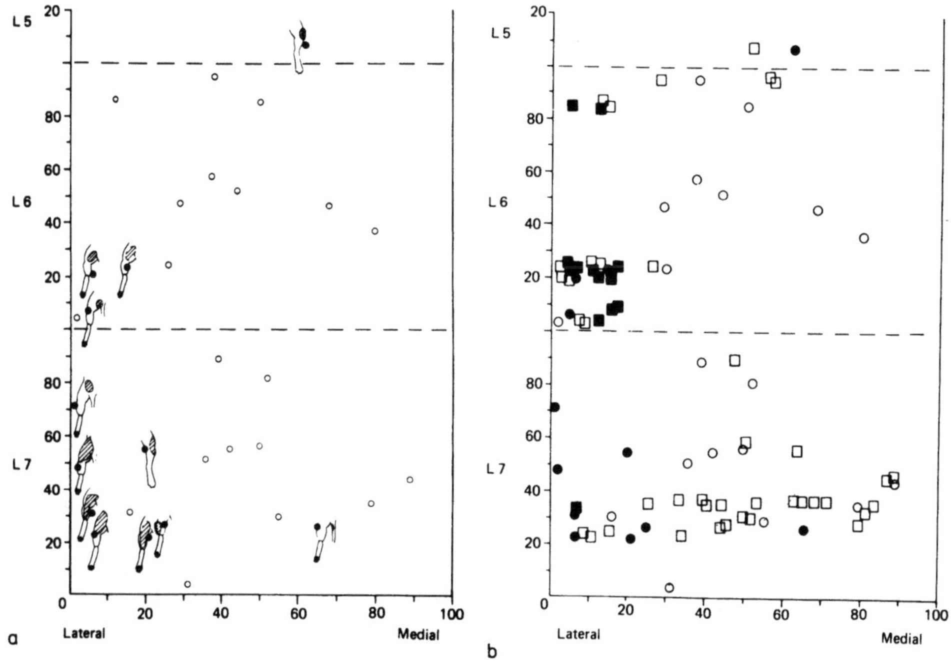

Fig. 4.16 a, b. Absence of somatotopically inappropriate receptive fields of SCT cells after chronic peripheral nerve section in the cat. **a** Plan view of L5–L7 dorsal horn showing locations and receptive fields of SCT cells intracellularly stained with HRP in dorsal horn ipsilateral to chronic transection of sciatic and saphenous nerves. **b** Same as **a** without drawings of leg, but also showing SCT cells which were recorded extracellularly. Segment lengths normalized for different animals. *Circles*, injected SCT cells; *squares*, extracellularly recorded SCT cells; *filled symbols*, cells with a receptive field; *open symbols*, cells lacking a receptive field. Chloralose anaesthesia. (From Brown et al. 1984, pp. 382 and 383)

neurons were isolated simply on the basis of their ongoing activity or response to tapping over the mechanically sensitive neuroma at the proximal stump of the cut nerve. The position of the cell body was determined for a number of each type of antidromically activated cell by intracellular staining with HRP. In cats which had survived chronic transection and ligation of the sciatic and saphenous nerves for periods of 19–55 days, the somatotopic organization of these neurons in the ipsilateral L6–L7 dorsal horn was examined and in all cases was found to be consistent simply with loss of afferent input to the region of foot and toe representation in the medial part of the dorsal horn (Fig. 4.16).

A similar conclusion was reached by Wilson (1987), who mapped the receptive fields of unidentified dorsal horn neurons throughout the L6 and L7 segments on both sides of the cord in chloralose-anaesthetized cats between 35 and 49 days after sciatic and saphenous nerve transection. In these experiments there was some indication that the number of cells responding to mechanical stimulation of unspecified, deep, sensitive mechanoreceptors (tapping the limb) had increased on the chronically deprived compared with the acutely deprived side of the cord. On both sides of the cord a few cells with proximal, *high-threshold*, cutaneous receptive fields were also found but these were situated deep to lamina V, and it could not be stated confidently that this kind of neuron was any more abundant after chronic than after acute nerve section. Dorsal horn neurons with very large, high-threshold receptive fields have been found in the deeper laminae of the dorsal horn in intact cats (Molinari, 1982), and the proximal high-threshold receptive fields could simply have represented truncated whole-limb receptive fields of these units. Wilson (1987) concluded that no unequivocal change of functional connectivity from low-threshold cutaneous mechanoreceptive primary afferents could be demonstrated across the dorsal horn after chronic peripheral nerve section. While it was felt that there might have been some reörganization of high-threshold receptive fields, and/or an increased number of cells responding to unspecified deep inputs, it was considered that this would require confirmation in a larger series of experiments.

In yet a further study, in cats anaesthetized with either chloralose or pentobarbital, Pubols (1984) examined the somatotopic organization of the dorsal horn in a region extending from the caudal third of L5 to the rostral third of L7 over a period of 1–98 days after sciatic and saphenous nerve transection. Statistical analysis of normalized measurements of the relative mediolateral position of microelectrode tracks indicated that postoperative survival time had no significant effect on the medial boundary of the representation of proximal hindlimb skin. While the most medial active track appeared to be significantly more medial under pentobarbital than under chloralose anaesthesia, this difference was minor compared with the magnitude of the shift of somatotopic boundaries described by Devor and Wall (1981a).

The possibility that there might be mediolateral shifts of the representation of high-threshold afferents has been borne out to some extent by a study of the behaviour of lamina I neurons following peripheral nerve section. Lamina I neurons receive input from cutaneous nociceptors and their receptive fields have sharp boundaries, permitting their somatotopic organization to be easily studied (Sect. 4.2.4) (Cervero et al. 1976). Hylden et al. (1987) made use of this fact and mapped the receptive fields of lamina I neurons throughout the L7 segment of cats after acute or chronic sciatic neurectomy. In these experiments recordings were performed under chloralose, supplemented with small doses of pentobarbital. All the cells responded to electrical stimulation of the sciatic nerve at latencies consistent with A∂ or A∂ plus C fibre input, and this criterion was used in isolating single units. The receptive fields of these cells formed a well-defined

Fig. 4.17 a–c. Reorganization of receptive fields of lamina I neurons of the cat after chronic ▶ peripheral nerve section. Receptive fields of lamina I neurons recorded from **a** control cats or from cats with **b** acute or **c** chronic (> 2 months) transection of the ipsilateral sciatic nerve. Each figure is a plan view of the L7 dorsal horn. Lateral boundary of dorsal horn is represented by the ordinate, medial boundary was 1.0–1.4 mm away, depending on rostrocaudal level. Position of figurines indicate position of unit. *Open circles* represent cells lacking a receptive field. Each transverse row of figurines represents the results obtained in a series of electrode tracks in a single animal, and data from several cats are combined. No attempt was made to correct for curvature of the lateral edge of the dorsal horn. (From Hylden et al. 1987, pp. 344 and 345)

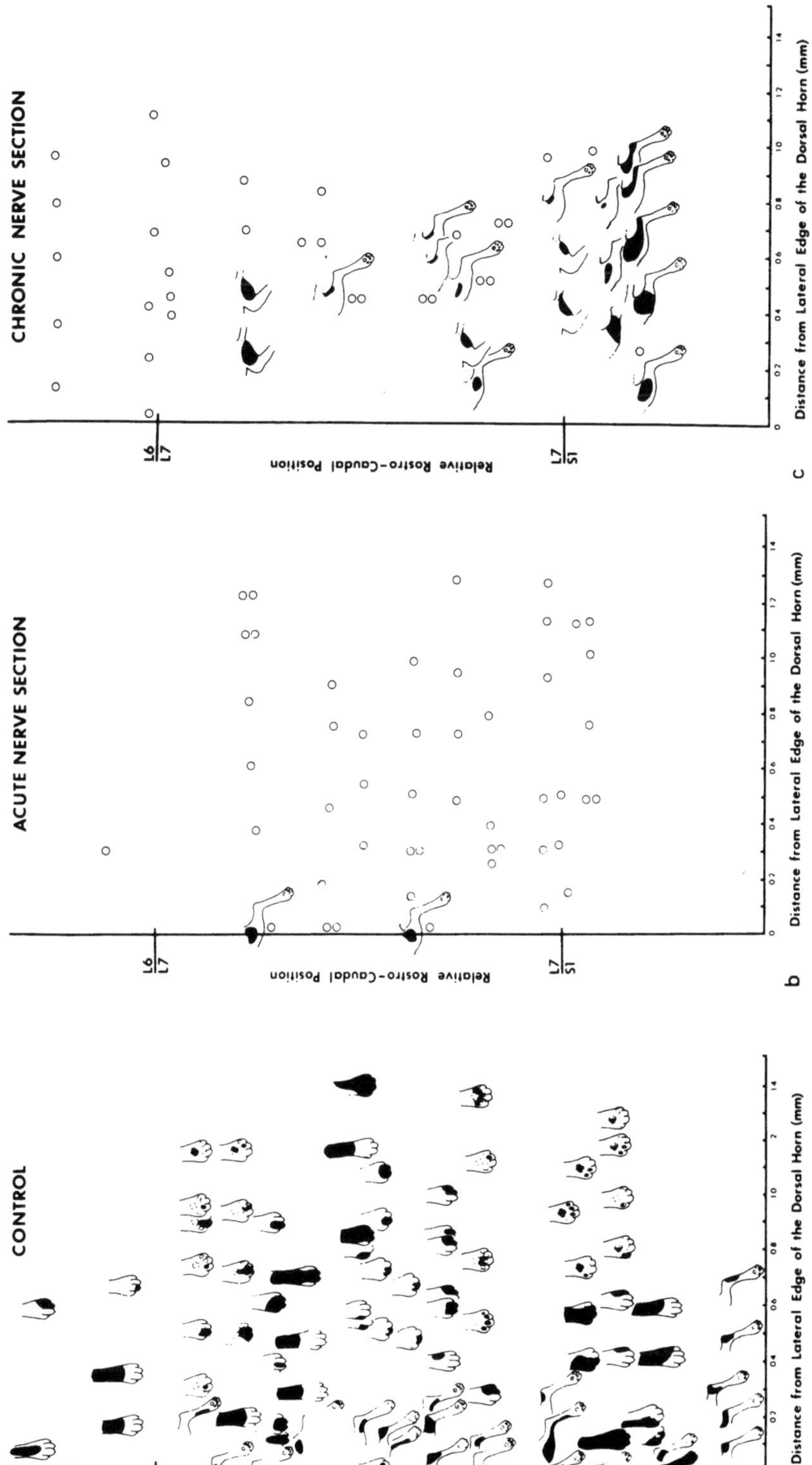
CONTROL
ACUTE NERVE SECTION
CHRONIC NERVE SECTION
L6
L7
L7
S1
Relative Rostro-Caudal Position
Distance from Lateral Edge of the Dorsal Horn (mm)
a
b
c

somatotopic map (Fig. 4.17a) which resembled the maps of the low-threshold receptive fields of dorsal horn neurons in laminae IV and V (Sect. 4.3.1).

In intact cats and cats after acute transection of the sciatic nerve, very few lamina I cells having receptive fields proximal to the knee were found, and, as expected these cells were all at the extreme lateral edge of the dorsal horn. After acute sciatic nerve transection all lamina I cells located more medially than this were found to lack a receptive field (Fig. 4.17b). However, when the dorsal horn was mapped in other animals 20–82 days after nerve section, cells with proximal receptive fields were found up to 1 mm more medially than they were in the acutely neurectomized animals (Fig. 4.17c). About 70% of the lamina I cells isolated still lacked any receptive field, but the remainder had a similar range of responsiveness to low- and high-threshold afferents to that found in intact cats. However, the inappropriate receptive fields lacked the well-defined edges of normal lamina I receptive fields, and most cells gave highly variably responses to repeated mechanical stimuli.

In these experiments the medial shift of the border of proximal limb representation was mainly confined to the caudal part of the L7 segment and rostral part of S1 (Fig. 4.17c). No shift was found in the rostral part of L7 and in the caudal part of the L6 segment. Moreover, the authors conceded that the reorganization they found was not a reliable phenomenon. In about one third of their chronically denervated animals no lamina I cells were found to have inappropriate receptive fields, whereas in other animals almost 50% of cells did so. The authors suggested that the inter-animal variability in the response to nerve section might help to explain the different results obtained in the various studies of reorganization of low-threshold receptive fields in the cat. However, there has never been any suggestion of such inter-animal variability in published investigations of the reorganization of lamina IV and V neurons in either the cat or the rat.

Two independent groups have described a different kind of somatotopic reorganization of dorsal horn neurons after chronic peripheral nerve section. While these studies differ in detail, they share the common feature that the new receptive fields acquired by cells which have lost their major input from the skin are on areas of skin represented rostral or caudal, rather than lateral or medial, of the deprived region. Markus et al. (1984) attempted to extend the findings of Devor and Wall (1981b) in the rat dorsal horn by examining the effect on dorsal horn somatotopy of transecting the sciatic nerve only, rather than both the sciatic and saphenous nerves. It was argued that this procedure should reveal competitive interaction between afferents in nerves innervating proximal skin, and those in the saphenous nerve, for providing new excitatory inputs to the neurons in the deprived sciatic nerve representation. In these experiments extracellular recordings were made from dorsal horn neurons in the medial part of the L3–S1 segments in rats anaesthetized with chloralose. The sciatic nerve had been transected and ligated 21 days earlier on one side, and on the other side the nerve was cut acutely. Spontaneous activity was generally lacking in these preparations, and presumably units were selected that responded to gross peripheral stimulation, although this is not stated explicitly. Acute transection of the nerve produced a region about 2 mm long in the medial part of the ipsilateral L4 and L5 segments in which neurons lacked a cutaneous receptive field. However, many cells in the corresponding region on the chronically denervated side had cutaneous receptive fields located in the peripheral territory of the saphenous nerve. On the acutely denervated side cells with receptive fields in the saphenous nerve territory were confined to a more rostral region, in the medial part of the L3 segment (Fig. 4.18a). An important point was that the receptive fields of cells in the reorganized sciatic nerve region were similar to those normally found in the saphenous nerve representation. For example, some neurons responded to hair movement or touch, while others responded to hair movement, touch and firm pressure. The receptive fields of these neurons were situated on the medial three digits and medial distal part of the foot, and in no case extended further proximally on the limb than those of cells in the control saphenous nerve representation (Fig. 4.18b). Thus over the 21-day postoperative period there had occurred a caudal spread of neuronal responsiveness to natural stimulation of the saphenous nerve territory of the order of 3000 μm into the deprived region of sciatic nerve representation. Remarkably, at the same time there was no observable spread of influence from afferents from the high, which terminate less than 500 μm away in the lateral part of the dorsal horn.

A second interesting feature of this study is that there was no apparent rostral expansion of the representation of receptive fields normally found caudal to the sciatic nerve representation. Cells in the medial part of the dorsal horn caudal to the sciatic representation normally have receptive fields on the thigh, perineum and tail, but there was apparently no tendency for cells in the caudal half of the deprived sciatic nerve representation to acquire such receptive fields. Instead, receptive fields in the saphenous nerve field were also found in this region (Fig. 4.18). Thus there appeared to be a marked directional asymmetry of whatever mechanism was responsible for the reorganization.

An important point to emerge from this work was that the caudal extension of responsiveness to natural stimulation of the saphenous nerve territory was not accompanied by any change in the extent over which cells responded to electrical stimulation of the saphenous nerve. Instead, it was found that the caudal boundary of the latter region extended an equal distance into the S1 segment on both sides of the cord. Markus et al. (1984) therefore considered that the somatotopic reorganization they demonstrated might be due to a strengthening of pre-existing projections, rather than sprouting. However, they were not prepared, on the available evidence, to argue that the responses to electrical stimulation in somatotopically inappropriate regions might be monosynaptically evoked.

On the basis that hair follicle afferent fibres have somatotopically inappropriate projections to regions of the dorsal horn rostral and caudal of the somatotopically appropriate region (Sect. 4.3.3) (Meyers and Snow 1984; 1986; Meyers et al. 1984; Snow and Meyers 1985), Wilson et al. (1987) postulated that nerve section might produce a rostrocaudal reorganization of somatotopy. In designing an experiments to test this theory they made use of information obtained from a detailed study of the somatotopic organization of unidentified dorsal horn neurons and SCT cells in the cat (Wilson et al. 1986). This study showed that the skin of individual toes is represented over a 3–4 mm long and 800 μm wide strip of dorsal horn (Fig. 4.5). Within this strip SCT cells and most other, unidentified, dorsal horn neurons have tactile

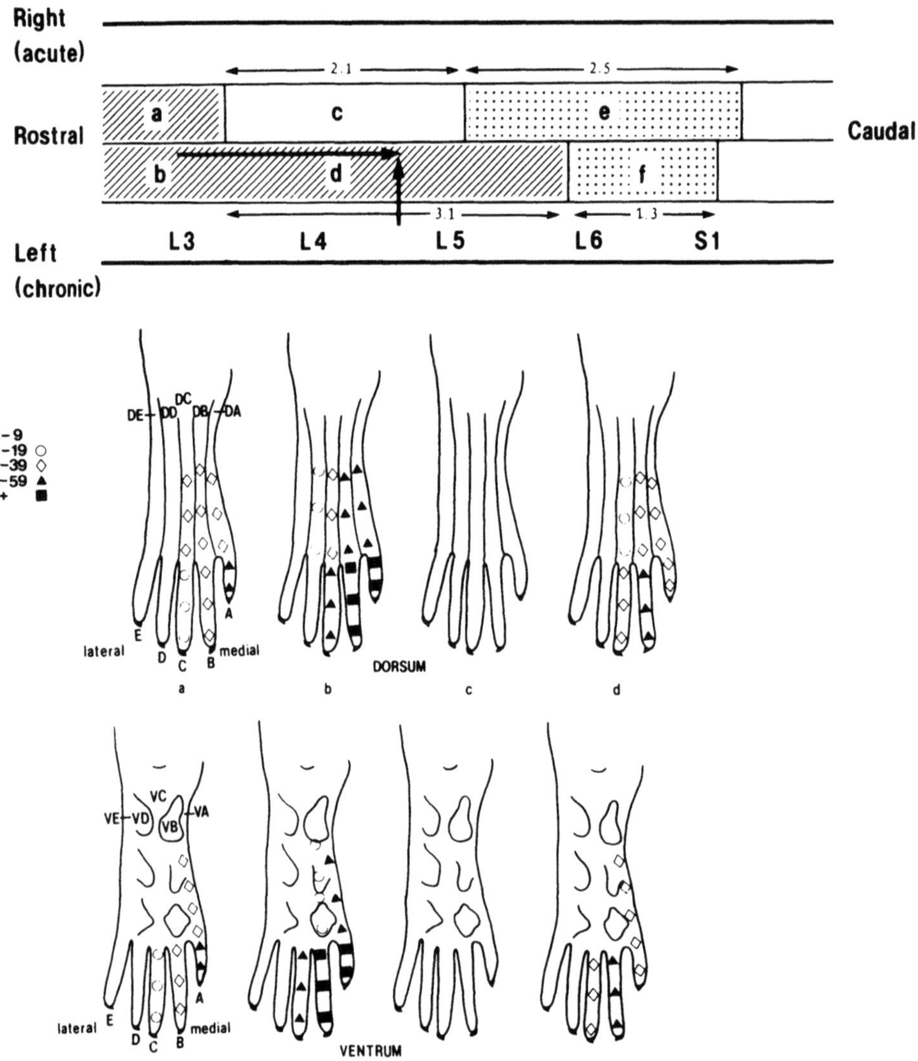

Fig. 4.18 a, b. Rostrocaudally directed reorganization of dorsal horn somatotopy in the rat. A dorsal view of dorsal horn showing somatotopic projection map of the saphenous nerve in the medial half of the dorsal horn obtained with natural stimulation of the hindlimb. *Thick horizontal lines* represent the lateral boundaries of the cord. The right side of the cord shows the map obtained after acute transection of the sciatic nerve, and the left side shows the map obtained after chronic transection. *L3–S1* are cord segments; *2.1*, *2.5*, *3.1* and *1.3* are micromanipulator measurements in millimeters. *Hatching* in zones *a*, *b* and *d* denote presence of neurons with saphenous mediated natural receptive fields on foot and toes. *Dots* in zones *e* and *f* denote natural receptive fields in the thigh, perineum and tail. Zone *c* (sciatic projection) contained no units with natural receptive fields. Note the caudal spread of about 3 mm of neurons responding to skin of the saphenous nerve territory on the *left* (side of chronic sciatic nerve section). **b** Distribution of receptive fields of dorsal horn neurons in zones *a–d* in **a**. Zones *a* and *c* are in the right side of the cord after acute sciatic neurectomy. Zones *b* and *d* are in the left side of the cord after chronic sciatic neurectomy.

receptive fields that are confined to a single toe and the boundaries between the representation of single toes are therefore well defined. Given this organization, Wilson and Snow (1987a) examined the effects of chronically and acutely denervating a single toe on neurons within the representation of that toe by transection and ligation of the digital nerves supplying the third digit of the hindlimb. Between 30 and 70 days later the corresponding nerves on the opposite side were cut acutely and, under chloralose anaesthesia, the receptive fields of SCT cells were mapped in the region of toe 3 representation on both sides. On the acutely denervated side it was found that a patch of SCT cells in the toe 3 region lacked receptive fields. Surrounding this patch were SCT cells whose receptive fields bordered the denervated skin but appeared to be truncated. Many of these neurons showed no obvious change in responsiveness or sensitivity to hair movement in the remaining part of the receptive field. However, a few SCT cells in this surrounding region gave sluggish, habituating responses to hair movement in very small regions adjacent to the denervated skin, and it was assumed that the latter cells had lost all but a very weak excitatory input from their original receptive field as a result of the nerve transections. It was concluded that there is no reorganization or unmasking of new receptive fields immediately after (i.e. within a few hours of) nerve section (Wilson and Snow 1987a, 1990b; Snow and Wilson 1989).

Owing to the bilateral symmetry of the somatotopic map, the rostral and caudal boundaries of the region where cells lacked a receptive field could be transposed to the chronically denervated side to establish the boundaries of the original toe 3 representation. When recordings were made from single SCT cells within this region, it was found that the vast majority gave brisk responses to light tactile stimulation of the skin around toe 3. In most cases, this receptive field was situated on the nearest innervated skin that normally would have been represented either rostral or caudal of the SCT cell although in a few instances receptive fields were situated on the proximal and still innervated part of toe 3. Particularly marked was an invasion of the rostral half of the former toe 3 region by receptive fields on the skin of the medially adjacent toe, toe 2 (Fig. 4.19). The time-course of this reorganization varied markedly in different animals, but in most cases the reorganization could be reliably observed after about 6–8 weeks (Wilson and Snow 1987b). It was concluded that the chronic peripheral nerve section leads to a spatial reorganization of the receptive fields of SCT cells which receive their major input from the transected nerve, such that they acquire new receptive fields on the nearest innervated skin, but that the reorganization shows a marked tendency to occur in a rostrocaudal direction, rather than across the mediolateral axis of the dorsal horn. Similar, though more extensive, rostrocaudally directed reorganization has been described following denervation of one or two toes in neonatal cats (Snow and Wilson 1989; Wilson and Snow 1988b) (see Sects. 4.6.3 and 4.6.4).

◀ **Fig. 4.18 a, b** (*continued*). The feet are divided into proximodistal strips, and the *symbols* denote the percentage of neurons which responded to the skin of the particular toe or strip of the foot on which they are sited. Dorsal views of the foot are at the *top*, and ventral views at the *bottom*. Note that the dorsal views of the foot for zones *b* and *d* are drawn as right legs to simplify comparison, but were in fact left legs. (From Markus et al. 1984, pp. 31 and 32)

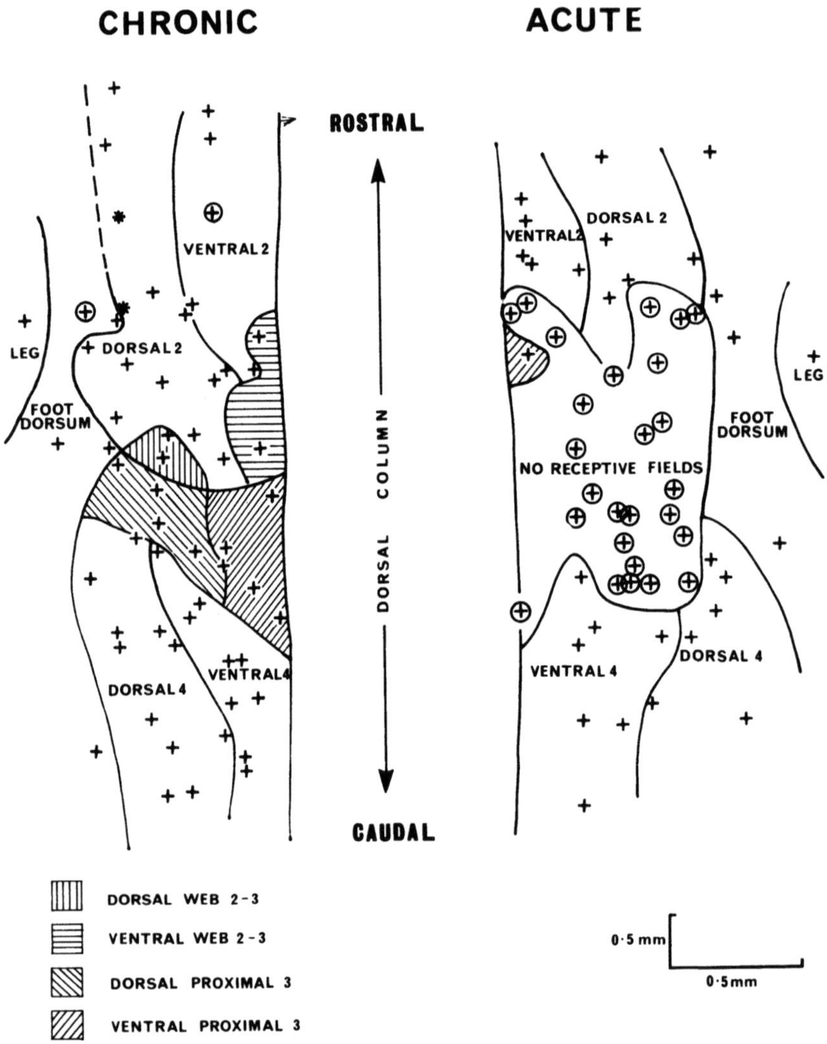

Fig. 4.19. Reorganization of somatotopy in the dorsal horn of the cat after chronic denervation of a single toe. Horizontal view of spinal cord of the L6 and L7 segments showing contrasting effects of chronic and acute transection of digital nerves of toe 3 on the light tactile receptive fields of SCT cells in the region of toe 3 representation. Combined results from 3 similar experiments. The *crosses* represent electrode tracks in which the receptive field of a SCT cell was mapped. *Circled crosses* are SCT cells which failed to respond to light tactile stimulation anywhere on the hindlimb. The numbers *2–4* refer to the corresponding digits, and the *web* refers to the skin between adjacent digits. *Dorsal, ventral* and *proximal* refer to corresponding regions of digital skin. Note the patch of dorsal horn ipsilateral to the acutely denervated toe in which SCT cells lacked a receptive field. In the corresponding region on the chronically denervated side SCT cells respond to gentle stimulation of regions of nearby skin, particularly that of toe 2. Chloralose anaesthesia. (From Wilson and Snow 1987a, p. 811)

Peripheral Nerve Crush. In the course of the experiments of Devor and Wall (1981b) in the rat, the intriguing observation was made that proximal receptive fields did not appear among cells in the medial dorsal horn if the sciatic nerve was simply crushed, rather than transected and ligated (the saphenous nerve was cut and ligated in both cases). Thus following nerve crush, the medial dorsal horn remained in a similar state to that seen immediately after nerve transection, with a very low incidence of cells having receptive fields. This was despite the fact that the time for regeneration of the crushed nerve was much longer than the week or so that was claimed to be necessary for somatotopic reorganization after nerve section. In rats examined 48 and 54 days after crushing the nerve, after sufficient time had elapsed for regeneration, tracks in the medial dorsal horn again contained many cells with receptive fields on the toes and foot.

Overview of Findings. There would appear to be no simple explanation for the different results obtained in all these studies of somatotopic organization of dorsal horn neurons following peripheral nerve lesions (for detailed critiques of the technical aspects of these studies see Brown et al. (1984) and Pubols (1984)). If the newly developed receptive fields are due to activation of relatively weak afferent inputs, one would anticipate that they might be heavily influenced by the excitability of dorsal horn neurons, and hence by the type of acute preparation used to demonstrate them. Devor and co-workers have argued that the crucial difference between their positive and other workers' negative evidence for mediolateral reorganization in the cat lies in their exclusive use of decerebrate low-spinal animals in which, owing to the absence of anaesthetic drugs and tonic descending inhibition, the excitability of dorsal horn neurons is maximal (Sect. 4.5.1). This explanation is unsatisfactory for several reasons: (1) Lisney (1982, 1983) claimed to replicate the results of Devor and Wall (1981a) in chloralose-anaesthetized cats which were not spinalized; (2) a similar pattern of mediolateral reorganization was found in the dorsal horn of the rat under barbiturate anaesthesia (Devor and Wall 1981b; Fitzgerald et al. 1985); (3) rostrocaudal reorganization has been demonstrated after chronic nerve section in both rats and cats under anaesthesia (Sect. 4.4.3). In relation to point (3), the reorganization described by Basbaum and Wall (1976), by Mendell et al. (1978) and by Sedivec et al. (1983) in the partially deafferented dorsal horn in anaesthetized preparations (Sect. 4.4.1) following chronic dorsal rhizotomy might reasonably be interpreted as occurring in a rostrocaudal direction, although this point was not made explicitly by these authors. In all these experiments, responses appeared in the partially deafferented region of dorsal horn to stimulation of skin normally represented either caudal or rostral to the dorsal rhizotomized segment.

Difficulties in the estimation of depth or mediolateral position of recording loci may also contribute to an erroneous impression of changed somatotopy in some of these experiments (for discussion see Wilson 1987; Pubols 1984), and it is perhaps significant that reorganization was not found in the experiments of Brown et al. (1983b, 1984), where the position of cells were most accurately determined by intracellular staining. However, Devor and Wall (1986) have vigorously defended the accuracy of their anatomical methods for determination of recording loci in both cats and rats. Different interpretations of somatotopic reorganization may also derive from the use of different intensities of mechanical stimulation in mapping receptive fields. It is obviously essential to be consistent and clear about the kind and intensity of stimuli used to map cutaneous receptive fields if the results obtained by different investigators are to be compared, and it is even more important to avoid the unintentional activation of deep receptors, or worse still, of the mechanically sensitive nerve endings in the neuroma of a transected nerve (Sect. 2.2.1).

A final possible source of error in such experiments, which may contribute to an artefactual reorganization of somatotopy, is the use of low impedance microelectrodes that certainly permit the activity of cells distant (e.g. > 200 μm) from the electrode to be monitored. Unless recordings are made only from well-isolated units very close to the electrode tip there will be an inevitable tendency for the borders of a denervated region of the somatotopic map to appear to shift in the direction of the region lacking input. This effect should, of course, apply equally to acutely denervated controls and chronically denervated animals (Pubols 1984).

Owing to the difficulty experienced by workers attempting to confirm some of the findings described above, there is now a tendency for investigators of plasticity in the dorsal horn of adult mammals to develop their own models. This is unfortunate, because it means that the controversy surrounding some of these experiments will probably never be settled. Though attempts have been made to find an anatomical basis for the hypothetical weak, or ineffective, synaptic connections between cutaneous primary afferent fibres innervating proximal skin and medial dorsal horn neurons (Devor et al. 1986), other groups interested in spinal somatotopic plasticity are pursuing models based on reorganization in the rostrocaudal direction, for which the anatomical basis, in terms of somatotopically inappropriate, primary afferent collateral arborizations are already well established (Sect. 4.3.3).

4.5 Mechanisms Underlying the Somatotopic Reorganization of Dorsal Horn Neurons

Owing to the controversial nature of the lesion-induced alterations of somatotopic organization of dorsal horn neurons (Sects. 4.4.1 and 4.4.3) it is difficult to propose even a multifaceted model to explain all the positive results. This problem is compounded by the two distinct kinds of somatotopic reorganization that have been reported in the dorsal horn, one occurring predominantly in a mediolateral direction, while the other is rostrocaudally directed. Despite these differences, suggestions of the likely mechanisms of reorganization may be reduced to only two major options, which are universally applicable throughout the somatosensory system. The new, somatotopically inappropriate receptive fields of cells in chronically deprived regions of dorsal horn must result either from a strengthening of pre-existing, but weak or ineffective, excitatory inputs to the cells, or from the anatomical formation of new connections by collateral sprouting and synaptogenesis, whether by primary afferent fibres, interneuronal or propriospinal axonal projections of dorsal horn neurons, or other systems, or by morphological changes in the dendritic trees of postsynaptic neurons. If ineffective inputs exist, they could either be the kind that normally form a subliminal fringe, or they could belong to a special class of potential inputs that normally do not influence the neuron in any detectable way. Many studies have been performed to demonstrate the existence and potential importance of one or the other of these alternatives in relation to somatotopic or other forms of plasticity, but the result is a morass of conflicting observations. In this review we will present the more important observations relevant to somatotopic reorganization, and attempt to assess their likely contribution to an overall understanding of the changes of somatotopy that follow chronic dorsal rhizotomy or peripheral nerve lesions. For an earlier review of plasticity in a range of spinal systems, see Mendell (1984).

As previously noted by Devor (1984b, 1987), the theoretically possible ways that the efficacy of existing weak inputs may be strengthened are legion, and include changes in the geometry or biophysical properties of the postsynaptic neuron, changes in the presynaptic component of the synapse and modulation of inhibitory inputs to the cell. Obviously, the receptive field of a neuron in a polysynaptic pathway may also be changed in the absence of any overt pre- or postsynaptic changes to that particular neuron. Such considerations have widespread applicability to neural plasticity. In terms of plasticity of the dorsal horn it is necessary to enquire not only whether detailed changes such as these do in fact occur after lesions (and at the present time there is little or no available evidence on any of these points), but whether the basic neural circuitry or phenomenology on which such hypotheses are based actually exist in the dorsal horn. First, if, as originally suggested by Merrill and Wall (1978), alterations of dorsal horn somatotopy are dependent upon the strengthening of normally ineffective afferent connections, we need to enquire what direct evidence there is for the existence of such connections in the dorsal horn systems in which somatotopic reorganization has been demonstrated, what is the nature of these connections and how does their organization relate to the observed patterns of somatotopic alterations? Second, if the strengthening of normally ineffective connections is responsible for changes of somatotopy after lesions, then how might this be brought about? Finally, what is the evidence that collateral sprouting and formation of new synaptic connections is involved in somatotopic reorganization, and how might this occur?

4.5.1 Physiological and Pharmacological Evidence for the Existence of Normally Ineffective Afferent Connections

Many physiological studies have shown that the receptive fields of dorsal horn neurons are not necessarily fixed, but can vary in size or even location in the absence of any *irreversible* changes of input, such as those caused by lesions. It is therefore clear that, under a given set of conditions, all possible excitatory inputs to a neuron are not necessarily demonstrable by natural stimulation of the skin. The most widely adopted method of demonstrating excitatory inputs to dorsal horn neurons from afferents supplying skin distant from the receptive field has been the use of electrical stimulation of afferents.

Electrical Stimulation of Afferents. In decerebrate, spinal or Dial/urethane-anaesthetized cats Merrill and Wall (1972) recorded extracellularly from the axons of a type of lamina IV neuron which, judging by their properties, were most probably type I SCT neurons (Brown and Franz 1969). By serially cutting dorsal rootlets it was shown that the receptive field of these cells usually depended on the integrity of afferents running in a single dorsal rootlet (Wall 1960). However, electrical stimulation of adjacent dorsal rootlets over a length of two segments could also evoke activity in the cell. It was concluded that there were afferents, capable of firing the cell monosynaptically, that could be transected without affecting the size or sensitivity of the receptive field. In a second experiment, two adjacent roots, stimulation of either of which fired a cell at a latency considered to be compatible with monosynaptic excitation, were alternately reversibly blocked by anodal current. When one root was blocked the cell's receptive field disappeared, but blocking the other root had no discernible effect on the receptive field. This was taken to mean that there were Aß afferents that were monosynaptically connected to the cell but which were ineffective in making the cell fire when activated only by natural stimulation of the skin. Merrill and Wall (1972) suggested that primary afferents might make two distinct kinds of monosynaptic connections. The first kind is highly effective in exciting the cell during natural stimulation. The second kind has no obvious effect on the cell's firing when activated by natural stimulation, but when activated in sychrony with many other similar inputs (as, for example, by electrical stimulation) can cause the cell to fire impulses.

It is difficult to interpret the findings of Merrill and Wall (1972) in terms of current knowledge of the functional connections underlying the organization of the receptive fields of SCT cells. For example, Brown and Noble (1982) showed, for a sample of intracellularly stained pairs of Aß hair follicle afferent fibres and SCT cells, that there were contacts between them if and only if the afferent fibre innervated skin within the firing zone of the cell (the region of skin from which spike discharges could be evoked by natural stimulation, see Fig. 4.20). If the ineffective monosynaptic connections to individual SCT cells proposed by Merrill and Wall (1972) are so few in number that they may not be demonstrated by such high resolution anatomical techniques it is difficult to imagine how the synchronization of afferent discharge produced by electrical stimulation could produce a strong monosynaptic excitatory influence on the cell. The most likely explanation of the apparently ineffective connections demonstrated by Merrill and Wall (1972) is that these inputs were not monosynaptic but rather were disynaptic in nature. It is known that SCT cells receive powerful di- or polysynaptic excitatory inputs from Aß cutaneous afferents (Hongo and Koike 1975; Brown et al. 1987). Like the monosynaptic inputs, these polysynaptic inputs normally originate mainly from within the firing zone of the receptive field (Hongo and Koike 1975; Brown et al. 1987), but it is possible that electrical stimulation of many afferents might cause convergence from over a wider area. The difficulty of demonstrating unequivocally monosynaptic connections in the vertebrate CNS (Berry and Pentreath 1976), particularly in the absence of intracellular recordings, due to the wide variation in rise-time of the EPSP, and uncertainties about the time of arrival of afferent impulses in individual fibres, may have been underestimated by Merrill and Wall (1972).

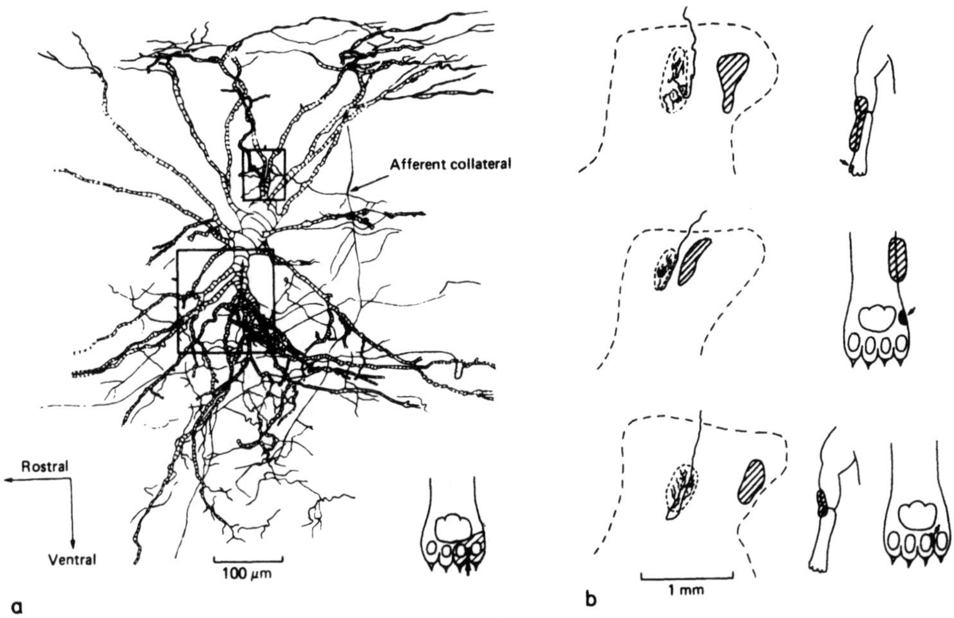

Fig. 4.20 a, b. Relation between receptive field location and connectivity of pairs of HRP-filled Aß hair follicle afferents and SCT cells in the cat lumbosacral dorsal horn. **a** Reconstruction from serial parasagittal sections of a SCT cell and part of a hair follicle afferent fibre collateral from which it received synaptic contacts. Most of the contacts were contained within the two rectangles. The receptive field of the afferent (*arrowed spot*) was in the centre of that of the SCT cell (*hatched*). **b** Relative mediolateral positions (*left*) of hair follicle afferent terminal arborizations and SCT cell dendritic trees (*hatched*) for 3 stained pairs where there was no overlap between receptive fields (shown on *right*). For each pair the afferent arborization is medial to the SCT cell and its receptive field (*arrowed*) is distal to that of the cell (*hatched*). (From Brown and Noble 1982. pp. 80 and 82)

Other studies of the responses of dorsal horn neurons to electrical stimulation of afferents are of direct relevance to plasticity, for they provide information about the distribution and proportions of dorsal horn cells responding to distant inputs. If new receptive fields appearing in regions of the dorsal horn partially deafferented by lesions result from a strengthening of ineffective inputs, then one would expect to find a correspondence between the distribution of such inputs and the reorganization of somatotopy. In decerebrate spinal cats, Devor et al. (1977) examined the distribution of dorsal horn neurons responding to electrical stimulation of afferents in a distant dorsal root. The stimulated afferents usually comprised a fraction of L3 or L4 dorsal root, and responsive cells were sought in the dorsal horn of the L7 and S1 segments. Typically, these cells were found in a rostrocaudally oriented band occupying laminae I–V in the lateral region of the dorsal horn. Many of the responsive cells had a cutaneous receptive field on the hindlimb, and, as expected from the lateral position of the cells, their receptive fields were usually situated proximally on the leg, in some cases including skin of the L4 or L3 dermatomes. However, the receptive fields of the great majority were clearly outside the L3 and L4 dermatomes, and many cells had small receptive fields on the foot or toes. In four cats in which a S1 dorsal rootlet was stimulated and the L4 dorsal horn was searched, responsive neurons were again found in the lateral half of the dorsal horn. The natural receptive fields of these cells were on the lower abdomen, flank or back, and none extended to the S1 dermatome. The distribution of responsive cells in these experiments appears to exceed the anatomically determined distribution of L3 or L4 dorsal root afferents, but S1 afferents have been shown to give rise to collaterals as far rostral as the L4 segment (Sect. 4.3.3). Measurements of response latency were made, but again were of little use in determining whether the responses to electrical stimulation were unequivocally monosynaptically evoked. In most cases they probably were not, for they were abolished by a moderate dose of barbiturate, and failed to follow repetitive stimulation at frequencies above 10–25 Hz. Thus these experiments may have produced the unsurprising result that some cells with somatotopically appropriate receptive fields were activated monosynaptically by the electrical stimulus, and the remainder responded polysynaptically. However, it is impossible to be sure of this because the relationship between cell position and dorsal root projections cannot be established in experiments of this kind.

Independent studies in both the cat and the rat have also revealed dorsal horn neurons responding to electrical stimulation of afferents which do not normally contribute to the cell's receptive field. In chloralose- or pentobarbital-anaesthetized cats electrical stimulation of the sural nerve was found to evoke responses from some dorsal horn neurons in the L6 and L7 segments having natural receptive fields outside the cutaneous territory of the sural nerve (Pubols et al. 1986). Most of these cells had characteristically long, variable latencies to sural nerve stimulation, whereas cells responding with a shorter, more stable (and therefore possibly monosynaptic) latency usually had a sural receptive field. Cells lacking a response to sural nerve stimulation generally did not have a natural receptive field in the sural nerve territory. Surprisingly, 10% of cells with a natural receptive field in sural nerve territory were not driven by sural nerve stimulation, a clear example of natural stimulation actually being *more* effective than electrical stimulation in demonstrating excitatory inputs. A second curious feature of this study was that none of a sample of 31 identified PSDC cells showed any sign of having ineffective inputs from the sural nerve, and conversely, no PSDC neurons which failed to respond to sural nerve stimulation were found to have a sural receptive field. The fact that relatively ineffective inputs could be demostrated for unidentified dorsal horn neurons of laminae III–V, but not for identified PSDC neurons, is significant in two respects. First, it indicates that different classes of dorsal horn neuron may have differing proportions of normally ineffective afferent input from distant skin, and second, it shows that the presence of an extensive excitatory subliminal fringe, which PSDC cells possess (Brown and Fyffe 1981), does not necessarily correlate with activation by electrical stimulation of afferents that innervate skin well outside the natural receptive field. It is therefore not clear what kind of excitatory inputs are being demonstrated by electrical stimulation of nerves.

In the chloralose-anaesthetized rat, Markus and Pomeranz (1987) found that, as expected, electrical stimulation of the saphenous nerve activated dorsal horn neurons in the medial part of the L2 and L3 segments, where cells have natural receptive fields in the peripheral territory of the saphenous nerve (medial edge of the foot). However, responses were also elicited from cells in the medial part of L4–L6, where cells normally respond to natural stimulation of the cutaneous territory of the sciatic nerve (remainder of the foot). At a superficial level of analysis (latency and number of spikes), there was no difference between the electrically evoked responses of cells in L3 and those in L5. This finding is of interest for two reasons. First, the caudal distribution of relatively ineffective projections from the saphenous nerve matches the region in which new receptive fields in cutaneous saphenous nerve territory appear following chronic sciatic neurectomy (Markus et al. 1984) (Sect. 4.4.3). Second, the presence of responses to natural stimulation of skin supplied by the saphenous nerve could be induced in the same region of dorsal horn by drugs which vary the excitability of dorsal horn neurons (see below).

Drugs Affecting Neuronal Excitability. An obvious way to demonstrate the presence of normally ineffective afferent input is to examine the effect on single neurons of drugs which are known to enhance the efficacy of synaptic transmission. The excitability of dorsal horn neurons can be raised *in vivo* by the microiontophoretic application of excitant amino acids, such as glutamate, or by blocking inhibitory inputs with antagonists of GABAergic transmission, such as bicuculline or picrotoxin, or with the glycinergic antagonist strychnine. Other drugs, such as the convulsant 4-aminopyridine (4-AP) (Thesleff 1980), facilitate both excitatory and inhibitory synaptic transmission by a presynaptic action when applied either locally onto neurons or systemically. This substance has also been found to affect receptive fields of neurons in the DCN (Sect. 5.4.5).

Microiontophoretic application of glutamate to wide-dynamic-range SCT neurons in decerebrate cats caused an increase in excitability and expansion of both high- and low-threshold regions of the cutaneous receptive field (Fig. 4.21d) (Zieglgansberger and Hertz 1971). Glycine and GABA produced a depression of activation from the skin, and a corresponding shrinkage of the receptive field. The response to peripheral stimulation could be completely blocked with these inhibitory amino acids. In both cats and monkeys anaesthetized with urethane-chloralose, intravenous injection of subconvulsant doses of the GABA antagonist picrotoxin caused a moderate expansion of the low-threshold receptive fields of wide-dynamic-range lamina V neurons (Fig. 4.21a, b). At the same time, however, the larger high-threshold region of the receptive field disappeared (Yokota and Nishikawa 1982).

Semba et al. (1985) observed expansion of the cutaneous receptive fields of neurons in laminae III–V of the chloralose-anaesthetized cat following intravenous injection of 4-AP. The degree of expansion was related to the dosage of the drug, but was never greater than about 50% (Fig. 4.21c). Because there was no measurable increase of the response evoked by controlled mechanical stimulation of the centre of the receptive field, this effect was thought to be due to a selective excitatory action on subliminal inputs from the periphery of the receptive field. However, dorsal horn neurons also have inhibitory inputs from regions of skin within their excitatory receptive fields (Brown et al. 1987; Noble and Short 1989) and any additional effects of the drug on these inputs could lead to results which are difficult to interpret.

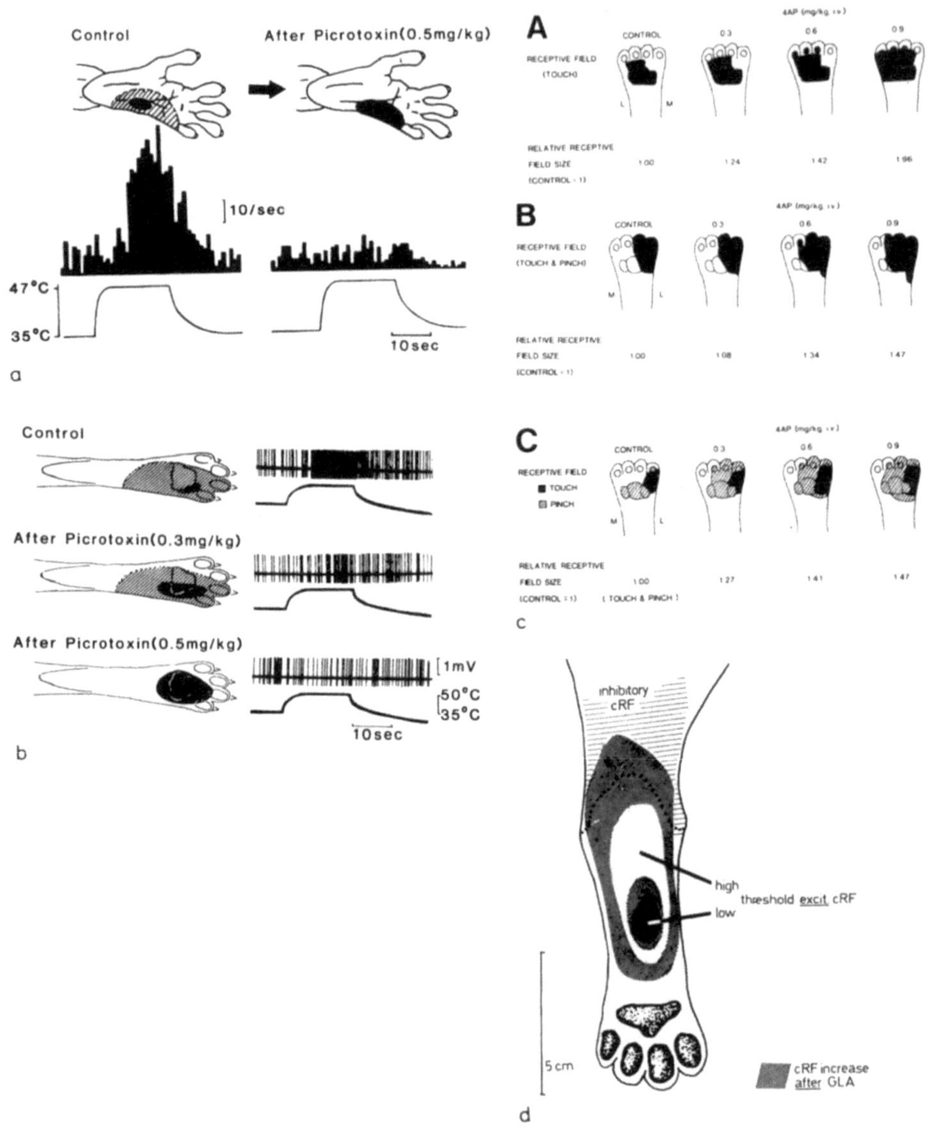

Fig. 4.21 a–d. Effect on the receptive fields of dorsal horn neurons of drugs that influence synaptic transmission, **a** Effect of intravenous administration of the specific GABA antagonist picrotoxin on the response characteristics of a wide-dynamic-range dorsal horn neuron in the monkey. The receptive field (*upper*) initially included a low-threshold area (*filled*) surrounded by a high-threshold area responsive to pinprick (*hatched*). After picrotoxin the low-threshold field expanded, and the high-threshold field disappeared. Note also the loss of the response to noxious heating (*lower*). **b** As in **a** but for a wide-dynamic-range dorsal horn neuron in the cat. **c** Expansion of the receptive fields of 3 dorsal horn neurons (*A*, *B*, *C*) in the cat following intravenous administration of 4-aminopyridine. For the wide-dynamic-range neurons (*B* and *C*) both low- and high-threshold fields expanded. **d** Expansion of the cutaneous receptive field (*cRF*) of a wide-dynamic-range SCT cell in the cat dorsal horn following iontophoretic application of the excitatory amino acid, L-glutamate (GLA).

In the decerebrate or barbiturate-anaesthetized cat, local application of 4-AP caused the receptive fields of most of a sample of lamina IV dorsal horn neurons recorded in the medial region of the lumbosacral dorsal horn (Saade et al. 1985) to increase in size. The expansion ranged from 50%–300% (mean 150%) and occurred in a predominantly proximal direction. In most cases expansion was not associated with any obvious change of mechanical threshold. In 15 cells, before application of the drug, electrical stimulation within the receptive field produced a response with a latency of around 5 ms. Stimulation of the skin 1 cm away was ineffective in firing the cell, even with a stimulus strength twice the threshold for activation from the receptive field. After application of the drug, electrical stimulation from the same distant point, which was now within the expanded receptive field, evoked a response at a latency of 5 ms. It was concluded that 4-AP had increased the effectiveness of normally ineffective excitatory inputs from afferents or interneurons.

These results suggest that inputs that were too weak to excite the cell to threshold were able to do so after transmission was enhanced by the application of 4-AP. However, it should be noted that the findings with electrical stimulation appear to contradict the earlier findings of Merrill and Wall (1972), who claimed that many lamina IV neurons could be excited by electrical stimulation of skin outside their cutaneous receptive field. This was originally taken to be evidence for the existence of relatively ineffective connections. According to Saade et al. (1985), the ineffective inputs could be detected by electrical stimulation only after drug treatment. It was suggested that the relatively small effect of 4-AP on distal receptive fields, compared with the large shift of somatotopic boundaries reportedly occurring after chronic peripheral nerve lesions (Sect. 4.4) might be due to the fact that the drug also increases inhibitory transmission. In contrast to 4-AP, bicuculline had little or no effect on the size of receptive fields. This presumably means either that the cells were not normally under significant tonic GABAergic inhibition, or that the increased excitability produced by the GABA antagonist was still insufficient to bring the ineffective inputs to threshold. The latter explanation seems more likely, because the drug did increase spontaneous firing and the number of spikes evoked by electrical stimulation of the receptive field.

It is of interest to compare the findings of Saade et al. (1985) in the cat with a similar study in the rat, where both the distribution of relatively ineffective excitatory inputs (as evinced by electrical stimulation of peripheral nerves) and the reorganization of somatotopy after peripheral nerve transection, have been found to proceed in a rostral to caudal, rather than in a mediolateral, direction (Markus et al. 1984) (Sect 4.4.3). Markus and Pomeranz (1987) found that dorsal horn neurons in the L4–L6 segments which lost their receptive fields after acute transection of the sciatic nerve, but responded to electrical stimulation of the saphenous nerve

◀ **Fig. 4.21 a–d** (*continued*). Both high-threshold (*white*) and low-threshold (*black*) receptive fields expanded. Note the inhibitory region (bordering the excitatory receptive field). (**a, b** From Yokota and Nishikawa 1982, pp. 262 and 261; **c** from Semba et al. 1985, p. 399; **d** from Zieglgansberger and Herz 1971, p. 115) (see above), acquired a natural receptive field in saphenous nerve territory after intravenous injection or topical administration of 4-AP or picrotoxin. Strychnine was usually ineffective. The drug-induced receptive fields were located on the medial toe, medial border of the paw, heel, ankle and leg. The appearance of these somatotopically inappropriate receptive fields was accompanied by increased responsiveness to electrical stimulation of the saphenous nerve, but there was no further caudal spread of the region from which responses to electrical stimulation could be obtained. It appears from the data, however, that these effects were not always obtained. Thus 4-AP was only effective in 4 out of 13 rats. Indeed, the most consistent effect was that produced by the GABA antagonist picrotoxin, which caused responses to natural stimulation of saphenous nerve territory to spread caudally about 2 mm, but only in 5 out of 18 rats. However, this distance was still considerably less than the spread reportedly occurring after chronic sciatic neurectomy (Fig. 4.18a).

Descending Inhibition. Since the earliest microelectrode recordings from dorsal horn neurons were made (Taub 1964; Wall 1960, 1967) it has been known that the functional properties of many dorsal horn neurons may be profoundly affected by tonic or evoked activity in pathways descending from the brain. In some studies it has been found that the size of the receptive field and the nature of the adequate stimulus of dorsal horn neurons depends on activity in descending inhibitory pathways. In themselves, these observations provide evidence for the existence of excitatory inputs which are labile, in addition to the stronger inputs whose function is evident even when cell excitability is depressed.

In decerebrate cats, Wall (1967) found that cells that he characterized as typical of lamina IV in the L7 segment increased their rate of spontaneous firing during reversible spinalization brought about by cooling the spinal cord at T 13 or L1, and often changed their response attributes from low-threshold mechanoreceptive to wide-dynamic range. However, the size of the cells' receptive fields was unchanged. Brown (1970, 1971) characterized identified SCT cells in decerebrate cats with and without a similar functional spinalization, and likewise found that the receptive field size of most cells was unaffected by the cold block. Nevertheless, SCT cells were found to be subject to descending inhibition from the dorsolateral funiculi, the ventral funiculi, and dorsal funiculi rostral to a transection of the tract (Brown et al. 1973a, b, and from the SI, SII and motor cortex (Brown and Short 1974; Brown et al. 1977c). Thus the absence of any effect on receptive field size reflects the minimal subliminal fringe adjoining the receptive fields of SCT cells (Brown et al. 1987).

In contrast, the lamina V neurons of Wall (1967) showed a marked expansion of their receptive fields during reversible spinalization. The expansion was often accompanied by a decrease in threshold at the centre of the receptive field, or the appearance of low-threshold responses in regions responding only to strong stimuli in the decerebrate state (Fig. 4.22a). Similar results were obtained for lamina V by Hillman and Wall (1969), who also noted that reversible spinalization blocked inhibitory responses evoked from the skin (Fig. 4.22b), and suggested that this might be due to the descending and afferent inhibition sharing a common segmental mechanism. For cells in deeper laminae (lamina VI type cells) which in the

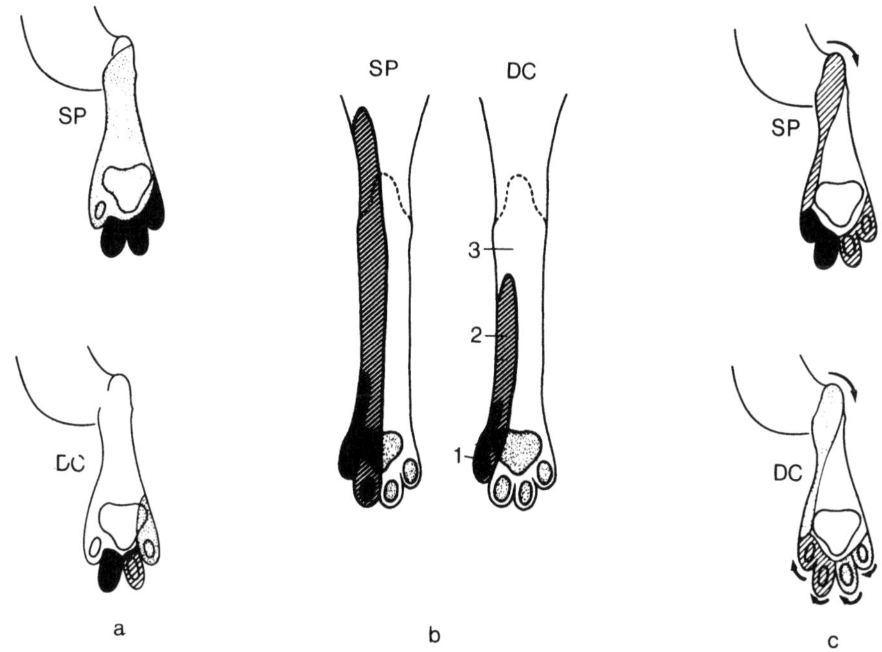

Fig. 4.22 a–c. Effect of blockade of tonic descending inhibition on receptive fields of dorsal horn neurons in the lumbosacral spinal cord of the decerebrate cat. The receptive field for each cell is shown before (*DC*) and after (*SP*) cooling the spinal cord rostrally. **a** A wide-dynamic-range neuron of lamina V responding to brush (*black*), touch (*hatched*) and pressure (*stippled*). **b** A wide-dynamic-range neuron with an inhibitory region bordering the excitatory receptive field. This cell was excited by brush and pinch in region *1*. In region *2* it was excited by pinch and inhibited by brush, and in region *3* it was inhibited by brush, touch and pressure. After cooling the cord, regions *1* and *2* expanded, and the inhibitory region disappeared. **c** A cell in lamina VI responding to both cutaneous inputs (key as in **a**) and joint movement (*arrows*). After cooling the cord responsiveness was enhanced to cutaneous inputs and decreased to deep inputs. (**a**, **c** From Wall 1967, pp. 413 and 417; **b** from Hillman and Wall 1969, p. 288)

decerebrate state responded to deep as well as cutaneous inputs, reversible spinalization converted the responsiveness of the cell from primarily kinaesthetic to primarily cutaneous, often with an expansion of the receptive field or the appearance of new areas of skin from which responses to gentle tactile stimuli could be obtained (Fig. 4.22c) (Wall 1967).

Effect of Anaesthesia. Wall (1967) noted that small intravenous doses of barbiturate in decerebrate cats decreased the spontaneous firing of dorsal horn neurons. Wide-dynamic-range lamina-V-type neurons were more severely depressed than lamina-IV-type neurons, and, as noted above, the expansion of their receptive fields during spinalization was blocked. Other investigators have found profound effects of anaesthetics on the receptive fields of neurons of lamina IV. For example, De Jong and Wagman (1968) found that 2% halothane progressively suppressed the

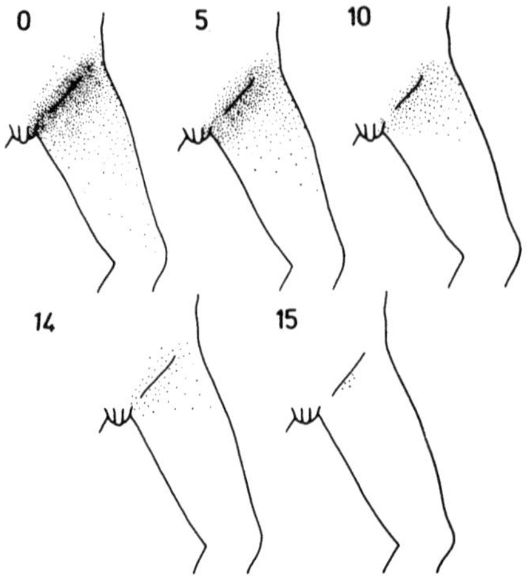

Fig. 4.23. Effect of anaesthesia on receptive field of a dorsal horn neuron. Receptive field of a cell in lamina IV of an unanaesthetized, decerebrate monkey. The cell responded to light pressure, the density of *stippling* indicating the strength of response. *Numbers* show time (min) after beginning administration of 2% halothane. The borders or the receptive field progressively contracted, and the field ultimately disappeared completely. (Modified from De Jong and Wagman 1968, p. 355)

responses of neurons in lamina IV of the decerebrate monkey to both natural stimulation of the skin and electrical stimulation of peripheral nerves. Depression of lamina IV neurons was accompanied by a progressive shrinking of the receptive field, sometimes leading to its complete disappearance (Fig. 4.23). Chloralose, barbiturates, N_2O and diethyl-ether are known to enhance presynaptic inhibition (Eccles et al. 1963; Miyahara et al. 1966), and Besson et al. (1968) found that a range of anaesthetics, including halothane, pentobarbital, thiopentone, and chloralose all increased presynaptic inhibition and decreased activation of dorsal horn neurons from primary afferents. Barbiturates also severely depress neurons in the superficial dorsal horn, and they have been reported to also suppress the spontaneous fluctuations of receptive fields of lamina II neurons (Sect. 4.5.2) (Dubuisson et al. 1979).

Clearly, many excitatory inputs to certain types of dorsal horn cells are masked in the anaesthetized state. However, more recent reports suggest that the response of some dorsal horn neurons to nociceptive inputs may also be *enhanced* by low doses of barbiturate (Hori et al. 1984; Collins and Ke Ren 1987). This effect may be due to blockade, by the drug, of systems that regulate the activity of dorsal horn neurons, possibly through an action on release of putative modulatory transmitters such as 5-HT or norepinephrine whose release is known to be inhibited *in vitro* by pentobarbital (Haycock et al. 1977; Richter and Jackson 1980).

4.5.2 Spontaneous Changes of Receptive Fields

When the low-threshold cutaneous excitatory receptive field of a typical SCT cell is repeatedly mapped over a long period of time in the absence of any other manipulation of the animal, in most cases it is found to be constant in size and position. If the cell exhibits a wide-dynamic range, the high-threshold regions of the receptive field also remain constant. Any surrounding regions of skin from which inhibition can be evoked also tend to be fixed. However, such constancy of receptive field size and organization is apparently not the case for all types of dorsal horn cells. Devor and Wall (1981a) claim to have recorded from single wide-dynamic-range neurons in the region of lamina V in the L6–L7 dorsal horn of the unanaesthetized decerebrate-spinal cat for periods of up to 5 h. During this time there were spontaneous fluctuations of excitability of the cells and sometimes changes in the borders of the receptive field, often involving 30% or more of the total area. In the receptive fields of some cells there were also variations in the regions from which high- and low-intensity mechanical stimuli were effective (Fig. 4.24a). Similar fluctuations have also been reported for PSDC neurons in cats anaesthetized with chloralose (Brown et al. 1983a). PSDC neurons have a wide-dynamic-range and complex receptive fields (Brown and Fyffe 1981). However, unlike the unidentified cells of Devor and Wall (1981a) the receptive fields of the PSDC cells only showed expansion after natural or electrical stimulation of the receptive field, or electrical stimulation of the dorsal columns. One further difference between the two sets of observations was that Devor and Wall (1981a) noted that brief repetitive electrical stimulation of the receptive field caused the field to shrink dramatically in size for a period of many seconds. In neither case is it clear which part(s) of the afferent fibre spectrum was being activated by the electrical stimuli. However, it seems likely that these effects were due to the activation of fine afferents, since in other studies, tetanic stimulation of Aß fibres has been found to have little or no long-lasting effect on the receptive fields of dorsal horn neurons (Sect. 4.5.3).

In contrast to these reports of relatively minor spontaneous receptive field changes in cells of the deep dorsal horn, some workers have described large fluctuations in the size of receptive fields of cells in the superficial dorsal horn (laminae I–III). While there remains much controversy about the functional properties of dorsal horn neurons in these superficial laminae (Sect. 4.2.4), most workers who have studied these cells have noted long-lasting changes of excitability, reflected in the presence of after-discharges or habituation, following cutaneous stimulation (Light et al. 1979; Fitzgerald 1981; Wall et al. 1979; Cervero et al. 1979; Dubuisson and Wall 1980; Dubuisson et al. 1979; Bennett et al. 1980; McMahon and Wall 1983).

Dubuisson et al. (1979), recording with low-impedance tungsten microelectrodes in decerebrate cats, made the surprising discovery that the receptive fields of a large proportion of superficial dorsal horn neurons underwent large variations in size in the absence of any intentional stimulation. The largest expansion observed in this study was a receptive field on the foot and toes which expanded proximally to include the entire back of the leg (Fig. 4.24b). In different cells, both high- and low-threshold fields were susceptible to expansion and contraction in this way. The fluctuations of receptive field were not related to the spontaneous firing of the cells,

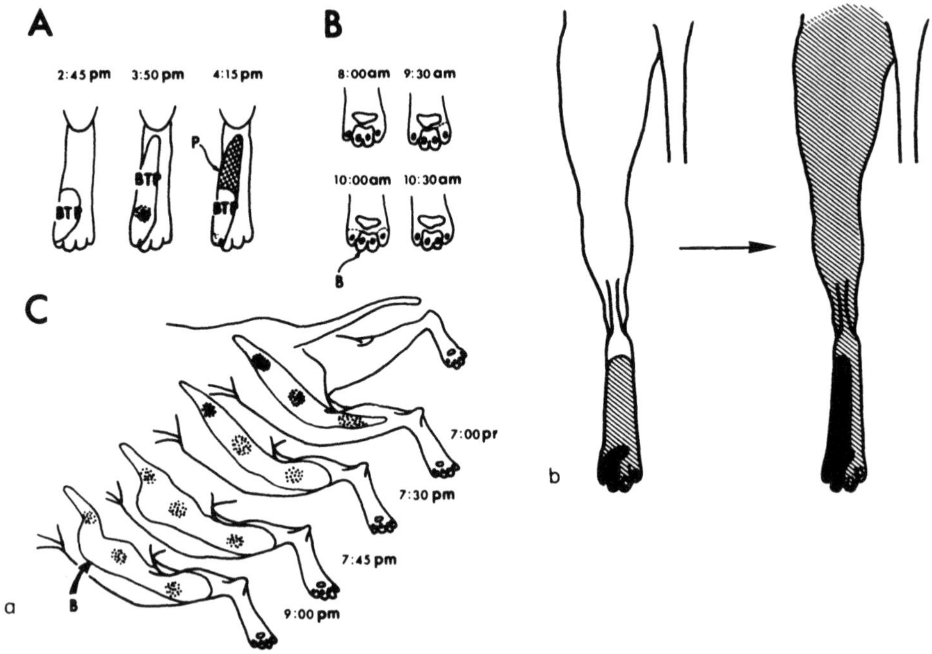

Fig. 4.24 a, b. Spontaneous fluctuations of receptive fields of dorsal horn neurons in unanaesthetized, decerebrate cats. **a** Spontaneous fluctuations in the receptive fields of 3 dorsal horn neurons (*A, B, C*) of the "lamina 5 type" (Wall 1967). The time of each observation is indicated. *Stippling* indicates areas giving strong responses to movement of individual hairs. *Abbreviations*: *BTP*, effective stimuli brush, touch and pressure; *B*, effective stimulus brush; *P*, effective stimulus pressure. **b** Maximal spontaneous expansion of the receptive field borders observed for a lamina II neuron. The brush sensitive region (*hatched*) expanded from an area on the foot to cover the whole caudal part of the leg. The pinch sensitive region (*black*) expanded simultaneously. (**a** From Devor and Wall 1981b, p. 279; **b** modified from Dubuisson et al. 1979, p. 377)

as one might expect if they were dependent on changes of neuronal excitability. Moreover, since the fluctuations were also found in spinalized animals, they could not have been due to variations in inhibitory or facilitatory drive from descending systems. Significantly, the fluctuations were abolished by a moderate intravenous dose of pentobarbitone, which caused the receptive field to contract and stabilize, and any after-discharge to disappear. Similar *amoeboid* receptive fields have been reported for 40% of a sample of lamina I projection neurons in the unanaesthetized decerebrate spinal rat (McMahon and Wall 1983). In view of our current lack of detailed knowledge of the microcircuitry of the substantia gelatinosa, it is possible to suggest only one segmental mechanism, modulation of PAD, by which spontaneous changes of receptive fields could occur in the absence of excitability changes. While these intriguing observations suggest that there is a large degree of plasticity inherent in the normal functional connectivity between afferent fibres and dorsal horn neurons, it should be noted that spontaneous changes of receptive field have not been reported in either anaesthetized or unanaesthetized animals by other workers who have recorded from the small neurons of lamina I and II using high-impedance glass microelectrodes (Light et al. 1979; Cervero et al. 1979).

4.5.3 Plasticity of Receptive Fields Induced by Afferent Activity

Studies of the basis of the prolonged and painful hypersensitivities to peripheral stimulation that often develop following peripheral injuries in man have led to investigation of abnormalities which can markedly change receptive fields of dorsal horn neurons.

Effect of Peripheral Tissue Injury. Dorsal horn neurons have been shown to alter their receptive fields in response to a peripheral injury. In monkeys anaesthetized with chloralose and pentobarbital, spinothalamic neurons increase their responsiveness to noxious heat, noxious cold and to innocuous tactile stimuli after noxious heating of the skin (Kenshalo et al. 1979, 1982). Provided the noxious thermal stimulus was applied for long enough the response to innocuous mechanical stimuli applied outside, as well as inside, the thermally damaged area of the cell's receptive field, was enhanced. Thus the less sensitive peripheral areas of the receptive field of a wide-dynamic-range neuron could become more sensitive to mechanical stimulation in the presence of increased C-fibre inputs. Although the detailed mechanisms of the enhanced responsiveness are still unknown, this observation fits well with clinical observations of secondary hyperalgesia following peripheral injury.

McMahon and Wall (1984b) examined the effect of a punctate burn placed on the skin just outside the receptive field of L4–L5 lamina I projection neurons in the decerebrate spinal rat. The majority of these cells showed slow fluctuations of their receptive field. When a small area of skin 5–15 mm outside the maximum observed extent of the receptive field was severely burned there was an initial reduction of responsiveness to mechanical stimuli and a shrinkage of the receptive field in some cells. After about 5 min, receptive fields began to expand in the direction of the damaged skin, and in most cases continued to expand or remained expanded for as long as observations on a single cell could be made (over 1 h) (Fig. 4.25a). This long-lasting receptive field enlargement was abolished if the peripheral nerves supplying the foot were treated with capsaicin, indicating that the expansion was linked to activity in C fibres. The area of skin from which the cell could be activated by transcutaneous electrical stimulation of cutaneous nerve branches was also increased (Fig. 4.25b), suggesting that central mechanisms, rather than changes in nociceptor thresholds, were responsible for the observed expansion of receptive fields.

Chronic pathological conditions related to articular or cutaneous diseases have also been shown to modify the responsiveness of dorsal horn neurons to peripheral stimuli (Menetrey and Besson 1982; Dubner et al. 1989). Plasticity of cutaneous receptive fields has been found in studies in decerebrate spinal rats rendered polyarthritic by injection of Freund's complete adjuvant (FCA). Menetrey and Besson (1982) found that the responsiveness of neurons in both the superficial (laminae I and II) and deep (laminae IV–VI) dorsal horn was increased to

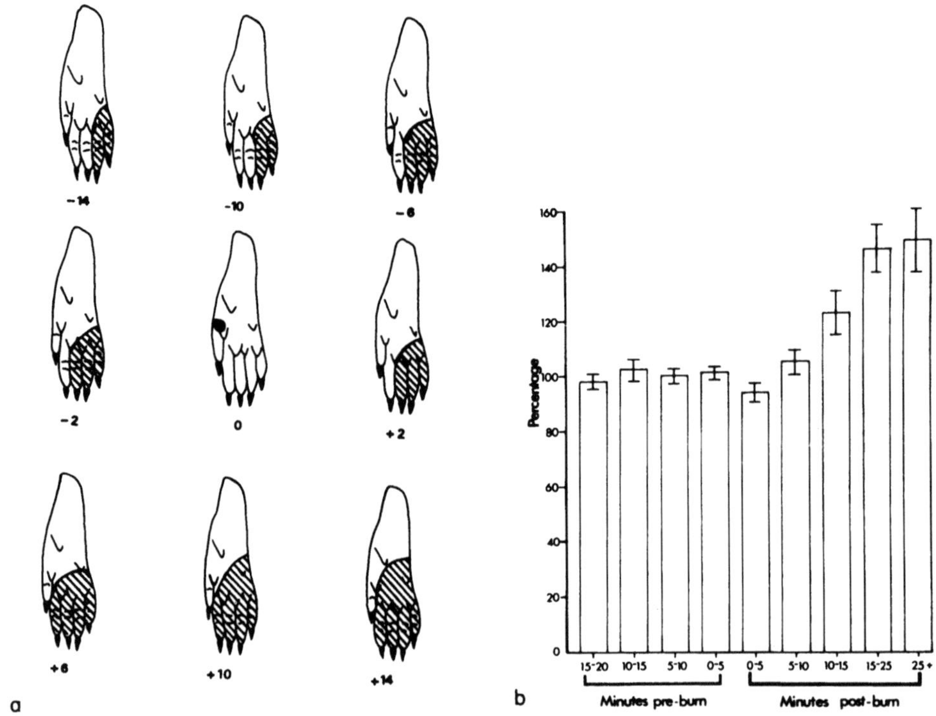

Fig. 4.25 a, b. Effect of skin injury on the receptive fields of lamina I dorsal horn neurons in the unanaesthetized, decerebrate rat. **a** Effect of a punctate burn on the receptive field of a single lamina I neuron. Each hindpaw figurine shows the size and location of the receptive field (*hatched*) at different times before the burn (negative times) and afterwards (positive times). Figures show times (min). Position of burn is indicated on time 0 drawing. After a short latent period, the receptive field expanded towards the site of the burn. **b** Effect of punctate burns on the receptive field size (mean and SE) of a population of 20 lamina I neurons at different times after the injury. For each unit receptive field sizes were converted to percentages of the average size before injury. The first significant increase of receptive field size occurs 10–15 min after injury. (From McMahon and Wall 1984, pp. 240 and 241)

stimulation of oedematous areas of skin. The increased responsiveness was manifested by an increased background firing, and a decrease in threshold, such that both neurons that normally would have responded only to noxious inputs and those with a wide-dynamic range now responded best to gentle mechanical stimulation. Receptive fields on the affected part of the limb were abnormally large, although in an earlier study (Menetrey et al. 1977) no convincing evidence was found for a *generalized* expansion of receptive fields on the affected hindlimb.

Dubner et al. (1989) have recently described clear differences in receptive field size of lamina I cells in the rat following injection of FCA into the paw. Compared to control rats, these animals had many lamina I neurons with large receptive fields covering the entire surface of the injected paw. Other abnormal findings included neurons with discontinuous receptive fields that also responded to joint movement or stimulation of deep tissue. There was no change in threshold of any of these neurons, all lamina I cells in both control and injected rats being nociceptor-specific. Primary nociceptive afferents were unaffected by the treatment, and thus this study lends further support to the idea that a central mechanism underlies plasticity of receptive field size of neurons in the superficial dorsal horn after peripheral tissue injury.

Repetitive Activation of C-Fibres. Evidence for a direct involvement of C fibre activity in plasticity of the receptive fields of dorsal horn neurons has come from a number of studies of the hindlimb flexor reflex in decerebrate spinal rats. The reflex discharge of flexor motoneurons evoked by electrical stimulation of an ipsilateral peripheral nerve at C fibre strength, or by the application of a pinch to the skin, is markedly enhanced for a prolonged period following repetitive stimulation of a peripheral nerve (Wall and Woolf 1984). The facilitation is produced by activation of afferents other than those involved in the reflex (heterosynaptic), and is therefore not due to a direct effect of repetitive activity on transmission at C fibre terminals. When the conditioning activity is evoked in a cutaneous nerve, the facilitation lasts a few minutes, but after stimulation of a muscle nerve the effect may last for more than an hour (Wall and Woolf 1984). Several lines of evidence point to the facilitation being dependent on repetitive activation of unmyelinated afferents. First, the effect is produced by electrical stimulation only at a strength suprathreshold for C-fibres (Wall and Woolf 1984). Second, facilitation is also produced by the application of mustard oil, which stimulates unmyelinated but not myelinated afferents (Woolf and Wall 1986), by burning the skin, or by chemically induced inflammatory conditions (Woolf and McMahon 1985). Third, the facilitation is reduced by the topical application of capsaicin to the nerve, which blocks some of the central actions of C fibres (Sect. 2.2.6). The facilitation does not results from an enhanced excitability of the flexor motoneurons themselves, because a monosynaptic reflex evoked in the same motoneurons is unaffected (Cook and Woolf 1985). Neither is it likely that excitability changes in the afferents responsible for the test reflex are responsible, because the reflex evoked by stimulation of a contralateral nerve is also facilitated. It follows that the most likely source of altered excitability lies in the dorsal horn interneurons on the polysynaptic flexor reflex pathway.

To investigate this possibility, recordings were made from neurons in various laminae of the L4–L5 dorsal horn of decerebrate spinal rats, and the effect of repetitive stimulation (20 s at 1 Hz) of either the sural or the gastrocnemius-soleus (GS) nerve on the size and responsiveness of their receptive fields was determined (Cook et al. 1987). Following C fibre strength conditioning of the GS nerve the receptive fields of almost 60% of neurons increased on average to a maximum of approximately four times their original size, for a period of about 45 min. Conditioning of Aß fibres had no such effect, while conditioning at A∂ stimulus strength produced much smaller expansions of shorter duration (Fig. 4.26a). In a few neurons that originally responded only to pinching of the skin, the expansion of the receptive field was accompanied by the appearance of responses to light tactile stimuli (Fig. 4.26b), and the responsiveness of some wide-dynamic-range neurons to standard light tactile stimuli also increased. However, expansion of the

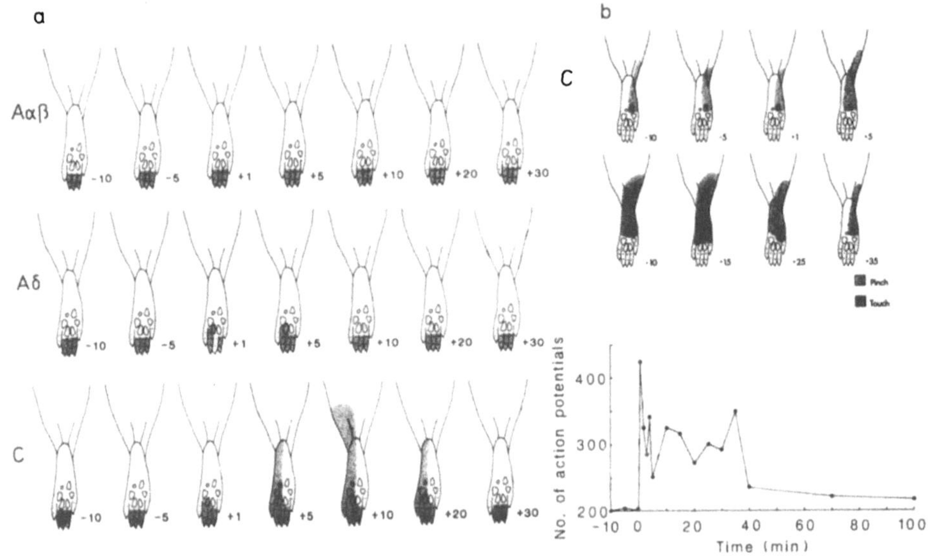

Fig. 4.26 a, b. Effect of repetitive electrical stimulation of C fibres on the receptive fields of dorsal horn neurons in the unanaesthetized, decerebrate rat. **a** Effect of graded strengths of conditioning stimuli (1 Hz for 20 s) applied to the gastrocnemius-soleus nerve on the size of the receptive field of a dorsal horn neuron. The pinch-sensitive field of the neuron is shown by the *shaded area* on each drawing. The number beside the drawing is the time (min) relative to the conditioning stimulus (delivered at time 0). At *left* the conditioning stimulus strength is indicated by the types of afferents recruited by the stimulus. Activation of A $\alpha\beta$ fibres alone had no effect. Inclusion of A∂ fibres caused a slight transient expansion of the receptive field, but when C fibres were also activated, there was a marked and long-lasting expansion of the receptive field. **b** Upper drawings as for **a**, but this cell also developed a low-threshold touch region in its receptive field. Graph shows the number of impulses evoked in the cell (*ordinate*) by a standard 3 s pinch stimulus to the medial edge of the foot at different times after the conditioning stimulus (*abscissa*). (From Cook et al. 1987, p. 152)

receptive field was not necessarily associated with any change in responsiveness to a standard stimulus. These experiments show that the exteroceptive properties of dorsal horn neurons may be influenced not only by lesions of afferent pathways, but also by previous afferent inputs. The receptive field expansions observed appear to be rather modest compared to those which would be required to explain the shifts of receptive fields in the medial dorsal horn following peripheral nerve or dorsal root lesions (Sect. 4.4), and similar effects of C fibre conditioning on the size of low-threshold receptive fields of dorsal horn neurons have not been unequivocably demonstrated.

While it is not known whether any of the dorsal horn neurons whose properties were claimed to be altered by the conditioning stimuli in these experiments were actually on the reflex pathway to flexor motoneurons, their increased excitability does afford an explanation of the facilitatory effect of C fibre activation on the flexor reflex. Repetitive activity in C fibres in fact has a dual facilitatory effect on the reflex activation of flexor motoneurons, and the two phases of facilitation are affected differently by chronic peripheral nerve transection. Several investigators have noted in a variety of species the phenomenon of *wind-up*, where the C fibre-evoked discharge of a dorsal horn neuron, or of flexor motoneurons, increases markedly during the application of the first few volleys in a low-frequency train (Mendell and Wall 1965; Mendell 1966; Wagman and Price 1969; Fitzgerald and Wall 1980; Woolf 1983; Cervero et al. 1984; Wall and Woolf 1986; Woolf and King 1987). Repetitive C fibre stimulation can also temporarily increase the response of dorsal horn neurons to inputs in myelinated afferents (Schouenborg and Sjolund 1983). Single or repetitive stimulation of the intact sciatic nerve in decerebrate spinal rats at C fibre strength produced a similar wind-up-like after-discharge lasting several seconds in flexor motoneurons. When the central stump of a sciatic nerve sectioned 12–14 days earlier was stimulated, this after-discharge was found to be significantly greater and more prolonged (Wall and Woolf 1986).

Repetitive stimulation of the intact sciatic nerve at C fibre strength was found to have a long-lasting (> 1 h) facilitatory effect on the flexor reflex evoked by pinching the toes similar to that induced by conditioning stimulation of the GS nerve (see above). After transection of the sciatic nerve the reflex was, of course, reduced due to partial denervation of the foot. Between 12 and 14 days later, the reflex was still reduced, and in sharp contrast to the increased short-term wind-up facilitation, the long-term facilitation was greatly reduced and curtailed. Because chronic nerve section had such different effects on the two phases of facilitation, Wall and Woolf (1986) suggested that the two facilitatory effects depended on different mechanisms. They further suggested that the late facilitation might depend on the modulatory function of neuropeptides, which are known to be depleted from the terminals of C fibres in the dorsal horn by 12–14 days after nerve section (Sect. 4.5.5).

In contrast to these reports of long-term facilitatory effects of repetitive C fibre activity on dorsal horn neurons, other workers have found either no effect or a long-lasting depression of both A and C fibre transmission to dorsal horn neurons. For example, in decerebrate spinal cats prolonged high-intensity, low-frequency stimulation of a peripheral nerve has been found to produce an *inhibition* of the late component of the flexion reflex that persisted for many minutes after the end of the conditioning stimulation (Chung et al. 1983). Brown et al. (1975) examined, in decerebrate spinal cats, identified SCT cells having a slowly adapting pressure-sensitive component to their receptive field. These cells responded with a discharge to electrical stimulation of C fibres, but pure C fibre conditioning volleys evoked in nerves in which myelinated fibres were blocked by dc polarization failed to produce either facilitation or inhibition of either the A or C fibre-evoked response. Recordings from single antidromically identified spinothalamic tract (STT) neurons in both superficial and deep laminae of the dorsal horn in monkeys have shown that these also may undergo a long period of heterosynaptic or homosynaptic inhibition as a result of high-intensity repetitive electrical stimulation of a peripheral nerve (Chung et al. 1984a). Whether the animals were anaesthetized (chloralose and pentobarbital), decerebrate or decerebrate spinal made little difference to the results of these experiments. In a subsequent study (Chung et al. 1984b) it was noted that the inhibition of C fibre activation of STT cells was in fact dependent mainly on repetitive activation of A∂ afferents, and that Aß and C fibre activation contributed much less to the phenomenon.

Even in the decerebrate spinal rat, Woolf (1983) noted that repetitive stimulation (2 Hz) of the sural nerve produced a powerful and prolonged heterosynaptic inhibition of C fibre responses evoked from the common peroneal nerve in laminae V and VI neurons. The reason for these conflicting observations of the effect of C fibre activation is not readily apparent, because different experiments have involved different species of animals, different nerves, and a variety of conditioning protocols. Furthermore, various types of dorsal horn neuron receiving excitatory input from C fibres may react differently to C fibre tetanization. Wall and Woolf (1984) suggested that *facilitation* is produced by brief (several seconds) tetani, and *inhibition* by much longer (several minutes) tetani. However, this explanation does not seem to hold for the heterosynaptic inhibition of STT cells in the cat, since Chung et al. (1984a) monitored the response of these cells to a test stimulus during the conditioning, and did not find an early facilitation preceding the long-lasting inhibition, though the early wind-up type of facilitation had been noted in other similar experiments (Chung et al. 1983).

Recording intracellularly from wide-dynamic-range neurons in lamina III–V of the decerebrate spinal rat, Woolf and King (1987) found that the long-duration facilitation following repetitive C fibre inputs could be partly accounted for by a prolonged depolarization, which in some neurons could become large enough to inactivate the somadendritic spike. Thus the basis of the apparent inhibition observed in some experiments could be excessive depolarization. Other workers have reported finding a decreased terminal excitability of C fibres in the cat, indicating that presynaptic hyperpolarization of these terminals might also be involved in the inhibition via mechanisms that have no known parallel in A fibres (Hentall and Fields 1979, 1980).

While the effects of prolonged C fibre inputs may differ in different acute preparations, there is little doubt that they are capable of producing, on at least some dorsal horn neurons, excitatory or inhibitory effects that outlast the C fibre activity. Because these effects are due, at least partly, to depolarization of dorsal horn neurons, repetitive C fibre activity is able to affect inputs from myelinated afferents and hence bring about expansion of both high- and low-threshold receptive fields of neurons having subliminal inputs.

4.5.4 Involvement of Unmyelinated Afferents in the Somatotopic Reorganization After Peripheral Nerve Injury

A series of observations of changes in the dorsal horn of the rat which follow chronic peripheral nerve transection led Wall (for reviews see Wall and Devor 1982; Wall and Fitzgerald 1982; Wall 1984, 1987) to postulate that the somatotopic reorganization of the low-threshold receptive fields of medial dorsal horn neurons (Sect. 4.4.3) might be induced mainly as a result of damage to the unmyelinated (C fibre) afferents of the transected nerve. According to this reasoning, whatever the underlying central mechanism of the long-term somatotopic reorganization might be, the fact that the change of somatotopy is induced by the peripheral nerve transection implies that it is initiated either by the *lack* of some influence *normally provided by the healthy nerve*, or by the *presence* of factors signalling damage, *provided by the damaged nerve*. Consideration was given by Wall and his co-workers to a number of factors that might be responsible for transmitting information about nerve transection to the spinal cord in either of these two cases. One obvious possibility was the chronic absence of normal afferent activity, but this was ruled out because it was claimed that mediolateral reorganization in the rat did not occur after either chronic blockade of conduction in the peripheral nerve (Wall et al. 1982), or following crush, rather than transection, of the nerve (Sect. 4.4.3) (Devor and Wall, 1981b).

The main reason for implicating C fibres in somatotopic reorganization was that most of the morphological and neurochemical changes which had been described in the dorsal horn following chronic peripheral nerve section involved the terminals of these afferents in the substantia gelatinosa (lamina II). These changes include: (i) the depletion of neuropeptides, such as substance P and CCK, (ii) the depletion of FRAP, and (iii) transganglionic degeneration of primary afferent terminals (Sect. 4.5.5). At the same time (10–14 days after nerve section) that these changes occurred at central C fibre terminals, most aspects of myelinated afferent function in the transected nerve appeared to be unchanged. Thus following a volley in the cut sciatic nerve in the rat, both the size of the maximal A wave evoked in the nerve, and the size of the afferent volley in the dorsal columns were normal. Also unaffected by chronic neurectomy were the size of A fibre-evoked volleys ascending in postsynaptic pathways, such as the dorsolateral funiculus, and the PAD evoked on roots distant from those containing the transected afferents (Wall and Devor, 1981).

Despite the apparent constancy of excitatory transmission from chronically transected A fibres, Wall (1982) reported a decline of PAD evoked both by and within the transected afferents. In apparent agreement with this finding, between 7 and 14 days after transection of the sciatic nerve the A fibre-evoked inhibition of both A and C fibre-evoked excitation of lamina V neurons of spinal rats (Sect. 4.2.5) was reported to be significantly reduced (Woolf and Wall 1982). Although the synaptic mechanism of the A-on-C inhibition of lamina V neurons is unknown, on the basis of earlier studies of C fibre terminal excitability (Hentall and Fields 1979; Fitzgerald and Woolf 1981) Woolf and Wall (1982) suggested that the inhibition was at least partially brought about by presynaptic depolarization of C fibre terminals in lamina II. Similarly, there is some evidence to indicate that the A fibre-evoked PAD in A fibres is mediated by lamina II interneurons (Wall 1962). Woolf and Wall (1982) thus proposed that the decline of A fibre-mediated inhibition of lamina V neurons following peripheral nerve section might be related to the various structural and neurochemical alterations in lamina II described above. They further proposed that the decline of A fibre mediated inhibition following peripheral nerve transection might lead to the expansion of receptive fields in the medial part of the dorsal horn, thus linking the changes in C fibres following nerve section to the reorganization of tactile receptive fields. Further evidence for a role of C-fibres in reorganization was the reported abolition of somatotopic reorganization by chronic application of NGF to the central cut end of the nerve (Sect. 4.4.3) (Fitzgerald et al. 1985). This finding implicates C fibres in the response to nerve injury, because NGF is thought to be essential for the metabolism of these afferents (Sect. 2.2.2). A final piece of indirect evidence for the involvement of C fibres came from observations of the effects of nerve crush. According to Wall and co-workers, if the peripheral nerve were simply crushed rather than cut, the depletion of substance P from lamina II was much reduced and delayed, there was no decline in PAD (Wall and Devor 1981; Wall 1982) and there was no development of proximal receptive fields by cells in the medial dorsal horn (Devor and Wall 1981b) (Sect. 4.4.3).

In order to obtain more direct evidence in support of these ideas, it was necessary to segregate the effects of damage to peripheral C fibres alone, from those due to simultaneous damage to both A and C fibres. Such a separation of effect was achieved with capsaicin, which has a specific neurotoxic effect on C fibres. Wall (1982), using the excitability testing technique (Wall 1958), claimed to find a severe reduction in the ability of apparently normal A fibres to generate PAD in neonatally capsaicin-treated rats, which lack around 95% of afferent C fibres (Sect. 2.2.6). This result clearly strengthened the argument for a sustaining influence of afferent C fibres on A fibre-evoked presynaptic inhibition of A fibres, and therefore also in the control of the low-threshold receptive fields of dorsal horn neurons, but has itself proved to be controversial (see below).

Logically, the next step was to examine the receptive fields of dorsal horn neurons in animals with selective damage to peripheral C fibres, to determine whether they were comparable to those in the reorganized region of dorsal horn after nerve section. In neonatally capsaicin-treated rats (Sect. 2.2.6), the lower part of the limb was denervated *acutely* by the transection of the sciatic and saphenous nerves, and the medial dorsal horn of segments L4 and L5 was searched for cells that responded to stimulation of skin proximal to the knee (Wall et al. 1982a, b). The animals were either anaesthetized with pentobarbital or decerebrated and all were spinalized to maximize the excitability of dorsal horn neurons. It was reported that a significantly greater proportion of cells were encountered with proximal receptive fields in the capsaicin-treated animals than in control rats, and the increase was comparable with that found after chronic nerve transection (Sect. 4.4.3). The receptive fields of these cells were on the upper leg and in a few cases extended onto the tail and over the midline to the contralateral limb. However, when evaluating the significance of this result the controversial nature of the evidence for mediolateral reorganization following peripheral nerve section must be considered (Sect. 4.4.3).

Capsaicin applied topically to peripheral nerves of adult rats severely reduced the C fibre-evoked excitation of dorsal horn neurons in the absence of either any permanent block of C fibre conduction or of alteration of the ultrastructure of the dorsal horn (Wall and Fitzgerald 1981). The central terminals of C fibres in the treated nerve are nevertheless depleted of substance P, CCK and FRAP (Ainsworth et al. 1981), indicating that the capsaicin had produced a blockade of trophic support from the periphery similar to that occurring after nerve transection. The effect of this treatment on the receptive fields of dorsal horn neurons in the rat has been investigated in three studies, two of which concentrated on cells in laminae IV–VI of the dorsal horn (Fitzgerald 1982; Wall et al. 1982b) while the other was aimed specifically at projection neurons of lamina I (McMahon et al. 1984).

In decerebrate spinal rats Fitzgerald (1982) and Wall et al. (1982b) recorded from neurons in laminae IV–VI of the L4/L5 dorsal horn with receptive fields in the cutaneous field of the sciatic nerve between 7 and 21 days after capsaicin treatment of the ipsilateral or contralateral sciatic nerve. While the capsaicin treatment did not affect the relative proportions of neurons responding to light tactile stimulation or to noxious pinch, the receptive fields of cells ipsilateral to the treated nerve were larger than normal. Thus many more cells had receptive fields which extended onto skin proximal of the sciatic nerve territory, and fewer cells had receptive fields confined to the foot and toes. On the side contralateral to the treated nerve there was also a significant but smaller tendency for receptive fields to be larger than normal. In addition to these changes, many cells (about 40%) on both sides of the cord showed an unusual excitatory receptive field on the distal part (toes and foot) of the contralateral limb. However, responses from these abnormal contralateral receptive fields were almost obtained by pinching or noxious heating of the skin, and not by light tactile stimulation.

McMahon et al. (1984) recorded from lamina I neurons in the medial region of the L4/L5 dorsal horn of rats which had one sciatic nerve treated with topical capsaicin 11–22 days earlier. Normally, the receptive fields of these cells extend over part of the foot, usually involving 2 or 3 toes and their associated foot-pads (McMahon and Wall 1983). After capsaicin treatment about half of these neurons ipsilateral to the treated nerve had considerably larger receptive fields than normal. These receptive fields often extended proximal to the foot onto the lower leg, and in some cases there were two main parts to the receptive field, one of normal size on the foot and the other on the leg, connected by a thin strip of sensitive skin along the medial or lateral edge of the foot (Fig. 4.27). There was a large decline in the proportion of cells responding at C fibre latency to electrical stimulation of their receptive fields, indicating that the capsaicin treatment had effectively interfered with transmission at the central terminals of C fibres in the sciatic nerve. The authors suggested that, like the receptive fields of cells in the deeper lamina, those of lamina I neurons due to inputs in myelinated afferents might normally be restricted by some action of C fibres supplying the same region of skin.

Certain features of Wall's arguments in favour of a role for C fibres in the control of A fibre-mediated inhibitions are unfortunately inconsistent with findings from other laboratories. First, there is the question of whether crushing the peripheral nerve really has no effect on PAD. Horch and Lisney (1981b) examined excitability changes at the central terminals of single normal or regenerating sural nerve A fibres following a conditioning A fibre volley in the sural or sciatic nerve in pentobarbital-anaesthetized cats. One month after either transection or crush of the sural nerve, very few regenerating sural nerve fibres showed any sign of PAD following the conditioning volley. During the succeeding months, PAD reappeared in crushed afferents earlier than it did in transected fibres, but recovery was still incomplete 6 months after crush and 9 months after transection. Whatever the underlying reason for this decline of PAD of damaged afferents, there seems little doubt that, in the cat at least, the short-term effects of crushing or cutting the nerve on presynaptic inhibition are identical.

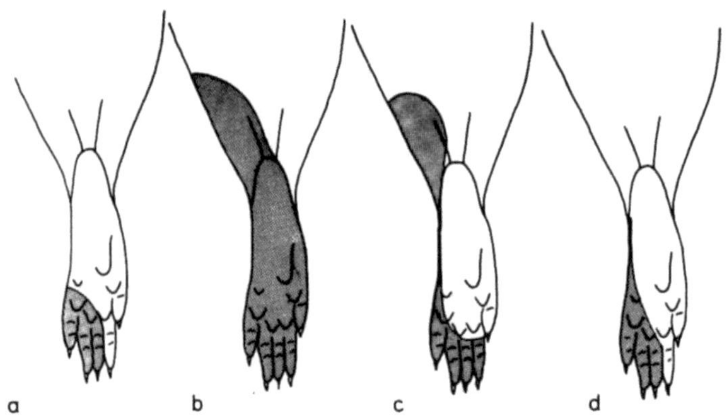

Fig. 4.27 a–d. Effect of unilateral topical application of capsaicin to the sciatic nerve on the receptive fields of lamina I neurons in the unanaesthetized decerebrate rat. **a** Typical receptive field on untreated side. **b, c** Examples of the kinds of expanded receptive fields commonly found on the capsaicin-treated side. **d** Typical normal-sized receptive field on the treated side. Note the unusual narrow lateral strip of responsive skin in **c** and **d**. (From McMahon et al. 1984, p. 397)

Second, other workers have been unable to confirm some of the observations of Wall and his associates on the effect of neonatal capsaicin on dorsal horn physiology in the rat. Initially it was found that the PAD generated in A fibres by activity in A fibres of the same or other nerves was unaffected by the virtual absence of C fibres (Cervero and Plenderleith 1984). The lack of effect of neonatal capsaicin treatment on PAD was shown both by measurements of the dorsal root potential and, as previously used by Wall (1982), by measuring excitability changes of central terminals. It was also independent of whether the animals were anaesthetized with barbiturate, or with urethane (Cervero and Plenderleith 1984). It should be pointed out that Wall's (1982) results were obtained in only two animals, and it is possible, using the excitability testing technique, to miss the true focus of PAD with the stimulating microelectrode in the dorsal horn in individual experiments (M. B. Plenderleith, personal communication).
Despite the unchanged PAD of A fibres, Cervero and co-workers nevertheless found some functional abnormalities of dorsal horn neurons of rats treated neonatally with capsaicin. These changes were mostly confined to neurons receiving convergent input both from the few remaining C fibres, and from Aß fibres. In control animals about one-half of the neurons driven by an Aß fibre volley in the sural nerve could also be driven by a volley in C fibres (Cervero et al. 1984; Cervero and Plenderleith 1985, 1987). In capsaicin-treated animals this proportion was reduced, but only by 50%. Since there is an approximately 95% reduction in the number of C fibres in these animals, either compensatory terminal sprouting of surviving C fibres occurred, or the convergence of C fibres onto dorsal horn neurons is normally so great that a large number of C fibres would have to be destroyed to achieve a significant fall in the number of postsynaptic cells receiving C fibre input. Sprouting has been discounted in this situation because there is also an approximately 90%

decrease in the number of synaptic glomeruli in lamina II, to which C fibres contribute their synaptic terminals (Ribeiro da Silva and Coimbra 1984). However, there is no reason to believe that axon sprouts may not form different, simple terminals, or that local axonal sprouting of interneurons with input from surviving C fibres could not make up the deficit of C fibre inputs (Sect. 4.5.5).

In the experiments of Cervero and his colleagues the responses of dorsal horn neurons to natural stimulation of their cutaneous receptive fields did not differ fundamentally in the capsaicin-treated rats from those in control rats. In both cases, about one-third of cells were mechanoreceptive, one-third were multireceptive and the remainder were nocireceptive. While there was a significant increase in the size of receptive fields, this change was restricted to neurons excited by C fibre volleys (Fig. 4.28), which account for only about 20%–30% of dorsal horn neurons in neonatally capsaicin-treated rats. The increase of receptive field size was not remarkable, amounting to approximately 100%, and since receptive fields of neurons in the medial dorsal horn are normally small, the capsaicin treatment led to an increase in the proportion of medium-sized, rather than very large, receptive fields. Thus, whereas in normal animals most receptive fields were confined to the toes, in capsaicin-treated animals they were located mainly on the foot and ankle, rather than involving more proximal skin. Cervero et al. (1984) concluded that there had been a shift of somatotopic boundaries in the medial part of the dorsal horn, such that there were more cells encountered with receptive fields on the foot and ankle, where fields are normally medium-sized, and fewer cells with the typically small receptive fields found on the toes. This rearrangement of somatotopy could be a direct result of the proximal expansion of receptive fields of dorsal horn neurons. However, there were also significantly fewer cells than normal with receptive fields on the leg in these studies (Cervero et al. 1984), and this is more difficult to reconcile with the results of Wall et al. (1982b) who found large numbers of neurons with proximal receptive fields in capsaicin-treated rats after acute sciatic and saphenous neurectomy. One possible explanation of the differences between these studies is that the receptive fields in the experiments of Cervero et al. (1984) were constricted by the use of barbiturate anaesthesia (Sect. 4.5.1). However, some of the rats studied by Wall and co-workers were also under pentobarbital anaesthesia, and this apparently did not prevent a marked increase in the proportion of cells with proximal receptive fields.

According to Cervero and Plenderleith (1985, 1987) tonic descending inhibition of C fibre-evoked responses is reduced in neonatally capsaicin-treated animals. That segmental inhibitory mechanisms are also adversely affected in the capsaicin-treated animals is shown by the fact that many more cells than normal, including some that lack a C fibre input, exhibit a temporary expansion of their receptive field repetitive activation of C fibres (Sect. 4.5.3) (Cervero et al. 1984). However, the absence of any *steady-state* enlargement of the receptive fields of neurons driven only by A fibre volleys indicates that the effect of neonatal capsaicin on a cell is somehow dependent on the presence of the C fibre input. One possibility is that the lamina II interneurons, which probably relay C fibre input to deeper wide-dynamic-range neurons, themselves have expanded receptive fields in capsaicin-treated rats. There is currently no physiological evidence in support of this proposal. However, there is

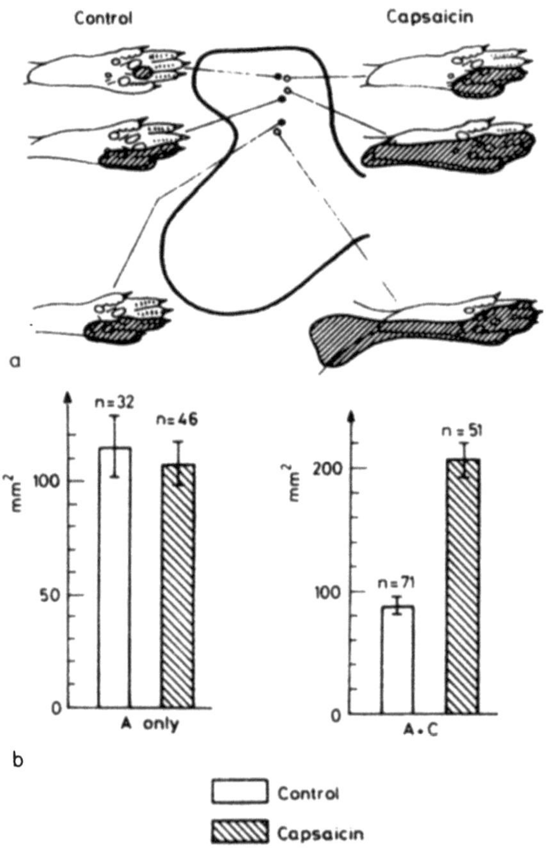

Fig. 4.28 a, b. Effect of neonatal capsaicin treatment on the receptive fields of rat dorsal horn neurons. **a** Cutaneous low-threshold receptive fields (*hatched*) of 3 neurons at similar locations in the dorsal horn (indicated on transverse section) and having excitatory input from both A and C fibres in control and neonatal capsaicin-treated rats. In the capsaicin-treated animal neurons have larger receptive fields. **b** Mean (± SE) of areas of cutaneous receptive fields of dorsal horn neurons responding to activation of A fibres only (*left*) and both A and C fibres (*right*) in control rats and in rats treated neonatally with capsaicin. Note different scales on the two graphs. (From Cervero and Plenderleith 1987, p. 397)

anatomical evidence from both HRP- and tritiated proline-labelling of dorsal root ganglion cells that neonatal capsaicin treatment in the rat may induce the sprouting of myelinated afferent fibres into lamina II, where they could form abnormal synaptic contacts with partially deafferented interneurons (Nagy and Hunt 1983). Recent Golgi studies of the dorsal horn of capsaicin-treated rats have shown that there is indeed an invasion of lamina II by arborizations of myelinated primary afferent fibres that, on the basis of their distinct collateral morphology (Sect. 4.2.3), appeared to be derived from both hair follicle and type I afferents (Beal and Knight 1987). Obviously it would be useful to have more information concerning the functional properties of lamina II neurons in these animals, before any firm conclusion can be reached about the mechanisms underlying the observed receptive field expansion. Furthermore, it is apparent that the dorsal horn of the neonatally capsaicin-treated rat does not simply display the functional effects of a lack of C fibres, but rather a whole range of adaptations (Cervero and Plenderleith 1987) to the removal of these afferents at a time when there is still a remarkable degree of structural plasticity of primary afferent projections (Sect. 4.6.3).

A further problem with Wall's proposal that damage to C fibres causes reorganization of A fibre inputs concerns the use of the word *expansion* to describe the effect of chronic nerve section on the receptive fields of cells in the medial dorsal horn. While it is acceptable that receptive fields might undergo expansion as a result of changes in the balance of excitatory and inhibitory inputs, the original descriptions of mediolateral reorganization by Devor and Wall (1981a, b) and Lisney (1983) clearly show that the receptive fields undergo *translocation* rather than simple *expansion*. This issue was recently raised by Devor (1987), who noted that the proximal reorganized receptive fields in the medial dorsal horn often did not extend down the leg as far as the denervated skin. In other words, they were distinct, proximal receptive fields, bordered distally by normally innervated skin. It is patently unreasonable to discuss such translocated receptive fields as if they were hugely expanded distal receptive fields with the distal component removed by peripheral denervation. Nevertheless such a description seems to be increasingly used in reviews of the question of mediolateral reorganization and the possible underlying mechanisms (Wall 1987).

C Fibres and Somatotopic Reorganization in the Trigeminal System. Some evidence for a possible role of C fibres in the control of somatotopy has come from studies of the trigeminal system. Between 7 and 15 days after denervation of the tooth pulp in adult cats there are significant changes in the functional properties of neurons in SpVo. The majority of these neurons are of the low-threshold mechanoreceptive variety, and their receptive fields form the well known *inverted-head* somatotopic map of the orofacial region (Fig. 3.3b) (reviewed by Darian-Smith 1973; Dubner et al. 1978). In intact animals neurons in the dorsal part of the nucleus have receptive fields localized within the mandibular region. Further ventrally, receptive fields are located within the maxillary region, and at the base of the nucleus most receptive fields are within the ophthalmic division. Medially, most neurons have perioral or intraoral receptive fields and laterally, receptive fields are found on the lateral aspect of the face (Fig. 4.29b) (Darian-Smith et al. 1963). One week after tooth pulp extirpation there is an expansion of mechanosensitive receptive fields within the division of the nucleus representing the denervated teeth. Thus, for example, after removal of mandibular tooth pulp, neurons in the mandibular (dorsal) division of the nucleus have receptive fields which include both mandibular and maxillary regions (Fig. 4.29a) (Hu et al. 1986). The deterioration of the somatotopic organization of low-threshold mechanoreceptive fields in these experiments is all the more remarkable because only a very small part of the total afferent input to the affected region is removed by the surgery, but this fraction consists almost entirely of unmyelinated or small myelinated nociceptive afferents (Sect. 2.3.4). The change of somatotopy is accompanied by a reduction in the

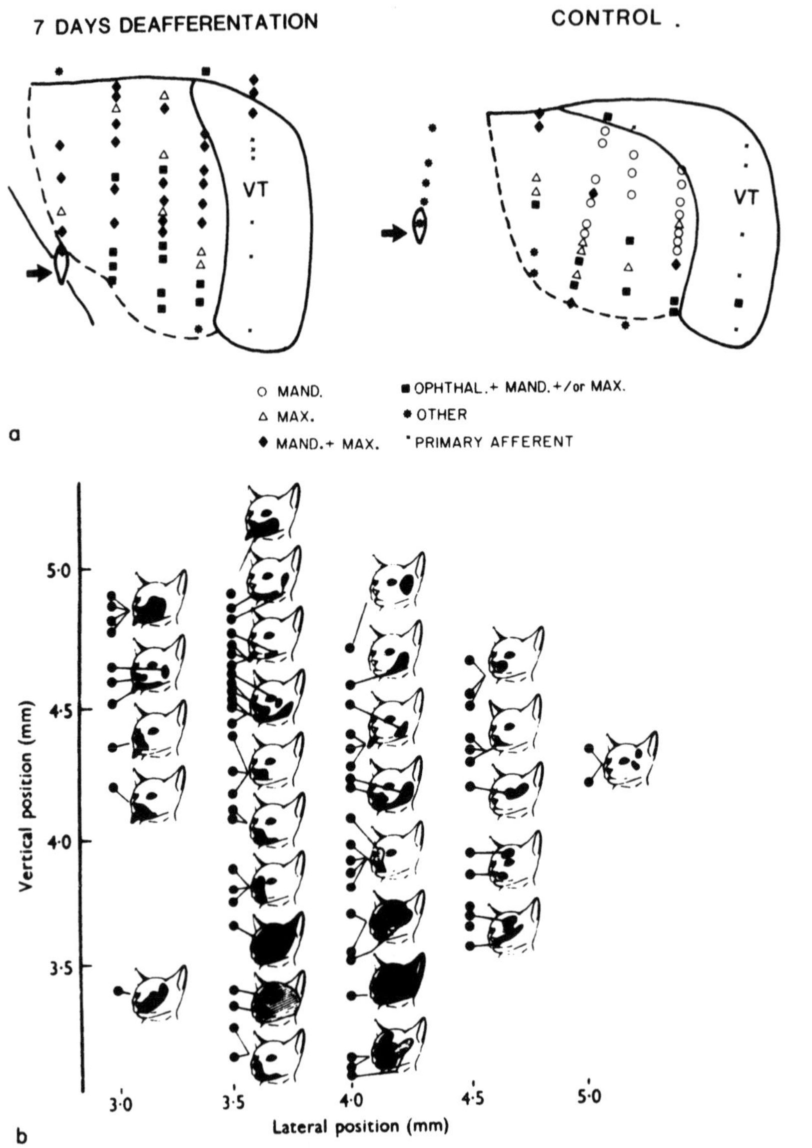

Fig. 4.29 a, b. Reorganization of somatotopic organization in trigeminal nucleus tractus spinalis oralis (SpVo) following chronic ipsilateral tooth pulp removal. **a** Representative reconstructions of somatotopic organization of SpVo in a control animal (*right*) and in a cat in which the ipsilateral mandibular coronal tooth pulps had been removed 7 days earlier (*left*). Each *column of symbols* represents the loci of single neurons recorded in a single electrode track. The extent and location of the orofacial mechanoreceptive field for each neuron is indicated by the *symbols in the key*. SpVo neurons with an ophthalmic receptive field and reticular formation neurons with large receptive fields are indicated by *asterisks*. Primary afferent fibres are indicated by *crosses*. Note in the partially deafferented nucleus the lack of neurons with a localized mandibular receptive field and the large increase in the number of neurons with a receptive field spanning both mandibular and maxillary regions. *Abbreviations: MAND*, mandibular; *MAX*, maxillary; *OPHTHAL*, ophthalmic; *VT*, spinal tract of trigeminal nerve.

proportion of neurons which can be driven by electrical stimulation of areas of the orofacial region outside those containing their receptive field, suggesting that the reorganization might indeed be due to an unmasking of pre-existing but ineffective inputs. Postsynaptic ultrastructural changes (Gobel 1984) and degeneration (Westrum et al. 1976; Gobel and Binck 1977) also occur in the brainstem following tooth pulp extirpation, and these are discussed below (Sect. 4.5.5).
Interestingly, while degenerative changes also occur in the brainstem after natural defoliation of the primary teeth in kittens, this form of denervation apparently has no effect on somatotopic organization of neurons in SpVo (Hu et al. 1987). The naturally occurring "lesion" differs from surgical pulp extirpation in that both large- and small-diameter periodontal afferents degenerate, in addition to the pulp afferents. Possible explanations for the lack of reorganization in the kitten are either (1) that the deafferentation produced by the natural defoliation is a gradual process, allowing adaptive processes in the brainstem to adjust to whatever changes, trophic or otherwise, occur in the afferent input, or (2) that the central portions of the afferents may not, in fact, degenerate in the kitten (Fried 1982; Westrum et al. 1984).

4.5.5 Sprouting of Primary Afferent Fibres and Other Neurons as a Basis of Somatotopic Reorganization

The possibility that sprouting of the central processes of intact primary afferent fibres or the axons of central neurons may be induced by dorsal root transection or by peripheral nerve injury, has sometimes seemed an attractive explanation both for the reorganization of somatotopic maps and for the functional recovery of sensori-motor co-ordination which have been observed following such lesions. Although collateral sprouting by intact axons in response to partial deafferentation of neurons has been described in many parts of the adult brain (Cotman et al. 1981) the question of its presence in adult mammalian spinal or trigeminal somatosensory systems is controversial. Thus, while sprouting has been promoted by some workers as the primary mechanism underlying functional plasticity in spinal somatosensory systems (Goldberger and Murray 1978, 1985), its existence has been vigorously denied by others (Micevych et al. 1986; Brown 1987).

Dorsal Rhizotomy and Cordotomy. The proposal that axonal sprouting may occur in the adult spinal cord following dorsal roots section was first made by Liu and Chambers (1958) (see also McCouch et al. 1958) in their well-known experiments which were also the first to provide evidence for collateral sprouting of axons anywhere in the adult CNS. Utilizing a Nauta silver technique to stain degenerating axons, these workers found that, following extensive dorsal rhizotomy in the adult cat, an intervening intact (spared) dorsal root appeared both to expand its

◀ **Fig. 4.29 a, b** (*continued*). **b** Typical single-unit receptive fields of SpVo recorded in 5 electrode tracks through the nucleus in the pentobarbital-anaesthetized cat. The most lateral track passes through the VT. (**a** From Hu et al. 1986; **b** from Darian-Smith et al. 1963, p. 151)

rostrocaudal projection within the spinal cord and to increase the density of its projection to existing regions of termination. In the example they used to illustrate this phenomenon, the most rostral extension of a spared L7 dorsal root's projection to the dorsal horn increased from L1 to T8–9, 277 days after transection of dorsal roots T10–L6 and S1–Ca7. At the same time, the projection to Clarke's column appeared to expand rostrally by about 3 segments, and that to the intermediate nucleus of the spinal grey by about 4 segments. The extent of the projection to the ventral horn remained unchanged (Fig. 4.30). Similar changes in the projection of dorsal root afferents were reported after chronic partial deafferentation of the spinal grey by transection of descending pathways (the corticospinal tract).

Kerr (1972) examined the possibility of primary afferent sprouting in the spinal division of the TNC of the adult cat. This nucleus had previously been shown to receive projections from both cervical and trigeminal roots in both the macaque (Kerr et al. 1968) and the cat (Kerr 1972). Kerr (1972) made use of the fact that, in the cat, the caudal portion of the spinal tract and nucleus of the Vth nerve, and the terminals of primary afferents of C1–C3 roots overlap throughout the dorsal horn of the C1 segment, as well as in some areas of the spinal division of the TNC. The trigeminal root was accordingly sectioned on one side and 11.5–24 months later dorsal roots C2 or C1–C3 were cut bilaterally. No evidence of sprouting of the cervical root projections on the side of the lesion was found by the Nauta method. Indeed, Kerr noted that "the density of degeneration was more often slightly greater on the side corresponding to the normal trigeminal system...." (Kerr 1972). He concluded that sprouting of primary afferent fibres did not occur within the trigeminal system, and suggested that the lack of sprouting might, in fact, be of advantage to the animal, because "the relative lack of plasticity in adult complex nervous systems, while presenting a major obstacle to phenomena of repair may, conversely, prevent unfortunate sequelae of anomalous neural connections" (Kerr 1972) (compare Sect. 8.6).

Goldberger and Murray (1974, 1978, 1982) and Murray and Goldberger (1974) repeated the experiment of Liu and Chambers (1958) with slight modifications. To investigate the effects of removal of descending projections on the projection of lumbosacral dorsal root axons in the cat, Murray and Goldberger (1974) partially hemisected the cord at low thoracic levels, sparing the dorsal funiculus. In six cats which survived longer than 10 months the L5 or L6 dorsal roots were then sectioned and the resulting degeneration in the lumbosacral cord was examined with variations of the Nauta method. On the previously hemisected side both the density and extent of staining were increased. The greatest increase in density was found in the central portion of laminae V, VI and VII, and the regions of expanded input were found medially in lamina VI and both medially and laterally in lamina VII. To examine the effect of dorsal rhizotomy on the projection of a spared root, Goldberger and Murray (1982) unilaterally cut all dorsal roots below T5, sparing L6. In some of the cats the L6 dorsal root was then cut bilaterally 18 or 24 months later. In others the cord was transected at L1 18–24 months after the initial rhizotomy, in order to determine whether the dorsal rhizotomy had caused an expansion of the projection of fibre systems descending to the spinal cord. It was found that the spared L6 dorsal root ipsilateral to the lesion appeared to have

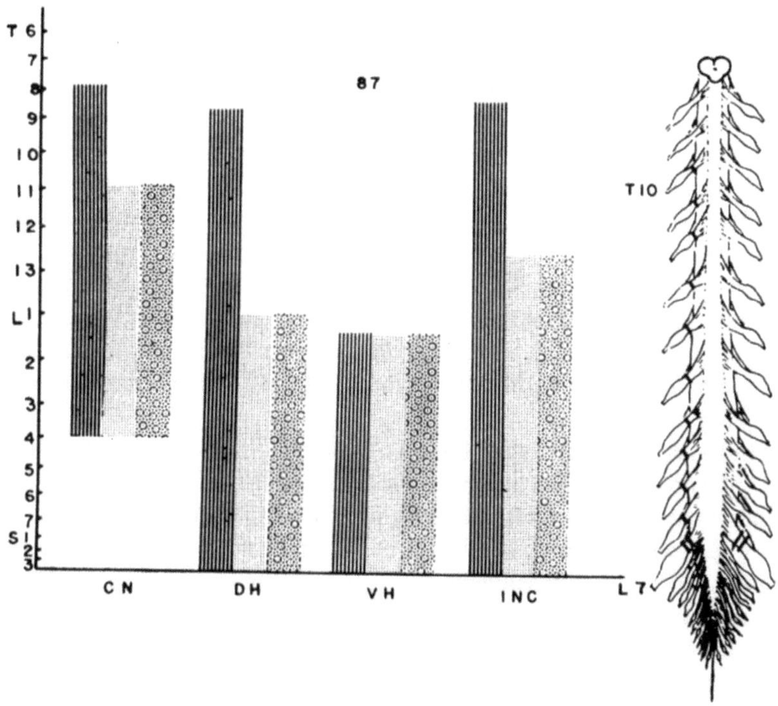

Fig. 4.30. Evidence for long-range sprouting of afferent fibres of an intact dorsal root in the spinal cord of the cat following extensive dorsal rhizotomy. *Right*, diagram showing the experimental procedure. The dorsal roots cut on one side in an initial operation (T10–L6 and S1–Ca7) are shown by a *thin line* across each. In a second operation, 227 days later, both L7 dorsal roots (marked by a *double line*) were cut. *The cat was sacrificed 4 days later and the cord stained by the Nauta technique. Left*, graph showing the rostrocaudal distribution of silver stained axon degeneration in the grey matter of the two sides of the cord resulting from the second operation. The distribution of the L7 root in a normal cat is also shown for comparison. Segment numbers are on *left*. *Vertical hatching*, distribution of L7 dorsal root on rhizotomized side in a single animal; *dots*, distribution of L7 dorsal root on intact side; *flowers*, distribution of L7 dorsal root in an unoperated animal. *Abbreviations*: *CN*, Clarke's nucleus; *DH*, dorsal horn; *VH*, ventral horn; *INC*, intermediate nucleus of Cajal. (From Liu and Chambers 1958, p. 49)

increased the density of its spinal projection (1) to the lateral part of the dorsal horn, mainly in the L6 segment (2) to the zona intermedia in all segments up to T10, and (3) to Clarke's column as far as the T9 segment. In addition, the density of projection of L6 afferents to the basal region of nucleus gracilis of the DCN also appeared to be increased (Sect. 5.3.2). However, in contrast to the report of Liu and Chambers (1958), there was no evidence for the appearance of *aberrant* projections.

It is not clear from the account of Goldberger and Murray (1982) whether sprouting of descending systems was investigated at this time in animals with complete unilateral lumbosacral dorsal rhizotomies, or only in animals with an L6 spared root (see above). However, in an earlier study, Goldberger and Murray (1974) transected dorsal roots T12 or L1 through S1 or S2 then 12 months later transected the cord at L1 or L2. Following the second lesion there was an increased density of degeneration on the same side as the dorsal rhizotomy, particularly in the nucleus proprius, in the base of the dorsal horn and in the zona intermedia of segments L2–L7. Degeneration was also apparent in regions in which none could be found on the control side, including the lateral part of the ventral horn, the medial base of the dorsal horn and Clarke's column (Goldberger and Murray 1974). However, despite the extensive deafferentation of the superficial dorsal horn that must have been produced by the dorsal rhizotomy, no evidence of sprouting was seen dorsal to lamina IV. In animals with a spared L6 root investigated in the later study (Goldberger and Murray 1982) the descending projections were bilaterally symmetrical, indicating that any sprouting of descending systems that may occur is suppressed by the presence of even a much reduced primary afferent projection.

Goldberger and Murray (1982) were at pains to point out the possible problems in the interpretation of increases in the density of staining caused by shrinkage of the denervated neuropil after extensive dorsal rhizotomies. In general, the amount of shrinkage was related to the degree of deafferentation of the segment, and ranged as high as 20% in the L6 segment deprived of all primary afferents. However, despite the increased density of staining in the lateral part of the dorsal horn (see above), shrinkage was absent from the L6 segment in cats with a spared L6 dorsal root. Thus the pattern of putative sprouting did *not* correspond in every case to the pattern of shrinkage. These workers concluded that, in adult cats following dorsal rhizotomy or rostral partial spinal hemisection, intact dorsal roots did *not* expand their projections rostrocaudally to invade new terminal regions, but did appear to increase the density of their projection to existing terminal regions (Goldberger and Murray 1974, 1978, 1982; Murray and Goldberger 1974). According to these authors the lack of sprouting in the trigeminal system in Kerr's (1972) work is compatible with their own suggestions. Thus collateral sprouting of primary afferent or other projections may be precluded in highly topographically organized regions by the overriding action of mechanisms of neuronal specificity (a second example being the apparent lack of sprouting by afferents in the cell nest region of the DCN, Sect. 5.3.2). However, we are not convinced that adequate allowance for *localized* shrinkage in the lateral region of L6 dorsal horn that receives primary afferents only from the transected roots could have been made in these experiments. Further shortcomings of the double-degeneration protocol are outlined below.

Stelzner et al. (1979) examined the response of dorsal root axons to rostral spinal hemisection in weanling (21–24 days old) rats. The hemisection was made at T6–T8, and 5–12 months later the pattern of degeneration in the L5–S1 segments of lumbosacral cord resulting from bilateral transection of dorsal roots caudal to L4 was examined with the Fink–Heimer technique. No detectable difference in either the distribution or density of degeneration could be seen when the hemisected and control sides were compared. The authors considered that their negative findings, compared with the positive signs of sprouting obtained by Murray and Goldberger (1974) following spinal hemisection in the adult cat, might be due in part to the fact that the distance (number of segments) between the initial hemisection and the subsequent rhizotomy was greater, and that consequently fewer propriospinal connections to neurons in the projection field of the rhizotomized roots would degenerate. Moreover, because all lumbosacral dorsal roots were cut, changes in the projection of individual roots might be masked. An additional problem with this protocol is that, in order to lesion the corticospinal tract in the rat it was necessary to damage the dorsal funiculus, thus cutting the ascending branch of dorsal root axons. Although this did not apparently causes severe retrograde degeneration of dorsal root ganglion cells, the authors suggested that it may have interfered with the sprouting of their central terminals.

The apparent conflict between the results of the above studies is all the more surprising because they all utilized similar double-degeneration protocols. However, the suitability of degeneration techniques for the demonstration of sprouting has since been questioned on several grounds by workers who have been unable to detect the phenomenon by alternative anatomical methods (Rodin et al. 1983; Rodin and Kruger 1984; Micevych et al. 1986). These technical limitations will only be briefly considered here, but include the following: (1) The resulting degeneration may produce spurious changes in the staining properties of non-neuronal cells, e.g. glia (Matthews and Kruger 1973) or of dendrites of neurons undergoing transsynaptic degeneration (De Olmos et al. 1981). (2) The staining properties of intact axons in regions chronically deprived of primary afferents may be changed in such a way as to increase their resistance to suppressive silver impregnation. As indicated by Micevych et al. (1986) this can lead to normal intact projections being visualized under conditions of reduced histological suppression designed to show degenerating axons. For example, Goldberger and Murray (1982) noted that after a reduction of the degree of suppression of the silver stain, a greater distribution of degeneration was found after spinal transection in their later experiments (Goldberger and Murray 1982) than in their earlier series (Goldberger and Murray 1974). In particular, degeneration became apparent on the control side in Clarke's column, and in laminae I and III. The projection to Clarke's column had previously been thought to be aberrant (Goldberger and Murray 1974), and therefore the same could possibly be said of the apparent degeneration in the dorsal horn. While Liu and Chambers (1958) were clearly aware of some of these possible pitfalls with the double-degeneration technique, their study is further flawed by the absence of adequate controls, and it is possible that the 277 or 303 days they allowed for removal of all debris from the first rhizotomy was insufficient. In view of all the uncertainties involved in the double-degeneration method it would seem hazardous to reach firm conclusions concerning the sprouting of primary afferent fibres (or descending systems) on the basis of this type of evidence alone.

Neuroanatomical tracing techniques have also been widely employed to study possible sprouting of primary afferent fibres. The projection of the chronic L6 spared root was studied autoradiographically in cats after injection of the L6 dorsal root ganglion with tritiated amino acid (Goldberger and Murray 1974) and the projection of L5 and L6 dorsal roots were studied by the same method in partially spinally hemisected animals (Murray and Goldberger 1974). These experiments provided somewhat weak support for the findings of these workers with the double-degeneration method (see above), but the technique was later abandoned owing to the difficulty of obtaining uniform and symmetrical labelling of a pair of dorsal root ganglia.

Aberrant projection of dorsal root afferents to the cat spinal dorsal horn after chronic spinal hemisection has also been demonstrated by labelling dorsal roots with HRP (Beattie et al. 1986). In cats HRP was applied bilaterally to cut dorsal rootlets both rostral and caudal to a chronic (6 weeks to 3 months) spinal hemisection at T13–L2. Two types of apparently aberrant dorsal root axons were observed from 1 to 7 segments caudal to the transection: (1) in some cats, coarse collaterals of axons in the dorsal funiculus appeared to diverge from their normal dorsoventral trajectories and traverse laminae I and II in a lateromedial direction, (2) smaller straight fibres radiated from the apex of the dorsal horn in an atypical fashion, often diverting from their typical laminar distribution in the dorsal horn. Both kinds of atypical axons possessed swellings along their course, and were only rarely found either in normal cords or contralateral to the lesion. Unfortunately, since the dorsal funiculi were cut in these experiments, it is not possible to say whether the atypical collaterals represent axonal sprouting in response to partial deafferentation of the grey matter, or whether they represent a growth response of the collaterals of injured fibres. Nevertheless, these are important and unique observations in relation to anatomical plasticity in the spinal cord of the adult mammal and are worthy of further investigation.

Arguably the most compelling anatomical evidence to date against the idea that primary afferents can undergo sprouting in partially deafferented regions of adult mammalian spinal cord has been provided in studies of the projection fields of peripheral nerves to the rat spinal cord with anterograde, transganglionic HRP labelling (Rodin et al. 1983). This technique had been successfully used previously to study the normal termination fields of peripheral nerves in the spinal cord of the cat (Koerber and Brown 1980, 1982) (Sect. 4.3.3). Rodin et al. (1983) examined the projection field of the sciatic nerve in the rat spinal cord 2 weeks to 3 months after removal of 2 to 5 lumbosacral dorsal root ganglia surrounding a spared root (either L4 or L5, which together contain nearly all the sciatic nerve's primary afferent neurons in the rat). Unlike the cat, where the cutaneous distribution of hindlimb nerves can vary widely, the projection fields of hindlimb nerves in the rat are normally highly consistent between animals (Rodin et al. 1983). Controls were therefore different animals in which the HRP was applied to the nerve on the day of the ganglionectomy. In these experiments no consistent differences were found in the intraspinal projection of sciatic nerve afferents between the experimental and control animals (Fig. 4.31). In particular, there was no evidence that intact sciatic afferents invaded new spinal segments, or that their projection density (as judged by the density of the HRP reaction), distribution along or across the dorsal horn or in the zona intermedia of either the spared or adjacent segments, or in Clarke's column, was affected by the ganglionectomy. Rodin et al. (1983) concluded that there was no anatomical basis for the electrophysiological plasticity demonstrated in earlier experiments (Sect. 4.4). Their explanation for the difference between their results and those of earlier studies employing the double-degeneration methods (see above) was the greater reliability of the anterograde HRP method, due to the fact that the enzyme, being strictly intra-axonal, labels only processes

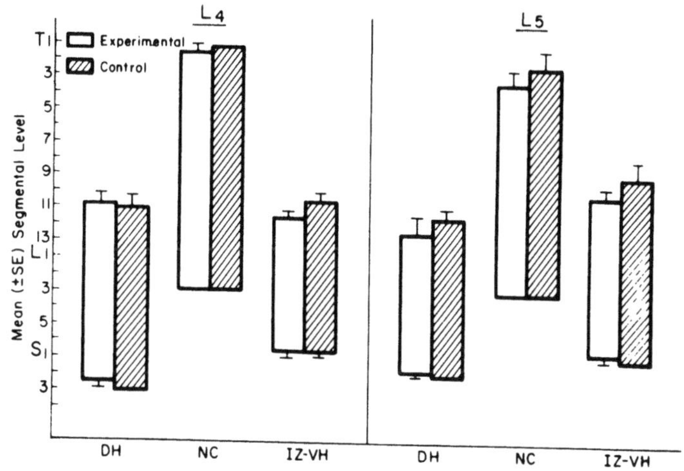

Fig. 4.31. Absence of long-range axonal sprouting in rat spinal cord following dorsal root ganglionectomy. Diagrams showing the average rostral and caudal boundaries (± SE) of the sciatic nerve projection, determined by transganglionic transport of HRP, to the dorsal horn (*DH*), Clarke's nucleus (*NC*) and intermediate zone-ventral horn (*IZ-VH*) in operated and control preparations with the L4 or L5 dorsal roots spared. (From Rodin et al. 1984, p. 189)

of primary afferent fibres. However, it remains possible that the anterograde HRP method is for unknown reasons incapable of adequately labelling newly sprouted axon collaterals. Similar studies also failed to show any sign of sprouting of sciatic nerve afferents after chronic spinal hemisection (Rodin and Kruger 1984).

A positive result obtained by the use of anterograde HRP is exemplified by studies of the plasticity of bladder reflexes in the cat. In neonatal kittens activation of pudendal nerve afferents by tactile stimulation of the perineum initiates micturition, but in normal adult cats micturition is inhibited. In adult cats with chronic (6–8 week) post–T12 spinal transection the reflex reverted to the neonatal response. In both neonatal and chronic spinal animals the pudendal nerve afferents as revealed by transganglionic transport of HRP had a substantially expanded distribution within the sacral spinal dorsal horn, when compared with normal adult cats. The increased projection was most evident in the contralateral dorsal horn, where there was normally only sparse labelling, and in the lateral and ventral part of the ipsilateral dorsal horn, a region containing preganglionic bladder neurons (Thor et al. 1982). This apparent sprouting of the adult projection to partially deafferented regions previously occupied by the afferents during development, and the accompanying functional plasticity, may represent a useful model system for studying the limitations imposed on the sprouting of intact primary afferent fibres. The factors underlying these limitations presumably also govern the contribution of sprouting by primary afferents or other axons to the reorganization of somatotopic maps in partially deafferented regions of the somatosensory system.

In addition to the studies discussed above, evidence of sprouting by primary afferent terminals has been looked for (1) by electrophysiological means, (2) by counting axons in the dorsal roots, (3) by ultrastructural studies of the density of central synapses, and (4) by studies of the anatomical plasticity of putative neurotransmitters or associated enzymes in the dorsal horn. Brinkhus and Zimmermann (1983) examined various electrophysiological parameters of dorsal horn neurons between 11 and 45 days after dorsal rhizotomy, but could find no evidence of changes in neurons located in a dorsal rhizotomized segment, compared with those in an adjacent intact segment. In adult cats dorsal rhizotomy was performed on either the L5 root alone, the L6 root alone, or the L5, L6 and L7 roots combined. The cord dorsum potential evoked by electrical stimulation of hindlimb nerves, or of intact dorsal rootlets just caudal to the rhizotomized region was recorded in acute experiments. The negative (N) component of the cord dorsum potential is related to current flow during synaptic activation of dorsal horn neurons, and the positive (P) component is related to the presynaptic depolarization of the terminals of primary afferent fibres (primary afferent depolarization). It was suggested that changes in connectivity between primary afferent fibres and dorsal horn neurons resulting from the chronic lesion should be reflected in characteristic changes in the distribution of N and P waves. In fact, both waves were found to have the same magnitude, time-course and longitudinal distribution on the lesioned side of the cord as on the opposite control side, where all dorsal roots were intact. However, owing to the extensive overlap of dorsal root projections to the lumbosacral dorsal horn (Sect. 4.3.3), transection of a single dorsal root probably would not produce sufficient deafferentation across the whole width of the dorsal horn to bring about significant changes in the N wave at any given level.

In a series of studies Coggeshall and co-workers used the electron microscope (Hulsebosch and Coggeshall 1981a, b, 1983a, b) to examine the effect of partial deafferentation of the spinal cord, by either extensive dorsal rhizotomy or cord hemisection, on the number of axons in selected intact dorsal roots in both neonatal and young adult rats. In adult rats the cord was hemisected at thoracic level, and 6 months later the number of axons in the three dorsal roots rostral and caudal (roots CR 1–3 and CA 1–3, Fig. 4.32a) of the hemisected segment were counted (Hulsebosch and Coggeshall 1983a). These counts were compared with those obtained from the corresponding roots on the contralateral unoperated side. There was a statistically significant fall in the number of myelinated axons on the hemisected side, amounting to about 8% for all six roots combined, and 14% for the adjacent roots only (roots 1 CR and 1 CA in Fig. 4.32a). While there was also an approximately 8% fall in the number of unmyelinated axons in the six roots, this was not statistically significant. The slight decrease of myelinated dorsal root axons was attributed to retrograde degeneration of some dorsal root ganglion neurons resulting from the severance of their central processes in the spinal cord, similar to that occurring after peripheral nerve transection (Sect. 2.2.1). Thus there was no evidence of sprouting of primary afferent fibres at the level of the dorsal root following spinal hemisection in the adult rat.

To avoid the complicating effects of damage to central processes of primary afferents on sprouting following partial deafferentation of the cord, Hulsebosch and Coggeshall (1981b) then used a spared root preparation similar to that used by Liu and Chambers (1958) in the adult cat. In rats aged 30 days, three consecutive dorsal roots above and below a spared low thoracic dorsal root were transected

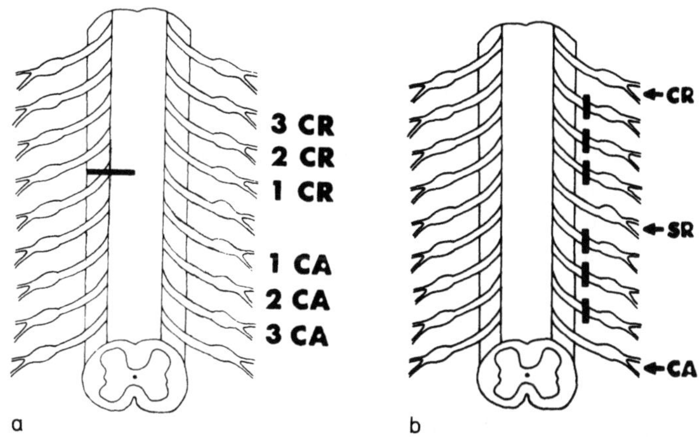

Fig. 4.32 a, b. Schematic representations of operative techniques used to demonstrate sprouting of dorsal root axons in rat spinal cord following spinal hemisection or dorsal root section. **a** Chronic spinal hemisection is shown by *solid bar* on the left side of cord. Axons were counted in the 3 roots rostral (*1*, *2* and *3 CR*) and caudal (*1*, 2, and *3 CA*) of the hemisection. **b** Three dorsal roots both rostral and caudal to a spared root (*SR*) are cut chronically (*vertical bars*). Axons were counted in the spared root, in the rostral adjacent intact root (*CR*) and the caudal adjacent intact root (*CA*). For further details see text. (**a** From Hulsebosch and Coggeshall 1983a, p. 188; **b** from Hulsebosch and Coggeshall 1983b, p. 78)

(Fig. 4.32b). After a survival time of 30 days axons were counted in the spared root (root SR in Fig. 4.32b) and in the cranial and caudal adjacent intact roots (roots CR and CA in Fig. 4.32b), as well as the corresponding roots on the contralateral, intact side. There was found to be no significant difference in the number of myelinated axons between corresponding roots of the two sides. However, there were 18% more unmyelinated axons in the spared root and 22% more unmyelinated axons in the cranial adjacent intact root (root CR in Fig. 4.32b) on the rhizotomized than on the intact side, and these differences were highly significant ($P < 0.05$ for SR, $P < 0.01$ for CR). Though admitting that their results might equally reflect a *reduction* in the number of unmyelinated axons on the unoperated side, the authors preferred to interpret them as indicating an increase on the rhizotomized side. The sprouting response appears to be more vigorous in young than in aged rats, and reaches its peak by about 1 month after the lesion (Hulsebosch and Coggeshall 1983a). The lack of sprouting in the caudal adjacent intact root was considered to be due to the relative paucity of caudally directed projections of primary afferents fibres (Brown 1981).

A different, if indirect, approach to the possibility of sprouting in the adult cord was taken by Murray and Goldberger (1986), who argued that an absence of sprouting in a partially deafferented zone of the dorsal horn should lead to a permanent decrease in the total population of synapses in that region. Further, they argued that the replacement of synapses must be considered a prerequisite for sprouting to be of functional importance in behavioural recovery following lesions of the nervous system. In adult cats they unilaterally transected dorsal roots L1–S2, and between 2.5 and 6 days (acute survival) or 3–15 months (chronic survival) later, stereological and morphometric analyses were performed with the electron microscope on lamina II of the L6 dorsal horn, and on Clarke's nucleus in the L3 segment. In both situations, there appeared to be a complete replacement of lost synaptic terminals (reactive reinnervation), such that the total number of terminal profiles per cross-section remained constant, at around 3×10^4 for lamina II and 1×10^4 for Clarke's nucleus. In lamina II, at chronic survival times there was a significant decrease in the mean area of synaptic profiles, and a reduction in the mean number of synaptic contacts made by each profile. Murray and Goldberger (1986) ascribe these changes to a permanent loss of virtually all the multisynaptic *central* or *scalloped* terminals, which are thought to originate largely or entirely from primary afferent fibres, in addition to a loss of some *smaller* terminals. The degenerated terminals are replaced by an equivalent number of small terminals that make fewer synaptic contacts. Unoccupied postsynaptic densities were not observed, but there was a statistically significant fall in the number of synaptic contacts in the chronic animals, compared with normal controls. The question arises whether the methods used were of sufficient sensitivity to detect the denervation of lamina II neurons which occurs in this situation, particularly as there was no observable transient reduction in terminal count at acute times. Murray and Goldberger (1986) explain this observation by suggesting that the reactive reinnervation occurs so rapidly that no loss of terminals may be seen. In support of this interpretation, they quote a parallel study of reinnervation of the goldfish optic tectum following a crush lesion of the optic nerve (Murray and Edwards 1982). In the latter case, there *was* a detectable decrease in the number of terminals prior to reinnervation by regenerating optic nerve axons, after which the normal complement of terminals was regained. It should be noted, however, that there are some puzzling features concerning the degeneration of primary afferent fibres in lamina II of the cat following dorsal rhizotomy. First, Heimer and Wall (1968), using the Fink–Heimer technique, found that terminal degeneration was scattered and irregular in lamina II of the cat, whereas in the rat it was fairly evenly distributed across the width of the dorsal horn. In addition, Ralston (1968) reported that most sections of lamina II of the cat L7 and S1 dorsal horn revealed no detectable synaptic degeneration whatsoever in the electron microscope between 24 h and 10 days after L5–S3 dorsal rhizotomy, while, in the rat, Heimer and Wall (1968) observed many degenerating boutons in lamina II at 30 and 48 h after section of only one or two dorsal roots. These degenerating boutons often appeared to form the central part of a rosette-like arrangement of multiple postsynaptic structures, suggesting that they corresponded to the central or scalloped terminals. The latter were never observed to degenerate in the cats studied by Ralston (1968), though they occasionally did so in the feline material studied by Murray and Goldberger (1986). It would appear from these observations that, due to the more uniform distribution of degeneration in the rat dorsal horn following dorsal rhizotomy, this might be the preferred species on which to perform stereological studies of the kind described by Murray and Goldberger (1986). In contrast to the results of Murray and Goldberger (1986) for Clarke's nucleus of the cat, in 12-week-old rats whose sciatic nerves had been crushed on postnatal day 1 there was reported to be a 36% reduction in the number of synapses per unit volume of Clarke's nucleus (Bondok and Sansone

1984). Whether this is due to a species difference, to a difference between dorsal rhizotomy in the adult and nerve crush in the neonate, both of which cause degeneration of primary afferents (Chap. 2) or whether either of these studies is in error for technical reasons, remains to be determined.

The intra-axonal injection of HRP has been used extensively to study the normal morphology of primary afferent collaterals in the dorsal horn, trigeminal system, and DCN (Sects. 4.2.3 and 5.4.2). The technique enables the finest branches of collaterals and their synaptic boutons to be visualized in the light microscope, and it would therefore seem to be a method of choice to examine the possibility of sprouting of primary afferent fibres in partially deafferented regions of these systems. Unfortunately, in this regard, the technique has its limitations. The morphology of primary afferent fibre collaterals is complex and it can therefore be difficult to determine whether a particular collateral is normal, or has undergone some degree of terminal proliferation. Hallas et al. (1987) applied the technique to the adult rat trigeminal system, after partial deafferentation of SpVc by medullary trigeminal tractotomy. The operation severed the central axonal processes of a large number of trigeminal sensory neurons projecting to the maxillary region of the nucleus. Two months later physiologically identified Aß trigeminal primary afferent fibres were injected with HRP, and their terminal morphology reconstructed. Caudal to the lesion, intact (i.e. non-lesioned) vibrissa, guard hair, and other hairy skin afferents gave rise to normal-looking arbors in topographically appropriate regions of SpVc and the C1 dorsal horn, and it was inferred that no sprouting had occurred in response to the lesion. The authors state that the arbor size (area), number of collaterals, and number of boutons per collateral all fall within the normal ranges determined for these types of afferents in the medullary dorsal horn by Jacquin et al. (1986). Unfortunately, only preliminary data are presently available and, with the possible exception of mystacial vibrissa afferents, this is largely unconvincing. While the collaterals of mystacial vibrissa afferents were found to terminate in highly circumscribed longitudinally orientated columns in medial dorsal horn (Sect. 3.1, Figs. 3.3 and 3.4) those of guard hair afferents had highly variable morphology, both within and between individual afferents. It is therefore doubtful whether relatively minor degrees of terminal proliferation could be detected in the latter. The lack of sprouting of mystacial vibrissa afferents is not surprising, given the highly isomorphic organization of the terminals of vibrissa afferents (Section 3.1). The lesion used by Hallas et al. (1987) leaves primary afferent input to the vibrissa representation relatively undisturbed. Perhaps a more effective stimulus for sprouting of the mystacial vibrissa afferents would be partial deafferentation of the mystacial vibrissa representation by removal of afferents supplying some vibrissae.

The protagonists of sprouting have argued that intraspinal sprouting in regions partially deafferented by lesions may underlie changes in the levels of certain putative neurotransmitters or intraneuronal enzymes in the dorsal horn that have been observed at different times after dorsal rhizotomy. The argument hinges mainly around the question of whether substance P(SP) levels in the dorsal horn exhibit plasticity following dorsal root transection. In a series of immunochemical studies (Tessler et al. 1980; Tessler et al. 1981; Tessler et al. 1984) it was claimed that SP levels in the dorsal horn initially fell following dorsal rhizotomy, but later recovered significantly. This was first shown immunohistochemically in the L6 dorsal horn of cats which had undergone complete unilateral lumbosacral dorsal rhizotomy. In normal cats SP was distributed in all laminae of the spinal grey, but the heaviest concentration was in laminae I and II, where the staining was intense, and aggregated into coarse, irregular clumps of granules. By 4 days after dorsal rhizotomy there was a noticeable decrease in the intensity of SP staining in the dorsal horn. The apparent depletion of SP was maximal at about 10–11 days postoperatively, although it was not completely abolished, particularly from the most superficial part of lamina I. Subsequently, SP levels appeared to recover. The intensity of staining increased up to about 30 days postoperatively, by which time the distribution of SP resembled that seen in control animals, though the staining intensity remained reduced, and the reaction product appeared finer and more regular than normal (Tessler et al. 1980). It was suggested that the partial recovery of SP might be due to the sprouting of SP-containing axons, and further experiments were devised to investigate the possible sources of these axons. By placing surgical lesions in the cord it was shown that the recovery of SP did not depend on the integrity of fibres extrinsic to the segment of dorsal horn, including long and short descending pathways, short ascending pathways, contralateral dorsal root afferents, and ipsilateral ventral root afferents. The destruction of interneurons by the intraspinal injection of kainic acid completely eliminated the recovered SP, indicating that the return of SP depended on the presence of SP-containing dorsal horn neurons (Tessler et al. 1981).

A similar time-course of SP depletion and recovery was found in the cat following unilateral lumbosacral ganglionectomy by Micevych and co-workers (Micevych et al. 1986; Micevych 1986). By 21 days postoperatively, it became difficult to distinguish between the staining on the normal and lesioned sides. In addition, these workers noted a parallel decline and recovery of immunoreactivity for a second primary afferent neuropeptide, cholecystokinin octapeptide (CCK). Unlike the recovered SP staining in the experiments of Tessler and co-workers, the staining observed by Micevych et al. (1986) had a similar appearance to that in control dorsal horn, and in one cat examined 6 years after unilateral ganglionectomy the immunohistochemical appearance of SP and of CCK on the two sides of the cord was indistinguishable (unpublished observations of Micevych and Kerr, quoted in Micevych 1986). Further evidence that the recovered neuropeptides were present in intrinsic dorsal horn neurons was obtained by performing mid-line myelotomy, thoracic cord transection, or intrathecal capsaicin injection; none of these procedures affected the recovery of SP or CCK, though intrathecal injection of capsaicin had earlier been shown to deplete SP and CCK associated with primary afferent fibres in the dorsal horn (Micevych 1983).

Further quantitative studies in the rat, utilizing radioimmunoassay (Tessler et al. 1984) appeared to confirm the qualitative immunohistochemical observations. Unilateral dorsal rhizotomy reduced the concentration of both SP and somatostatin (SS) in the dorsal quadrant of the cord 10–14 days postoperatively. The concentration of SS, which is also found in some neuronal cell bodies, did not recover, indicating that perhaps, unlike SP-containing neurons, these cells were not involved in any sprouting response. While the recovery of SP and CCK in the dorsal horn would certainly lend support to the idea that synapse replacement might occur in partially deafferented region by sprouting of interneurons, some other workers have not observed the late recovery of SP levels. For example, Jessel et al. (1979) reported that after dorsal rhizotomy in the rat the SP level began to fall by 7 days, but reached its lowest level by about 28 days, and underwent no further observable change by 57 days. These differences could of course be ascribed to a species difference. Perhaps in the rat any sprouting role is taken over by non-SP-containing neurons. Alternatively, as suggested by Micevych et al. (1986), the results may reflect a unique ability of the cat dorsal horn to reorganize following dorsal rhizotomy, but there is currently no physiological evidence to lend support to this suggestion (Sect. 4.4.1).

Recently, it was reported that the mean neurotensin content of the L6 dorsal horn in the cat, 21 days after ipsilateral lumbosacral ganglionectomy, was significantly higher, by about 65%, than in either normal controls, 2- and 7-day postoperative animals or contralateral L6 dorsal horn of the same animals (Howe et al. 1987). This increase occurred in the absence of any decrease at earlier postoperative times. This observation is compatible with the theory that interneurons sprout, because, unlike SP- and CCK-containing neurons, neurotensin-immunoreactive fibres are absent from the dorsal root ganglia of the cat (Lundberg et al. 1982). Unfortunately, it is not known whether the increase in SP or neurotensin occurs in neuronal perikarya, in axon terminals, or in both. Neither is it possible to distinguish in these immunochemical studies between an increase in the number of neuropeptide-containing terminals (sprouting) and an increase in the neuropeptide concentration of existing terminals. Moreover, there is a great disparity between the time-course of synaptic replacement in lamina II, which according to Murray and Goldberger (1986) is almost instantaneous, and the much slower time-course of the changes in SP or neurotensin levels. That neuropeptide levels in the dorsal horn are capable of much more rapid change in response to dorsal root section, has been demonstrated by a significant increase in the intensity of immunoreactivity for the opioid peptide, dynorphin, only 24 h after ipsilateral dorsal rhizotomy in the rat. However, in this case at least part of the increase occurred in neuronal perikarya, rather than terminals (Cho and Basbaum 1987).

If it is accepted that the recovery and/or increase of levels of neurotransmitter candidates generally is indicative of sprouting after dorsal rhizotomy, then there is also evidence for the sprouting of descending serotonergic systems, which originate in the medulla, into the dorsal rhizotomized spinal cord of the adult cat. This was suggested by Goldberger and Paige (1986), who found that the area fraction of lamina II occupied by 5-HT-immunoreactive axons and varicosities was increased by about 50% ipsilateral to a lumbosacral dorsal rhizotomy. Following chronic spinal hemisection at T13 the lumbosacral 5-HT-immunoreactivity, which initially fell in both laminae I and II, returned to nearly normal levels in lamina II. This recovery was considered to be due to homonymous sprouting of the descending serotonergic pathways contralateral to the lesion, which overlap the ipsilateral lesioned projection. Unfortunately this preliminary report contains no information about the time-course of these changes or the nature of the data, but plasticity of these descending tracts might well contribute to the reactive reinnervation considered by Murray and Goldberger (1986) to occur in lamina II.

Using a bilateral thoracic spared-root preparation (Fig. 4.28b) (Hulsebosch and Coggeshall 1983b), McNeill and Hulsebosch (1987) noted that in rats in which one side of the cord had undergone dorsal rhizotomy 30 days earlier (chronic) and the opposite side only 4 days earlier (acute), the intensity and distribution of FRAP staining in the segment with intact innervation was greater on the side of the chronic lesion. Since FRAP is known to be a marker for a subpopulation of small-diameter primary afferent fibres and their central terminals in the dorsal horn, the increase in FRAP activity on the chronically rhizotomized side might be consistent with sprouting of the terminals of unmyelinated primary afferent fibres in the spared root. These additional terminals might originate, at least partially, from the increased number of unmyelinated axons present in the chronic spared root (Hulsebosch and Coggeshall 1983b). A major problem with all such attempts to compare the levels of transmitters or enzymes in chronically deafferented and normal tissue is common to the double-degeneration techniques discussed above. Inadequate controls exist for possible non-specific effects of long-term deafferentation on the staining or antibody-binding characteristics of the tissue. Until this problem is recognized and effectively dealt with the positive results may only be taken as suggestive of sprouting, whether by primary afferent fibres, by propriospinal neurons or by interneuronal systems.

Peripheral Nerve Injury. There is at present no unequivocal evidence that central axonal sprouting, either of primary afferent fibres or of postsynaptic neurons, is involved in any of the variety of anatomical and physiological responses to peripheral nerve injury which have been described in spinal or trigeminal somatosensory systems of mature mammals (Sect. 4.4). However, this is not surprising when one considers the controversial nature of some of the reported physiological changes, and the relatively few direct attempts to detect the presence of sprouting in animals with chronic peripheral nerve injury. As discussed in Chap. 2, peripheral nerve transection in adult mammals, at least in the medium term, may cause the death and degeneration of relatively few axotomized primary afferent fibres (Sect. 2.2). Therefore, if partial deafferentation of central neurons and the resultant loss of synaptic terminals is the most potent trigger for some kind of reactive reinnervation, it would be surprising if collateral sprouting were to occur to a significant degree after peripheral nerve section. If, on the other hand, central terminals of axotomized, but surviving, primary afferent fibres are withdrawn due, for example, to lack of a peripheral trophic influence, then some degree of reactive reinnervation might occur in the absence of extensive sensory cell death.

Transection of a peripheral nerve in the adult mammal results in a series of progressive morphological changes in primary afferent terminals, collectively referred to as *transganglionic degenerative atrophy* or more commonly *transganglionic degeneration*. Such changes have been observed in the superficial dorsal horn of the rat spinal cord following sciatic nerve transection (Csillik and Knyihar 1975; Knyihar and Csillik 1976; Csillik and Knyihar 1978; Csillik and Knyihar-Csillik 1981) and in SpVc of the cat after tooth pulp extirpation (Gobel and Binck 1977) or trigeminal nerve lesions (Arvidsson 1979). According to Csillik the normal structure of some terminals may undergo shrinkage associated with an increase in electron density, followed by glial engulfment. However, in other terminals an electron-lucent type of change may occur in which "sheet-like ultraterminal branches" (Knyihar and Csillik 1976) are formed. These, together with synapses and dendritic material, form whorl-like structures which have been termed axonal labyrinths. The entire process is considered to be a limited and unsuccessful effort at regeneration. The presence of axonal labyrinths is characteristic of the later stages of transganglionic degeneration, when there might be as many as 300 per mm^2 in sections of the substantia gelatinosa (Csillik and Knyihar-Csillik 1981).

These structures in the substantia gelatinosa (lamina II) appear to be formed from the terminals of unmyelinated afferent fibres. Although the central processes of myelinated afferent fibres also undergo some degree of ultrastructural alteration (Knyihar and Csillik 1976), it is not clear whether they also take part in the formation of the labyrinth-like structures. Similar changes have been observed in the absence of any signs of nerve axon degeneration 10 days after the application of inhibitors of axoplasmic transport (Vinca alkaloids) to the sciatic nerve (Csillik 1977; Csillik and Kyihar-Csillik 1981). Thus the axonal labyrinths might represent the final stage of the central terminals of those axotomized sensory neurons which, following nerve transection, undergo transganglionic changes but do not actually die.

Despite these degenerative changes, many central terminals of axotomized sensory neurons appear to retain an essentially normal ultrastructure. This has been shown in experiments in which HRP was applied to the central stump of the superficial radial nerve in adult cats 1 month or 3 months after transection of the nerve. Of the many primary afferent terminals in the C6 through C8 dorsal horn which were labelled transganglionically with HRP, all had normal ultrastructure. Moreover, the size distribution of labelled neuronal perikarya in the C6 through C8 dorsal root ganglia was the same as in the contralateral control ganglia (Sugimoto and Gobel 1982). This is not surprising, however, as the neurons which retain the ability to transport HRP are also likely to retain sufficient levels of axonal transport to maintain a normal central terminal arbor and to prevent atrophy of the cell body.

Since most sensory neurons appear to survive peripheral nerve transection, an important question is whether the surviving axotomized neurons maintain their synaptic contact with postsynaptic neurons, or whether they are lost, thus allowing reactive reinnervation of vacated postsynaptic sites to take place (Murray and Goldberger 1986). Clearly, if reactive reinnervation can occur only at sites vacated by degenerating primary afferent fibres, its scale will be severely restricted after peripheral nerve section, compared with dorsal rhizotomy.

Gobel (1984) examined the effects of peripheral nerve injury on postsynaptic structures in laminae I and II of the medullary dorsal horn following extirpation of the tooth pulps of all the mandibular teeth in adult cats. By 30 and 60 days postoperatively, cavitation of small dendrites ($< 2\,\mu m$ diameter) was seen in laminae I and II. This condition preceded the destruction of the affected dendrites. Cavitation occurred despite the presence of synaptic connections from a variety of axon terminals, including scalloped and dome-shaped terminals, and small fragments of dendritic debris were often found to be still in contact with axodendritic synaptic terminals. Many synaptic glomeruli in lamina II, major synaptic sites between primary afferent fibres and second-order neurons, displayed empty spaces where small dendritic profiles once occupied their scalloped depressions. Similar dendritic cavitation was seen in small ($< 2\ \mu m$ diameter) and medium-sized (3-4 μm diameter) dendrites of laminae I–IV of the dorsal horn in the C7 segment of adult cats between 2 weeks and 2 months after section of the superficial radial nerve (Sugimoto and Gobel 1984). Thus, while peripheral nerve section may induce transsynaptic degenerative changes in dendrites, there is no ultrastructural evidence that it causes postsynaptic sites on healthy dendrites to be vacated. On the contrary, the severance of synaptic connections on the degenerating dendrites appears to occur relatively late in the degenerative process.

To date very few studies have sought to examine directly any possible relationship between sprouting and somatotopic changes. Seltzer and Devor (1984) made one such attempt to determine whether sprouting of afferents from the thigh into the medial dorsal horn might be responsible for the mediolateral reorganization of dorsal horn somatotopy observed by Wall and co-workers in the cat and rat (Sect. 4.4.3). The projection of axons and terminals of the sciatic nerve and of "thigh nerves" (the posterior cutaneous nerve of the thigh, the pudendal nerve, the perineal nerve and branches of the gluteal nerves) to the lumbosacral dorsal horn of adult rats was examined by means of transganglionic labelling with WGA-HRP after injection of the lectin into the crushed nerves. In normal control animals and on the unoperated side of animals with unilateral chronic sciatic and saphenous nerve transection, the thigh nerves were found to project to the lateral one-third to half of the dorsal horn, at the same rostrocaudal level as the sciatic nerve, which projected to the medial region. After chronic (4–12 days) transection and ligation of either the sciatic nerve alone or both the sciatic and saphenous nerves there was no indication of any medial expansion of the thigh nerve terminal region in any of the laminae of the dorsal horn.

Recent anatomical investigations have indicated that, under certain conditions, intraspinal sprouting of the afferent terminals of C fibres may occur following peripheral nerve damage in the adult rat. In the first of these studies Molander et al. (1988) labelled the spinal terminals of saphenous nerve afferents transganglionically with HRP in adult rats in which the sciatic nerve had been transected 8 weeks earlier. If the saphenous nerve was crushed at the time of the sciatic nerve transection, the area of labelling in lamina II of the dorsal horn increased by about 13%. In the second study, LaMotte et al. (1989) examined the saphenous nerve projection by transganglionic labelling with HRP and WGA-HRP 4 months after destruction of ipsilateral sciatic nerve afferents with pronase (Sect. 2.2.6). The area of the saphenous projection to the superficial dorsal horn on the treated side was increased by about 26%. The zone of termination of sciatic afferents, as determined by transganglionic labelling, borders that of saphenous afferents medially and laterally, as well as caudally (Devor and Claman 1980; Fitzgerald 1985; Swett and Woolf 1985; Molander and Grant 1986) (Sect. 4.3.3). Molander et al. (1988) found that the expansion of labelling was greatest at the caudal end of the saphenous nerve projection, and, in fact, their data show no clear evidence of significant mediolateral expansion of saphenous afferents. LaMotte et al. (1989) claimed to find a small degree of mediolateral expansion, but only into a fringe region of the saphenous projection that was lightly labelled on the control (unoperated) side. By comparison, the caudal expansion occurred into a level that, on the control side, was devoid of labelling.

Unfortunately, owing to the vagaries of the technique, it is difficult to make direct comparisons between the results of different transganglionic tracing studies. Thus, for example, while some workers have reported a significant overlap between the "normal" sciatic and saphenous terminal fields in lamina II of the rat dorsal horn, others have claimed that there is no overlap (Sect. 4.3.3). Consequently, it is difficult to assess the significance of the small mediolateral changes in the extent of labelling reported by LaMotte et al. (1989). However, as will be shown below, the relatively large degree of rostrocaudal expansion of labelling in these experiments may be explained on the basis of localized sprouting of sparse, somatotopically inappropriate collaterals of saphenous afferents. It is therefore not necessary to postulate long range sprouting of primary afferent fibres to explain these results. It is also important to note that these experiments do not provide information about sprouting of large diameter myelinated fibres, which would probably not be demonstrable by HRP or WGA-HRP labelling, at least in the rat.

Indirect studies where the effect of peripheral nerve section on the levels of various transmitter candidates or intra-neuronal enzymes have been investigated, point to a lack of central sprouting by intact primary afferent fibres following peripheral neurectomy. After sciatic nerve section, FRAP in the dorsal horn in rats is severely depleted from the central terminal regions of peripherally axotomized nerves. The subsequent demonstration that FRAP was contained within many primary afferent sensory neurons and their terminals within the substantia gelatinosa (Coimbra et al. 1970; Knyihar and Gerebtzoff 1973; Knyihar and Csillik 1976; Knyihar-Csillik et al. 1986) indicated that the depletion was a result of changes within the axotomized primary afferent neurons. This phenomenon was subsequently used to map the distribution of the trigeminal nerve's projection to the rat substantia gelatinosa (Rustioni et al. 1971) and to map the three-dimensional terminal distribution of various hindlimb nerves in the rat lumbosacral spinal cord (Devor and Claman 1980). In intact rats, lamina II of the dorsal horn stained intensely for FRAP, and transection of the sciatic nerve caused the virtual elimination of FRAP activity from U-shaped region extending through L2–L5 in he medial part of lamina II. This depletion was already detectable by 2 days, and was complete by 4 days after nerve section. This size and shape of the area of FRAP depletion did not change subsequently for up to 6 months, indicating that the region of sciatic nerve termination was not invaded by sprouting FRAP-containing terminals of afferents from intact nerves. A late (> 12 months) return of FRAP to the depleted region in these experiments was considered by the authors to be due to a recommencement of transport of FRAP by some of the injured axons, rather than sprouting, because the filling-in of the depleted region occurred evenly over the whole region, rather than moving in as a wave from the edges. This late return of FRAP has not been observed in a more recent investigation (Shehab and Atkinson 1986a). Perhaps a more likely explanation for the late return of the enzyme in the experiments of Devor and Claman (1980) is that there was some undetected regeneration of fine, FRAP-containing afferents at these very long postoperative times.

That there is no gross sprouting of primary afferent fibres into terminal regions of axotomized nerves may also be deduced from the behaviour of some neuropeptides after nerve transection. All neuropeptides that have been found in primary afferent neurons, including SP, SS, CCK, vasoactive intestinal polypeptide (VIP), angiotensin II, and bombesin, are located in the superficial dorsal horn, where each has a similar distribution (for a review see Salt and Hill 1983). Substance P is depleted from the terminal region of a transected peripheral nerve within the superficial dorsal horn in a similar way to FRAP (Barbut et al. 1981). In a quantitative study, the SP content of the rat lumbar dorsal horn fell by about 75%–80% after sciatic nerve transection, and remained low for at least 2 months (Jessell et al. 1978). Recently it was found that depletion of not only SP, but also SS and CCK begins about 7 days after sciatic nerve section in he rat, and is maximal by 15 days (Shehab and Atkinson 1986a). The region of depletion of all three neuropeptides was identical to that of FRAP depletion. Since these workers examined animals as long as 12 months postoperatively, and did not describe any late resurgence of neuropeptide levels, it would seem that there is no recovery during this time. Interestingly, the behaviour of VIP was quite different from that of the other neuropeptides. The initial control levels of this neuropeptide in the dorsal horn were quite low, but began to increase 10 days after sciatic nerve section, reaching a maximum by 15 days and remaining at the same level thereafter. The spatial distribution of increased VIP levels was, again, identical to that of FRAP, SP, SS and CCK depletion. Further, the increase was dependent on the integrity of the dorsal roots, showing that the source of the VIP was sensory neurons in the dorsal root ganglion, and not spinal neurons (Shehab and Atkinson 1986b). However, the absence of any recovery of SP, SS, and CCK at long postoperative times, like the behaviour of FRAP discussed above, implies a lack of gross sprouting by intact primary afferent fibres containing these markers into the depleted region. The lack of recovery, after peripheral nerve section, of SP and more particularly SS, which is found in a large number of terminals of spinal interneurons in the rat (Hokfelt et al. 1976; Priestley et al. 1982) indicate further that these intrinsic systems also do not sprout. The increase in VIP levels might be interpreted as being due to sprouting of intact VIP-containing afferents into the terminal region of the transected nerve. However, Shehab and Atkinson (1986a) give two reasons for discounting this possibility: (1) the increase in VIP-immunoreactivity occurred uniformly throughout the region of sciatic nerve termination, and (2) section of other nerves which terminate in the same segments as the sciatic did not prevent the increase in VIP, which then occurred also in the region of projection of the other transected nerves. About 40% of small and medium-sized L5 dorsal root ganglion cells become heavily immunoreactive for VIP after section of the sciatic nerve and other nerves terminating in the same segments (saphenous, femoral, posterior or lateral cutaneous nerves of the thigh, pudendal and perineal nerves and cutaneous nerves of the lower back). In addition, terminal areas of the sciatic nerve in nucleus gracilis of the DCN contain VIP-positive terminals 28 days after sciatic neurectomy, whereas the normal nucleus gracilis contains no VIP (Shehab and Atkinson 1986a). Thus the most likely explanation for the VIP response is that it is due to a change in peptide expression in injured sensory neurons, rather than sprouting of intact VIP-containing afferents.

The rostrocaudally directed somatotopic reorganization of the receptive fields of SCT cells in the dorsal horn of the adult cat following chronic denervation of a single toe (Sect. 4.4.3) (Wilson and Snow 1986), together with the known rostrocaudal distribution of the somatotopically inappropriate collaterals of Aß hair follicle afferents (Sect. 4.3.3) (Meyers and Snow 1984; Snow and Meyers 1985), led Wilson and Snow (1987) to suggest that the altered somatotopy might result, at least in part, from changes in the somatotopically inappropriate collaterals of large-diameter primary afferent fibres. Electrophysiological studies, in which signal averaging was used to record extracellularly from the collateral arborizations of single afferents in somatotopically identified regions of the dorsal horn, have indicated that at least some of the somatotopically inappropriate collaterals are invaded by action potentials (Meyers et al. 1984; Meyers and Snow 1986). However, their apparent lack of synaptic boutons suggests that under normal circumstances they are unlikely to contribute to the integrative activity of local dorsal horn neurons (Wilson and Snow, 1990a). In the case of the SCT cell, which is the only type of functionally identified dorsal horn neuron whose primary afferent input has been studied in detail, there is convincing evidence from intracellular recordings that any excitatory subliminal fringes are not only very small and adjacent to the receptive field, but are displaced in a proximo-distal direction from the firing zone of the cell (Brown et al. 1987). Furthermore, both electrophysiological (Brown et al. 1987) and anatomical (Brown and Noble 1982) studies of pairs of SCT cells and single Aß hair follicle afferents with overlapping or non-overlapping receptive fields have indicated that subliminal excitatory effects are unlikely to be mediated monosynaptically from the somatotopically inappropriate collaterals of Aß hair follicle afferents. Synaptically inactive, somatotopically inappropriate projections of this kind would not of course be detected by electrical stimulation of afferents (Sect. 4.5.1). Following peripheral nerve injury, however, localized proliferation and synaptogenesis could occur in the somatotopically inappropriate collaterals, possibly as a result of the degeneration of a small proportion of axotomized afferents (Sect. 2.2.1). The somatotopically inappropriate collaterals would then be able to evoke excitatory actions on the partially deafferented dorsal horn neurons. Whether these excitatory actions are evoked monosynaptically or polysynaptically in any particular type of dorsal horn neuron, such as the SCT cells, is of little consequence for the location of the new receptive fields. Abnormal inhibitory receptive fields might also be anticipated to develop as a consequence of sprouting of primary afferent fibres, but these have not yet been documented. There is as yet no direct anatomical evidence of localized sprouting of the rostrocaudally distributed, somatotopically inappropriate collateral arborizations of primary afferent fibres in the mammalian dorsal horn following peripheral nerve transection *in the adult*. However, Snow and Wilson (1989), using intra-axonal injection of HRP, have obtained direct evidence for such sprouting in the dorsal horn of cats which have undergone denervation of either one or two digits in the first postnatal week (Sects. 4.6.2 and 4.6.3), and they have made the suggestion that a similar phenomenon may occur, although perhaps to a reduced extent, in the adult cord. The recent anatomical evidence of "sprouting" in the adult rat dorsal horn described above (Molander et al. 1988; LaMotte et al. 1989) also lends support to this hypothesis. While the increased area of labelling in these studies may well reflect some new growth of saphenous terminals into somatotopically inappropriate regions of dorsal horn, the greater caudal expansion observed might result from an increase in the density of preexisting somatotopically inappropriate projections of saphenous afferents that are normally too sparse to be detected with transganglionic tracing techniques at the light microscope level.

Effect of Lesions on the Morphology of Dorsal Horn Neurons. Unfortunately, little emphasis has been placed on possible anatomical changes of postsynaptic neurons as a direct cause of plasticity in the dorsal horn. Golgi studies of dorsal rhizotomized segments of the cat dorsal horn indicated only that there might be atrophy of some dendrites (Brown et al. 1979), but this was not borne out in a light-microscopic study of SCT cells intracellularly injected with HRP (Brown et al. 1983). In the latter study it was found that partially deafferented SCT cells maintained a relatively normal somatic and dendritic morphology following dorsal root section, although there were minor signs of an abnormal extension of dorsal dendrites into lamina II. At the same time, however, there were no signs of reorganization of the SCT cell's receptive field (Sect. 4.4.1). In contrast, Sedivec et al. (1986) claim to have found definite morphological changes in SCT cells in a chronically and partially deafferented segment of cat spinal cord. SCT cells stained by intracellular injection of HRP had dendrites with significantly larger diameters than normal, though their length was unchanged. Sedivec et al. (1986) suggested that SCT cells in deafferented cord may undergo limited growth and possible dendritic reorganization. This is potentially an important observation in relation to plasticity, and deserves further attention. However, while morphological changes in dorsal horn neurons may well play a crucial role in mechanisms of somatotopic plasticity in the adult mammal, there is currently little direct evidence to implicate this kind of structural plasticity in the reorganization of receptive fields following lesions.

4.6 Plasticity of the Developing Dorsal Horn

A major goal of developmental neurobiologists is to understand how specific connections are formed between sets of neurons during ontogeny. The dorsal horn of the mammalian spinal cord has been considered by some to be a useful area for such studies, because in some species with a short gestation period, such as the rat, there is appreciable development of synaptic connectivity after birth. For example, behavioural studies of the results of mid-thoracic spinal transection in infant rats have shown an increased capacity for recovery of some hindlimb postural reflexes when the lesion is made between P 0 and P 15 (Weber and Stelzner 1977). In contrast, a lesion made after this time caused permanent impairment of these functions. The difference between neonatal and adult lesions is thought to be related to the development of descending pathways to the dorsal horn, which do not attain their mature form in the lumbosacral cord until P 15 (Gilbert and Stelzner 1979). Moreover, in the substantia gelatinosa (lamina II) of the rat both interneurons and afferent input also appear to undergo substantial maturation during the postnatal period (Bicknell and Beal 1984; Fitzgerald 1985b), and there is an increasing body of evidence that this part of the dorsal horn system is able to undergo considerable plastic changes in response to lesions made during the immediate postnatal period.

4.6.1 Development of Dorsal Horn Neurons and Primary Afferents

Neurogenesis begins in the rat spinal cord on G 11 and is complete by G 16. The birth of neurons proceeds in a ventrodorsal direction such that the neurons of the substantia gelatinosa are formed last (Nornes and Das 1972). There is a rostrocaudal gradient in the maturation of the spinal cord and the dorsal root ganglia, such that more rostral segments mature earlier than more caudal segments (Hamburger and Levi-Montalcini 1949; Nornes and Das 1974; Lawson 1979). Lumbosacral dorsal root ganglion cells are generated between G 11 and G 15 in the rat (Lawson et al. 1974). On G 12 bundles of nerve fibres from the sensory ganglia and ventral roots leave the spinal cord and combine to form the spinal nerves. By G 13 many fibres have entered the base of the limb bud, which at this stage consists solely of mesenchymal cells. Differentiation of the limb begins on G 14, and by G 15 bundles of nerve fibres project into the limb, with some running distally (Comans and Hendry 1983). By G 17, many hindlimb afferents are responsive to peripheral stimulation (Fitzgerald 1987b).

Transganglionic HRP studies have shown that primary afferent fibres send collaterals into the grey matter 2 days after their central axonal processes have entered the dorsal columns (Smith 1983). In the thoracic cord, primary afferents begin to form terminal arborizations in the deeper laminae around G 17–18 and the branching pattern of these afferents is similar to that of low-threshold cutaneous myelinated afferents in the adult animal. Substantial development of afferent input to the superficial laminae (laminae I–III) appears to occur postnatally. At the time of birth, fibres in lamina IV have dorsally directed branches that penetrate the deeper part of lamina III, and over the succeeding 4 days some of these extend up to lamina II. Initially, the results of transganglionic HRP studies indicated that afferent projections to laminae I and II in the rat might develop only postnatally, although it was also suggested that the fibres might be too fine before this time to be resolved by this light-microscopic method (Smith 1983). Transganglionic labelling of sciatic nerve afferents on P0 with WGA-HRP revealed dense afferent projections to lamina I and II by P2 (Fitzgerald and Swett 1983) and systemic capsaicin treatment on P2 caused terminal degeneration in this region within a few hours (Jancso and Kiraly 1978). Subsequently however, Fitzgerald (1987a), using transganglionic transport of WGA-HRP in rat embryos, found that fine afferents grow into the lumbar dorsal roots at G 19. By G 19.5 they have penetrated the dorsal horn and formed sparse terminations in the outer part of the substantia gelatinosa (lamina IIo). By G 20 they have reached the inner part of the substantia gelatinosa (lamina IIi) and have increased in density, and by the time of birth the labelling in laminae I and II is comparable to that seen in the first postnatal week.

The development of C fibre input to the dorsal horn in the rat has also been studied by specific immunocytochemical staining for SP and by histochemical studies of the enzyme FRAP, both of which are markers for separate populations of unmyelinated sensory fibres (Nagy and Hunt 1982). Substance P first appears in spinal neurons at G15, but does not appear in sensory neurons of the dorsal root ganglia and their processes in the dorsal columns or dorsal horn until G17–18 (Senba et al. 1982; Peckel et al. 1982). Postnatally, SP-containing fibres are concentrated in lamina I on P1. In lamina II their density increases over the first week, after which it resembles that in the adult (Fitzgerald and Gibson 1984). FRAP staining appears in the cord postnatally. On P1 FRAP is already present in lamina II but staining is weak until P5 or P6 (Mattio et al. 1981; Fitzgerald and Gibson 1984).

Primary afferent fibres project to the dorsal horn in a precise somatotopic pattern that is reflected in the somatotopic organization of the receptive fields of dorsal horn neurons (Sect. 4.3.3). Experiments involving labelling of peripheral nerves with HRP in neonatal rats have shown that a miniature version of the normal adult central projection pattern of peripheral nerves is present from the time that afferents first invade the dorsal horn (Smith 1983; Fitzgerald and Swett 1983). Thus, unlike the visual system, in the somatosensory system there is no evidence that afferents initially make diffuse projections that are withdrawn at a later developmental stage.

A possible exception to this is the primate dorsal horn. Knyihar et al. (1978), using electron-microscopic and tritiated thymidine autoradiographic analysis of the fetal rhesus monkey dorsal horn, found that afferent fibres initially make synaptic contact with cells in the border region of the primordial dorsal horn at G31, before the permanent neurons of the dorsal horn have been born. By G50, most of these border cells were found to be degenerating, and eventually all disappeared. It is not known whether the primary afferent axons that make these initial contacts all degenerate, or whether some of them form new synapses with newly generated dorsal horn cells. No temporary synaptic relationships of this kind have been found in the primordial spinal cord of birds or rodents.

The axonal and dendritic development of neurons of the substantia gelatinosa of the neonatal rat have been studied in Golgi impregnated tissue by Bicknell and Beal (1984). These workers found that neurons whose axons entered the white matter (projection and propriospinal neurons) underwent maturation prenatally by a process of dendritic growth and branching. In contrast, neurons with axons that remain within the grey matter (intrinsic non-projection neurons) mature postnatally by a process involving an intermediate stage (star-shaped cell) in which the cell forms a large number of short beaded dendrites that radiate from the cell body. This intermediate stage then undergoes a further maturation in which the dendrites are rearranged into the adult pattern. These changes were observed as late as P 20. It would therefore appear that fundamental changes of neuronal connectivity probably occur postnatally among interneurons of the substantia gelatinosa, although it is not clear how these changes might relate to the demonstrated functional changes in dorsal horn neurons. One possible contributory factor to the expanded receptive fields reported by Fitzgerald (1985b) in the neonatal rat dorsal horn (see below) might be a delayed development of the organization of segmental inhibitory mechanisms.

Certainly, descending inhibition appears to be weak in the neonatal rat as evinced both by a lack of descending axonal projections to the dorsal horn (Gilbert and Stelzner 1979), and by the rapid recovery of postural reflexes following mid-thoracic spinal transection (Weber and Stelzner 1977). However, some dorsal horn neurons in the neonatal rat do have inhibitory as well as excitatory cutaneous receptive fields (Fitzgerald 1985b), and therefore inhibitory interneurons must be operative at this time.

4.6.2 Functional Plasticity in Development

From G17 in the rat a flexor withdrawal reflex may be evoked by pinching the skin of the hindlimb (Angulo-y-Gonzalez 1932; Windle and Baxter 1936; Narayanan et al. 1971), suggesting that some nociceptive afferents are in place and making connections with dorsal horn neurons on reflex pathways. Fitzgerald and Gibson (1984) studied the development of the flexor reflex in neonatal rats lightly anaesthetized with urethane and found that while a response to pinch and noxious heat was present (and even exaggerated) from birth, no response to the irritant chemical mustard oil could be obtained until P 10–11. Development of neurogenic oedema produced by mustard oil or by antidromic activation of sciatic nerve C fibres was likewise delayed. There are, of course, many possible reasons for the delayed development of these functions, both of which are mediated by polymodal nociceptors (Kenins 1981). The most parsimonious explanation would seem to be that the delayed onset of chemical responsiveness is due to a delayed development of chemosensitivity by polymodal nociceptors. However, Fitzgerald and Gibson (1984) appear to favour the alternative suggestion, that the central connections of all sensory C fibres are not fully functional until the second week of life. This explanation implies, of course, that the earlier-appearing flexor reflex responses are in fact mediated via developing A fibres. Owing to the slow myelination of A fibres in the neonate this question is unfortunately not easily resolved by considerations of latency and threshold. Fitzgerald (1985b) went on to examine the responses of dorsal horn neurons to natural and electrical stimulation of the skin in rats aged 0–15 days, and made observations which also suggested that the maturation of C-fibre input is largely a postnatal process. She found that on days P0–P3 both background and evoked activity was low for neurons throughout the dorsal horn in rats anaesthetized with urethane. Most cells with cutaneous receptive fields responded only to pinching the skin. Measurement of response latency to electrical stimulation of the skin at an intensity twice the threshold for evoking either A or C fibre volleys in the dorsal root showed that until P7–8, responses to both A and C fibre inputs could be obtained only from cells in superficial laminae (laminae I–III). In deeper cells (laminae IV–VI), only A fibre responses were obtained at these early times. Convergence of inputs from A and C fibres onto single dorsal horn neurons (found only in the superficial laminae during

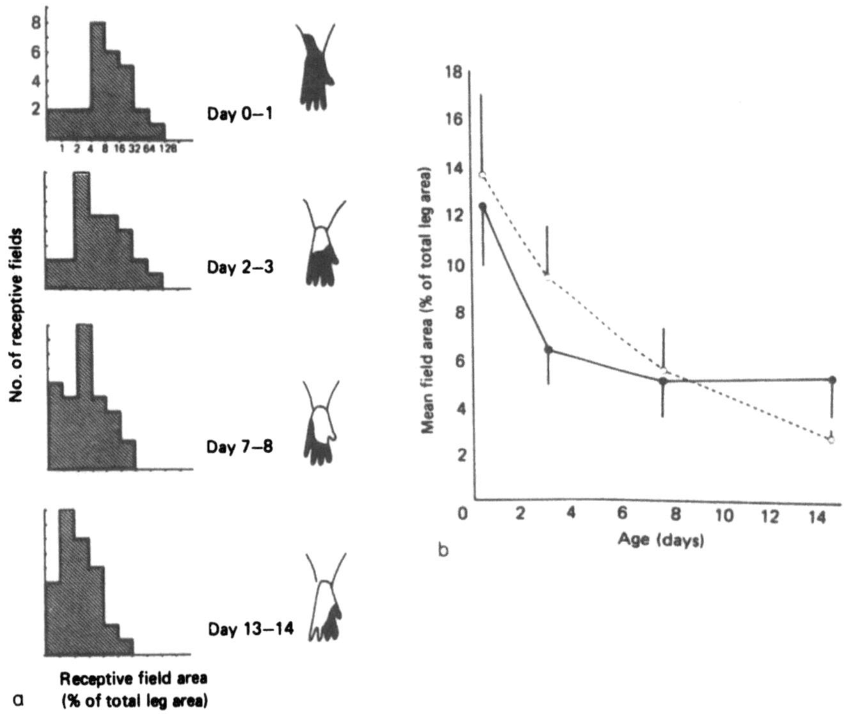

Fig. 4.33 a, b. Receptive field size of dorsal horn neurons in the early postnatal rat. **a** Histograms of the distribution of receptive field areas of dorsal horn neurons in neonatal spinal cord. Receptive fields are expressed as a percentage of total hind-limb area to allow for rapid growth (note logarithmic scale). Note the gradual decrease in receptive field size over the first 2 postnatal weeks. The figurines indicate typical receptive fields. **b** Graphical representation of change in mean ($\pm$ SE) receptive field size with age from P 0 to P 14–15. *Filled circles* indicate results from cells in superficial laminae (I–III), and *open circles*, cells in deep laminae (IV–V). (From Fitzgerald 1985b)

the first week) was rare at birth, but became correspondingly more common with age. A third important observation was that the average size of cutaneous receptive fields decreased markedly over the first two weeks of life (Fig. 4.33). The shrinkage of receptive fields was particularly marked for cells of the deep dorsal horn. A final unusual feature of some dorsal horn neurons in the neonate was the presence of a marked after-discharge following a single stimulus, or *wind-up* of the response to repeated stimulation (Sect. 4.5.3). This effect is comparable to the exaggerated response and wind-up of the flexor reflex also reported to occur in neonatal rats (Fitzgerald and Gibson 1984; Stelzner 1971). Both are considerably reduced by the end of the first postnatal week.

There is some anatomical evidence that the terminal arborizations of large-diameter myelinated afferent fibres in lamina III undergo considerable postnatal development in the cat (Scheibel and Scheibel 1968b). Furthermore, postnatal maturation of substantia gelatinosa neurons of a similar nature to that described for the rat by Bicknell and Beal (1984) has also been reported to occur in the cat (Falls and Gobel 1979). Nevertheless, a recent study of the somatotopic organization of the light tactile receptive fields of dorsal horn neurons in the chloralose-anaesthetized neonatal cat (Wilson and Snow 1988a) failed to show significantly larger receptive fields than those found in the adult. We assume that the difference between the cat and the rat is due to the difference of gestation period (cat 62 days, rat 22 days), although it is also possible that the high-threshold receptive fields of dorsal horn neurons, which Wilson and Snow (1988a) did not examine, might be larger in the neonate, especially if they are associated with a different set of dorsal horn cells.

4.6.3 Somatotopic Reorganization Following Neonatal Peripheral Nerve Lesions

When either toe 3 alone or toe 3 and toe 4 were denervated in kittens during the first postnatal week, and the somatotopic organization of SCT cells in the toe representation in the lumbosacral dorsal horn mapped under chloralose anaesthesia about one year later, Wilson and Snow (1988b) discovered a marked somatotopic reorganization throughout the deprived region of the toe representation. Many SCT cells which would normally have had a receptive field on the denervated skin had new receptive fields on skin normally represented rostral or caudal of their location in the cord. Where both toes 3 and 4 were denervated, SCT cells having receptive fields on the skin of toe 2 were found up to 5 mm caudal to the normal caudal boundary of the toe 2 representation (Snow and Wilson 1989; Wilson and Snow 1988b, 1990b). This is the largest shift of somatotopic boundaries recorded in the dorsal horn following peripheral nerve section. A large proportion of SCT cells in the reorganized region were found to have two or more small receptive fields, situated in many cases on both of the immediately adjacent unoperated toes. This rostrocaudally directed reorganization is thus similar to, but more extensive than that seen after chronic denervation of a single digit in the adult cat (Wilson and Snow 1987a) (Sect. 4.4.3). Because of the large rostrocaudal shifts of somatotopic boundaries, the authors considered that the reorganization in both cases might result from local proliferation and synaptogenesis, at the rostrocaudally distributed, somatotopically inappropriate collaterals of Aß primary afferent fibres (Figs. 4.13 and 8.2) (Meyers and Snow 1984) (Sects. 4.4.3, 8.3 and 8.4). Owing to the relatively greater attrition of primary sensory neurons following peripheral nerve section in neonates than in adults (Sect. 2.2.1), there is likely to be a more extensive deafferentation of dorsal horn neurons when peripheral nerves are chronically transected in the neonate. This could lead in turn to an enhanced synaptogenesis in the somatotopically inappropriate collaterals and on the basis of competition for synaptic space could account for the large number of split receptive fields in the deprived region following neonatal nerve section (Wilson and Snow, in preparation). Preliminary evidence from intra-axonal staining with HRP has strongly indicated that the suggested kind of morphological changes may indeed occur in neonatally nerve-sectioned kittens (Fig. 4.35) (Sect. 4.6.4).

4.6.4 Anatomical Plasticity of Neonatal Afferent Projections

Effect of Neonatal Capsaicin. The observations of Nagy and Hunt (1983) showing sprouting of myelinated afferent fibres from lamina III into lamina II after neonatal capsaicin treatment in rats have been discussed in relation to the enlargement of receptive fields of dorsal horn neurons (Sect. 4.5.4). In a more recent study employing Golgi staining in neonatally capsaicin-treated rats, Beal and Knight (1987) found that both ascending and descending collateral branches of large myelinated afferent fibres exhibited abnormal arborizations in lamina II. A dorsal expansion of coarse myelinated afferents in the dorsal horn of neonatally capsaicin-treated rats was also found by Rethelyi et al (1986), who labelled primary afferent fibres with HRP applied to transected lumbar dorsal roots. These workers further noted that flameshaped arborizations in the lateral part of the dorsal horn reached only to the inner part of lamina II, whereas a different class of coarse afferent in the medial part of the dorsal horn sprouted as far as the grey-white border. This mediolateral difference was ascribed to differences in afferent innervation of hairy and glabrous skin. Electron microscopic examination of the sprouted arborizations revealed that they formed large numbers of complex synapse-like structures in the "foreign" neuropil. Sprouting has also been reported by Shortland et al. (1990), who intraaxonally injected Aß hair follicle afferents (HFAs) with HRP in the lumbar dorsal horn of neonatally capsaicin-treated rats. Of a sample of 14 HFAs injected in these animals, the proportion of collateral arborizations that entered lamina II was 71%, compared with only 27% of collaterals of 24 HFAs in normal rats. It therefore seems that the neonatal removal of unmyelinated fibres may provide the stimulus necessary to elicit sprouting of nearby terminations of myelinated afferent fibres. One possibility is that afferents which survive capsaicin treatment have access to a greater supply of NGF or other trophic material from the periphery. Alternatively, the presence of a normal C fibre innervation of lamina II may block the dorsal expansion of myelinated afferents by occupying all available synaptic sites in this lamina.

Structural Plasticity of Afferent Fibres Following Chronic Neonatal Nerve Section. In contrast to the mounting evidence against long-range axonal sprouting in the dorsal horn of adult mammals (Sect. 4.5.5) it appears that primary afferent fibres in neonates have a variable ability to sprout in the dorsal horn following peripheral nerve lesions. The first report to show this difference between adult and neonate was that of Kerr (1975), who studied the effect of chronic trigeminal root section in kittens in the first postnatal week on the projection of cervical root afferents to the spinal division of the TNC. Kerr (1975) found that compared with the situation following root section in the adult (Kerr 1972) (Sect. 4.5.5) there was a mildly increased degeneration of cervical afferents in the C1 dorsal horn, and in SpVi on the chronically denervated side, particularly in kittens whose trigeminal roots were evulsed during the first three postnatal days. However, he concluded that primary afferents of the upper cervical roots have little potential for sprouting even at early stages.

Where sprouting of C fibres has been studied, the results have been more dramatic. In the rat, Fitzgerald (1985a) noted that transection of the sciatic nerve on days

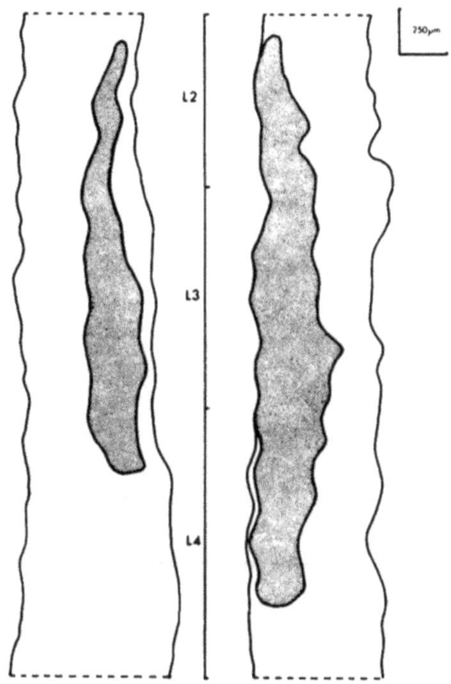

Fig. 4.34. Distribution of saphenous nerve terminals in a horizontal view of lamina II of the rat dorsal horn, revealed by retrograde transport of WGA-HRP. The *wavy vertical lines* represent the medial and lateral boundaries of the lamina II grey matter. The *central vertical line* represents the midline, and boundaries of segments are shown. *Stippled area* shows the extent of terminal labelling in the dorsal horn. On the *left* is shown the control labelling in a normal rat. On the *right* the sciatic nerve was sectioned on P 1. Note the apparent mediolateral and caudal expansion of the saphenous nerve projection to lamina II. (From Fitzgerald 1985a)

P 0–P 3 caused a significant spread of the central arborizations of the saphenous nerve in lamina II as indicated by transganglionic labelling with WGA-HRP. The projection region of the saphenous nerve was enlarged both mediolaterally and caudally in the direction of the projection zones of the superficial peroneal, tibial and sural branches of the sciatic nerve (Fig. 4.34). The central projection of C fibres in the sciatic nerve to lamina II of the neonatal rat has also been examined by the method of FRAP depletion following nerve transection (Sect. 4.5.5). As in the adult rat, following crush or transection of the sciatic nerve on P1 FRAP staining disappears from the region of dorsal horn occupied by sciatic nerve afferents. However, after the neonatal lesion the region of FRAP depletion shrinks with time (Fitzgerald and Vrbova 1985). The region of FRAP depletion does not disappear completely. Maximal shrinkage of the area of dorsal horn from which the enzyme is depleted following sciatic nerve section is to about 23% of normal, and is achieved between P 12 and P 30. The shrinkage of the region of FRAP depletion is due to a filling-in of the sciatic terminal region by a thinner, less dense, line of FRAP staining. This new FRAP appears to result from the sprouting of adjacent nerve terminals, since it is unaffected by resection or recrush of the regenerated sciatic nerve. A second major effect of the neonatal nerve transection is to produce a considerable stunting of growth of the ipsilateral dorsal horn at the level of the sciatic nerve projection, presumably due to either a loss of trophic material normally provided by the sectioned nerve, or a loss of afferent terminals, or both factors. Peripheral nerve transection in neonatal rats has been shown to result in considerable and rapid loss of axotomized sensory neurons (Sect. 2.2.2). Thus, unlike nerve section in the adult, neonatal nerve section could be expected to produce significant deafferentation of neurons in the nerve's central projection region. In fact, owing to the lack of overlap in the dorsal horn between the central projections of different peripheral nerves (Sect 4.3.3), such a lesion would produce a more profound deafferentation of dorsal horn neurons than that which follows a limited dorsal rhizotomy. However, in addition to the deafferentation of neurons, there must also be other factors which encourage long-range axonal sprouting in neonates, because dorsal rhizotomy *per se* does not appear to lead to such sprouting in the adult cord (Sect. 4.5.5). These factors are unknown, but are likely to be similar to the trophic factors that affect the collateral sprouting of primary afferent fibres in the periphery (Sect. 2.3.2). The functional consequences of the neonatally induced sprouting of C fibres have not yet been investigated.

Structural Plasticity of Somatotopically Inappropriate Collaterals of Aβ Afferents Following Neonatal Peripheral Nerve Section. As described above (Sect. 4.6.3) chronic denervation of digit 3 alone or both digits 3 and 4 of the hindlimb of the neonatal cat leads to a reorganization of the receptive fields of SCT cells throughout the affected part of the toe representation (Snow and Wilson 1989; Wilson and Snow 1988b, 1990b). Cells which would normally have receptive fields on the denervated digits are found to have fields on the adjacent innervated digits. Recent studies, using intraaxonal injection of HRP, of the morphology of somatotopically inappropriate collaterals (Sect. 4.3.3) of Aß hair follicle afferents with receptive fields on toe 2 or toe 5 have indicated that some collaterals terminating in the somatotopically reorganized region exhibit more complex branching, and bear many more synaptic boutons, than would normally be expected for collaterals arborizing at such large distances rostral or caudal to the somatotopically appropriate region (Fig. 4.35) (Snow and Wilson 1989). Normally, the terminations of somatotopically inappropriate collaterals of primary afferent fibres do not have synapses (Snow Wilson, 1990; Wilson and Snow, 1990a) (Sect. 4.3.3). Thus these collaterals clearly constitute a *potential*, rather than a *weak*, source of excitatory input to neurons of the dorsal horn. Indeed the somatotopically inappropriate afferent projections form the *only* anatomical basis of potentially effective afferent projections that has been positively identified in the dorsal horn to date. The existence of such synapseless terminations should perhaps be considered in any attempt to explain alterations of somatotopy consequent to chronic lesions in afferent pathways at any level of the neuvaxis (Fig. 8.2) (Sects. 8.2 and 8.3) (Snow and Wilson, 1990).

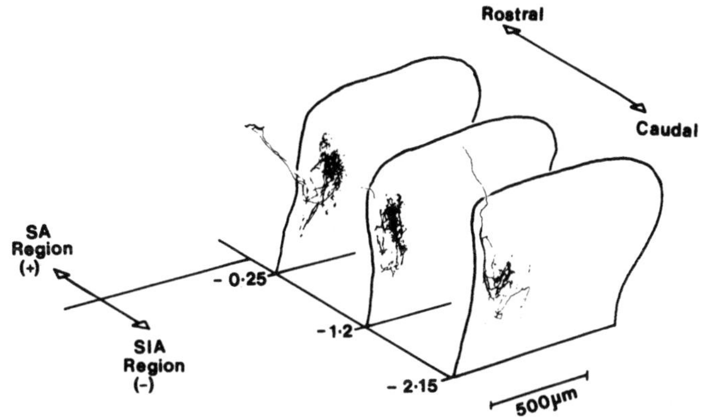

a

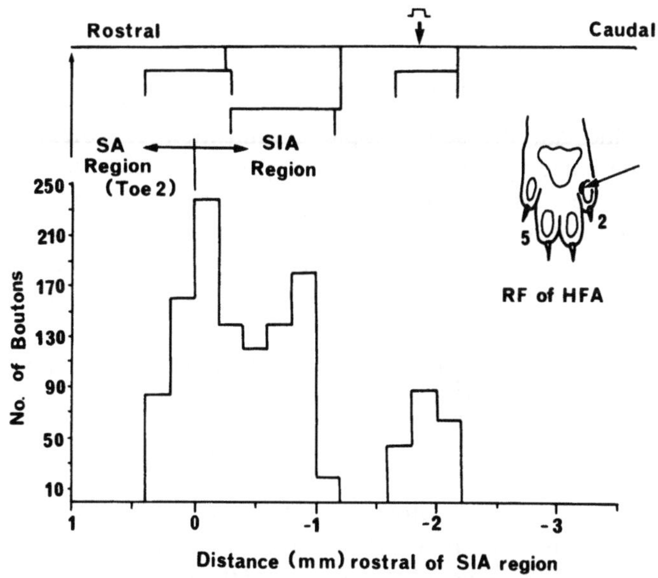

b

Fig. 4.35 a, b. Anatomical plasticity of primary afferent projections following neonatal peripheral nerve transection and ligation in the cat. **a** Reconstructions, in the transverse plane, of the collateral arborizations of a single Aß hair follicle afferent fibre with a receptive field on toe 2, injected intra-axonally with HRP, in an adult cat in which the digital nerves to toes 3 and 4 had been cut and ligated in the first week of life. To enable accurate localization of the somatotopic boundaries between the toe 2 region, the toe 3 and 4 region, and the toe 5 region, the digital nerves of toes 3 and 4 were cut acutely on the opposite side at the time of the recording. Mapping the receptive fields of SCT cells on both sides of the cord showed that the stained collaterals were all within the region that originally would have contained the representation of toe 3. However, SCT cells in this region and the more caudal toe 4 region were found to have reorganized receptive fields on toes 2 and 5. The axon was injected (*down arrow*) about 2 mm caudal to the boundary between the original toe 2 and toe 3 representations (cf. Fig. 4.5, 4.11 and 4.19). Note the complexity of the most caudal collateral of this axon compared with the somatotopically inappropriate collaterals of the afferent shown in Fig. 4.11a (*f–i*). **b** Density of synaptic *boutons* visible in the light microscope. Note that although all 3 collaterals arborized in the somatotopically inappropriate region, they nevertheless give rise to large numbers of boutons, even at a distance of 2 mm from the somatotopically appropriate region (cf. Fig. 4.11b). (Modified from Snow and Wilson 1989)

4.7 Major Conclusions

The dorsal horn of the spinal cord has a clear somatotopic organization, and receives input from primary afferent fibres as well as from a number of descending pathways that are amenable to experimental manipulation. This organization has resulted in the dorsal horn becoming a major area for studies of plasticity. Unfortunately, despite the large number of experiments performed and the confident claims of many workers that they have a model reorganizing system there is really very little evidence that shows unequivocally that dorsal horn neurons develop new connections following lesions such as dorsal rhizotomy, peripheral nerve transection or cordotomy. Owing to these problems, it is difficult to reach definitive conclusions regarding plasticity in the dorsal horn. However, with this reservation in mind, the following statements may be made concerning the present state of our knowledge.

(1) Somatotopic Organization of Primary Afferent Terminals in the Dorsal Horn

In the cat, the somatotopic organization of dorsal horn neurons is largely determined by the organization of primary afferent projections to the dorsal horn. However, while traditional neuroanatomical techniques have failed to show significant somatotopically inappropriate primary afferent projections, studies using intra-axonal injection of tracers have revealed afferent collaterals projecting to somatotopically inappropriate regions rostral and caudal of the somatotopically appropriate region. These somatotopically inappropriate collateral arborizations are much simpler than those in the somatotopically appropriate region, and have few or no synaptic boutons.

(2) Immediate "Unmasking" Effects of Lesions

Neither dorsal rhizotomy nor peripheral nerve transection has been shown to have any *immediate* effect on the functional properties of dorsal horn neurons, other than the removal of all or part of their peripheral receptive fields. There is thus no evidence for an immediate "unmasking" of latent excitatory inputs similar to that reported for neurons of the DCN, thalamus and SI cortex.

(3) Long-Term Effects of Lesions

Dorsal Rhizotomy. Experiments to investigate the effects of chronic dorsal root section on dorsal horn somatotopy have produced mixed results.

(a) One week after transection of all dorsal roots caudal to L3, except for S1, unidentified dorsal horn neurons in the region of the L4/L5 junction in the cat acquired new, low-threshold tactile receptive fields in the dermatomes of either the intact lumbar roots or the spared S1 root. The proportion of cells with such somatotopically inappropriate receptive fields then increased over subsequent weeks. Intracellular recording from dorsal horn neurons in the rhizotomized segments indicated that they had developed new excitatory inputs from afferents in intact roots, and that some of these new inputs might be via monosynaptic connections from primary afferent fibres.

(b) In a single attempt to evaluate the effect of chronic dorsal rhizotomy on identified SCT cells and other neurons in the lumbosacral dorsal horn of the cat whose locations were carefully checked by intracellular injection of HRP, no evidence could be found of somatotopically inappropriate receptive fields.

(c) Chronic dorsal rhizotomy caused an *increase* in the proportion of dorsal horn neurons responding only to noxious stimuli. This increase was particularly marked when compared with the acute state, where the proportion of such cells was below normal. The reduction following acute lesions may reflect the limited rostrocaudal distribution of A∂ and C nociceptive afferents compared with that of Aß mechanoreceptive afferents. It was tentatively suggested that the increased responsiveness to noxious inputs might result from sprouting of C afferents.

Suggested reasons for the differing results in (a) and (b) include: (i) the initial, positive results were obtained in unanaesthetized, decerebrate spinal cats, in which dorsal horn neurons have high excitability, whereas the negative results were obtained in anaesthetized animals, where excitability may be lowered; (ii) both the segments examined and the degree of deafferentation (i.e. the number of dorsal roots cut) differed. It therefore remains possible that reorganization of excitatory inputs occurs following dorsal rhizotomy and that this can be detected only in heavily deafferented regions of the dorsal horn; (iii) partial deafferentation of dorsal horn neurons by dorsal rhizotomy may immediately cause some dorsal horn neurons to lose reponsiveness to intact afferent inputs. In the long term, these cells may recover their responsiveness to cutaneous stimuli. This may reflect the proliferation of the terminals of the intact afferents and a consequent increase in their excitatory influence over the partially deprived cells.

Spinal Cord Hemisection. Partial deafferentation of lumbosacral dorsal horn neurons by chronic ipsilateral spinal hemisection leads to an increase in the receptive field size of some wide-dynamic-range neurons with proximal receptive fields.

Peripheral Nerve Transection. Experiments designed to investigate changes of somatotopy in the adult mammalian spinal dorsal horn following peripheral nerve section have produced the most controversial results.

(a) Chronic denervation of the whole limb below the knee has been reported to lead to the appearance of light-tactile, proximal, receptive fields among neurons in the foot representation of the medial dorsal horn. Since proximal skin is normally represented in the far lateral part of the dorsal horn at the appropriate segmental level, this type of reorganization has been termed *mediolateral.* The phenomenon has been demonstrated in unanaesthetized, decerebrate spinal cats and in barbiturate-anaesthetized rats.

(b) Mediolateral reorganization of a more limited extent has also been reported to occur among lamina I neurons following chronic sciatic nerve transection in the cat. This was shown in anaesthetized preparations.

(c) Several independent attempts to replicate the original demonstration of mediolateral reorganization in the dorsal horn of the cat have failed but in none of these was the final recording performed in unanaesthetized decerebrate spinal animals.
(d) When the sciatic nerve is transected in the rat there is reported to be a gradual *caudal* spread of receptive fields located within the cutaneous field of the saphenous nerve into the region of dorsal horn representing the sciatic nerve field. The extent of the spread may be as much as 2–3 mm. In contrast to the findings outlined above, in these experiments there was no apparent medial spread of receptive fields on proximal skin. This rostrocaudal reorganization was demonstrated in anaesthetized animals.
(e) In the adult cat, following chronic denervation of a single digit, many antidromically identified SCT cells in the representation of that digit acquire new receptive fields on nearby innervated skin. Most of these new receptive fields are on the skin of the adjacent digits, which are normally represented rostral and caudal to the denervated digit. The spatial extent of this rostrocaudal reorganization was up to 3 mm, equivalent to the rostrocaudal extent of the representation of a single digit in the dorsal horn of the cat. This reorganization was demonstrated in anaesthetized animals.

(4) Mechanisms Underlying Reorganization in the Dorsal Horn

The controversy over somatotopic reorganization in the dorsal horn of adult rats and cats has resulted in different models being proposed to explain the fundamentally different results. However, all the models are based ultimately on only two fundamental hypotheses.

(A) Reorganization Is Due to Expression of a Class of Pre-existing, Ineffective Synaptic Inputs

(a) In some cases the existence of a subliminal excitatory fringe to the receptive field of dorsal horn neurons has been demonstrated by intracellular recording. Subliminal inputs are usually from small regions of skin adjacent to the firing zone of the receptive field.
(b) The existence of a class of somatotopically inappropriate inputs that do not normally contribute to the firing zone of neurons' natural receptive fields but differ functionally from those forming a subliminal fringe has been shown by: (i) neuron responses to electrical stimulation of afferents innervating skin distant from their natural receptive fields; (ii) in unanaesthetized animals receptive fields of lamina II neurons spontaneously undergo considerable changes of size; and (iii) the convulsant drug 4-aminopyridine (4-AP), which enhances synaptic transmission generally, causes the expansion of receptive fields when applied microiontophoretically to dorsal horn neurons.

(B) Synapseless Terminals, Sprouting and Synaptogenesis

(a) There is no good evidence that intact Aß primary afferents are able to sprout over relatively long distances in the dorsal horn. The results of early anatomical studies utilizing selective silver staining of degenerating axoplasm to show long-distance sprouting after dorsal rhizotomy, have not been supported by recent studies using tracer techniques. There may be a proliferation of unmyelinated C fibres after dorsal rhizotomy, as shown by an increase in their numbers in nearby intact dorsal roots. Indirect evidence of recovery of neuropeptide levels and synapse density in the deafferented dorsal horn indicate that interneurons may also sprout in lamina II.

(b) An alternative explanation of somatotopic reorganization based on local sprouting and synaptogenesis may involve a special class of primary afferent collaterals that do not normally make synaptic connections with dorsal horn neurons. Intra-axonal injections of HRP together with somatotopic mapping has shown that Aß hair follicle afferent fibres give off collaterals to somatotopically inappropriate regions of dorsal horn rostral and caudal of the somatotopically appropriate region, where their main collaterals arborize. The somatotopically inappropriate collaterals have been observed up to 3–4 mm from the boundary of the somatotopically appropriate region in the adult cat. However, beyond about 0.5 mm the collaterals are relatively simple in structure and somatotopically appropriate lack the synaptic boutons that are numerous in the somatotopically region. Due to their relatively simple arborizations these collaterals may be overlooked in studies of peripheral nerve projections utilizing tracer techniques. Following peripheral nerve section or dorsal rhizotomy it is suggested that these somatotopically inappropriate collaterals of intact afferents may undergo limited collateral sprouting and synaptogenesis, thus causing nearby dorsal horn neurons to develop new receptive fields. It is further suggested that the stimulus for such morphological changes may be the limited degree of deafferentation of dorsal horn neurons produced by the chronic nerve section.

(5) Plasticity in the Dorsal Horn Following Neonatal Lesions

C fibre input to the dorsal horn undergoes much development after birth in the rat, and is accompanied by a structural maturation of interneurons of the substantia gelatinosa. During the immediate postnatal period there is reported to be a reduction of the average receptive field size of dorsal horn neurons which might be related to the maturation of the C fibre input.

(a) In the rat during the immediate postnatal period the C fibre projection to lamina II shows a remarkable capacity to sprout into regions deafferented by neonatal nerve section or crush. Furthermore, myelinated afferent fibres appear to sprout into lamina II following partial deafferentation of the lamina by neonatal capsaicin treatment. However, in the cat, primary afferent fibres appear to show little or no ability no undergo long-range sprouting into neonatally deafferented regions of dorsal horn.

(b) When one or two toes of the cat hindlimb are denervated in the first postnatal week, and recordings made from SCT cells in the lumbosacral dorsal horn one year later, the cells are found to have developed receptive fields on the adjacent innervated digits, normally represented rostral and caudal to the denervated digit in the dorsal horn. This rostrocaudal reorganization is similar to, but more extensive than, that described following digit denervation in the adult cat, and is demonstrable in anaesthetized preparations. The change in somatotopic organization has been correlated with local proliferation and synaptogenesis in the somatotopically inappropriate terminations of Aß afferents innervating toes surrounding the denervated digit or digits. Synapse-free, somatotopically inappropriate afferent terminals in both neonatal and adult mammals might thus provide a substrate by which spatially constrained structural changes could bring about changes of somatotopic organization (Sect. 8.2 and 8.3).

5 Plasticity and the Dorsal Column Nuclei

5.1 Advantages of the Dorsal Column Nuclei for Studies of Plasticity

The dorsal column nuclei (DCN), nucleus gracilis and nucleus cuneatus, offer several advantages for studies of the effects of changing peripheral input on neuronal connectivity. In common with the other somatosensory regions reviewed here, the DCN contain a somatotopically organized representation of the body surface. In addition, a majority of neurons in the somatotopically organized region of the nuclei exhibit a high degree of specificity in relation to their receptor input. Thus changes of either somatotopic organization or the distribution of modality specificities, resulting from changes of input, may be indicative of functional or structural changes of connectivity within the DCN. Neurons in the somatotopically organized regions of the nuclei are directly contacted by clusters of terminals from large-diameter, myelinated, cutaneous, primary afferent fibres, as well as the terminals of a few fibres from other extrinsic sources. However, in other regions of the nuclei, afferent input enters over several anatomically discrete pathways, allowing the effect of interruption of one or more of these pathways on the anatomical organization and functional representation of the remaining intact inputs to be studied. Finally, a limited number of behavioural studies have shown that certain deficits of tactile sensibility may result from lesions of the dorsal columns and, in some of these cases, a substantial degree of time-dependent recovery has been demonstrated. Whether this recovery is dependent on changes of connectivity within the nuclei themselves, or whether it is due entirely to compensatory changes occurring in intact neural pathways is an important question which remains to be answered.

Despite these apparent advantages of the DCN for studies of somatosensory plasticity, there have to date been disappointingly few detailed studies of the effects of lesioning afferent pathways on the functional and somatotopic organization of the DCN. This may be due in part to the technical problems involved in extensively mapping the receptive fields of single neurons in the nuclei, which are situated on the dorsal surface of the medulla in a region of poor mechanical stability for electrophysiological recording. Moreover, where changes have been reported after deafferentation, the basis of the changes remains obscure.

5.2 Organization of the Dorsal Column Nuclei

The cytoarchitectonic and functional organization of the DCN has been described for a number of species, but most detailed descriptions are available for the rat, cat, monkey and raccoon. To date, only the rat and cat have been utilized in electrophysiological and morphological studies of somatosensory plasticity and thus in what follows we will direct attention to these species. It is, however, clear that the organization of the dorsal column-medial lemniscal system may be very different in other mammals, such as the primate and the raccoon, in which the somatosensory function of the distal-extremities is much more highly developed. Only recently have raccoons been considered for studies of the mechanisms underlying the synthesis and plasticity of somatotopic maps in the DCN (Rasmusson 1988). Findings derived from these experiments will be discussed in more detail in a later section of this chapter.

For many years the gracile and cuneate nuclei were regarded simply as somatosensory relays between primary afferent fibres entering the spinal cord and the ventrobasal (VB) complex of the thalamus. More recently, however, this view has given way to one of a more heterogeneous organization. The traditional view that the dorsal column-medial lemniscal system is concerned with the transmission of information regarding the location, spatial form, quality and temporal sequence of mechanical stimuli applied to the skin, and with the transmission of kinaesthetic information from the joints (Mountcastle and Darian-Smith 1968) is well known, but has been strongly challenged as several investigations have indicated that an intact dorsal column-medial lemniscal system may not be *essential* for the procesing of such inputs. An alternative proposal is that the system may instead have an *alerting* function in relation to subsequent processing by other somatosensory pathways. These controversies concerning the role of the dorsal column system in somatic sensation are beyond the scope of this review. For discussions of dorsal column function the reader is directed to Wall (1970), Wall and Noordenbos (1977), Wall and Dubner (1972), Beck (1976) and for reviews of various aspects of the dorsal column system see also Boivie and Perl (1975), Brown and Gordon (1977), Willis and Coggeshall (1978), Vierck (1984) and Berkley et al. (1986).

5.2.1 Cytoarchitectonics

Most authors agree that along the rostrocaudal axis there are three major cytoarchitectonic divisions in the DCN (Figs. 5.1a–c and 5.2a–c). The disposition of these three regions varies considerably in different species. In the DCN of the rat only two such divisions have been described. In all species the DCN can be divided into a reticular zone and a core region which are described in detail in a following section (Sect. 5.2.4).

In the cat, two principal varieties of neuron can be recognized in Golgi-stained tissue. Large (20–30 μm diameter) round cells with short, heavily branched, dendrites bearing many spines are found throughout the whole rostrocaudal extent of the nuclei. These cells are most numerous in the dorsal part of the middle third, where they occur in clusters of 8–10 cells, which have been termed *cell-nests*. The appearance of these neurons in Ramon y Cajal's (1952) classical Golgi studies of the cat and human DCN is shown in Fig. 5.3. In the region of the cell-nests the large round neurons are frequently arranged spherically around a relatively cell-free central region, with their dendrites projecting into and intertwining within the central zone (Ramon y Cajal 1952; Kuypers and Tuerk 1964; Ellis and Rustioni

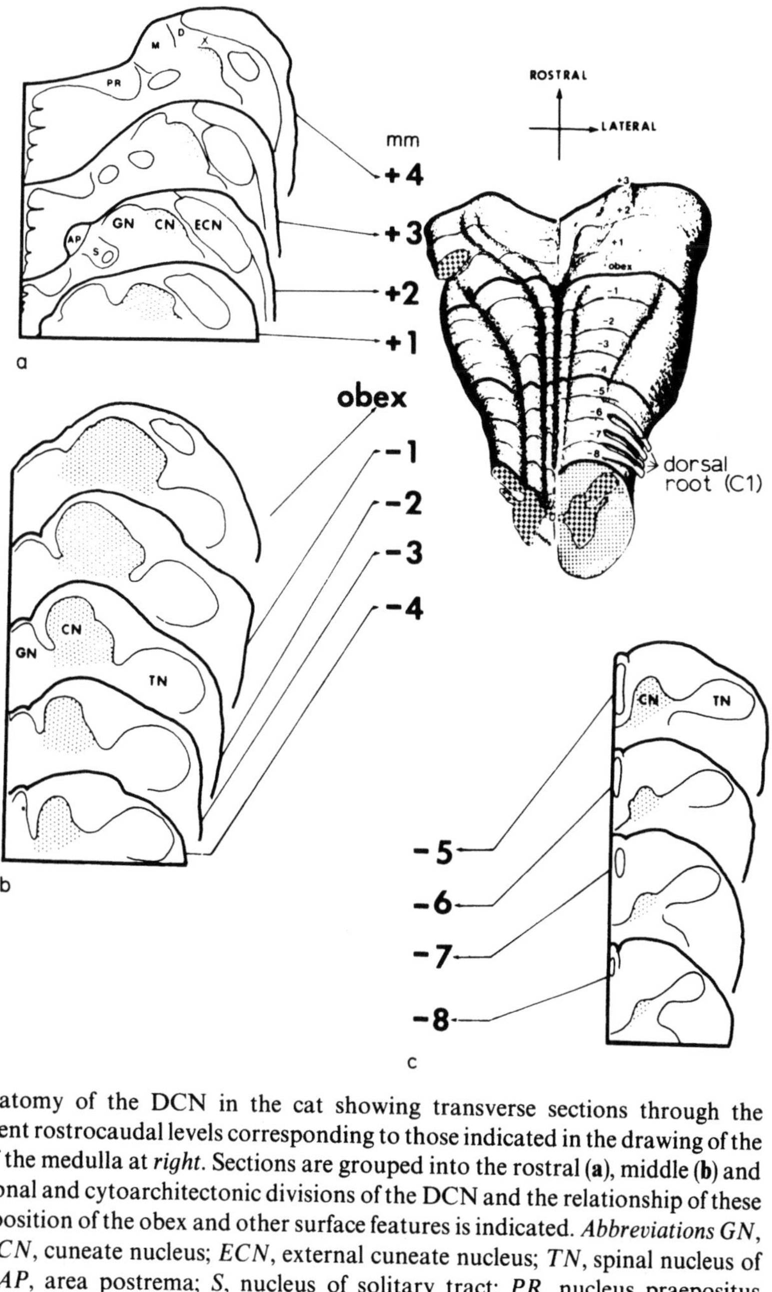

Fig. 5.1 a–c. Anatomy of the DCN in the cat showing transverse sections through the medulla at different rostrocaudal levels corresponding to those indicated in the drawing of the dorsal surface of the medulla at *right*. Sections are grouped into the rostral (**a**), middle (**b**) and caudal (**c**) functional and cytoarchitectonic divisions of the DCN and the relationship of these divisions to the position of the obex and other surface features is indicated. *Abbreviations GN*, gracile nucleus; *CN*, cuneate nucleus; *ECN*, external cuneate nucleus; *TN*, spinal nucleus of the trigeminal; *AP*, area postrema; *S*, nucleus of solitary tract; *PR*, nucleus praepositus hypoglossi; *M*, medial vestibular nucleus; *D*, descending vestibular nucleus, *X*, nucleus X (Modified from Cheema et al. 1983, p. 173)

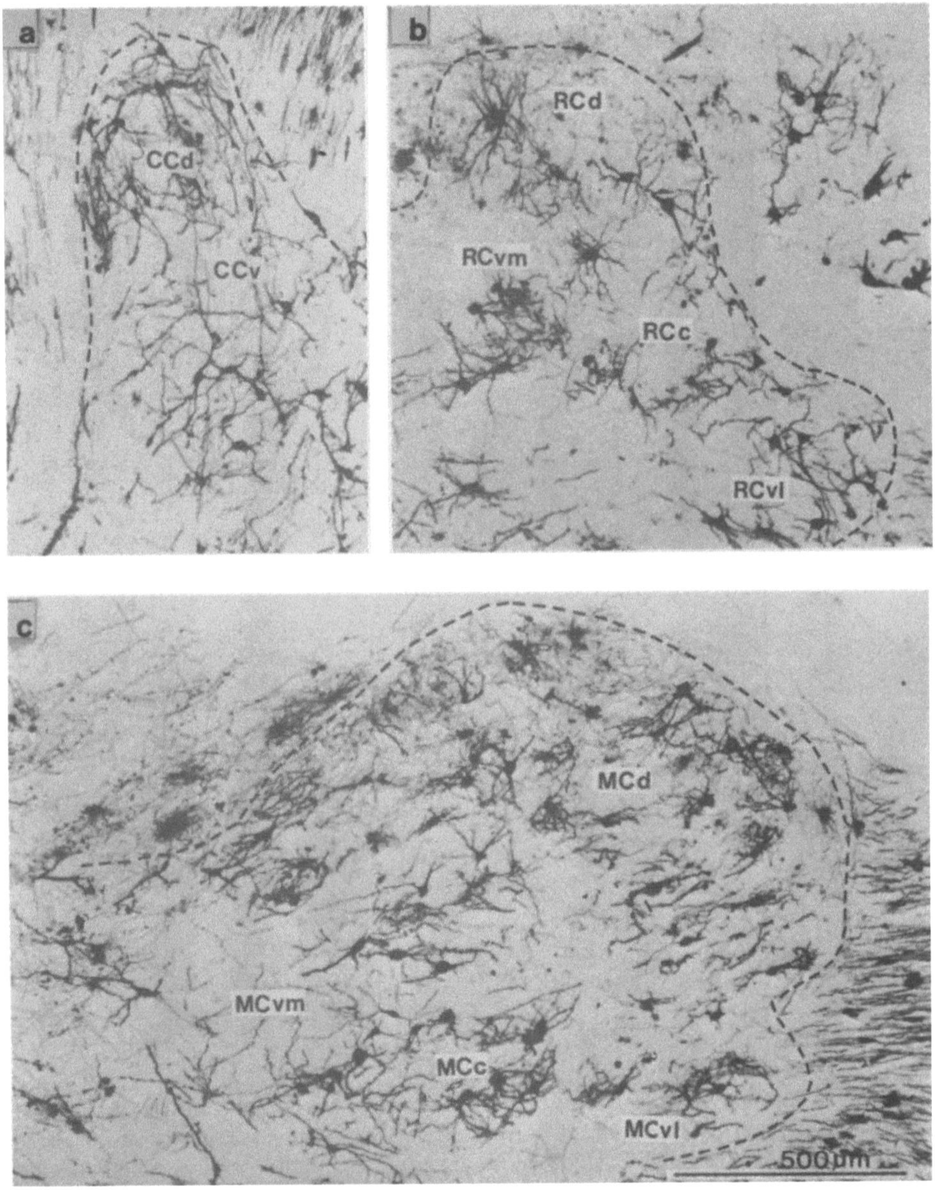

Fig. 5.2 a–c. Major cytoarchitectonic and functional regions of the cat cuneate nucleus (outlined in Fig. 5.1 and Table 5.1) as they appear in transverse, Golgi-stained sections. The dorsal, medial and lateral borders of the nucleus are outlined by the *dashed lines*. In each figure, dorsal is at the *top*, and medial to the *left*. **a** Caudal cuneate; **b** rostral cuneate; **c** middle cuneate. The cell-nests are located in the dorsal part of the middle cuneate (*MCd*). Abbreviations and distinguishing characteristics of the different subdivisions are described in Table 5.1. (From Cheema et al. 1983, p. 181)

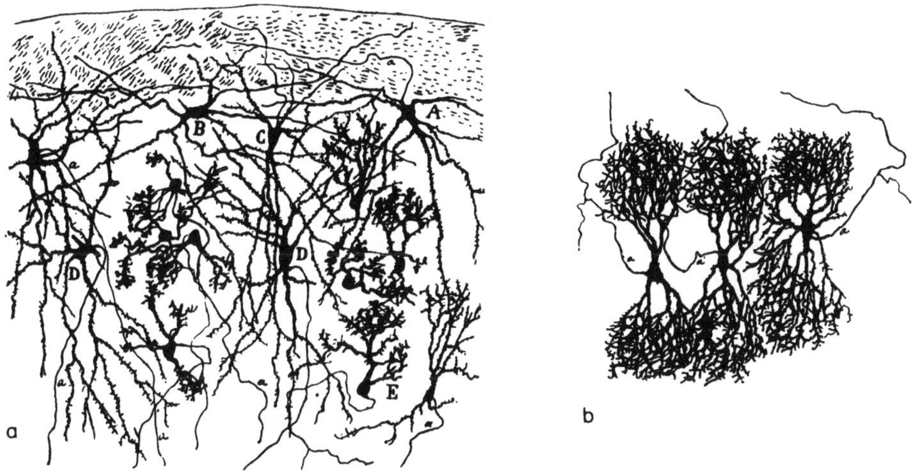

Fig. 5.3 a, b. Appearance of neurons in the cell-nest region of the DCN as shown in the early Golgi studies of Ramon y Cajal (1952). **a** Transverse section through the *nucleus cuneatus* of the kitten. The round neurons with short bushy dendrites labelled *E* are typical of this region. **b** Neurons in the cell-nests of the dorsal region of the cuneate nucleus of a human fetus. (From Ramon y Cajal 1952, pp. 903 and 902)

1981; Cheema et al. 1983). The second major cell type consists of small to medium-sized (10–20 μm diameter) neurons with triangular, multipolar or fusiform cell bodies, and characteristic long, radiating dendrites. These cells are most common outside the cell-nests region – the so-called *reticular* regions – although they are also found scattered between the cell-nests and along the dorsal rim of the nuclei (Kuypers and Tuerk 1964; Berkley 1975; Ellis and Rustioni 1981; Cheema et al. 1983). Particularly in the cuneate nucleus, most neurons in the rostral and ventral parts are of this type. At the caudal end of the nuclei many of the cells are triangular in shape and there are also fewer cell-nests (Taber 1961; Kuypers and Tuerk 1964; Hand 1966; Rustioni and Macchi 1968; Keller and Hand 1970; Berkley 1975; Ellis and Rustioni 1981; Dykes et al. 1982; Cheema et al. 1983). Recently a small region of large, darkly staining, uniformly distributed neurons extending in a continuous horseshoe-shaped band across the caudal ends of both the gracile and cuneate nuclei has been recognized. This region may be a special receiving zone for input from Pacinian afferents (Dykes et al. 1982) (Sect. 5.2.5).

The thalamic projection is most dense from the region of cell-nests (Gordon and Seed 1961; Berkley 1975; Blomqvist and Westman 1976; Blomqvist 1980; Ellis and Rustioni 1981; Berkley et al. 1986). Over 90% of neurons in the cell-nest region of the nuclei have been labelled after injections of HRP into the VB complex of the cat (Ellis and Rustioni 1981) and the monkey (Rustioni et al. 1979). Indeed it appears likely that all the neurons arranged in typical clusters throughout the dorsal two-thirds of the nuclei project to the VB complex. The few cells in this region that are not labelled after thalamic injections are of the small fusiform type, and are located at the periphery of the cell-nests (Ellis and Rustioni 1981).

In the rat, Nissl and Golgi studies of the DCN have revealed a somewhat simpler cytoarchitectonic organization than that found in the cat. In the cuneate nucleus, only two main zones, rostral and caudal, have been recognized (Basbaum and Hand 1973; Gulley 1973; Odutola 1977). The caudal zone contains mainly round cells. These are arranged in transversely-oriented slabs or *cellular bricks*, each of which is about 380 μm wide by 50 μm thick. Vertically, these slabs extend from a depth of about 400–850 μm, such that they almost reach the base of the nucleus. Each slab of neurons is surrounded by bundles of fibres running vertically and horizontally through the nucleus. Presumably these cellular bricks correspond functionally to the more spherical cell clusters present in the cat and other mammals. The rostral region of the nucleus contains scattered cells of several types, including round, spindle-shaped and multipolar cells (Basbaum and Hand 1973; Beck 1981). In the gracile nucleus of the rat, the rostrocaudal differences are less obvious than they are in the cuneate. Neither cell-nests nor slabs are evident in the caudal region of the gracile nucleus. However, while the neuronal perikarya are not arranged in any particular fashion, Gulley (1973) noted that the dendrites of these cells intertwined to form vertical columns. Cuneothalamic projection cells in the rat have been identified after injection of HRP into the contralateral VPL. Labelled cells were 16–20 μm in diameter, round or oval in shape and were distributed throughout the nucleus although they were more concentrated at the level of the obex (Feldman and Kruger 1980). Information is lacking concerning the projection of the neurons comprising the cellular bricks.

5.2.2 Ascending Afferent Pathways

The DCN receive ascending afferent input from three major sources: (1) ascending branches of primary afferent fibres in the dorsal funiculi, (2) non-primary afferent fibres also ascending in the dorsal funiculi, and (3) non-primary afferent fibres ascending in the dorsolateral funiculi.

Primary Afferents. Only a small proportion of the primary afferent fibres which enter the dorsal funiculi over dorsal roots send branches all the way up to the medulla. In the cat, only about 22%–23% of fibres ascending the fasciculus gracilis, and probably even fewer of those in the fasciculus cuneatus, reach the medulla (Glees and Soler 1951). Furthermore, not all functional classes of cutaneous afferent fibre project directly to the DCN, those that do so being nearly all of the rapidly adapting type. Among single afferent units recorded from the upper cervical dorsal columns in the cat, hair follicle afferents and other rapidly adapting afferents in the pads have been found most often (Yamamoto et al. 1956; Brown 1968; Petit and Burgess 1968; Burgess and Clark 1969; Uddenberg 1968a; Bromberg and Whitehorn 1974; Tracey 1980). Most afferent units supply hair follicles, with all physiological classes of hair follicle afferent except type D being well represented both in fasciculus cuneatus (Yamamoto et al. 1956; Brown 1968; Petit and Burgess 1968) and fasciculus gracilis (Uddenberg 1968a; Bromberg and Whitehorn 1974). Pacinian afferents and other rapidly adapting afferents from glabrous skin were also commonly found in these studies. The degree of projection from slowly adapting afferent fibres to the nuclei is less certain, due mainly to problems of identification of these afferents and possible contamination of samples of recorded fibres with postsynaptic dorsal column fibres. Of cutaneous, slowly adapting, mechanoreceptive afferents, type II but not type I have been found in reasonable numbers in fasciculus gracilis at the cervical level (Petit and Burgess 1968; Burgess and Horch 1973) (Table 2.3; Sect. 2.5.1). However, in one study of the fasciculus cuneatus in the cat both types of slowly adapting afferents were found (Bromberg and Whitehorn 1974). Large numbers of group I muscle afferents from the forelimb have also been found in fasciculus cuneatus in the cat (Uddenberg 1968a). Many of these slowly adapting afferents in the cat's fasciculus cuneatus, particularly those from muscle, probably terminate within the external cuneate nucleus. However, some certainly terminate within the main cuneate nucleus, as shown both by HRP injection of single, identified afferents (Sect. 5.4.1), and by the responsiveness of some cuneothalamic projection neurons to electrical stimulation of muscle nerves (see below). It has been suggested that in the squirrel monkey only rapidly adapting afferents of cutaneous origin reach the gracile nucleus (Whitsel et al. 1969); this may represent a further functional specialization in primates towards the processing by the dorsal column lemniscal system of information concerning mainly the phasic components of external stimuli or movements. It was generally believed that primary afferent fibres having axons with conduction velocities in the C and Aδ range (C low-threshold mechanoreceptors, polymodal nociceptors, thermoreceptors, and mechanical nociceptors, Aδ mechanical nociceptor and D hair afferents) do not ascend in the dorsal funiculi more than one or two segments (Brown 1968, 1973; Petit and Burgess 1968; Burgess and Perl 1973; Horch et al. 1976). However, the recent discovery, in the electron microscope, of unmyelinated primary afferent fibres running in the dorsal funiculi of the rat (Chung and Coggeshall 1985) must lead to a dramatic reappraisal of this dogma, particularly as many of these fine afferents have been shown to ascend in the fasciculus gracilis as far as the C3 segment, and probably terminate in the DCN. Surprisingly, at C3 level, almost one-third of axons in the fasciculus gracilis (excluding the corticospinal tract) are unmyelinated, and 80% of these are lost following an extensive (T8 and more caudal roots) ipsilateral dorsal rhizotomy. Many of these unmyelinated afferents also appear to contribute an ascending branch to the contralateral funiculus gracilis (Patterson et al. 1989). The demonstration of these unmyelinated projections leads, of course, to questions concerning their functional role, if any, in transmission through the DCN. In view of the controversial role proposed by some workers for unmyelinated afferents in the control of somatotopy within the dorsal horn and other regions of the somatosensory system (see Sect. 3.3.4 and Sect. 4.5.4) there is clearly an urgent need to elucidate the physiological status of these unmyelinated afferents.

Non-Primary Afferents. The presence of non-primary long afferent fibres in the cervical dorsal funiculi has been known since the physiological single-unit studies of Uddenberg (1968b) in the fasciculus cuneatus and Petit (1972) in the fasciculus gracilis. Anatomical degeneration studies soon confirmed the projection of these postsynaptic fibres in the dorsal funiculi to the DCN in the cat (Rustioni 1973, 1974) and the monkey (Rustioni 1977; Rustioni et al. 1979). In both species the non-primary afferent fibres in fasciculus cuneatus originate from cervical levels of the spinal cord, while those in fasciculus gracilis originate from lumbosacral segments. In both the cat and the rat the distribution of the cells of origin of the postsynaptic dorsal column (PSDC) projection is similar to that of spinocervical tract cells. In the rat both the PSDC pathway (Giesler et al. 1984) and the spinocervical tract (Giesler et al. 1978) originate from a band of neurons lying immediately below the lamina II/III border. In the cat, both pathways appear to originate mainly from cells in lamina IV and V (Rustioni and Kaufman 1977; Brown et al. 1980b; Bennett et al. 1983; Bennett et al. 1984; Brown and Fyffe 1981). In the monkey, as in the rat, the PSDC projection originates from cells situated immediately below lamina II (Bennett et al. 1983).

A third afferent pathway ascends to the DCN in the ipsilateral dorsolateral funiculus (Gordon and Grant 1982; Tomasulo and Emmers 1972; Hazlett et al. 1972; Dart and Gordon 1973; Rustioni 1973, 1974; Rustioni and Molenaar 1975; Nijensohn and Kerr 1975). It has been proposed that this projection consists of collaterals of spinocervical tract axons (Tomasulo and Emmers 1972). However, labelling of lumbosacral dorsal horn neurons in the rat (Giesler et al. 1984), cat (Rustioni 1973) and monkey (Rustioni et al. 1979) following injections of the retrograde tracer HRP into the gracile nucleus, is virtually abolished by transection of the dorsal funiculi, indicating that the number of dorsal horn neurons projecting to the nucleus via the dorsolateral funiculus is much smaller than the population of spinocervical tract neurons.

The distribution of both primary and non-primary afferent fibres to the DCN has been examined in many studies of degeneration after lesions of dorsal roots or afferent pathways and in studies employing anterograde transport of tracers. Although, in the cat, primary afferent fibres terminate throughout the whole of the DCN, the input to the cell-nest region is heavier than that to the reticular zones (Sect. 5.2.4). In this species, non-primary afferent fibres ascending in both the dorsal funiculi and dorsolateral funiculi have similar termination patterns within the nuclei. Those from the cervical and thoracic segments terminate mainly in the rostral region, and, to a lesser extent, the base of the cuneate nucleus. Fibres arising from segments below T6 terminate in corresponding regions of the gracile nucleus (Rustioni 1973, 1974; Rustioni and Dekker 1974; Gordon and Grant 1982). In the rat, the distribution of the terminals of non-primary afferents in the DCN, as determined by anterograde tracing, has been found to correspond more closely with that of the terminals of primary afferent fibres. Moreover, there is anatomical evidence that, in this species, both types of afferent may terminate on individual thalamic projection neurons, and that the distribution of non-primary afferent fibres is somatotopically organized (Patterson et al. 1989). In addition, spinal PSDC neurons appear to be less concerned with input from nociceptors in the rat (Giesler and Cliffer 1985) than they are in the cat (Brown and Fyffe 1981, see Sect. 4.5.2). Such interspecies differences might have important consequences in relation to the central effects of deafferentation, both within the DCN themselves, and at higher levels of the somatosensory pathways.

5.2.3 Responses of Neurons to Natural Stimulation

The receptive properties of neurons in the DCN have been investigated by means of single and multiunit recording in a number of mammals, including the rat, cat, monkey, tree squirrel and raccoon. Most information is available for the cat, and since this animal has also been used in various studies of plasticity, it will be considered in most detail. It is generally agreed that most, but not all, cuneate and gracile neurons respond to a variety of types of primary afferents that have in common a sensitivity to the dynamic phase of mechanical displacement. Thus the majority of cells behaves as though they receive excitatory input from a single class of primary afferent fibre.

Cutaneous Inputs. Single-unit studies in the cat have generally revealed a preponderance of low-threshold, cutaneous inputs, with cells responding either to hair movement or to light touch of the skin. Brown, et al. (1974), recording from axons in the medial lemiscus, found examples of the hair-sensitivite neurons responding solely to movement of the tylotrych hairs. Similarly, using constant-velocity, mechanical stimuli, Golovchinsky (1980) found cuneate neurons which behaved as though they were driven by different individual subsets of the group II afferents innervating guard hair follicles (G 1, G 2 or intermediate types). Other cells behave as though they are excited only by slowly adapting type II afferents, responding selectively to claw movement, pressure on the claw base, or stretch of the skin. Units capable of following vibratory stimuli at 100–300 Hz have also been found and are thought to receive input exclusively from Pacinian corpuscles (Gordon and Paine 1960; Kruger et al. 1961; Perl et al. 1962; Gordon and Jukes 1964; Brown et al. 1974; Angaut-Petit 1975b; Janig et al. 1977; Golovchinsky 1980). Low-threshold inputs from glabrous skin are well represented in the cat. In all, three main functional classes of cuneate neurons having input from tactile receptors in the foot-pads have been described, of which two gave rapidly adapting and the other slowly adapting responses. In response to controlled sinusoidal mechanical stimuli, neurons in each of these three classes behaved as though they received input from one and only one of the three known types of low-threshold mechanoreceptors found in glabrous skin. In the case of the dynamically sensitivity neurons these receptor types are the Pacinian corpuscles and Meissner (or Krause) corpuscles, whereas the behaviour of the statically sensitive neurons suggested that they receive input exclusively from the known slowly adapting receptor of glabrous skin, the Merkel cell complex (Bystrzycka et al. 1977; Douglas et al. 1978). Unequivocal type I-like responses have not been found in some single-unit studies of feline dorsal column neurons (Janig et al. 1977; Golovchinsky 1980). Brown et al. (1974) described type I receptor-like responses of a small number of medial lemniscal axons, but their sample may have included axons from sources other than the DCN.

Deep Inputs. Cells responsive to stimulation of deep receptors have also been found in the DCN by most workers. These include units responsive to joint movement, to deep pressure or to activation of muscle afferents. Many neurons in the cat's cuneate nucleus are activated by electrical stimulation of group I muscle afferents in forelimb nerves. Of these, more than 50% are antidromically activated from the medial lemniscus, and are therefore probably cuneothalamic relay cells. A few receive convergent group I inputs from several muscle nerves, but the majority are excited from only one. Many also receive excitatory input from group II muscle afferents. Cells responding to group I muscle input are preferentially found deep in the nucleus, below those with cutaneous receptive fields (Rosen 1967, 1969a, b; Rosen and Sjolund 1973a, b). Of special interest in relation to possible plasticity in the cuneate nucleus is the fact that though they never possessed a cutaneous receptive field, a few muscle Group I activated cells could be weakly driven by electrical stimulation of a cutaneous nerve. Further evidence of modality segregation in the DCN is discussed below (Sect. 5.2.5).

Inhibitory Receptive Fields. Though it is widely believed that afferent inhibition plays an important part in the synthesis of excitatory receptive fields in the somatosensory system, the precise function of this inhibition is poorly understood. As observed in the spinal cord, the thalamus and the SI cortex, an electrical or natural cutaneous stimulus applied within the excitatory receptive field of a neuron evokes a brief excitatory response that is followed by a period of strong inhibition. In the DCN this strong inhibition lasts from 50 to 150 ms. The time-course and strength of this inhibition may be measured by studying the inhibition of a second stimulus applied within the excitatory receptive field at various times after the first. In cats under light barbiturate anaesthesia (a condition that enhances significantly afferent inhibition of DCN cells) Janig et al. (1977) have shown that inhibition is maximal from the centre of the excitatory receptive field and decays steeply with distance from the centre. Inhibition was still observed when the first stimulus was applied outside the excitatory receptive field but its decay with distance from the edge of the excitatory receptive field had a milder slope. For hair-sensitive neurons with distal receptive fields the gradients of afferent inhibition across and outside the excitatory receptive field were much less steep than the gradient of excitation. This results in a region around the excitatory receptive field from which inhibition may be elicited (Fig. 5.4 a, c). The inhibition from within the excitatory receptive field is thus identical to the *in-field* inhibition described in the thalamus and the SI cortex (Fig. 6.7; Sect. 6.3, 7.3.1 and 8.5.1) while the inhibition from the skin around the excitatory receptive field constitutes what is classically known as *surround* inhibition.

In neurons thought to be driven by Pacinian afferents the spatial gradients of inhibition and excitation were more equal and surround inhibition was consequently reduced (Fig. 5.4 b, d) (Janig et al. 1977). This is presumably true also of the touch-pressure sensitive neurons described by Gordon and Paine (1960), Gordon and Jukes (1964) and Perl et al. (1962) as also having large receptive fields lacking inhibitory surrounds. As inhibition is most effective when it is generated from the centre of the excitatory field, resting, or incidentally evoked, discharge in afferents supplying this area might effectively inhibit any weak excitatory inputs from afferents supplying skin regions outside the excitatory receptive field. Such weak inputs may be those onto distal dendrites that project onto the termination regions of afferents supplying skin outside the normal excitatory receptive field of the neuron or they might be derived from sparse ramifications of afferents that innervate skin

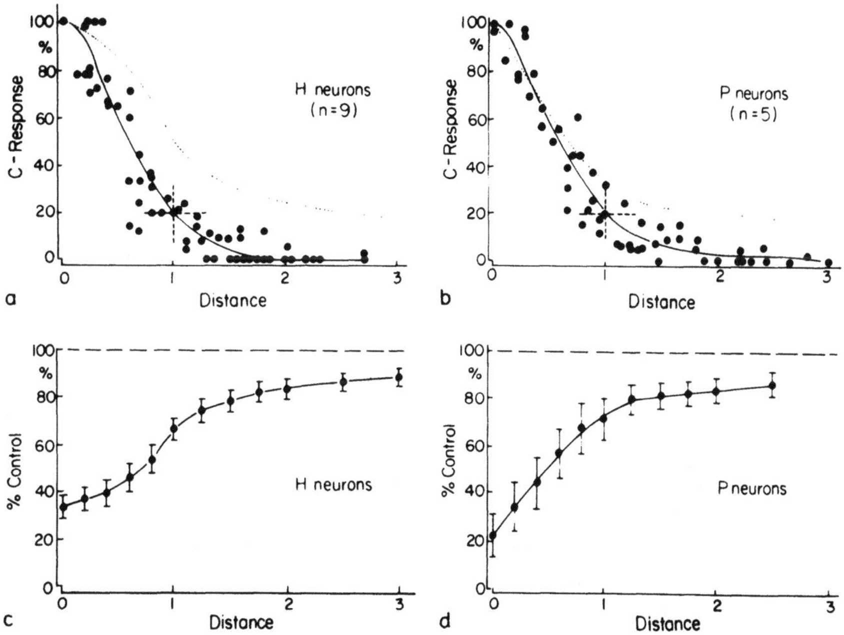

Fig. 5.4 a–d. The basis of *surround* and *in-field* inhibition in neurons of the cat cuneate nucleus. Averaged spatial distribution of excitation and inhibition in hair sensitive (*H*) neurons (**a**,**c**) and Pacinian-sensitive (*P*) neurons (**b**,**d**). The profile of the excitatory RF was determined by applying a brief pulse of air (supramaximal at the centre) at different points along a line passing through the RF centre. A conditioning stimulus also consisting of a brief pulse of air was applied at points on the skin at different distances from the centre of the excitatory RF, where a test pulse was given 16–22 ms later. For each neuron the distance scale is normalized such that unit distance is the distance from the RF centre at which the excitatory response to a standard pulse of air (*filled-circles*) falls to 20% of that at the centre. In **a** and **b** the *ordinates* are the conditioned response, expressed as a percentage of the unconditioned (control) response. The *dotted lines* in **a** and **b** show the spatial distribution of the inhibition with normalized inverted ordinate scales. *Vertical bars* in **c** and **d** indicate + and − S.E. Note that the inhibition produced by the conditioning stimulus is maximal from the centre, and decays to the border of the RF. The spatial gradients of afferent inhibition within and outside the RF of H neurons (**a**) is much flatter than that of excitation, producing a clear inhibitory surround. For P neurons the excitatory and inhibitory profiles are more similar (**b**), and surround inhibition is thus weak or absent. (From Janig et al. 1977, p. 830)

outside the neuron's receptive field and that primarily project to a distant focus in the DCN (Sect. 5.4). While the presence of afferent inhibition in the DCN is well established there is still some confusion concerning the source of the inhibition and its effectiveness under different conditions. Andersen et al. (1970) considered that in barbiturate-anaesthetized cats primary afferent depolarization and afferent inhibition in the cuneate nucleus was mainly produced by activation of rapidly adapting hair afferents. However, Bystrzycka et al. (1977) argued that most if not all afferent inhibition in the cuneate nucleus of decerebrate unanaesthetized cats was dependent on activity in Pacinian afferents. In cats lightly anaesthetized with halothane,

Golovchinsky (1980) found that it was almost impossible to demonstrate afferent inhibition of cuneate neurons from outside the excitatory receptive field, although he did note signs of inhibition from within the excitatory receptive field of some neurons.

5.2.4 Core and Shell Organization

It soon emerged from some of the first microelectrode studies of the DCN that there might be functional differences between the different cytoarchitectonic zones (Gordon and Paine 1960; Gordon and Kuypers 1964). This aspect of nuclear organization was later emphasized by Gordon and Jukes (1964), who concluded that the gracile nucleus in the cat was organized into two functionally different zones, corresponding to (1) the cell-nest region in the dorsal part of the middle of the nucleus, and (2) the reticular zone encompassing the rostral and basal regions of the nucleus. According to this scheme, in the middle, superficial, cell-nest zone cells have small, somatotopically organized receptive fields with inhibitory surrounds, and a large proportion send their axons to the contralateral VB complex via the medial lemniscus. In the reticular zones cells have large receptive fields which lack inhibitory surrounds and somatotopic organization is absent. Only a small proportion of cells in these regions project to the VB complex. In general, cells in this region do not respond to a single class of receptor and it is possibly these neurons that are responsible, at least in part, for the widefield properties of some VB neurons (Sect. 6.3). While different patterns of functional organization have been described in other studies (Kruger et al. 1961; Perl et al. 1962), it is possible that the differences are due mainly to the susceptibility of responses in the rostral region to anaesthesia or to the animal's physiological status (Gordon and Jukes 1964).

More recently, combined anatomical and physiological studies of the DCN in the cat (Cheema et al. 1983) have confirmed and extended earlier work showing that the cuneate nucleus in the cat may be divided into the following 3 major regions: (1) the region between 0 and 4 mm rostral of the obex (the rostral cuneate, RC) (Fig. 5.1a), (2) the region between 0 and 4 mm caudal of the obex (the middle cuneate, MC) (Fig. 5.1a), and (3) the region between 4 and 8 mm caudal of the obex (the caudal cuneate, CC) (Fig. 5.1c). Within each of these three main divisions further subdivisions have been recognized on the basis of combined functional, cytoarchitectonic and connectional criteria. Thus the rostral and middle regions have each been subdivided into dorsal (RCd, MCd), central (RCc, MCc), ventromedial (RCvm, MCvm) and ventrolateral (RCvl, MCv) parts while the caudal region has been subdivided into a dorsal (CCd) and a ventral (CCv) part. This extreme caudal region of the nucleus, between 6 and 8 mm caudal of the obex, has been designated the spinal region (CCs). Table 5.1 (Cheema et al. 1983), provides a convenient summary of the distinguishing features of these subdivisions and Fig. 5.2 shows the disposition of these areas within the 3 main divisions of the nuclei.

A contemporary view of the organization of the mammalian DCN is thus that it consists of a *core* region, comprising the region of cell clusters in the dorsal part of the middle of the nuclei, and a *shell* region comprising the reticular regions of the

Table 5.1. Subdivisions of the rostral, middle, and caudal parts of the cuneate nucleus based on (1) Nissl and Golgi material; (2) distribution of thalamic projection neurons (TPNs); (3) receptive field location and submodality; and (4) distribution of corticofugal afferent terminations. (After Cheema et al. 1983)

Rostral cuneate	
Dorsal (RCd)	Mainly large and small neurons; large neurons display many long radiating dendrites; few TPNs; neurons respond predominantly to rotation of forelimb joint; dense corticofugal terminations
Central (RCc)	Mainly small and medium neurons; few impregnated in Golgi material; few TPNs; mainly large cutaneous receptive fields on arm; sparse corticofugal terminations
Ventromedial (RCvm)	Mainly large neurons with long radiating dendrites; deep and cutaneous receptive fields around perioral region, jaw and trunk; sparse corticofugal terminations
Ventrolateral (RCvl)	Mainly small and medium neurons with few dendritic processes usually directed ventrolaterally; fusiform TPNs; cutaneous receptive fields on neck and mainly from deep receptors at the shoulder region; no corticofugal terminations
Middle cuneate	
Dorsal (MCd)	Medium neurons arranged in clusters; small neurons in between clusters; large number of TPNs; discrete cutaneous receptive fields on digits; sparse corticofugal terminations
Central (MCc)	Mainly medium and large neurons with long radiating dendrites; few TPNs; deep and cutaneous receptive fields on arm; cortical projections denser from rostral to caudal
Ventromedial (MCvm)	Sparsely populated region; few TPNs; both cutaneous and deep receptive fields on trunk; moderate corticofugal projections
Ventrolateral (MCvl)	Many small and medium neurons; variety of dendritic shapes in Golgi material; cutaneous receptive fields but also some deep input from the neck area; corticofugal termination absent
Caudal cuneate	
Dorsal (CCd)	Very small number of neurons, mainly medium and large with a variety of dendritic shapes; mainly cutaneous receptive fields on arm; very sparse corticofugal terminations
Ventral (CCv)	Mainly small neurons; sparse TPNs; large receptive fields on arm and trunk areas and neurons driven by deep input preferentially; very heavy corticofugal projections
Spinal (CCsp)	Sparsely populated region at extreme caudal end of cuneate nucleus; receptive fields not determined; dense corticofugal projections

rostral and caudal poles, the base, and the dorsal rim of the nuclei. The core region contains a well-organized somatotopic representation of the body which is largely given over to the distal extremities of the limbs. Neurons in the core region respond almost exclusively to primary afferent fibres of a particular modality. They have small receptive fields and project their axon to the contralateral VB complex where they are responsible for the classical *lemniscal* properties of cells in the VPM and VPL (Sect. 6.3) (Figs. 6.1b and 6.2b). Most cells in the VPL and VPM project to areas 1 and 3b of the SI cortex giving rise to the precise somatotopic representation of the body surface across each of these areas (Figs. 6.1, 6.2, 7.1 and 7.14) (Sects. 7.2.1 and 7.3.1). In contrast, within the shell region a somatotopic representation of the body is lacking. Neurons in the shell region receive multimodal input mainly from non-primary afferents in the dorsolateral funiculi. They have large receptive fields which are usually derived at least in part from deep tissues and generally are centred on proximal parts of the body. Although there are fewer neurons in the shell region that project to the VB complex it nevertheless seems likely that they are at least in part responsible for the widefield properties of the so-called *m* cells that have been well documented in both the VB complex and the SI cortex of the cat (Sects. 6.3 and 7.3.3).

The shell region may be regarded as the phylogenetically more primitive of the two systems. Presumably the core system has developed more recently in response to the high adaptive value of an increased acuity of somatoaesthesia, especially at the distal extremities of the limbs (Dreyer et al. 1974). Most studies of plasticity in the DCN following lesions of afferent pathways deal with changes that are presumably restricted to the somatotopically organized core system. However, it is important to take into account the dual nature of nuclear organization when interpreting such changes both in the DCN and at higher levels of the neuraxis (Chaps. 6 and 7). Generally this has not been done.

5.2.5 Somatotopic Organization

The main features of the somatotopic organization of the mammalian cuneate and gracile nuclei were first established many years ago in the cat by use of the evoked-potential technique (Kuhn 1949). The forelimb region is represented in the cuneate nucleus, and the hindlimb region more medially in the gracile nucleus. The somatotopic organization of the mammalian DCN has since been investigated in several species by means of electrophysiological mapping techniques employing various kinds of microelectrode. In addition to the rat, cat and raccoon, which are considered here in some detail, maps have also been obtained in the tree squirrel (Ostapoff et al. 1983), the sheep (Woudenberg 1970) and the opossum (Hamilton and Johnson 1983). Unfortunately, detailed mapping experiments have not yet been performed in any species of primate. While there are many features common to all species it is evident that the details of the organization not only vary considerably between different animals, but also depend on the mapping technique used. With the exception of a single study in the decerebrate, unanaesthetized cat (Cheema et al. 1983), all detailed electrophysiological mapping studies of the somatotopic organization the DCN have been performed in animals that were deeply anesthetized with barbiturates.

Cat. Therman (1941) should probably be credited with the first electrophysiological studies of the mammalian cuneate nucleus, but the first study aimed at determining the somatotopic organization of the nucleus was that of Kuhn (1949), who recorded potentials in cats anaesthetized with barbiturate, evoked by natural stimulation of the body surface in a grid of penetrations with a coarse wire electrode. He noted that maximal potentials to hair and skin stimulation were recorded from levels caudal to the obex, while more rostrally, maximal potential were obtained in response to stimulation of deep structures (muscle stretch, joint movement, deep pressure). Kuhn (1949) summarized his findings regarding the somatotopic organization of the gracile and cuneate nuclei, together with the spinal division of the trigeminal nuclear complex, thus ".... the ipsilateral surface of the cat is represented within the caudal portions of the DCN as an inverted figure of the animal with the tail pointed dorso-caudally, extremities dorsally, and the face and mouth situated laterally. The figure is considerably distorted to allow greater volume representation of those parts of the body with the greatest innervation density. The dorsal portions of the body are represented in continuity from tail caudally to neck more rostrally across the ventral portions of the responsive areas" (Kuhn 1949).

The results of later physiological studies were in general agreement with these findings (Johnson 1952; Winter 1965). In cats under deep barbiturate anaesthesia Kruger et al. (1961) found that the receptive fields of hair-sensitive and touch-sensitive neurons in the gracile and cuneate nuclei and spinal division of the trigeminal nuclear complex were such that the entire body surface was represented across all transverse planes between 4 mm rostral, and 8 mm caudal, of the obex (Fig. 5.5a). Unlike Kuhn (1949) these workers found no evidence that cells responding to deep inputs were located rostral to cells responding to cutaneous inputs. Kruger et al. (1961) did, however, report that cutaneous and deep modalities were distributed such that cutaneous inputs predominated in regions containing a representation of the distal extremities whereas deep inputs were found more often in the representation of the trunk and proximal parts of the limbs.

The uniform and rather straightforward somatotopic organization proposed by Kruger et al. (1961) has not been confirmed by some subsequent studies of the DCN in the cat. Perl et al. (1962) examined the receptive fields of neurons in the gracile nucleus of decerebrate, unanaesthetized cats and emphasized the modality specificity shown by gracile neurons (Sect. 5.2). Although these workers did not map the somatotopic organization in detail certain of their observations suggested a more complex organization than that reported by Kruger et al. (1961). For example, the receptive fields of hair-sensitive neurons varied widely in average size, from around 50 mm^2 on the tips of the digits, to 2000 mm^2 on the flank, while the receptive fields of touch-pressure-sensitive neurons were generally much larger, varying between about 400 mm^2 on the digits and 11000 mm^2 on the flank. Furthermore, neurons with these two different modality specificities were not uniformly distributed along the rostrocaudal axis of the gracile nucleus. Neurons with hair-sensitive fields were most commonly encountered more than 1 mm caudal of the obex, while those having

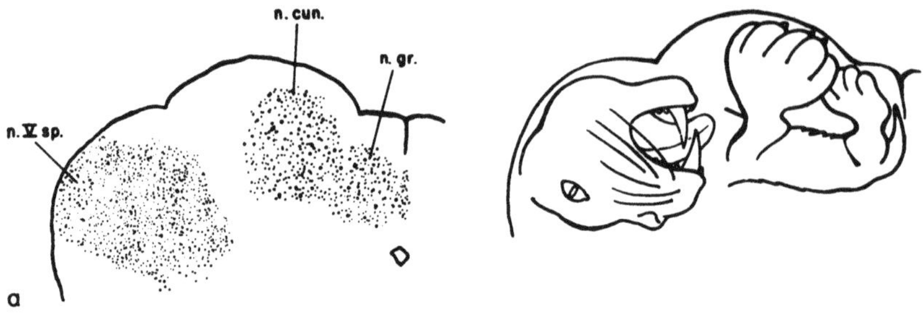

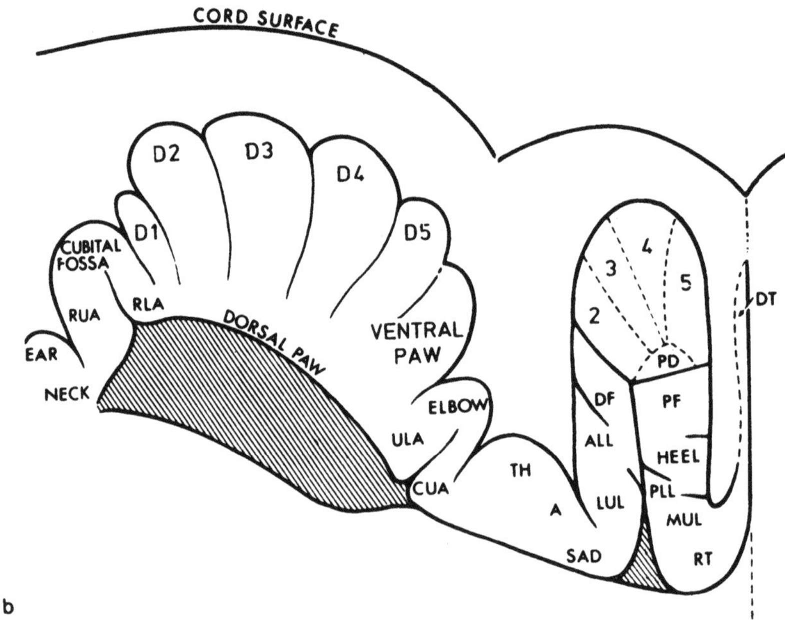

Fig. 5.5 a, b. Somatotopic organization of the DCN in the cat. **a** Early impression of the somatotopic organization of neurons in the gracile nucleus (*n.gr.*), cuneate nucleus (*n.cun.*) and subnucleus caudalis of the trigeminal nuclear complex (*n. V sp.*) The representation in the gracile and cuneate nuclei resembles an inverted "feliculus", or view of the animal's body, with the tail most medially in the gracile nucleus, followed by the hindlimb more laterally and the forelimb occupying the cuneate nucleus. The representation of the trunk occupies a relatively small ventral region of the nuclei. According to Kruger et al. (1961), this somatotopic organization is the same at all rostrocaudal levels of the DCN (and spinal trigeminal nucleus). **b** More detailed "feliculus" of the cuneate and gracile nuclei obtained by mapping multiunit RFs at points in a 100 μm × 100 μm grid in a single transverse plane through the middle cell-nest region of the nuclei. Note the split representation of the forelimb, with the radial aspect represented laterally, and the ulnar aspect medially in the cuneate nucleus. Between the two is a large region containing neurons responding to deep input (*shaded*). *Abbreviations*: *DT*, distal tail; *RT*, root of tail and groin; *MUL*, medial upper leg; *PLL*, posterior aspect of lower leg; *PF*, proximal plantar foot; *PD*, foot-pad; *DF*, dorsal foot; *ALL*, anterior aspect of lower leg; *LUL*, lateral upper leg; *SAD*, saddle and lumbar back; *A*, abdomen and perineum; *TH*, thorax; *CUA*, caudal upper arm and shoulder; ULA, ulnar aspect of lower arm; *RLA*, radial aspect of lower arm; *D1* to *D5*, digits 1 to 5; *RUA*, radial aspect of upper arm. (**a** From Kruger et al. 1961, p. 343, **b** modified from Millar and Basbaum 1975, p. 287)

fields on the digits were found 2–4 mm caudal of this landmark. Touch-pressure-sensitive neurons were found in the region 0–2 mm caudal of the obex, where they usually had receptive fields on the distal part of the limb. They were also often found in a region 5–6 mm caudal of the obex, where their receptive fields included proximal parts of the limb and trunk and often extended onto the distal portions of the limb. Although the rostral region of the nucleus was not explored in these experiments it is clear that the information reported is at odds with the observations of Kruger et al. (1961).

Gordon and co-workers (Gordon and Paine 1960; Gordon and Jukes 1964) examined the responses of gracile neurons in a variety of preparations of which most were deeply anaesthetized with barbiturate. Some of their findings have been touched upon previously (Sect. 5.2). Like Perl et al. (1962) they also noted the high degree of modality specificity of gracile neurons, the most common types being hair-sensitive and touch-pressure-sensitive cells. Both types were found throughout the length of the nucleus, but touch-pressure cells were most common in the rostral section and in the deep part of the middle region (Fig. 5.6). Although the size of the receptive fields of both types of cell varied widely the actual size was to some extent determined by the postion of the receptive field along the proximo-distal axis of the limbs. The receptive fields of touch-pressure cells were much bigger, on average, than those of any other type of neuron and in keeping with their position in the basal region of the nucleus most touch-pressure-sensitive neurons had proximal receptive fields on the upper leg or trunk. In the rostral third of the nucleus the receptive fields of both hair- and touch-pressure-sensitive neurons were generally much larger than in the middle third. Although the somatotopic organization of the nucleus was not mapped in detail, an organization similar to that described by Kruger et al. (1961) was clearly evident in the middle region (corresponding to a region between about 2 and 5 mm caudal of the obex), where hair-sensitive cells having small receptive fields with inhibitory surrounds were commonly found in clusters (Fig. 5.6a). In the rostral region there was no sign of somatotopic organization and units with receptive fields on the proximal part of the limb were sometimes recorded at the same dorsoventral position as cells with distal receptive fields (Fig. 5.6b). Gordon and Jukes (1964) also found a concentration of cells with Pacinian input in a restricted region of the gracile nucleus between 3 and 6 mm caudal of the obex. Many of the cells in this region appeared to respond to activation of Pacinian corpuscles within the abdominal cavity. Moreover, some of these neurons also had well defined cutaneous receptive fields on the hindpaw or lower leg.

Millar and Basbaum (1975) mapped the middle part of the cuneate and gracile nuclei in cats anaesthetized with Dial. The receptive fields of multiunit clusters were tested at intervals of 100 μm along vertical microelectrode tracks. At each recording locus, between 1 and 4 cells were simultaneously recorded, and where multiunit activity was present, all the recorded cells were found to have receptive fields on overlapping or adjacent parts of the body. However, to improve the resolution of the map, only receptive fields of cells with spikes $>100\,\mu$V were actually mapped. Receptive fields were classified according to whether they were superficial (cutaneous) or deep, but no attempt was made in this study to distinguish different sensory modalities. In the cuneate nucleus the somatotopic organization was

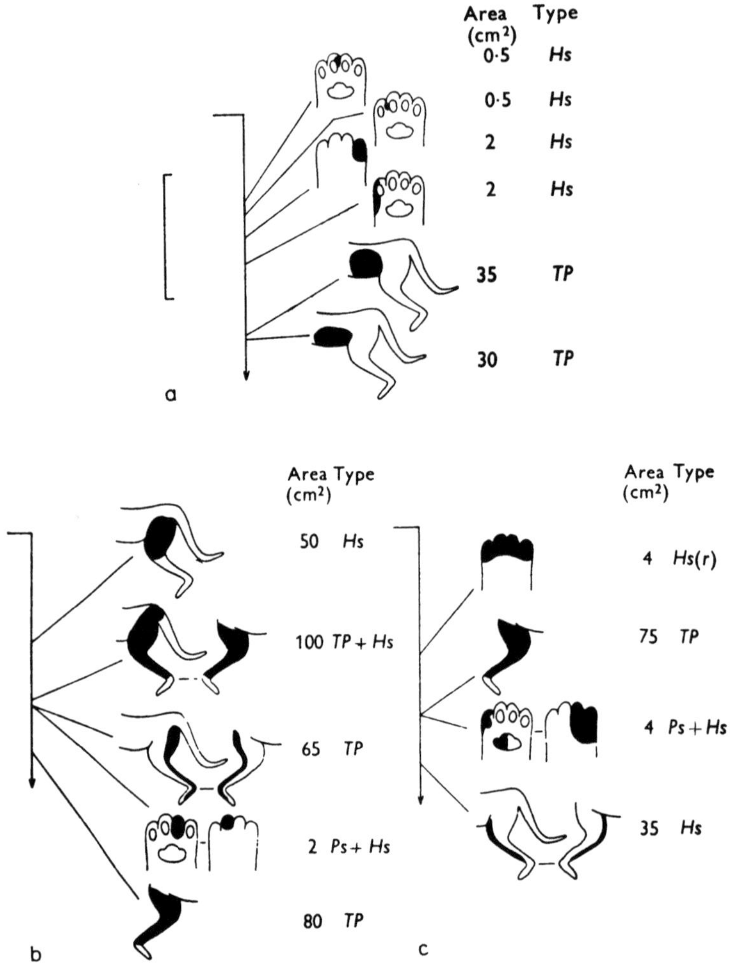

Fig. 5.6 a–c. Somatotopic organization in the gracile nucleus of the barbiturate-anaesthetized cat. Sequences of single units, and the characteristics of their RFs, in 3 typical vertical microelectrode penetrations through the cuneate nucleus in the cell-nest (**a**) and rostral reticular (**b**, **c**) regions. The dorsoventral movement of the electrode in each track is shown by the *arrow*. *Scale bar* = 1 mm. Table at *right* gives the size and submodality type of the RF for each cell. *Abbreviations*: *Hs*, hair-sensitive; *Ps*, pad-sensitive; *TP*, touch-pressure-sensitive; *r*, cell with poorly defined properties. Note the much larger RFs of TP than Hs neurons at both levels of the nucleus, the presence of TP cells in the ventral part of the track through the middle region, the somatotopic organization in the track through the middle region, and its absence from the rostral tracks. (From Gordon and Jukes 1964, p. 270)

essentially the same at different rostrocaudal levels between 1 and 3 mm caudal of the obex (Fig. 5.5b). Within this region the position of receptive fields changed in an orderly manner acrosss the nucleus, with the most lateral toe represented most medially. The representation of the digits was massively enlarged when compared to that of the rest of the paw and the limb. In the most superficial part of the nucleus, cells with receptive fields on the toe pads were found dorsal to cells with receptive fields on the dorsal surface of the digits. A further point of interest was the discovery of a split representation of the foreleg. Whereas the radial surface of the lower foreleg, the cubital fossa, and the rostral part of the upper foreleg were represented laterally in the nucleus, the ulnar side of the lower foreleg, the elbow and the caudal part of the upper foreleg were represented medially. Between these representations, in the base of the nucleus, was a large region containing neurons with deep receptive fields (Fig. 5.5b). Although the rostral reticular zone was not explored in these experiments, the somatotopic organization of the nucleus was the same at all rostrocaudal planes examined. In the gracile nucleus the somatotopic organization was more variable than in the cuneate. For example, certain parts of the hindlimb had greater representation at some rostrocaudal planes of the cell-nest region than at others. The representation of the hindlimb was split mediolaterally in a similar fashion to that of the forelimb in the cuneate nucleus. The representation of the toes was very erratic and in some animals cells with receptive fields on one of the digits were not found, a feature that is perhaps reflected in the variability of somatotopic representation of the digits at the level of the SI cortex and that may well relate to directed attention and/or usage of a particular part of the skin surface (Figs. 7.2 and 7.21; Sects. 7.2.4, 7.4.1 and 7.5.1). In the dorsal part of the cuneate nucleus regions were often encountered where the background noise was low and, as suggested earlier by Gordon and Jukes (1964), these quite regions presumably corresponded to the regions between the cell clusters. As the electrode traversed these silent regions, there were often sudden large jumps of receptive field from the toes to area of the lower foreleg. The abrupt changes in receptive field location suggested that in the DCN the body surface might be represented not as a continuous map where the centres of receptive fields move smoothly across the surface of the body, but as a map in which neurons are collected into small groups within which each neuron has a receptive field on the same region of skin. Shifts of receptive fields are thus related to transitions from one group of cells to another, rather than from cell to cell.

In 10- to 14-week-old kittens anesthetized with Dial the general somatotopic organization of the cell-nest region of the feline cuneate nucleus was confirmed by Kalaska and Pomeranz (1982). Recording from single units, they found that receptive fields tended to progress from distal to proximal along the dorsoventral axis (Fig. 5.14a). Across the mediolateral axis receptive fields shifted from the ulnar to the radial aspect of the forelimb. There was a disproportionately large representation of the distal part of the limb, cells in the dorsal and central parts of the nucleus all having receptive fields on the paw. In the base of the nucleus cells were hard to drive, and usually responded to movement of joints or squeezing of muscles.

Recent studies in the adult cat have confirmed and extended these rather simple views of somatotopic organization, and have produced a more complex picture consistent with the view that different somatosensory modalities and submodalities might be relayed and processed separately at all levels of the somatosensory system up to and including the somatosensory cortex (Dykes 1983). Cheema et al. (1983) examined the responses of cuneate neurons in different cytoarchitectonic regions of the cuneate nucleus in decerebrate, unanaesthetized preparations. Units in the dorsal part of the middle region (MCd, Table 5.1) were activated exclusively from small cutaneous receptive fields on the foot or digits. In the deeper regions of the middle section (MCvm and MCc, Table 5.1) receptive fields were generally larger, and could extend from the digits to the shoulder. Many cells in this region had deep inputs, and responded to movement of wrist, elbow or shoulder joints. Units in different subdivisions of the rostral region (RC) either responded to deep inputs from proximal parts of the limb or had large cutaneous receptive fields covering the whole forelimb. Within individual subdivisions of the nucleus there might be large shifts of receptive field, but responses were uniformly cutaneous or deep. In the caudal part of the nucleus (CC) units had cutaneous receptive fields on the paw or digits, but unlike those in the middle region, these fields were large, covering several digits, or stocking-like, in that they extended from the digits to the elbow.

In cats anesthetized with barbiturate, Dykes et al. (1982) recorded multiunit activity at a large number of recording sites in the dorsal part of the medulla, many of which were confirmed histologically to be within the DCN or subnucleus caudalis of the spinal nucleus (Sp Vc) of the TNC. Multiunit receptive fields were mapped every 50 or 100 μm along each track. Particular attention was paid to the modality (cutaneous or deep) and submodality (slowly adapting, rapidly adapting) of units. Deep responses were thought to arise mainly from muscle receptors, rather than joints or other deep structures. Very few recording sites within the nuclear complex showed no response at all to peripheral mechanical stimulation, and responses to both cutaneous and deep modalities were rarely found at the same site. Almost all sites within the *external* cuneate nucleus responded to deep modalities, whereas the great majority of sites in the Sp Vc responded to cutaneous modalities. In the main cuneate and gracile nuclei almost 60% of recording sites yielded units with cutaneous receptive fields while 25% yielded units with deep receptive fields and about 13% revealed units responding to Pacinian input. A small proportion of sites showed responses to other modalities including joint and high-threshold mechanoreceptors and a significant proportion of sites (12.5% of sites overall) were activated by somatic stimuli which could not be classified as clearly affecting either deep or cutaneous receptors (responses to tapping). When the cuneate and gracile nuclei were divided into rostral, middle and caudal thirds, using the level of the obex as the dividing line between rostral and middle thirds (Fig. 5.1), there were significantly more cutaneous sites in the middle third and more sites responding to deep input in the rostral third. There were also significant differences in the dorsoventral axis. In the rostral third 90% of sites in the basal half responded to deep input, whereas in the dorsal half sites with deep or cutaneous input were equally common. In contrast, the dorsal half of the caudal two-thirds was occupied mainly by cutaneous sites, and the basal half contained a mixture of cutaneous and deep sites. All cutaneous responses in the rostral region were slowly adapting, whereas those in the middle and caudal regions could be either slowly or rapidly adapting. In the cuneate and external cuneate nuclei responses to deep inputs were mostly slowly adapting, but in the gracile nucleus nearly all responses, whether of deep or cutaneous modality, were rapidly adapting. The distribution of Pacinian responsive sites through the nuclei was also not homogeneous. None were found in the external cuneate or SpVc, and in the gracile and cuneate nuclei they were mostly encountered in the caudal third, where they appeared to coincide with a cytoarchitectonically discrete zone containing large, darkly staining cells (Sect. 5.2.1).

The most remarkable finding of Dykes et al. (1982) was the clear demonstration that small regions within the gracile and cuneate nuclei are devoted to the representation of a single afferent modality. Neurons in each of these small regions thus respond to activity in only a single functional class of primary afferent fibre. Physically, individual modality specific regions consisted of rostrocaudally oriented cylinders of neural tissue. In the case of the cutaneous modality regions that were found in the dorsal half of the middle of the nuclei, the cylinders had roughly the same dimensions as the cell clusters (approximately 210 μm wide by 325 μm long). A microelectrode traversing a typical single modality region could either encounter successive receptive fields located on the same part of the body, or record a smooth sequence of shifting but overlapping receptive fields. When breaks in a smooth sequence of receptive fields occurred they always coincided with a change in modality or submodality (Fig. 5.7a). Dykes et al. (1982) noted that the receptive field sequences obtained in rostrocaudally oriented electrode tracks could also undergo gradual and abrupt changes in location or modality at intervals of a few hundred micrometers and that the same receptive field was never found at more than one place within each track (Fig. 5.7a). This observation is difficult to reconcile with the idea that the mediolateral somatotopy is uniform throughout the rostrocaudal extent of the cell-nest region. Moreover, it suggests that if the terminal arborizations of individual primary afferent fibres line up in the rostrocaudal axis, they may not all terminate in somatotopically equivalent regions. Thus the body may be represented in the DCN not in a single, continuous, somatotopic map lacking discontinuities, but as a gradient of gradually shifting receptive fields which is interrupted at boundaries between populations of cells subserving different modalities. For each afferent modality Dykes et al. (1982) found small regions representing nearly all body parts (Table 5.2). Thus essentially, there are somatotopic maps in each modality represented within the nuclei, and owing to the relatively large dimensions of the modality-specific regions, these maps are not likely to be in register with one another.

Raccoon. Like the representation of the vibrissae in the TNC of rodents, the representation of the forepaw of the raccoon in the cuneate nucleus may be ideally suited to investigations of plasticity. This is because in the cuneate nucleus of the raccoon there are clear morphological specializations related to the representation of the volar skin of the digits. The cell-nest region of the raccoon cuneate nucleus is organized primarily into longitudinally oriented cellular subnuclei separated from each other by fibrous laminae. These subnuclei extend rostrocaudally as thin, irregular columns of cells running the entire length of the nucleus, and each receives the projection from the volar surface of an individual digit (Johnson et al. 1968). Within each subnucleus (Fig. 5.8a, b), receptive fields are confined to the one digit and are somatotopically organized. Evidently, the subnuclei represent a highly specialized development of the cell-nest, or core, region of the DCN, related to the enhanced somatosensory function of the forepaw in this mammal (Johnson et al. 1968).

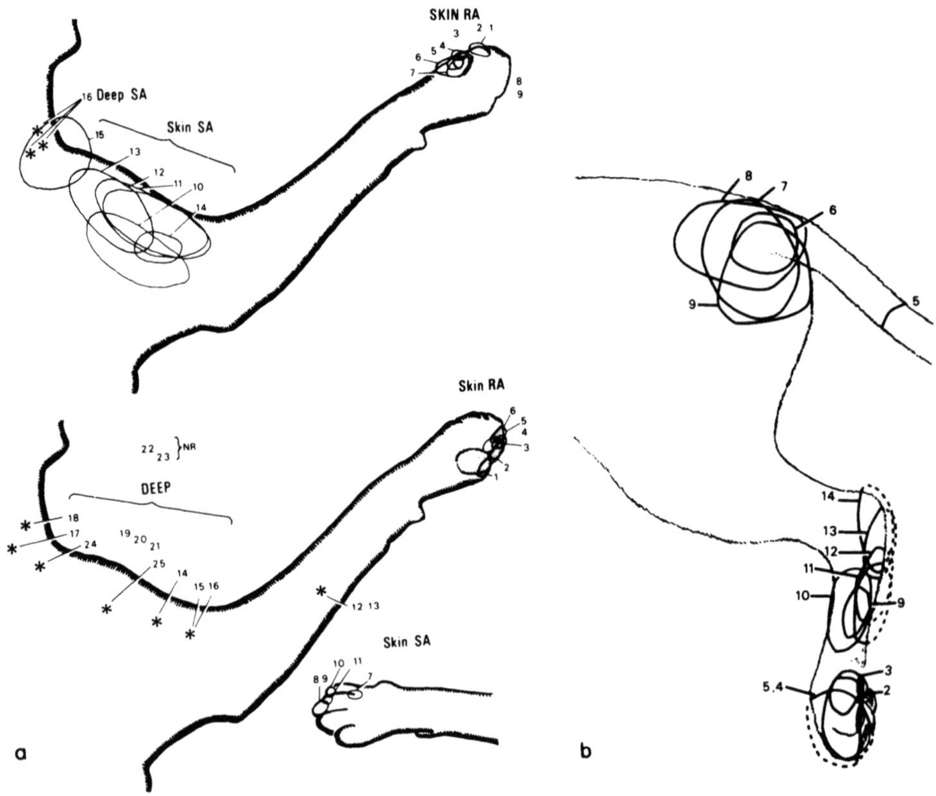

Fig. 5.7 a, b. Relation between submodality segregation and somatotopic organization of neurons in the DCN of the barbiturate-anaesthetized cat. **a** Receptive field sequences in two long, rostrocaudally directed, horizontal penetrations, through the cuneate nucleus. The *numbers* referring to each receptive field show the order in which each neuron was encountered in 100 μm steps throughout each track. Note the gradual shift of receptive field centre over distances of several hundred micrometers, and that large shifts in receptive field location, correlated with changes in response modality (*skin, deep*) or submodality (*skin RA, skin SA*) can occur over relatively small distances (e.g. 100 μm). **b** Receptive field sequences every 100 μm throughout a rare vertical penetration through the caudal part of the gracile nucleus where a large jump in RF location occurred in the absence of any change in receptor modality or submodality. The *numbers* have the same significance as in **a**. All the receptive fields are cutaneous rapidly adapting. A sequence of receptive fields on the paw (200–500 μm) was followed by a sequence on the tail (500–900 μm) then a sequence on the hind foot (900–1400 μm). Dual multiunit receptive fields, on the tail and the hind foot, could be recorded for a distance of about 150 μm. Note the large jump in RF location from base of the tail to the foot over this relatively small distance. *NRF*, no receptive field. (From Dykes et al. 1982, pp. 405 and 404)

Table 5.2. Representation of body parts in different submodality regions of dorsal column nuclei in the cat. (Dykes et al. 1982)

	Skin			Deep		
Deep	SA	LV	RA	SA	RA	Pacinian
First digit	+	+	−	−	−	+
Second digit	+	+	+	+	+	+
Third digit	+	+	+	+	+	+
Fourth digit	+	+	+	+	−	+
Fifth digit	+	+	+	+	−	+
Paw	+	+	−	+	+	+
Wrist	+	+	+	+	+	+
Forearm	+	+	+	+	+	+
Upper arm	+	+	+	+	+	+
Shoulder	+	+	+	+	+	−
Neck	+	+	+	+	+	−
Thorax	+	+	+	+	+	+
Abdomen	+	+	+	+	−	+
Upper hind leg	+	+	+	+	+	+
Lower hind leg	−	+	−	+	+	+
Hind Foot	+	+	+	+	+	+
Hind digits	+	+	+	+	+	+
Tail	+	+	+	−	+	+

Abbreviations: SA, slowly adapting; *RA*, rapidly adapting; *LV*, low velocity (D-hair) rapidly adapting.

Rat. Relatively few details are available concerning the somatotopic organization of the DCN in the rat. This is particularly unfortunate because this species has been frequently used in experiments to investigate the effects of lesions of afferent pathways. Notably lacking is precise information about the size of receptive fields, the representation of different sensory modalities, the detailed representation of the digits, and the effect on the somatotopic organization of the transverse clustering of neurons (Sect. 5.2.1). In an early microelectrode study, McComas (1963) examined the response of single cells in the gracile nucleus of urethane-anaesthetized rats to mechanical stimulation of the plantar skin of the hindpaw. Cells isolated within 1 mm of the rostal tip of the nucleus had much larger receptive fields on average than those isolated in the next 1 mm. Cells situated more caudally also had larger fields. These results are qualitatively similar to those reported for the cat gracile nucleus by Gordon and co-workers, and indicate that the rat gracile nucleus may be organized in a similar manner. Unfortunately, no attempt was made to map the detailed somatotopic organization or functional properties of neurons in the three different regions of the nucleus. However, it was noted that cells in the medial part had receptive fields on the tail and those in the lateral part had receptive fields on the flank, while the representation of the hindlimb occupied a relatively large

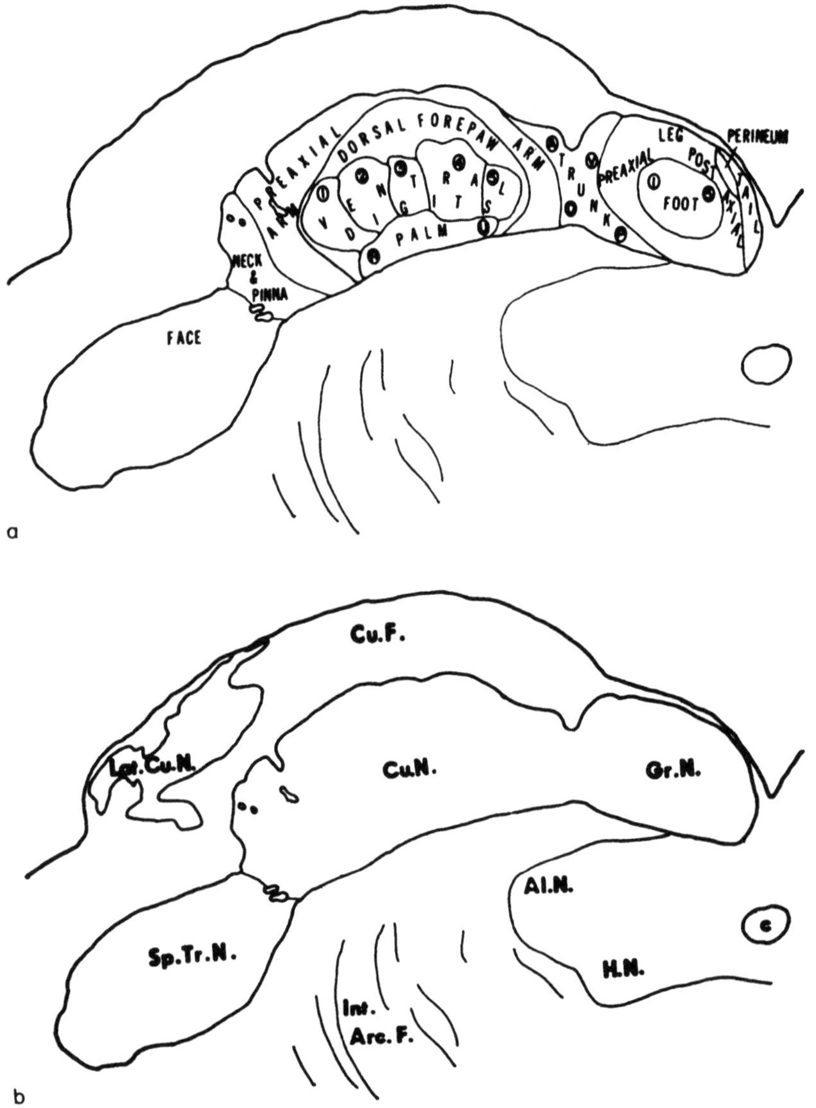

Fig. 5.8 a, b. Somatotopic organization of the DCN of the raccoon. **a** Somatotopic map of the left DCN superimposed on a transverse section through the nuclei at their widest extent. Corresponding neuroanatomical regions are shown in **b**. Note the relatively huge and separate representation of the ventral skin of each of the forepaw digits. *Cu.F.*, cuneate fasciculus; *Cu.N.*, cuneate nucleus; *Lat.Cu.N.*, lateral cuneate nucleus; *Gr. N.*, gracile nucleus; *Sp.Tr.N.*, spinal trigeminal nucleus; *Int.Arc.F.*, internal arcuate fibres; *A1.N.*, dorsal motor nucleus of X; *H.N.*, hypoglossal nucleus; *c*, central canal. (From Johnson et al. 1968, p. 42)

intermediate region. Some cells in the middle of the long axis had receptive fields smaller than $10\,mm^2$ but it is not clear whether any of these were restricted to only part of a single digit. A further omission in this study was that no attempt was made to determine the modality specificity of different neurons.

In rats anaesthetized with pentobarbital, Nord (1967) explored the medulla in dorsoventrally oriented tracks with gross wire electrodes. No attempt was made to isolate single units in this study, though they were sometimes recorded. Tracks were made between 3.5 mm caudal and 3.1 mm rostral of the obex, which would encompass the whole of the region explored by McComas (1963), and include both the caudal and rostral cytoarchitectonic regions described by Basbaum and Hand (1973) (Fig. 5.5). Relatively few tracks were made in the cuneate and gracile nuclei in this study, the majority passing through the face representation in adacent parts of the SpVc (Fig. 3.3a, b). In the DCN proper, multiunit receptive fields were frequently large, the smallest covering two digits plus part of the paw. In general, distally situated receptive fields were smaller than those situated more proximally. However, there was no other correlation between receptive field size and position of the recording site in the nucleus. It should be noted that the multiunit receptive fields in this study are somewhat artificial, because their boundaries were set by selection of an arbitrary minimum spike amplitude. The data shown in Fig. 5.9 illustrate the marked degree of expansion of such multiunit receptive fields as the selected spike amplitude was varied between $>100\,\mu V$ through $25–99\,\mu V$ to $<25\,\mu V$. This illustration taken from very early work, underscores the need for extreme caution when using multiunit activity to map receptive fields and determine somatotopy in the rat DCN—a warning that is particularly important when recordings are made in regions which have been partially deprived of their normal afferent input (Sects. 5.3 and 8.2.3).

The general picture of somatotopic organization that emerged from Nord's (1967) study was that cells in the medial region of the nuclei had receptive fields on the tail, while those in successively more lateral regions had receptive fields on successively more rostral parts of the body (Figs. 3.3a, b). Cells in the dorsal part of the nuclei had distal receptive fields, while those in the ventral region had proximal receptive fields. Almost any point on the skin appeared to be represented in the responses of a chain of cells stretching rostrocaudally throughout the length of the nuclei. Thus Nord (1967) concluded that within the DCN of the rat there is a continuous representation of the body surface which in the transverse plane of the medulla resembles a highly distorted and inverted view of the limbs, trunk and tail. The relative size of the representation of a particular body part is directly proportional to the peripheral innervation density of that part. Finally, in reporting on the somatotopic representation of the forelimb in the VPL of the rat, Angel and Clarke (1975) have cited an unpublished study by Berridge (1973) that shows that within the cuneate nucleus of the rat the forepaw is represented as a closed fist that stretches along the nucleus such that different parts are represented at different rostrocaudal levels. Interestingly, a similar representation has been described at the levels of the thalamus and the SI cortex in this species (Figs. 6.8 and 7.2b; Sect. 6.4).

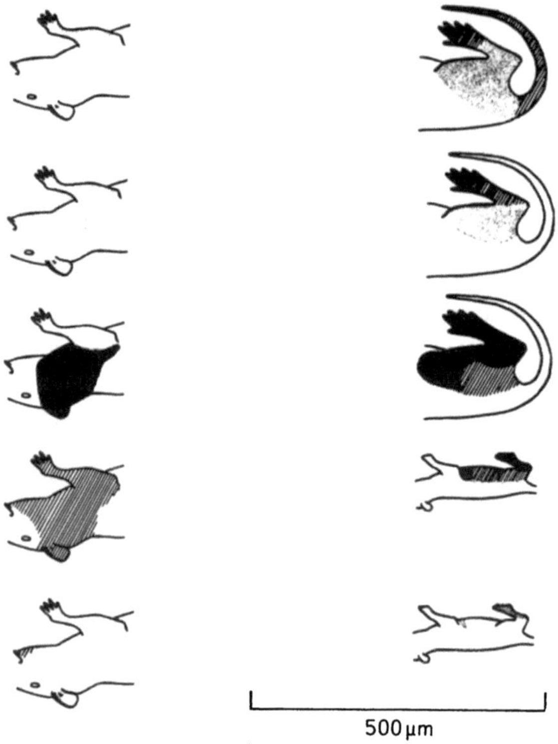

Fig. 5.9. Somatotopic organization in the DCN of the urethane-anaesthetized rat. Multiunit receptive fields found in two typical vertical microelectrode tracks through the middle region of the DCN 2 mm caudal to the obex. The track on the *left* passes through the cuneate nucleus and that on the *right* traverses the gracile nucleus. The rat figurines are inverted to emphasize the inverted representation of the body within the DCN. The *shading* indicates the extent to which multiunit RFs may vary when the minimum acceptable amplitude of the evoked activity is arbitrarily set at different levels. (*black*, $> 100\,\mu V$; *hatched*, 25–99 μV; *shaded*, $< 25\,\mu V$). (Modified from Nord 1967, p. 351)

5.3 Alterations of Inputs to the Nuclei

Attempts of demonstrate plasticity within the DCN following lesions of ascending pathways include descriptions of alterations of normal somatotopic organization, morphological changes in afferent projections and synaptology that might indicate neuronal sprouting and recovery from sensorimotor deficits induced by the lesion. In some cases the effects of lesioning afferent pathways have been shown to be dependent upon whether the lesion is made in the immature or the adult animal. In relation to somatotopic changes it is important to appreciate that very small movements of the recording site within the DCN can result in very large proximo-distal shifts in the receptive fields of neurons. A good example of this is illustrated in Fig. 5.7b and this should be consulted before reading the following section.

Three types of lesions have been performed and it is as well to keep in mind the essential differences between these lesions in terms of our current knowledge of central pathways and the chronic versus acute effects of such lesions. The lesions that have been studied include: (1) Cold block or section of the dorsal columns in which all inputs ascending in the dorsal columns are removed but inputs ascending in other pathways are presumably intact (Sect. 5.3.1); (2) Dorsal rhizotomy in which there is a permanent loss of inputs to the spinal cord and/or DCN from all primary afferents innervating the dermatomes of the severed roots (Sect. 5.3.2); (3) Peripheral nerve section and ligation in which there is a permanent loss of primary afferent input to the spinal cord and/or the DCN from a clearly circumscribed area of the body (Sect. 5.3.3).

5.3.1 Section of Ascending Pathways

Physiological and Anatomical Effects of Dorsal Column Lesions. As the primary effect of dorsal column lesions is to destroy most of the afferent input (including all the primary afferent input) to the DCN it is necessary to enquire whether there is any evidence of time-dependent changes in the nuclei which might themselves account at least partially for the observed behavioural recovery which has been described in the primate (Sect. 5.5).

Physiological Effects. As part of a study of the acute effects of dorsal root transection (Sect. 5.3.2), Dostrovsky et al. (1976) examined the effects of placing ice on the cord to effect a cold block of the dorsal funiculi at L4, on the receptive fields of single cells in the cat gracile nucleus. In a region of the middle cell-nest zone, intermediate between the major representation of the hindlimb and that of the abdomen, cold block caused 11 out of 40 cells (in 8 cats) to switch their receptive fields from the leg, foot or toes to skin on the abdomen. The newly appearing receptive fields were all of low threshold and electrical stimulation of the skin within the new receptive field gave response latencies comparable to those obtained on stimulation of the original receptive field. Despite the fact that cold block was used, no report was made of the reversibility of these shifts in receptive field. The authors interpreted these results to mean that, in addition to the powerful, high-safety factor, excitatory transmission from primary afferent fibres to postsynaptic neurons previously shown to exist in the nuclei (Amassian and de Vito 1957; Andersen et al. 1964), there are also normally weak or ineffective monosynaptic excitatory inputs from primary afferent fibres to neurons whose receptive fields are distant from those of the afferent fibres (Sect. 5.4). Clearly it would be informative to extend these experiments by using antidromically identified thalamic projecting neurons and monitoring the receptive fields before, during and after a thermostatically controlled cold block of the dorsal columns.

In order to assess the functional role played by non-primary afferents ascending in the dorsolateral funiculi the acute effects of transection of the dorsal funiculi at C4 have been investigated in the cat (Dostrovsky and Millar 1977). In both anaesthetized (Dial) or decerebrate cats there was a profound reduction in the proportion of cells in the cell-nest region of the gracile nucleus which could be driven by peripheral stimulation. Those cells which still responded were no longer driven by low-threshold stimuli, and their responses were weak and exhibited rapid habituation. Some of the cells which remained responsive to hair movement could only be driven by rapidly flicking the hairs. Pressure-receptive fields were frequently deep and units responding to joint movement were responsive only to large movements or movements at the extremes of joint freedom. Other differences from the normal nucleus included a lack of both somatotopic organization and surround inhibition (Dostrovsky and Millar 1977). Essentially similar findings were made by Dart and Gordon (1973), (in cats anaesthetized with either pentobarbitone or chloralose) following acute transection of the dorsal funiculi at C4. These authors explored the rostral reticular region as well as the cell-nest region of the DCN and selected for study those cells that responded to electrical stimulation of the dorsolateral funiculi. Of cells which were activated postsynaptically from stimulation of the dorsolateral funiculi, 30% were thalamic relay cells. Of these, 40% had peripheral receptive fields on the appropriate ipsilateral limb and most of these receptive fields were of the low-threshold cutaneous type. Other cells, including some relay cells, were inhibited by stimulation of the dorsolateral funiculi. These cells also had widefield, high-threshold, inhibitory receptive fields which usually involved more than one paw and were often contralateral or bilateral. Neither of these studies succeeded in distinguishing more completely the effects of dorsal funiculi transection on the properties of cells in the rostral and basal reticular regions, where the majority of postsynaptic afferents terminate, from those in the cell-nest region where primary afferent fibres provide the major input (Sect. 5.2.2). However both studies (particularly that of Dart and Gordon 1973) indicate that information from low-threshold cutaneous afferent fibres can still enter the medial lemniscus, albeit at a much reduced level, after acute transection of the dorsal funiculi.

To date there have been no electrophysiological studies of the DCN after long-term chronic transection of the dorsal funiculi, so that it is still unknown (1) whether the proportion of cells projecting to the thalamus that respond to inputs from the dorsolateral funiculi (or other intact pathways) changes with time, (2) whether the proportion of cuneothalamic relay cells with low-threshold receptive fields increases with time and, if so, whether their receptive fields become somatotopically organized, and (3) whether any chronic changes occur differentially in the functionally different regions of the nuclei. Information about such changes would appear to be essential to an increased understanding of the behavioural deficits and recovery that occur after lesions of the dorsal funiculi (Sect. 5.5).

Anatomical Effects. The question of whether sprouting of afferent fibres occurs in the DCN following massive deafferentation by transection of the dorsal funiculi is, of course, fundamental to the elucidation of the mechanisms underlying the behavioural recovery which occurs after such lesions. Available evidence to date points to the fact that intact pathways to the nuclei do not sprout following dorsal column lesions in adult animals. In the cat the pattern of degeneration in the DCN following a lesion of the dorsolateral funiculi is not changed when this lesion is preceded at least 13 months earlier by transection of the ipsilateral dorsal funiculi

(Rustioni and Molenaar 1975; Gordon and Grant 1982). This is not altogether unexpected for the dorsolateral funiculi project mainly to the rostral and basal reticular regions of the nuclei, whereas the dorsal column projection terminates mainly in the cell-nest region. Of the afferents in the dorsal funiculi, only the postsynaptic fibres distribute terminals in the rostral region. It remains possible that fine terminal or preterminal sprouting may have taken place but not been revealed in these double-degeneration experiments.

The only attempt to use ultrastructural stereology to assess terminal sprouting in the deafferented DCN was made in the rat, with curious results. Ganchrow et al. (1981) counted synaptic boutons on the somata and proximal dendrites of rat gracile neurons at different postoperative times after bilateral transection of the dorsal funiculi at T12. In the caudal part of the nucleus there were postoperative changes in bouton counts and these were most pronounced on the somatic portions of neurons. However, except for a decline in somatic boutons at 7 days postoperatively in the caudal part of the nucleus there was no statistically significant change from normal afferentation at any postoperative time. Thus there was not clear-cut deafferentation and reafferentation of dendritic profiles suggestive of terminal degeneration and renewal by intranuclear sprouting. Further studies are required to determine the nature of the oscillations in the numbers of boutons on the cell somata, since, in the DCN, primary afferent fibres probably make many more axodendritic than axosomatic synapses (Walberg 1966; Rustioni and Sotelo 1974). Three possible sources of new synaptic terminals in the deafferented cell-nest region might include the sparse corticofugal fibres which terminate there, the interneurons that are found between the cell-nests, or the axon collaterals of cells projecting to the thalamus, some of which form extensive terminal arborizations in the dorsal part of the nuclei (Fyffe et al. 1986).

It has been known for many years that anterograde transneuronal changes may occur in groups of neurons deprived of their major afferent input. These changes include cell atrophy, loss of dendritic branching, and in some cases cell death. In the gracile nucleus of monkey and man, there is not obvious cell loss following long-standing transection of the dorsal funiculi. There is, however, atrophy of neurons in the gracile nucleus of the monkey, amounting to a 25% reduction in cell body area 6 months after lesion of the dorsal columns (Loewy 1973). Furthermore, it has been shown recently that gracilothalamic relay neurons in the monkey fail to transport HRP retrogradely in the first two weeks after transection of the ipsilateral dorsal funiculi at thoracic levels. By 17–21 days after the lesion, many more neurons transport the enzyme, but labelled cells remain fewer than normal and the amount of enzyme transported by each neuron also appears reduced (Berkley and Vierck 1987). The reduced transport may be associated with a reduced level of neural activity in the deafferented cells—an explanation that was advanced previously for similar reductions of HRP transport by deafferented cat lateral geniculate neurons (Singer et al. 1977) and rat motoneurons (Peyronnard and Charron 1983). Electrophysiological evidence for an initial reduction followed by a recovery of levels of activity in gracile relay neurons after dorsal column transection is lacking, though changes of this kind in ongoing activity has been observed in neurons both in the deafferented external cuneate nucleus (Kjerulf et al. 1973) and spinal dorsal horn

(Loeser and Ward, 1967). Berkley and Vierck (1987) were impressed by the similarity of the time-course of the recovery of retrograde axonal transport of HRP in gracilo-thalamic neurons and the time-course of the early phase of recovery from sensorimotor deficits following dorsal column lesions (Sect. 5.5). Strictly speaking, however, there is currently no evidence of a functional relationship between the two phenomena. Finally, as discussed elsewhere (Sect. 6.5.2), there is to date no good evidence that lesion of the dorsal columns elicits sprouting in the VB complex of the thalamus.

5.3.2 Effects of Dorsal Rhizotomy

Electrophysiological Changes. The effect of partial deafferentation by transection of dorsal roots on the somatotopic organization of the DCN has been examined in the adult cat and rat. In the rat, the standard mapping technique used by Millar and Basbaum (1975) to investigate the somatotopic organization of the normal cat gracile and cuneate nuclei (Sect. 5.2.5) was used (Millar et al. 1976). During recording sessions cats were anaesthetized with Dial. Initially the somatotopic organization of cells in the gracile nucleus was studied after partial deafferentation by acute or chronic section of all but one of the dorsal roots supplying the hindlimb (all dorsal roots caudal to L3, except S1 or L7). One track was made every 100 μm across the middle of the gracile nucleus and a recording was made every 100 μm along each track. Points across the gracile nucleus were thus assigned to a part of the body, depending upon where units recorded at that point had their receptive fields. The total number of points that could be assigned to different body parts (e.g. toes, foot, etc.) were then taken to represent the extent of the representation of that part of the body surface within the nucleus. Points at which units had no discernible receptive fields were also noted (Fig. 5.10a). Immediately after section of all but one of the dorsal roots supplying the hindlimb only about one third of recording points in the gracile nucleus showed unresponsive units and there was a large decrease in the proportion of points with receptive fields on the toes, together with an increase in the proportion of points at which responses from the trunk were obtained (Fig. 5.10b). In addition, some individual neurons began to respond to afferents from two widely different areas of skin (e.g. toes and leg). These abnormal cells were usually located in the border regions between the representations of different body parts. Between 6 weeks and 8 months after root section further changes in somatotopy were observed. These included a large decrease in the population of unresponsive loci, a further increase in the proportion of grid points at which responses to the trunk were obtained (Fig. 5.10c, d) and an increased incidence of cells with split receptive fields, usually one on the trunk and the other on the foot or toes (Millar et al. 1976). Unfortunately it proved impossible to determine whether deafferented cells in the toe and foot regions had gained new inputs from the intact root, because of the wide variability in the representation of the toes and foot found in different animals.

In a second series of experiments on the Dial-anaesthetized cat, the somatotopic organization of the gracile nucleus was examined after more extensive deafferentation

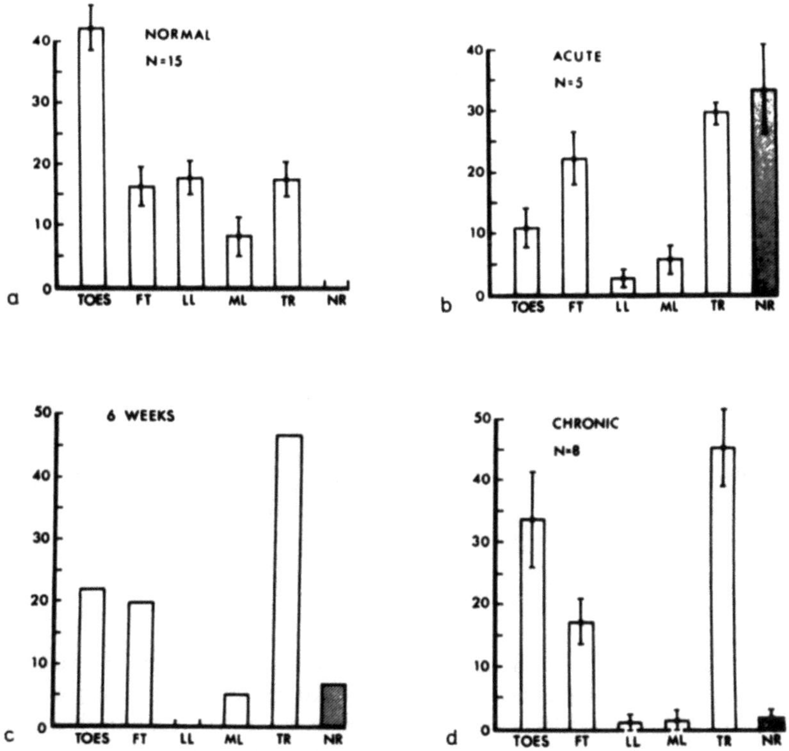

Fig. 5.10 a–d. Plasticity of the somatotopic map in the deafferented gracile nucleus of the adult cat anaesthetized with Dial. Each graph indicates the percentage of recording loci (mean and SD), in a 100 μm × 100 μm grid across a tranverse plane in the cell-nest region at which responses to various body parts could be recorded. **a** *NORMAL*, intact control animals; **b** *ACUTE*, immediately after cutting all dorsal roots caudal to L4 except S1; **c** *6 WEEKS*, a single animal with S1 intact 6 weeks after dorsal rhizotomy; **d** *CHRONIC*, animals with either S1 or L7 intact and surviving > 8 months after dorsal rhizotomy. *Abbreviations*: *FT*, foot and heel; *LL*, anterolateral leg; *ML*, posteromedial leg, perineum, and groin; *TR*, abdominal trunk; *NR*, spontaneously active but non-driveable cells. (From Millar et al. 1976, p. 662)

of the nucleus by acute or chronic section of all dorsal roots serving the hindlimb (all roots caudal to L3) (Dostrovsky et al. 1976). In these animals the number of recording points at which units had receptive fields on abdominal skin approximately doubled immediately after rhizotomy (Fig. 5.11). At many of the recording loci cells could be driven only by rapid flicking of the hairs in the receptive field, an uncommon feature in the intact nucleus. Despite this abnormality the receptive fields were described by the authors as having well-defined edges. Those cells which responded to light brushing of hairs that lay outside areas of skin normally represented in that region of the nucleus responded to electrical stimulation of the underlying skin at latencies which had the same distribution as those seen following electrical stimulation within the receptive fields in the normal cat. Unfortunately, it is not clear what proportion of those neurons from which latency measurements

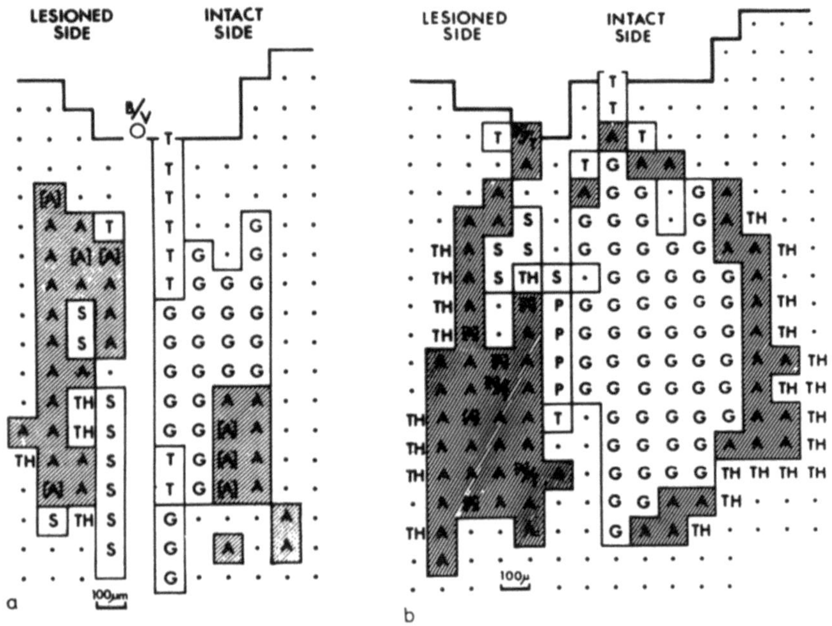

Fig 5.11 a, b. Somatotopic organization of the nucleus gracilis of the Dial-anaesthetized cat after acute and chronic dorsal rhizotomy. **a** Bilateral somatotopic map across the cell-nest region of the gracile nuclei in the cat, obtained from a grid of recording loci made at intervals of 100 μm in both transverse and dorsoventral axes. The part of the body surface represented at each locus was determined by mapping the receptive fields of multiunit activity. Receptive fields are grouped into 6 categories; *TH*, thorax; *A*, abdomen, lumbar trunk and back; *G*, hindlimb; *T*, tail; *P*, perineum; *S*, spontaneously active but non-driveable cells. *[ ]*, indicates that units could be driven only by rapidly flicking hairs. *Dots* represent points where cellular activity was absent. On the *left* (lesioned) side all dorsal roots caudal to L3 had been cut acutely, and the *right* (intact) side shows the normal representation of the skin in the nucleus. Note the regions on the deafferented side where neurons had no receptive fields, and the increased number of loci where responses to the abdomen were obtained (*cross-hatched*). **b** Same as **a** but in a different animal after chronic section of all dorsal roots caudal to L3 on the lesioned side for a period of 8 months. There are apparently fewer loci with "silent" cells and more loci giving responses to the abdomen than in **a**. (From Dostrovsky et al. 1976, pp. 485 and 490)

were made actually had somatotopically displaced abdominal receptive fields. The authors nevertheless concluded that removal of the major input to a neuron caused an *immediate unmasking* of a set of previously established connections from intact afferents that are normally ineffective (Sects. 5.3.1 and 5.4).

There are, of course, several other possible explanations for this unmasking phenomenon, some which were considered by Dostrovsky et al. (1976). First, cells having abdominal receptive fields might normally be present in small numbers in regions dominated by input from the hindlimb. Silencing of the cells with hindlimb receptive fields by limb deafferentation would then give a mistaken impression that the same region of the nucleus had switched from representing the limb to representing the adbomen. Second, there might be cells in the region which initially lack any receptive field whatsoever, but which acquire a trunk receptive field as a result of the deafferentation of surrounding hindlimb cells, and a consequent release from inhibition. This is of course merely an extension of the suggestion that receptive fields actually switch. Thirdly, because the spikes of individual neurons can be recorded from a distance it would be easy to record in the deafferented region the responses of distant cells that still receive inputs from intact dorsal roots (see Sect. 8.2.3). Clearly the first and third of the possibilities might give a false impression of the degree of spread of the abdominal representation into regions that originally represented distal skin. Taken at face value these experiments indicate that the normal somatotopic organization of the gracile nucleus in the Dial-anaesthetized cat represents the expression of excitatory connections from only those primary afferents that supply a small area of skin. Afferent inputs from skin outside these small areas are repressed. Unfortunately, subsequent experiments on intact cats have shown that the somatotopic organization of the cat DCN may not be as straightforward as previously assumed (Sect. 5.2.5). Further independent evidence for the existence of widespread excitatory afferent connections to individual cuneate and gracile neurons is discussed in Sect. 5.4.

While there is, in the cat, some evidence for an immediate change in the receptive fields of some gracile neurons following dorsal rhizotomy, it is less certain whether further changes occurs with the passage of time. In two Dial-anaesthetized cats the gracile nuclei were mapped 8 months after ipsilateral dorsal rhizotomy. The proportion of recording points with receptive fields on the abdomen was 17% and 27% on the intact side and 76% and 86%, respectively, on the deafferented side. However, Table 1 of Dostrovsky et al. (1976) shows that in seven intact Dial-anaesthetized cats the proportion of grid points where responses were obtained from the abdomen varied between 19% and 40%, and in the same animals after acute dorsal rhizotomy the range was 40% to 94%. These are insufficient grounds for believing that chronic deafferentation produces more pronounced changes in somatotopy than acute deafferentation. There are also doubts as to whether the mapping used permits the unequivocal demonstration of somatotopic reorganization. The authors attempt to compare, in only two animals, the number of points at which multiunit activity could be evoked by stimulation of a part of the body surface in the contralateral intact nucleus with the number of points in a nucleus having large areas of unresponsive cells. Since the activity of individual cells can easily be recorded from distances in excess of the distance between sampling points it is clear that the number of points at which no response will be recorded in the deafferented nucleus will be fewer than the number of points at which local cells have lost their receptive fields. The number of points at which cells with unaffected receptive fields will be recorded will appear to increase accordingly. This problem is likely to be exacerbated, following a chronic lesion, by atrophy of the nucleus. In conclusion, there seems little doubt that, in the cat, dorsal rhizotomy causes the receptive fields of some neurons in the affected part of the gracile nucleus to shift immediately to new areas of skin. However, there is little or no convincing evidence that further changes occur in the chronically deafferented state.

In the rat, McMahon and Wall (1983b) have tested the effects of chronic section of the dorsal roots that supply the hindlimb. In this study they transected the L4, L5 and L6 dorsal roots unilaterally and after 15–20 days they transected the spinal cord bilaterally at L3 immediately prior to making recordings in the gracile nuclei of both sides. Cord transection was utilized as a method of acute removal of all input to both nuclei from all dorsal roots below L3, permitting the effects of chronic dorsal root section to the compared with those of acute section. McMahon and Wall (1983b) claim that cord transection was selected over acute section of dorsal roots on the control side in order to avoid complications that might result from the bilateral asymmetry of dermatomal organization that is apparent in the rat. Differences between the chronically deprived side and the acutely deprived side were determined by making transverse rows of vertical microelectrode tracks across both the left and the right gracile nuclei in rats under urethane anaesthesia. In each row, tracks were separated by 50 μm and receptive fields were mapped at 50 μm intervals along each track. It was found that receptive fields located in dermatomes rostral to the cord transection were represented at more recording loci on the side ipsilateral to the chronic dorsal rhizotomy than on the side with intact dorsal roots. The apparent increase in the number of sites responsive to inputs from these dermatomes was more marked in the basal region of the nucleus, where the medial boundary of the responsive region was found to be about 200 μm more medial on the chronic than on the acute side (Fig. 5.12a). In contrast, there was little if any spread of responsiveness into the deafferented dorsal region where the digits are normally represented. Despite the earlier demonstrations from the same laboratory of immediate changes of somatotopy after acute dorsal rhizotomy in the cat (Dostrovsky et al. 1976), no detailed data are presented in this study to show the effect of acute cord transection alone, beyond the simple statement that there were signs of some minor and statistically non-significant immediate unmasking of previously unsuspected inputs. It should be emphasized that on methodological grounds this study cannot reveal the effects of chronic dorsal root section alone, for it compares the effects of chronic dorsal root section plus acute section of all other ascending pathways with the effects of acute section of all ascending pathways. One might ask, for instance, what are the effects of chronic dorsal root section on the influence of ascending tract neurons within the DCN – an influence which, whether normal or abnormal, will be removed following the bilateral cord section which is employed in the above experiments. In conclusion, the effects of acute and chronic dorsal rhizotomy on the somatotopic organization of the DCN in the rat merit further investigation.

Anatomical Changes. This section seeks to evaluate whether following dorsal rhizotomy there is any evidence for the sprouting of intact pathways into partially deafferented regions. Only two attempts have been made to determine whether sprouting of primary afferent fibres occurs in the DCN after chronic partial deafferentation by dorsal rhizotomy. In their original description of primary afferent sprouting in the adult cat, Liu and Chambers (1958) stated that, 281 days after sectioning all dorsal roots caudal of L3 except L7, primary afferents entering through the L7 dorsal root had increased the breadth of their projection to the

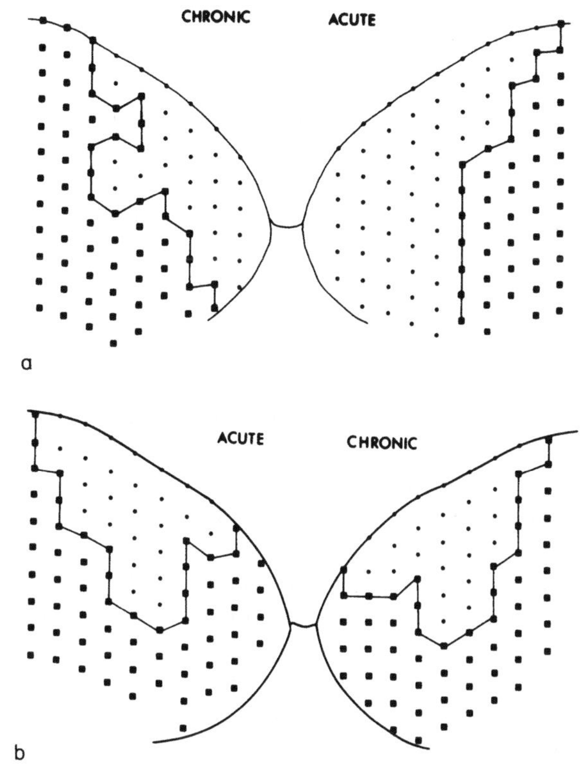

Fig. 5.12 a, b. The effects of dorsal rhizotomy or peripheral nerve transection on the somatotopic organization of the gracilis nucleus in the urethane-anaesthetized rat. Superimposed on the outline of the gracile nuclei in the mid-dorsal region of the medulla is a 50 μm × 50 μm grid of recording loci made in a single transverse plane across the middle part of the nucleus. At each recording locus, the electrode was stopped and the presence or absence of neuronal responses (with an amplitude > 50 μV) to natural stimulation of the skin determined. *Circles* indicate loci where no evoked activity was recorded; *squares* indicate points at which responses were obtained. **a** Bilateral map obtained from a rat with a chronic (17-day) L4, L5, and L6 dorsal rhizotomy on the *left* (*CHRONIC*) side and acute complete cord transection at L3. Responses to skin stimulation were recorded at significantly more points on the side of the chronic rhizotomy than on the acutely deafferented side (*ACUTE*). Due to the cord section, responses were obtained only from the flank and upper leg, implying that the chronic deafferentation had caused a dorsal and medial spread of the representation of the intact L3 dermatome. **b** Bilateral map obtained from a rat with chronic (15-day) transection of the sciatic and saphenous nerves on the *right* (*CHRONIC*) and acute transection of the same nerves on the *left* (*ACUTE*). There is no significant difference in the proportion of loci at which responses to skin stimulation can be obtained on the two sides. (From McMahon and Wall 1983b, pp. 201 and 203)

gracile nucleus. More recently, Goldberger and Murray (1982) unilaterally transected all lumbosacral dorsal roots except L6 and compared the density and distribution of degeneration in the gracile nucleus on each side after bilateral transection of the L6 dorsal root 18 or 24 months later. There was no sign of a spread of the projection from the spared root within the cell-nest region. There was, however, an apparent increase in density of degeneration in the basal and rostral reticular regions on the chronic side when compared to the control side (Fig. 5.13). This was taken to indicate that the projection within the basal reticular portion of the nucleus from the spared dorsal root had in fact increased as a result of the partial deafferentation. The report of Liu and Chambers (1958) was based on a single animal and both studies are subject to the same problems in relation to possible artifacts resulting from the usage of a double-degeneration protocol (see Sect. 4.5.5). Liu and Chambers (1958) actually state that shrinkage was especially marked in the nucleus gracilis after dorsal rhizotomy. However, because there was no sign of increased degeneration in the deafferented cell-nest region, where shrinkage was greatest, Goldberger and Murray (1982) argued that the enhanced density of degeneration in the reticular region was not due to shrinkage. Indeed, they suggested that sprouting of axons into deafferented regions might protect them from such shrinkage. The absence of sprouting in the somatotopically organized cell-nest region was, they suggested, due to the limitations on sprouting imposed by mechanisms of neuronal and somatotopic specificity. In general, sprouting may only occur within the territory in which an afferent axon already has some, albeit few, terminations. Owing to the more diffuse and overlapping projections from single dorsal roots to the reticular zones of the DCN, sprouting (and presumably synaptogenesis) in these regions might be detectable whereas this may not be the case within the afferent terminal fields of the cell-nest region.

Ultrastructural studies of the chronically deafferented nuclei do not support the idea that vacated synaptic sites are reoccupied by sprouting primary afferents. Thirty days after L3–S2 dorsal rhizotomy in the cat, many dark degenerating boutons can still be seen in the gracile nucleus, although they are not as numerous as they are at 48 h. Deafferented postsynaptic specializations are still plentiful, and often retain the remnants of presynaptic membrane of degenerated former boutons. Small boutons with flattened vesicles remain in close association with the deafferented postsynaptic sites, which remain plentiful as long as 6 months after the rhizotomy (Rustioni and Sotelo 1974). Similar changes have been noted in the lateral cuneate nucleus of the cat up to 16 days after C1–T7 dorsal rhizotomies (Kjerulf et al. 1973).

The possibility of axonal sprouting in the DCN following dorsal rhizotomy should be reinvestigated using more reliable methods, such as orthograde tracers or ultrastructural stereological techniques to count degenerating terminals. In any such study it is also important to distinguish between the different functional regions of the nuclei. At present it is difficult to assess the possible contribution to changes of somatotopy in the cell-nest region which might result from afferents sprouting only into the basal reticular region.

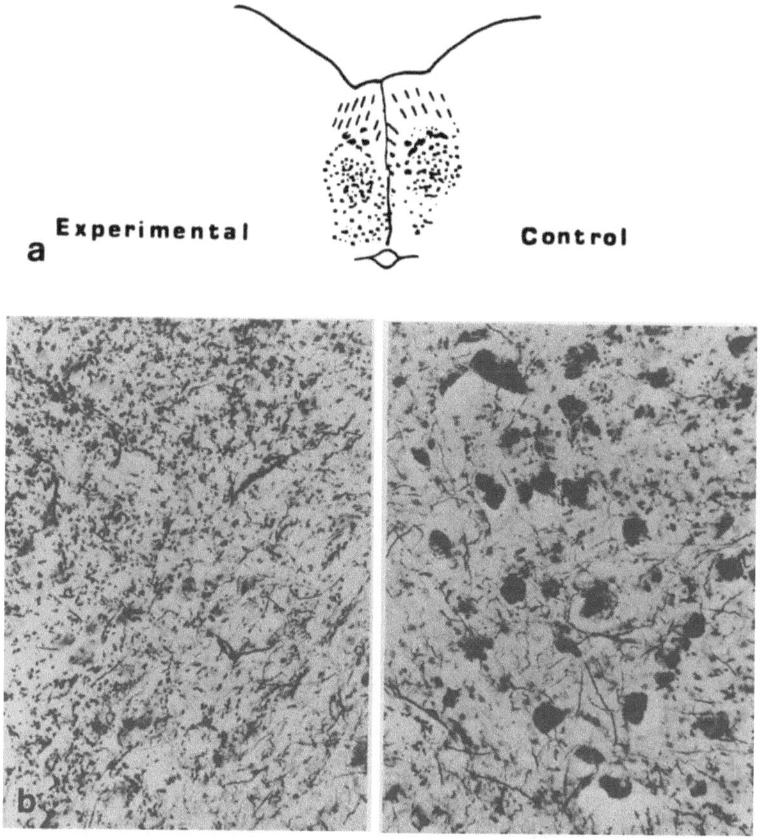

Fig. 5.13 a, b. Axon sprouting in the deafferented gracile nucleus. **a** Drawing showing terminal degeneration in gracile nuclei of both sides in the cat following bilateral transection of the L6 dorsal roots. On the experimental (*left*) side a lumbosacral dorsal rhizotomy, sparing L6, had been performed over 18 months earlier. There is evidence of heavier degeneration in the base of the nucleus from the L6 root on the chronically rhizotomized side than on the unoperated side, indicating that axons of the spared root may have undergone sprouting. **b** Photomicrograph of degeneration in the basal region of the gracile nucleus on the chronically deafferented (*experimental*) and unoperated (*control*) sides. (From Goldberger and Murray 1978, p. 87)

5.3.3 Peripheral Nerve Section

Only a single study has been made of the long-term effects of peripheral denervation on the functional organization of the DCN in the cat. In 10- to 14-week-old kittens which had undergone unilateral denervation of the paw at the age of 2–3 weeks there was strong evidence of a reorganized somatotopy (Kalaska and Pomeranz 1982). In this careful study the skin of the right forepaw was denervated distally by transection and ligation of the superficial radial, median, and ulnar nerves. Recordings were made from single units in the middle part of the cuneate nucleus in both normal and paw-denervated animals and it was found that in the latter many neurons in the dorsal part of the nucleus, where the digits are normally represented (Fig. 5.14a),

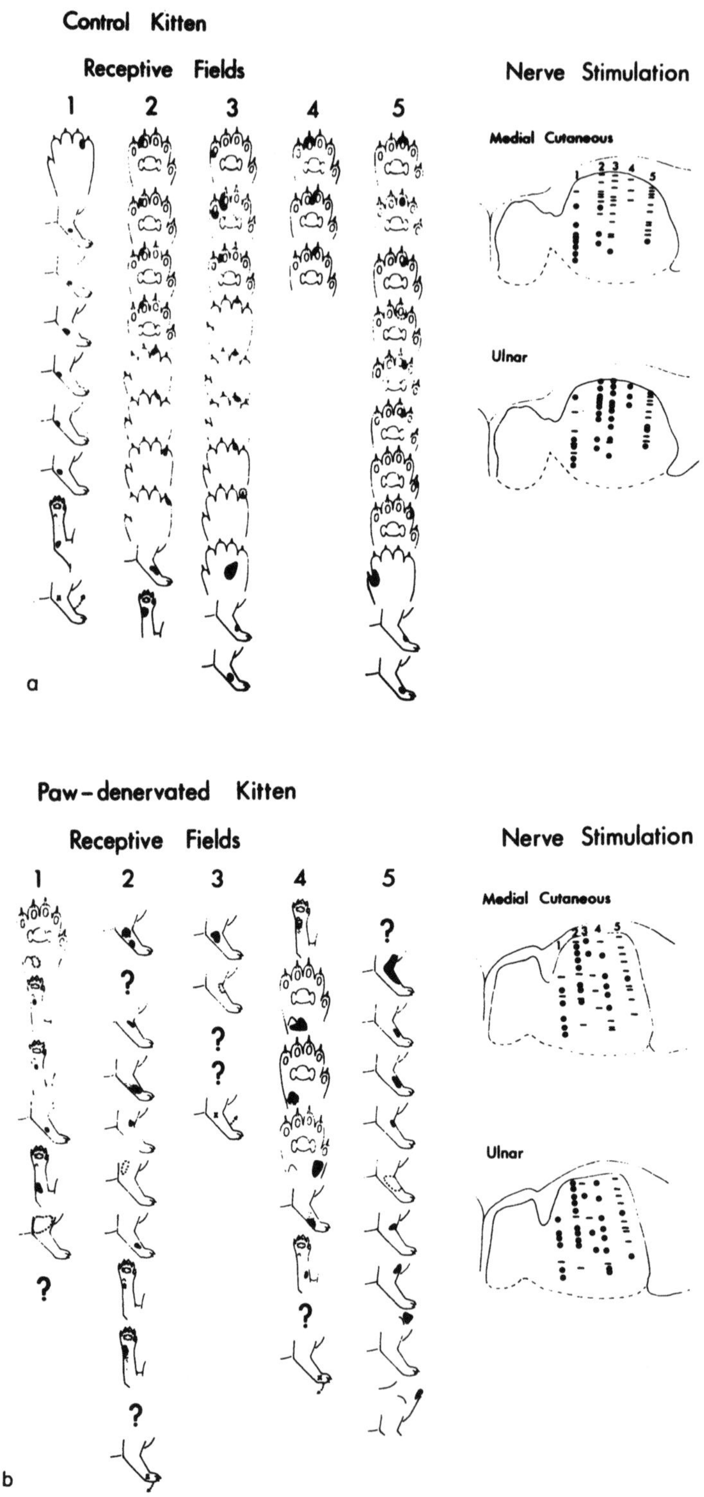
Control Kitten
Receptive Fields
1
2
3
4
5
Nerve Stimulation
Medial Cutaneous
1
2
3
4
5
Ulnar
a
Paw-denervated Kitten
Receptive Fields
1
2
3
4
5
Nerve Stimulation
Medial Cutaneous
1
2
3
4
5
Ulnar
b

responded to tactile stimulation of the wrist, forearm, or proximal parts of the body (Fig. 5.14b). This change could not be ascribed to a difference in sampling from those regions of the nucleus where proximal parts of the limb are normally represented because recording loci were accurately reconstructed in histological sections. A significant number of neurons in the dorsal region of the nucleus had no apparent receptive field, indicating that reorganization was, at best, incomplete. Most of the cells with somatotopically displaced receptive fields responded to brushing or light stroking of the hair or skin, although their thresholds were rather higher than normal. Moreover, the borders of their receptive fields were less well defined than normal, and a few had abnormally large receptive fields (Fig. 5.14b). The change of somatotopic organization was accompanied by the appearance, in the central region of the nucleus, of cells responding to electrical stimulation of the medial cutaneous nerve (Fig. 5.14b). Such responses are normally confined to the medial and ventral regions, where cells have receptive fields on the ulnar surface of the lower foreleg (Fig. 5.14a). These changes at the level of the cuneate nucleus are strikingly similar to changes which occur simultaneously in the somatotopic organization of SI cortex. Since the reorganization of SI cortex is more pronounced when the deafferentation is performed in the young (2- to 3-week-old) kitten than in the adult cat (Sect. 7.3.6) (Kalaska and Pomeranz 1979), it is possible that in the cuneate nucleus some degree of somatotopic reorganization might also occur if peripheral nerve section is performed in adult cats. This has not, however, been investigated. Unfortunately, no attempt was made to map the cuneate nucleus after acute denervation of the paw. It is therefore impossible to say whether, following paw denervation in the kitten, there are any immediate changes that might persist to adulthood or whether the changes observed at adulthood were entirely due to a time-dependent reorganization of afferent connections.

The effect of chronic peripheral nerve injury on the somatotopic organization of the gracile nucleus has also been studied in the adult rat (McMahon and Wall 1983). In these experiments the lower part of the hindlimb was totally denervated by transection of the sciatic and saphenous nerves, and the somatotopic organization of

◀ **Fig. 5.14 a, b.** Somatotopic reorganization in the cuneate nucleus of Dial-anaesthetized kittens after chronic denervation of the distal forepaw. **a** Single-unit receptive fields recorded from the cell-nest region of the right cuneate nucleus of a normal kitten aged 10–14 weeks are shown in the figurines on the *left*. Cutaneous receptive fields are in *black*, responses to joint rotation are indicated by a *cross*, *arrow* indicates the direction of movement. The histologically reconstructed position of each recording locus is shown on the *right*, along with the result of electrically stimulating the medial cutaneous (upper) or ulnar (lower) nerves. *Dots* indicate that the cell was driven by the stimulus, *dashes* indicate no response. **b** Same as **a** but in a kitten which had the right forepaw denervated by section of the superficial radial, median, and ulnar nerves at age 2–3 weeks. *Areas outlined by broken lines* indicate responses to gentle sustained pressure or light taps, *stippled areas* indicate weak responses to cutaneous stimuli over large areas of foream, *question marks* indicate units lacking any definable receptive field. Note the presence of receptive fields on the proximal part of the limb, and the responses to electrical stimulation of the medial cutaneous nerve, in the dorsal region of the nucleus where the digits are normally represented. (From Kalaska and Pomeranz 1982, pp. 39 and 40) the gracile nucleus was then mapped electrophysiologically by the same method used to investigate changes of somatotopy after dorsal rhizotomy (Sect. 5.3.2). Unlike the rhizotomized animal there was no apparent change in the number of unresponsive grid points in the nucleus 15 and 20 days after peripheral nerve section compared to the number immediately after nerve section (Fig. 5.12b). Thus in these experiments there was no sign in the adult rat of a spread of the representation of the upper leg or abdomen into the deprived region that once contained the representation of the denervated foot and toes. It remains possible that some change may have been measured if longer survival times had been examined.

5.4 Evidence for Ineffective Afferent Connections

Ineffective afferent connections within the DCN, or any other level of the neuraxis, could contribute in two ways to alterations in central connectivity. First, they could be rendered effective in firing the postsynaptic neuron by classical neurophysiological or neuropharmacological mechanisms. Second, they could participate in longer-term changes by sprouting new synapses and thereby increasing the effectiveness of presynaptic action potentials on postsynaptic cells. Although there is little good evidence for somatotopic reorganization in the DCN following partial deafferentation it is still instructive to simply examine the evidence for ineffective afferent terminations that might contribute to the changes in connectivity that several authors have claimed to demonstrate. Two distinct but possibly related types of change have to be considered: (1) immediate changes of somatotopy which occur after acute deafferentation by dorsal rhizotomy or cooling the dorsal funiculi, but not after transection of a peripheral nerve, and (2) possible further changes of somatotopy occurring at longer times after both dorsal rhizotomy and peripheral nerve section. For an explanation of the immediate changes, it is necessary to look at the mechanisms responsible for the synthesis of normal receptive fields of cuneate and gracile neurons to determine whether there might already be in place somatopically inappropriate afferent terminals which are only expressed after disruption of the normal input to neurons in part of the cell-nest region. In contrast, longer-term changes might result from a Hebbian kind of reinforcement of transmission at synapses made newly effective by an acute lesion and/or from axonal or terminal sprouting and synaptogenesis by axons from a variety of sources.

5.4.1 Projections of Dorsal Roots and Peripheral Nerves

Dorsal Root Projections. The projection pattern of individual dorsal roots to the DCN has been studied many times in the cat by means of degeneration techniques (for a review of early work see Keller and Hand 1970). Fibres from coccygeal, sacral and low lumbar roots project to the gracile nucleus, while roots rostral to and including L5 project to the cuneate nucleus. There is, however, some evidence that roots around the L4–L5 boundary send fibres to both nuclei – an observation which perhaps relates to the overlap between the dermatomes of adjacent dorsal roots and the precise somatotopic representation of skin in the receptive fields of cells in the cell-nest region (Sects. 5.2.3, 4 and 5).

Transection of a single cervical dorsal root is followed by axon and terminal degeneration in a rostrocaudally oriented band of the cuneate nucleus which occupies all rostrocaudal levels of the nucleus (Rustioni and Macchi 1968; Keller and Hand 1970). However, there is also some evidence for rostrocaudal differences in the projections of some dorsal roots to the gracile nucleus. Thus in the experiments of Rustioni and Macchi (1968) only L5 and S1 were found to project to the rostral pole, while the projection from L1 did not even reach the level of the obex. Their conclusion was based on observations in very few animals, and the Nauta staining technique which these authors used may not have allowed degenerating terminals to be detected (Keller and Hand 1970). However, this distribution of dermatomal projections could explain physiological observations of rostrocaudal differences of somatotopic representation (Sect. 5.2.5). In the mediolateral axis of the nuclei there is a rostrocaudal sequence of root projections, with more caudal roots projecting medially to more rostral roots (Fig. 5.15a, b). In addition, roots with distal dermatomes project more dorsally than those with proximal dermatomes. There are also sparse projections from almost all the lumbosacral and coccygeal roots which cross the midline to enter the reticular zone, but not the cell-nest region, of the contralateral gracile nucleus. Within the cell-nest region of the gracile nucleus of the cat, dorsal root section gives rise to foci of degenerating terminals that lie within the cell clusters. There is some overlap between adjacent dorsal roots which again is characteristic of the overlap in the dermatomes (compare parts a and b of Fig. 5.16). In contrast, within the reticular regions of the rostral and deep caudal zones the degeneration resulting from the section of a single dorsal root is more diffuse and there is greater overlap between the degeneration produced by lesions to adjacent dorsal roots.

In the cuneate nucleus of the rat the terminal degeneration patterns following transection of single dorsal roots have been studied using the Fink-Heimer technique (Basbaum and Hand 1973). The projection pattern was found to be different in the two cytoarchitectonic zones of the nucleus. In the rostral reticular region the projections were diffuse, but in the caudal region there were discrete *terminal field bands* corresponding to the position of the *cellular bricks* (Sect. 5.2.1) and separated by regions of much sparser degeneration. In both rostral and caudal regions of the cuneate nucleus the projections were topographically organized, with rostral roots projecting ventrolaterally and more caudal roots projecting dorsomedially. The intersegmental overlap was greater in the rostral than in the caudal region. However, there were also signs of rostrocaudal differences in projection in that more rostral roots, C3 and C4, projected mainly to the rostral region, while the most caudal root, T2, projected primarily to the caudal region. Intermediate roots, C5 to T1, projected to both regions. In summary, studies of the projections of whole dorsal roots have provided no evidence of overlap in the projections of adjacent dorsal roots that does not reflect the overlap of their corresponding dermatomes. From these experiments there is thus no evidence of somatotopically inappropriate afferent projections to the DCN.

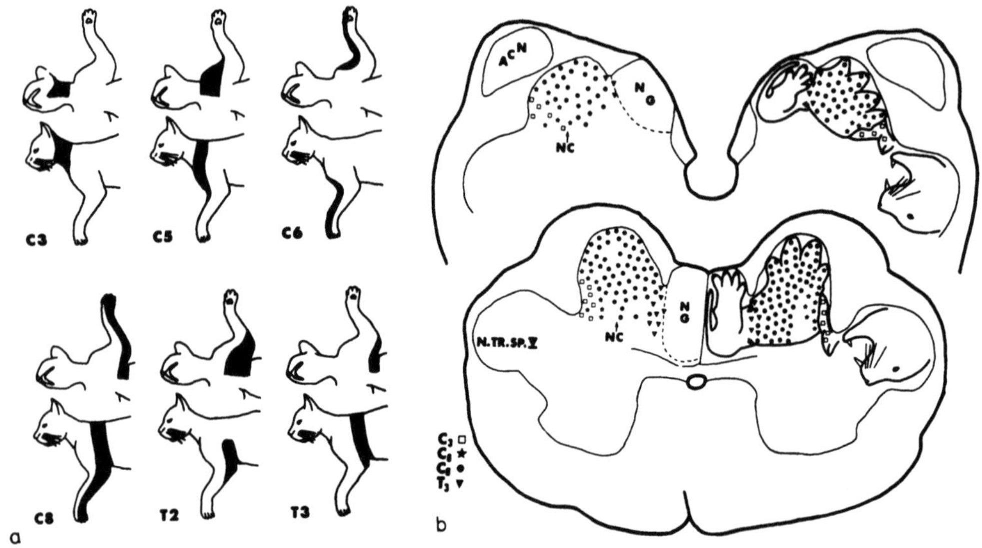

Fig. 5.15 a, b. Projection of cervical dermatomes to the cuneate nucleus of the cat, determined by the Nauta and Fink-Heimer degeneration techniques. **a** Peripheral distribution of selected cervical dermatomes. **b** Areas of axon degeneration in transverse sections at two different levels of the cuneate nucleus after transection of four of the cervical dorsal roots in **a**. The *upper section* is at the level of the rostral reticular region, and the *lower* passes through the middle cell-nest region. For comparison, on the *right* side of each section is shown the somatotopic map of the body representation in the medullary sensory nuclei, with the forelimb and neck drawn to match the experimentally determined dermatotopic projections. In the cell-nest region, the projection from each dermatome forms a rostrocaudal band through the length of the nucleus. More rostral roots project laterally to more caudal roots. In the dorsal, cell-nest region of the nucleus the projections are highly ordered, with little intersegmental overlap, but in the rostral and ventral reticular regions the projections are more diffuse. Note that a dermatotopic organization is still apparent in the rostral region. *Abbreviations*: *ACN*, accessory (external) cuneate nucleus; *NC*, cuneate nucleus; *NG*, gracile nucleus; *N.TR.SP.V*, spinal trigeminal nucleus. (From Keller and Hand 1970, pp. 7 and 9)

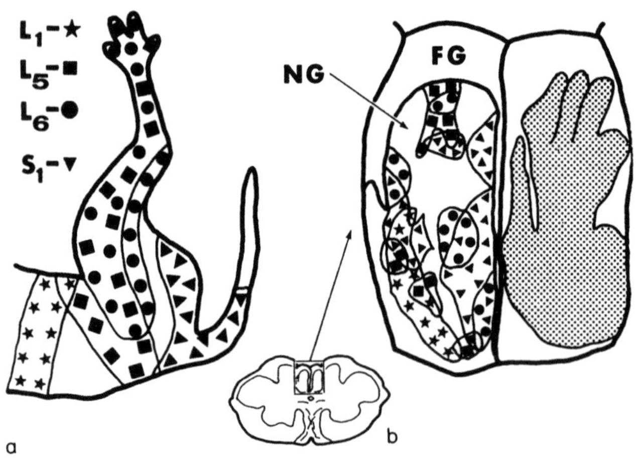

Projections of Peripheral Nerves. Transganglionic tracer techniques have now been extensively utilized to examine the projection patterns of individual peripheral nerves to the DCN. While the chosen nerves often supply rather large and overlapping areas of the periphery, these studies have revealed those details of the topographic organization of their terminals within the nuclei which may be important in relation to possible mechanisms of neural plasticity.

In a thorough study, Beck (1981) applied HRP to the radial, median, ulnar, musculocutaneous, superficial and deep radial nerves of the forelimb in the rat to investigate their projections to the cuneate nucleus. All six nerves were found to project to both rostral and caudal regions of the cuneate nucleus and although there was considerable overlap of their projection zones, there was nevertheless a clear mediolateral shift in the projection of each nerve with respect to any other. In general, the projection of each nerve corresponded well to the projection of the dorsal roots through which its afferent fibres enter the dorsal funiculi (Basbaum and Hand 1973). Unfortunately, in the rat, there is insufficient information on the detailed somatotopy of the DCN to enable conclusions to be drawn regarding the correspondence between the anatomy of nerve projections and the receptive fields of postsynaptic cells (Sect. 5.2.5).

In the cat, similar anatomical studies of the projections of forelimb nerves to the cuneate nucleus have produced results which may be inconsistent with the physiologically determined somatotopy. Using transganglionic labelling with HRP, Nyberg and Blomqvist (1982) examined the projection fields of a range of forelimb nerves, including the median, ulnar, musculocutaneous, medial cutaneous, lateral brachial and antebrachial cutaneous, superficial and deep radial, and palmar digital nerves. The deep radial nerve, and all other nerves having components from muscle, projected to the ventral, reticular region of the cuneate nucleus (Fig. 5.17). In contrast, pure cutaneous nerves did not project to this region, indicating a separation of cutaneous and deep modalities in the nucleus in agreement with most physiological data (Sect. 5.2.4). Both the modality separation, and the somatotopic organization, were lacking in the projections to the rostral pole of the nucleus. Afferents from the paw occupied a large dorsal part of the cell-nest region, with the ulnar component projecting medially, and the radial component projecting laterally. Nerves from the palmar side of the digits project more dorsally than those from the dorsal side. The authors concluded that cutaneous nerves project in a somatotopic pattern to this region of the cuneate nucleus. However, superimposed upon this somatotopically organized distribution was a second termination pattern,

◀ **Fig. 5.16 a, b.** Projection of lumbosacral dermatomes to the gracile nucleus of the cat, determined by the Nauta degeneration technique. **a** Peripheral representation of selected lumbosacral dermatomes on cat, inverted as in the gracile nucleus. **b** Terminal fields of dorsal roots of the same dermatomes as in **a** superimposed on a transverse section of the cell-nest region of the left gracile nucleus. In the *right* gracile nucleus is drawn a somatotopic map of the hind limb. *Inset* shows orientation of the area shown in **b**. Note correspondence between the projection of different dermatomes and their contribution to the somatotopic map. Overlap of dermatomes centrally parallels that in the periphery. *Abbreviations:* FG, fasciculus gracilis; NG, nucleus gracilis. (From Hand 1966, p. 140)

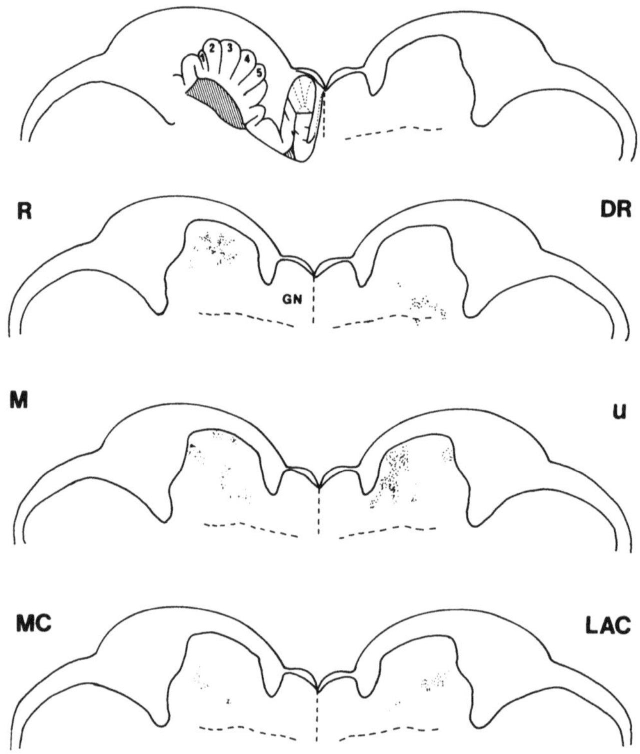

Fig. 5.17. Dual projection of peripheral nerves to the cuneate nucleus in the cat. Areas of terminal labelling in transverse sections of the medulla through the middle part of the cuneate nucleus 2 mm caudal of the obex after applying HRP to six different peripheral nerves. For comparison, the somatotopic map obtained from this region of the nucleus (Fig. 5.5b) is reproduced approximately to scale on one side of a similar section at the *top* of the figure. Note the clusters of labelling from the ulnar, median and superficial radial nerves associated with the representation of the digits, and the projection of the non-cutaneous deep radial nerve to the base of the nucleus. Most of the cutaneous nerves appear to project to two widely separated regions, but it is difficult, on this evidence, to decide whether any of these projections is likely to be somatotopically inappropriate. *Abbreviations*: *SR*, superficial radial nerve; *DR*, deep radial nerve; *M*, median nerve; *U*, ulnar nerve; *MC*, medial cutaneous nerve; *LAC*, lateral anterbrachial cutaneous nerve. (Modified from Nyberg and Blomqvist 1982, pp. 212, 214 and 216 and Millar and Basbaum 1975, p. 287)

in which medially projecting nerves also had a small projection to the lateral part of the nucleus, and *vice versa* (Fig. 5.17). This unexpected secondary projection may provide a source of somatotopically inappropriate afferent terminals but it has first to be shown that the secondary projections are indeed to regions where the receptive fields of cuneate neurons do not encompass those of the parent afferent fibres. This need not necessarily be the case for nerves supplying the forearm because as we have seen, this part of the limb is represented in both the medial and lateral regions of the cuneate nucleus (Fig. 5.5b). However, Nyberg and Blomqvist (1982) also describe a projection of the medial palmar digital nerve of digit 5 to the lateral parts of the cuneate nucleus in the cat. Clearly this projection is somatotopically inappropriate, at least for the somatotopic maps obtained in cats anaesthetized with barbiturates.

In order to understand the possible relevance to plasticity of the dual projections of some nerves to the DCN in the cat, it would be helpful to have information concerning the somatotopic and modality representation at the two projection zones of individual forelimb nerves. Likewise, any such correlation in the DCN of the rat must be accompanied by a detailed study of the somatotopic organization of the cellular brick region of the DCN in that species. Recent preliminary results of injections of tracers into the glabrous skin of the digits in the monkey showed that afferents from a single digit terminate only within a well-defined, cytochrome-oxidase-rich zone of the cell-nest region related to that digit (Florence et al. 1987).

To date, the only serious attempt to determine whether sprouting of primary afferent fibres may occur within the DCN following peripheral deafferentation has been performed in the raccoon. As we have seen above (Fig. 5.8, Sect. 5.2.5), the somatotopic organization of the cuneate nucleus in this animal is organized, at least within the middle, cell-nest region, such that the volar surface of each digit of the forelimb is represented in the receptive fields of neurons occupying a discrete, longitudinally oriented column, or subnucleus. Anterograde HRP tracing studies have shown that the terminals of primary afferent fibres innervating a single digit occupy a narrow (150–250 μm wide) longitudinal band confined within the appropriate subnucleus, and therefore do not overlap the terminals of afferents from adjacent digits (Rasmusson 1988). Amputation of the 5th digit, with ligation of the digital nerves to inhibit regeneration, did not lead within 2–4 months to any perceived change in the terminal distribution of afferents from the adjacent 4th digit (Rasmusson 1988). This is yet another clear example of the inability of primary afferent fibres in mature mammals to sprout into areas to which they do not normally provide some projection within highly somatotopically organized regions of the somatosensory pathways. There remains some doubt about whether sprouting may have occurred within the rostral reticular region of the cuneate nucleus in these studies, because the separation between the digital subnuclei is less clear at this level and there is partial overlap between the projections of afferents from adjacent digits. This lack of sprouting into new regions does not, of course, eliminate the possibility of changes in connectivity between primary afferents and cuneate neurons, because the dendritic domains of the latter may be extensive, as they are in the cat (see below, Sect. 5.4.3).

5.4.2 Projections of Single Afferent Fibres

Single Unidentified Afferent Fibres. Compared with our detailed knowledge of the termination of a range of large myelinated primary afferent fibres in the spinal dorsal horn (Brown 1981; Fyffe 1984; Maxwell and Rethelyi 1987), we have relatively little information concerning the terminations of afferent pathways in the DCN and their relationship to the somatotopic organization of postsynaptic neurons. The termination pattern of unidentified afferent fibres (both primary and non-primary) of the dorsal funiculi has been studied with Golgi techniques in the rat (Gulley 1973; Odutola 1977), cat (Ramon y Cajal 1952; Glees and Soler 1951; Valverde 1966) and monkey (Ferraro and Barrera 1935; Chang and Ruch 1947), and the general features are well known. On reaching the nuclei the dorsal column divides into bundles, the axons of which sweep rostrally and ventrally to branch and terminate among the neurons of the nuclei (Glees and Soler 1951). In the rat, individual axons run rostrally along the surface of the nuclei giving rise to 4–6 collaterals which descend vertically, bifurcate once or twice, then break into fine terminal branches that end in synaptic boutons. Irregularly shaped swellings, resembling *en passant* boutons, are also found at intervals along the entire length of the collaterals. Usually each parent axons tends to remain in a fixed parasagittal plane as it ascends through the caudal and middle regions of the gracile nucleus, thus giving rise to a narrow sheet of terminals throughout the length of the nucleus. On reaching the rostral region of the nucleus, the parent axon may deviate laterally or medially before giving off horizontally oriented collaterals which branch and then terminate. An occasional axon projects diagonally through the gracile nucleus giving off both dorsally and ventrally directed collaterals (Fig. 5.18) (Gulley 1973). In addition, some axons have been seen to project directly from the dorsal funiculus into the nuclei, where they terminate either diffusely or in discrete arbors (Odutola 1977). These relatively minor aberrant projections might form one basis of the large receptive fields observed in a minority of DCN neurons and thus the widefield neurons described in the VB complex and the SI cortex (Sects. 6.3, 7.3.3) (Dostrovsky et al. 1978). In the DCN these afferents may also have an important role in any somatotopic reorganization which might occur within the core region after lesioning of afferent pathways (Sect. 5.3).

Single Identified Primary Afferent Fibres. Recently, information concerning the termination pattern of functionally identified primary afferent fibres has been obtained in experiments employing the intra-axonal injection of HRP. Fyffe et al. (1986) succeeded in staining a sample of 17 afferents in the cat cuneate nucleus sufficiently well to visualize their collateral branches. Of these axons, 9 innervated hair follicles, 2 supplied slowly adapting type I receptors, and 6 were group Ia muscle afferents. The cutaneous afferents innervated skin of the paw and digits, and the muscle afferents probably originated in wrist extensor muscles. The stained portions of the axons could be traced into the middle and caudal divisions of the cuneate nucleus, between 0 and 8 mm caudal of the obex. Most axons gave off a series of 2–5 collateral branches at intervals of 100–1000 μm as they traversed the medulla in a caudal to rostral direction. However, two of the axons each had only a single detectable collateral. The cutaneous afferent fibres terminated mainly in the principal cell-nest region of the nucleus. Most of the hair follicle afferent (HFA) collaterals branched repeatedly, usually giving rise to a single focus of synaptic boutons that was about 100 μm wide and extended about 200 μm in the rostrocaudal axis (Fig. 5.19). However, in the case of one axon, a single terminal arborization extended about 400 μm rostrocaudally, and consisted of two separate foci, each 150–200 μm in length. Adjacent collaterals of individual axons had non-overlapping

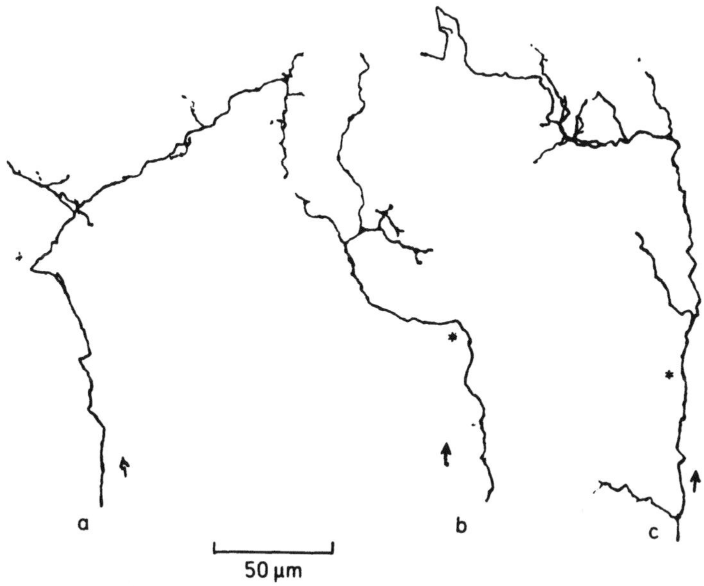

Fig. 5.18. Abnormal termination of some dorsal column fibres in the rat gracile nucleus. Three dorsal column fibres in Golgi-stained horizontal sections of the gracile nucleus in the rat showing unusual patterns of collateral branching. **a,c** are from the right and left gracile nucleus, respectively. These fibres bend sharply near the level of the obex (*asterisk*) and terminate in horizontal arborizations. Fibre (**b**) also bends near the obex, but gives rise to a more longitudinally oriented terminal arborization. *Arrows* indicate the direction of the mid-sagittal plane for each fibre. Owing to the rostrocaudal constancy of the somatotopic organization of the nucleus, such afferents are likely to give rise to terminals across a larger part of the somatotopic representation than that occupied by their receptive field. (Modified from Gulley 1973, p. 339)

terminal arborizations. For the slowly adapting type I afferents, the mediolateral extent of the foci of terminal arborizations in the cell-nest region appeared to be about twice that of HFAs. Two collaterals of one slowly adapting type I axon each produced a second focus of terminal arborizations distant from the main arborization in the cell-nest region. In one case, this was situated in the ventral part of the nucleus, while in the other it was in the dorsal region about 100 μm from the medial edge of the nucleus (Fig. 5.20).

Owing to the small sample it was not possible for the authors to conclude that the differences between the organization of the HFA and type I terminal arborizations were significant. However, the terminal organization of group Ia muscle afferents was quite different from that of the cutaneous afferents. Each muscle afferent fibre produced two or three collaterals, which coursed ventrally and rostrally through the nucleus, supplying terminals to two distinct regions: (1) the dorsal rim, a 50–100 μm thick layer along the lateral, medial and dorsal borders of the middle region of the nucleus, and (2) the ventral region of the nucleus (Fig. 5.21). Occasional side branches provided a few boutons to the cell-nest region, but all such arborizations had a limited rostrocaudal spread (usually

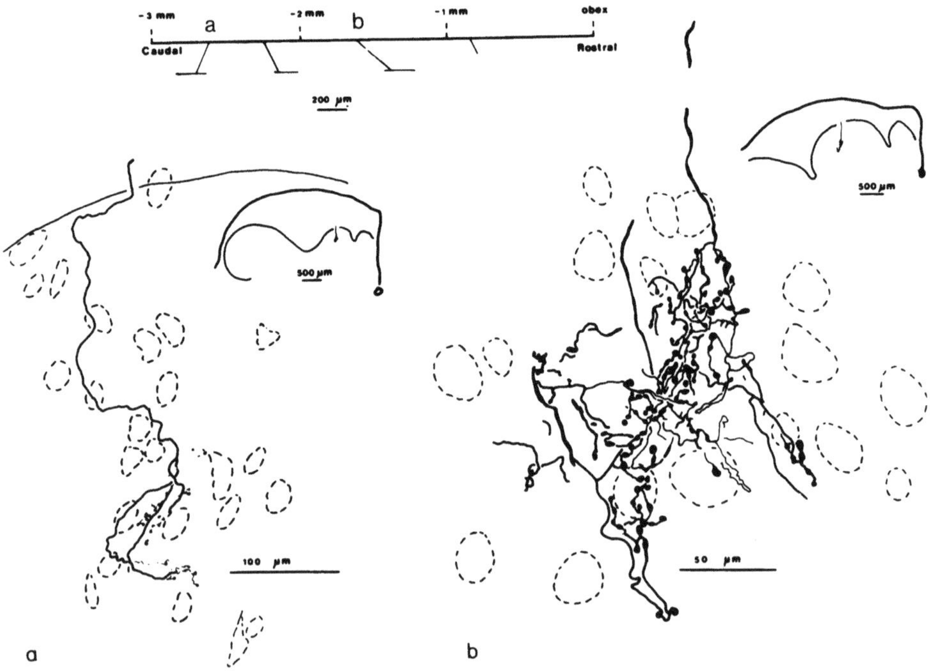

Fig. 5.19 a, b. Anatomy of the collaterals of a single identified Group II hair follicle afferent in the middle region of the cat cuneate nucleus, stained by intra-axonal injection of HRP. The diagram at *top* shows the position of well-stained collaterals and their terminal arborizations in the rostrocaudal axis. Reconstructions in a single transverse plane of the terminal arborizations of collaterals **a** and **b** are shown in detail below. Note the regular spacing of collaterals at intervals of 300–500 µm as the axon runs rostrally in the fasciculus cuneatus dorsal to the nucleus. Each collateral gives rise to a narrow, rostrocaudally oriented patch of terminal arborizations bearing synaptic boutons. The arborization of collateral **b** is clearly associated with the central region of a single cell-nest (*broken lines* indicate outlines of cell bodies). (From Fyffe et al. 1986a, p. 1269)

Fig. 5.21 a, b. Anatomy of the collaterals of identified Ia muscle afferent axons in the middle part of the cat cuneate nucleus, stained by intra-axonal injection of HRP, and reconstructed in a single transverse plane. In **a** terminals are located at the lateral border as well as in the ventral region of the cuneate nucleus. A thin medial branch also provides a few boutons to the cell-nest region. In **b** most of the terminals (shown at higher magnification at *arrow*) are in the ventral region of the nucleus, but a few are also given off dorsally in the cell-nest region. (From Fyffe et al. 1986a, p. 1273)

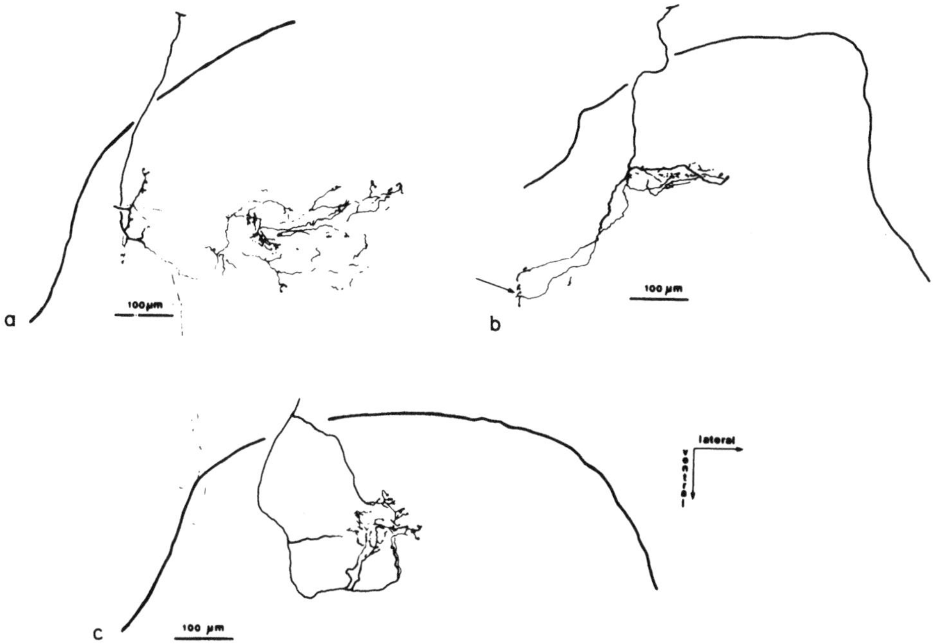

Fig. 5.20 a–c. Anatomy of the collaterals of identified type I axons in the middle part of the cat cuneate nucleus, stained by intra-axonal injection of HRP, and reconstructed in a single transverse plane. **a, b** are collaterals of a single axon issued at the caudal part of the middle of the nucleus. In **a** the collateral gives a few terminals to the medial edge of the nucleus before terminating in a larger arborization more laterally. *Arrow* in **b** indicates a few terminals of the second collateral in the ventral part of the nucleus. The collateral in **c**, from a different axon, produces a restricted terminal arborization in the centre of the cell-nest region. (From Fyffe et al. 1986a, p. 1273)

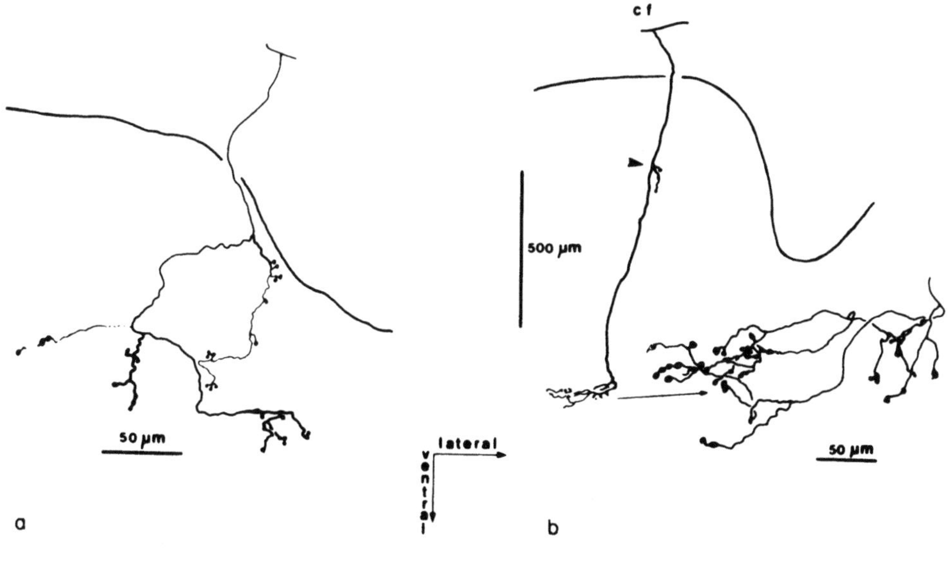

< 50 μm). Terminal boutons were much more numerous in the ventral region than in more dorsal regions, and the ventral arborizations extended up to 200 μm in both mediolateral and rostrocaudal axes. It is interesting that the projection of group Ia muscle afferents to the dorsal rim of the nucleus was not detected in any of the many functional studies of the responses of cells in the DCN, in which responses to muscle inputs were found only in the ventral region (Sect. 5.2.3). Since the terminal arborizations in the dorsal rim region are relatively sparse they may normally produce only subthreshold excitatory effects, and/or be functionally suppressed by pre- or post-synaptic inhibitory mechanisms, as suggested by Fyffe et al. (1986). Neurons in this region receive their major excitatory input from cutaneous afferents on the digits and the distal parts of the forepaw. As DCN cells also receive their major inhibitory inputs from the centre of their excitatory receptive fields (Sect. 5.2.3) perhaps removal of input from distal skin leads to some release from tonic inhibition and a consequent expression of previously sub-threshold inputs from the wrist extensors. In the long term, transganglionic degeneration of terminals might facilitate the proliferation of the terminals of intact afferents and so reinforce these previously ineffective connections. It is clearly important in elucidating the mechanism of plasticity in this situation to accurately determine whether apparently novel proximal receptive fields in the deafferented digit region of the nucleus are in fact cutaneous or due to inadvertent activation of muscle receptors.

5.4.3 Dendritic Spread of Cuneate Neurons

Intracellular staining of identified cuneothalamic projecting neurons in the barbiturate-anaesthetized cat (Fyffe et al. 1986b) has recently led to new insights into the synthesis of their receptive fields. This technique has revealed that the dendritic extent of these neurons may be much greater than indicated by the early Golgi studies. It is now estimated that the dendritic tree of an individual cell may be large enough to receive synaptic input from afferents that terminate up to 500 μm from the cell body. Thus instead of dendrites being confined, as once thought, to a single cell cluster, it is now apparent that many cells have dendrites that branch widely and extend into several nearby groups of neurons (Fig. 5.22). This was even true of neurons with small receptive fields on the distal part of limb. Intracellular recording showed that many of the cells had receptive field properties indicating input from more than a single type of receptor. The high degree of selectivity to a single receptor type and the small receptive fields seen in most extracellular recording (Sect. 5.2.3) means that, at least in anaesthetized animals, the inputs to these neurons are highly filtered so that the separation of modality and topographical localization of inputs are well preserved across the DCN. Finally, the large dendritic spread of some DCN cells is reminiscent of cells in the trigeminal nuclear complex of the rat where again afferents from particular receptors on a small area of skin project to a small focus across the transverse plane of the complex (Figs. 3.3 and 3.4; Sect. 3.1).

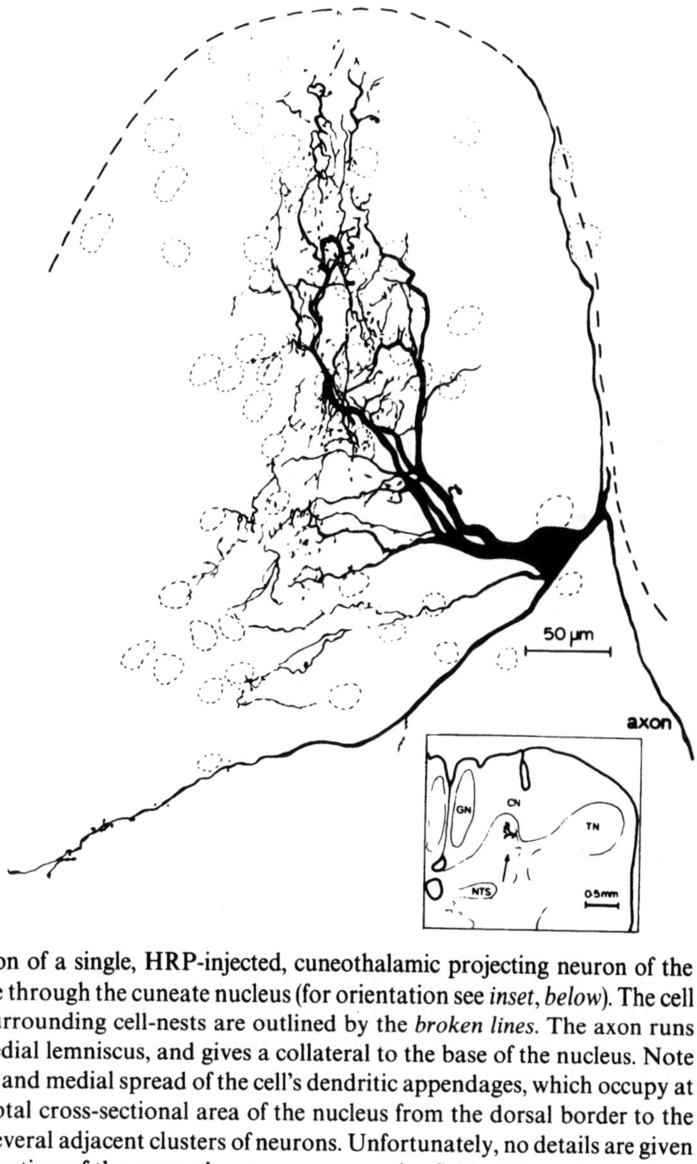

Fig. 5.22. Reconstruction of a single, HRP-injected, cuneothalamic projecting neuron of the cat in a transverse plane through the cuneate nucleus (for orientation see *inset, below*). The cell bodies of neurons in surrounding cell-nests are outlined by the *broken lines*. The axon runs ventrally to join the medial lemniscus, and gives a collateral to the base of the nucleus. Note the considerable dorsal and medial spread of the cell's dendritic appendages, which occupy at least one-third of the total cross-sectional area of the nucleus from the dorsal border to the base, and project into several adjacent clusters of neurons. Unfortunately, no details are given of the dimensions or location of the neuron's cutaneous receptive field. *GN*, gracile nucleus; *CN*, cuneate nucleus; *TN*, spinal trigeminal nucleus; *NTS*, nucleus of solitary tract. (From Fyffe et al. 1986b, p. 1291)

5.4.4 Electrical Stimulation and Widefield Neurons

Physiological evidence for the existence of ineffective afferent treminations in the DCN of intact, though barbiturate-anesthetized cats has been sought by applying percutaneous electrical stimuli to all four foot-pads while recording the activity of single neurons in the core region of the gracile nucleus (Dostrovsky et al. 1978). Cells were found that had a localized receptive field on the ipsilateral hindlimb and which were also excited by electrical stimulation of one or more of the other limbs. Neurons with both localized mechanoreceptive fields and widefield inputs to electrical stimulation were rare. Most responded to hair movement or touch within a localized receptive field on the ipsilateral hindlimb, and the widefield inputs detected were mainly from the ipsilateral forelimb or contralateral hindlimb. Few cells were excited from both these limbs and none were driven from the contralateral forepaw. While there is no doubt that these data reveal, in anaesthetized cats, the presence of inputs to gracile neurons which are not directly involved in the synthesis of their receptive fields, the authors' tentative conclusion that these inputs are probably weak, or ineffective, monosynaptic connections directly from primary afferent fibres is not well supported by their data. Ignoring those with very long (> 16 ms) latencies, the cells activated from the contralateral forepaw had minimal latencies 2–4 ms longer than those expected for cuneate neurons activated by electrical stimulation of the ipsilateral forepaw. Those cells activated from the contralateral hindpaw had minimal latencies which tended to be 2–3 ms longer than their latencies to stimulation of their natural receptive fields on the ipsilateral hindpaw. When the natural receptive fields of the cells were repetitively stimulated electrically at different frequencies, all cells were able to follow at > 30 Hz and many could even follow at 100 Hz. In contrast, only one cell was able to follow stimulation of its widefield input at 30 Hz, the remainder either showing failure of response or responding with very erratic latencies even at stimulus presentation rates as low as 1–2 Hz. In view of these findings it could easily be argued that the widefield inputs were in fact arriving over polysynaptic pathways. Nevertheless, Dostrovsky et al. (1978) tentatively suggested that the widefield connections might be monosynaptic for the following reasons: (1) the presence of degeneration in the contralateral gracile nucleus following lumbosacral dorsal rhizotomy (Rustioni and Macchi 1968) indicates that there are indeed direct contralateral projections to the nucleus; (2) the low following frequency and the increased latency might result from the blockade and partial blockade, respectively, of the action potential in the presynaptic arborization of an axon, especially if the component branches are of small diameter; and (3) the low following frequency and erratic latencies might be accounted for by assuming that the monosynaptic connections are on the very distal parts of dendrites. In the face of their own data these arguments are not convincing, particularly as the contralateral projection demonstrated by Rustioni and Macchi (1968) is to the ventral reticular region, not to the cell-nests. Thus while the experiments of Dostrovsky et al. (1978) indicate that there are connections from the skin to somatotopically inappropriate regions of the DCN, they do not provide any useful information about the pathways traversed by these inputs. The authors suggest that the widefield connections merely represent errors of connectivity from primary afferent fibres in the nucleus and that they are normally functionally suppressed by inhibitory mechanisms.

Finally, the excitation of some DCN neurons within the core region by electrical stimuli applied to regions well outside their normal excitatory receptive fields might well depend on the sychronous activation of weak somatotopically inappropriate connections from many afferent fibres to the distal dendrites of these neurons (Sect. 5.4.3). Alternatively, such responses could be mediated by the synchronous activation of many afferents that give rise to a few wide-ranging terminations onto the somal region of a neuron that under conditions of natural stimulation receives its effective input from an area of skin distant to that which is being electrically stimulated (Sect. 5.4.2). Whatever the mechanism, these neurons, together with those neurons within the reticular zones that project to the VB complex (Sect. 5.2.4) possibly contribute to the widefield properties of some cells within both the VB complex and the SI cortex of the cat (Sects. 6.3 and 7.3.3). Clearly, after partial deafferentation, these cells might still respond to some peripheral input giving the mistaken impression that there have been changes in functional connectivity to cells with confined receptive fields.

5.4.5 Pharmacological Alterations of Receptive Fields

Futher evidence for the existence of normally ineffective monosynaptic connections was later provided by Saade et al. (1982) who reported the effect of the convulsant 4-aminopyridine (4-AP) on the receptive fields of neurons in the cell-nest region of the pentobarbital-anaesthetized cat. This drug enhances transmission at many types of synapse in the CNS by a variety of mechanisms, both pre- and postsynaptic, of which the best known is a prolongation of the presynaptic spike. Transmission is thus enhanced at both excitatory and inhibitory synapses. Recordings were made from cuneate and gracile neurons and the 4-AP was either applied through the second barrel of a double-barrelled micropipette, or applied topically to the surface of the nuclei. All the tested cells had low-threshold receptive fields and responded to hair movement, touch-pressure, deep pressure, or hair movement and touch. The receptive field of most cells expanded after application of the drug but in a few cases the field contracted or remained the same size. Cuneothalamic cells and inter-neurons were affected in the same way. The magnitude of the receptive field expansion was variable, ranging from a spread from 2 toes to 3 toes for a distal receptive fields to much larger expansions for more proximal receptive fields, including a few which expanded across the midline. Initially, threshold, electrical stimuli in the receptive field produced a short latency burst of spikes, while the same stimulus applied outside the receptive field produced either no response or a variable response of longer latency. Application of the drug reduced the latency of the response to threshold, electrical stimulation within the original receptive field by 0.2–0.3 ms, and stimulation within the expanded part of the receptive field then produced brisk responses which had the same latency. These results support the notion that, at least under barbiturate anaesthesia, there is an excitatory subliminal fringe around the cutaneous receptive field of cuneate neurons. However, they do not make a strong case for this excitatory fringe being due to monosynaptic input from primary afferent fibres because the pathway is too long for measurements of overall latency alone, especially to threshold stimulation, to be a reliable indicator of monosynaptic excitation.

5.5 Recovery from Sensorimotor Deficits Following Dorsal Column Lesions

Much of the early literature concerning the behavioural effects of lesions of pathways afferent of the DCN relates to the important question of the central role played by the dorsal column-medial lemniscal system in somatic sensation (for references, see Wall 1970; Wall and Dubner 1972; Vierck 1973; Wall and Noordenbos 1977). A detailed discussion of this extensive and controversial subject is beyond the scope of this review. In relation to plasticity, the questions we are mainly concerned with are: (1) whether any reliable sensorimotor deficit can be attributed to acute transection of the dorsal funiculi, and (2) whether there is any observable time-dependent recovery from such a deficit. The evidence for physiological and/or anatomical changes that might explain any recovery from deficits resulting from lesions to the dorsal columns are dealt with in the preceding sections (Sects. 5.3 and 5.4).

Unlike dorsal rhizotomy, transection of the dorsal funiculi does not totally deprive the brain of somatosensory input from a particular area of the periphery. The degree of sensorimotor disability is thus much milder after dorsal column lesions because alternative pathways are still able to transmit topographical information to central somatosensory structures. Thus the ability to recognize the presence of a tactile stimulus remains intact (Frommer et al. 1977) and, contrary to generally accepted dogma, there appears to be, in the monkey, no measurable change in the threshold for touch or for two-point discrimination (Vierck 1982). In both the gracile and cuneate nuclei, transection of the dorsal funiculi undeniably produces an extensive deafferentation of neurons in the cell-nest region. Because their input comes mainly from rapidly adapting receptors (Sect. 5.2.2) these neurons are particularly suited to the transmission of information about the onset or offset of movement or of externally applied tactile stimuli. Moreover, because they have small, somatotopically organized receptive fields situated mainly on the distal part of the limb (Sect. 5.2.5) they are also well suited to play a role in the transmission of sensory information about the spatial relationships between moving parts of the body and objects in the immediate environment. When behavioural tasks were designed to test these specific aspects of sensorimotor function, it became clear that sensorimotor deficits do occur after transection of the dorsal funiculi (reviewed by Vierck 1982, 1984). It has been claimed that lesions of the dorsal funiculi in different species generally tend to produce sensorimotor deficits in areas related to any specialized use of the forelimbs and hindlimbs by that species (Beck 1976). For example in carnivores, the limbs are specialized for locomotion, and cats with lesions of the cervical dorsal funiculi slip and stumble when made to jump to a rotating platform (Melzack and Bridges 1971). A variety of behavioural deficits of forelimb performance have been demonstrated in primates with similar lesions. These include severe impairment of manual dexterity and fine control of the digits in species of monkeys with highly developed manual skills (Ferraro and Barrera 1934; Gilman and Denny-Brown 1966; Schwartz et al. 1972; Vierck 1975). Squirrel monkeys, for instance, though normally adept at catching flying insects, show an impairment of the ability to catch falling baits following dorsal column lesions (Beck 1976).

Ganchrow et al. (1980) rated the hindlimb performance of rats at different times before and after a bilateral lesion of the thoracic dorsal funiculi. The degree of postoperative behavioural impairment correlated fairly well with the extent of the lesion. Unfortunately, in rats the corticospinal tract runs in the ventral part of the dorsal columns and thus it is not clear to what extent damage of the corticospinal tract determined the observed deficits. Surprisingly, in this case the deficit was significantly greater 90 and 120 days after the lesion than immediately postoperatively, an observation which the authors attributed to rather vague changes that they have observed in bouton renewal and "neuronal excitability" at these late times (Sect. 5.3.1).

The most convincing evidence of recovery from behavioural deficits after dorsal column lesions comes from the work of Vierck (1982). Within 3 weeks of lesioning one dorsal funiculus in the primate, there was a rapid recovery of performance on motor tasks which require accurate placement of the ipsilateral hindlimb or precise orientation of a distal extremity. Recovery from somatosensory deficits was also demonstrated. To permit detection of these deficits monkeys were trained to discriminate between different-sized disks impressed on the glabrous skin of the hindlimb. In these animals, unilateral lesion of the thoracic dorsal column produced an immediate and substantial impairment in their ability to discriminate different-sized disks ipsilateral to the lesion (Vierck 1973). After about 80 days, there began a gradual improvement in this ability which returned toward preoperative levels over the succeeding 5 months. A subsequent bilateral lesion at a different level caused a loss of size discrimination in both feet. This deficit was less severe than the first and the difference was ascribed to conceptual learning. Thus, a large part of the recovery process after the first lesion could be accounted for by the animal developing fresh strategies for making the necessary discrimation, strategies which remained immediately available to the animal after the second lesion. As suggested by Vierck (1973) an example of this would be a switch from discrimination based primarily on edge detection, to one based primarily on measuring the area of the stimulus, achieved by integrating the number of slowly adapting units with overlapping receptive fields activated by the stimulus (Vierck 1973, 1982). Further improvement of performance after the lesion occurs as a result of rote learning (learning related to practice of the particular behavioural task). Improvement of performance due to rote learning proceeds at roughly the same rate postoperatively as preoperatively (Vierck 1982). However, even after a second dorsal column lesion there remains an early phase of rapid recovery lasting about 3 weeks which corresponds in duration to the early recovery from motor deficits. Because this early phase of recovery is of fairly constant duration in different animals, Vierck (1982) has argued that it represents a recovery of function (i.e. a functional, if not structural, reconstitution of input) amongst cells of the deafferented core of the DCN.

5.6 Plasticity of the DCN During Development

5.6.1 Effects of Prenatal Lesions

To date, relatively little use has been made of the projections from the limbs to the DCN to investigate the factors responsible for orderly development of projections in the somatosensory system. The main reason for this appears to be that the basic organization of these pathways is already present at birth in animals, such as the cat, in which most information is available concerning the normal organization of the DCN. For example, Ferrington and Rowe (1982) found that the highly specific connections from different types of glabrous skin mechanoreceptor onto individual cuneate neurons (Sect. 5.2.3) were already demonstrable in the 1- to 5-day-old kitten, at a time when the afferents have not yet acquired their mature physiological profiles.

Most available information concerning developmental plasticity of the mammalian dorsal column-medial lemniscal system has come from a small number of studies in the rat. In this species, sensory neurons of the C7-T1 DRGs that innervate the forelimb differentiate between G12 and G14. The central processes of these neurons reach the cuneate nucleus on G17, the time at which cuneate cells undergo their final mitoses (Altman and Bayer 1984). The peripheral processes of these afferents reach the forelimb on G16. The developmental timetable for the hindlimb innervation lags behind that of the forelimb by 1 day, hindlimb afferents thus reaching the gracile nucleus on G18 (see also Sect. 4.6.1).

In the rat DCN, a distinct segmentation of SDH activity has been observed on the day of birth (P0). As is well established for the rodent trigeminal system (Sect. 3.2), this segmentation reflects the pattern of afferent termination within the DCN. The two groups of large neurons that comprise the cuneate and gracile nuclei are separated by a narrow region containing fewer, and smaller, cells. Correspondingly, regions of high SDH activity within the two nuclei are separated by a region of low activity. Killackey and Dawson (1989) found that, in rats that had undergone complete forelimb removal on G17, this characteristic organisation was replaced by a continuous band of high cell density and of SDH activity that encompassed both nuclei. Forelimb amputation 1 day later, on G18, though causing a reduction in the size of the cuneate nucleus, did not produce this apparent fusion of the cuneate and gracile nuclei, although in both cases forelimb amputation would have destroyed a majority of forelimb afferents (Sect. 2.2.2). The fusion of the gracile and cuneate nuclei corresponds with an enlargement of the hindlimb representation in SI cortex (Sect. 7.5.3). Taken together, these results suggest that the removal of forelimb afferents before they are able to establish synaptic connections with cuneate neurons allows hindlimb afferents to invade the region between the gracile and cuneate nuclei (and possibly also part of the cuneate nucleus itself). However, once ordered synaptic connections are established between primary afferents and DCN neurons, removal of afferent input to the cuneate nucleus no longer results in expansion of hindpaw afferent projections.

Young of marsupials are born at a much earlier developmental stage than those of placental mammals, and they complete much of their development in the pouch,

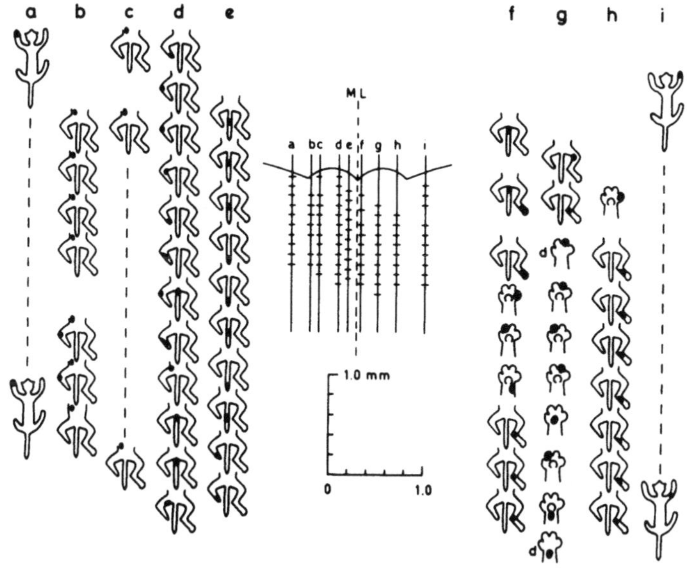

Fig. 5.23. Somatotopic organization of gracile nucleus in barbiturate-anaesthetized cats with congenital agenesis of one hindlimb. Receptive fields found in single and multiunit recordings in serial microelectrode tracks through the DCN. Each figurine shows the receptive field of a single cell or group of cells at the position marked in the corresponding track. Broken lines in tracks *a* and *i* show where receptive fields were obtained from the forelimb, and in track *c* indicate similar receptive fields through the entire track. **d** indicates deep receptive field. Note that the region on the *left* where the digits and distal limb would normally be represented is given over to the representation of the limb stump. The left hind foot of this cat was missing from the heel downwards, and the tibia and fibula were shortened. (From Schulz et al. 1981, p. 414)

where they are accessible to experimentation. In the neonatal American opossum (*Didelphis marsupialis virginiana*) the DCN are not yet formed at birth (Johnson et al. 1972). Removal of one hindlimb at this time results in a lack of development of the dorsal part of the gracile nucleus in which the somatotopic representation of the hindlimb is normally found (Johnson et al. 1972). Presumably the amputation causes the death of any primary afferent neurons innervating the hindlimb at birth (Sect. 2.2.2). Like the manipulations of the mystacial vibrissae of rodents (Chap. 3), these experiments indicate that during early developmental phases peripheral structures have an overwhelming influence on the proper formation of centrally located parts of the somatosensory system.

Cats with congenital agenesis of a hindfoot (possibly induced by teratogenic compounds) have been found to have a normally developed gracile nucleus. As we have already seen, development of the nucleus in the rat is disrupted by removal of hindlimb tissue at what, in the cat, would be an early stage of embryonic development (cat gestation period = approximately 62 days). When the nucleus was mapped electrophysiologically in adult specimens of these cats, it was found that cells in what was normally the dorsal part of the representation of the hindpaw had receptive fields on the leg stump (Schultz et al. 1981) (Fig. 5.23). These receptive fields were smaller than those normally found on this proximal part of the limb. Thus it appears that the degree of convergence of primary afferent fibres into postsynaptic neurons from the skin is determined by either the position of the afferent projection site within the nucleus or the position of the receptors along the proximo-distal axis of the limb stump, rather than the specific identity (e.g. toe versus calf) of the innervated skin. To some extent this supports the findings of Simons et al. (1984) on the vibrissae system which suggest that afferent fibres neonatally deprived of their correct receptive structures will still maintain an influence over central cells at the somatotopically appropriate location (Sect. 3.3.2). Finally it is fascinating to note that in humans with congenital absence of one arm two point discrimination on the distal part of the stump matched that of the palm of the contralateral, intact arm (Weinstein et al. 1964).

5.6.2 Effect of Neonatal Destruction of Unmyelinated Afferents

After neonatal capsaicin treatment the receptive fields of neurons in the gracile nucleus of the adult rat were reported to be very much larger than normal (McMahon and Wall 1983b). Capsaicin treatment of neonatal rats destroys about 95% of unmyelinated primary afferent fibres (Sect. 2.2.6). Since a large number of unmyelinated axons of primary afferent origin have been described in the dorsal columns at C3 (Chung and Coggeshall, 1985) (Sect. 5.2.2), it is not surprising that the nearly complete elimination of C-fibres should affect the synthesis of tactile receptive fields of DCN cells. Alternatively, such changes might be brought about by an enlargement of the receptive fields of those dorsal horn neurons which project to the nuclei (Sect. 4.5.4). Unfortunately, no data concerning the size of receptive fields of individual neurons are presented, the enlargement being deduced from an increase in the number of grid points at which cells responding to a particular body area could be recorded (Fig. 5.24). As might be expected, acute transection of the sciatic and saphenous nerves in animals treated neonatally with capsaicin produced fewer unresponsive grid points than normal. These results imply an important role for the non-primary afferent and/or C fibre input to the DCN in the control of somatotopic organization which is difficult to explain on the basis of current knowledge. However, owing to the paucity of data in this experiment (only four rats were studied), the results cannot be regarded as more than preliminary.

5.7 Major Conclusions

Studies of somatosensory plasticity at the level of the DCN are relatively few in number and in some cases they are flawed by their dependence on assumptions about the homogeneity of the somatotopic map within the nuclei of the cat and the rat. Experimental evidence has accumulated that show that such assumptions are incorrect. Using somatotopy as an indicator of connectional changes within the DCN requires some caution as small dorsoventral movements through the nucleus can result in large shifts in the receptive fields of neurons (Fig. 5.7b). This difficulty

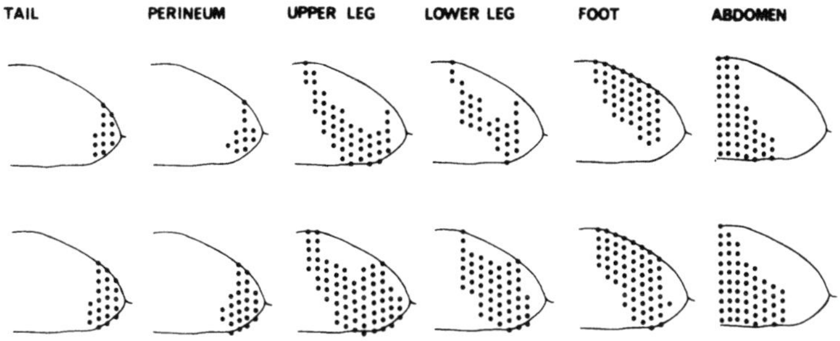

Fig. 5.24. Plasticity of the somatotopic map in the gracile nucleus of the urethane-anaesthetized rat after neonatal treatment with capsaicin (*bottom*; *top*, normal). In each diagram the outline of the left nucleus is shown in transverse section with the midline to the *right*. Superimposed on this is a 50 μm × 50 μm grid of recording loci (*dots*), made in a single transverse plane across the middle of the nucleus, at which evoked activity (amplitude > 50 μV) was recorded in response to natural stimulation of the appropriate part of the body. Note the greater number of loci at which responses to each body part were found in the capsaicin-treated animals. From these data, it was concluded that the capsaicin treatment had caused an increase of receptive field size of gracile neurons. (Modified from McMahon and Wall 1983b, p. 199)

could be overcome if antidromically identified projection neurons were used for studies of the acute effects of lesions. While there are undoubtedly changes in the somatotopic organization of the nuclei in adult mammals after lesions of afferent pathways, these changes have not been fully documented, and fundamentally important details which would aid in the interpretation of the changes, such as the size and the modality specificity of single-unit receptive fields are lacking. In the case of experiments which have been performed on the rat (and these unfortunately constitute the bulk of physiological investigations), there are additional problems of interpretation due to a relative lack of detailed information about the normal functional and somatotopic organization of the nuclei in this species, and to the increased likelihood of overestimating the true spread of evoked neuronal activity into deafferented regions due to the relatively small distances over which changes have supposedly occurred. For the same reasons, it is particularly difficult to distinguish between short-term functional changes and those occurring over longer times. These reservations notwithstanding, the following statements may be made at the present time, concerning plasticity in the nuclei.

(1) Somatotopic Reorganization After Blockade or Lesion of the Dorsal Columns. In the cat cold block of the dorsal columns at L4 causes a few cells to *immediately* develop new receptive fields on skin supplied by roots rostral of the block. Dorsal column lesions in the cat cause an immediate profound reduction of neurons with small somatotopically organized receptive fields on the digits. No physiological studies have been done of the long-term effects of such lesions, but it has been hypothesized that in primates functional (or even structural) reafferentation of the neurons may underly behavioural recovery from dorsal column lesions (see *(5)* below). While there is some indirect evidence for an

initial depression and later recovery of activity in deafferented neurons, this interesting speculation remains to be demonstrated directly. In the cat there is evidence that afferents from intact pathways, such as the dorsolateral funiculus, do not sprout into deafferented regions of the nuclei.

(2) *Somatotopic Reorganization After Dorsal Rhizotomy.* After deafferentation of the hindlimb in the cat there is an immediate spread of responses to abdominal skin into more dorsal regions of the gracile nucleus where cells originally had fields on the hindlimb. The observed extent of the expansion is of the order of 500–600 μm in the dorsal direction. This expansion is incomplete and cells in some areas remain unresponsive to natural stimulation of the skin. After a period of months there is some evidence of further expansion of responses to abdominal skin into the silent areas. A similar spread of responsiveness to abdominal skin occurs after chronic deafferentation of the hindpaw in the rat, but acute changes have not been well documented in this species.

(3) *Somatotopic Reorganization After Peripheral Nerve Section.* In the adult rat, neither acute nor chronic deafferentation of the hindpaw by transection of peripheral nerves, leads to expansion of the representation of innervated skin into the deprived region of hindpaw representation. In the cat, chronic denervation of the forepaw causes some cuneate neurons in the region of digit representation to acquire receptive fields on more proximal skin, mainly of the upper limb. However, this reorganization has been demonstrated only after nerve section in the 10- to 14-week-old kitten, and whether similar changes occur in mature adult cats is unknown. The acute effects of peripheral denervation have not been investigated in the cat.

(4) *Factor Underlying the Observed Changes. (a)* There is little doubt that the immediate reorganization of somatotopy which has been reported in the DCN following dorsal rhizotomy is to some unknown degree artefactual. When receptive fields of multiunit activity are mapped at fixed loci in nuclei there will inevitably be an *apparent* shift in the boundary between the representations of two regions if one of the regions is suddenly deafferented. The likely extent of this apparent shift is not predictable from published data. *(b)* Added to this there may be problems of interpretation due to the complexity of the somatotopic organization. Recent studies have indicated that there may be a somatotopic map for each different modality and spatially these maps are not necessarily in register. *(c)* Further immediate expansion of responsiveness to intact inputs after dorsal rhizotomy is undoubtedly due to a breakdown in the normal balance of afferent excitatory and inhibitory influences from the skin onto neurons of the DCN, but there is insufficient available information to predict accurately the likely extent of such expansion. While it is true that some single gracile neurons have been observed to switch their receptive fields from the foot to the abdomen on reversible blocking of conduction in the dorsal columns, these neurons were located in boundary regions intermediate between the representation of the two areas. There are no equivalent data on switching of receptive fields by individual neurons with small receptive fields on the digits, which occupy the great majority of the cell-nest region of the DCN, but many of these would have to undergo a similar change of receptive field to explain the

observed degree of expansion of the abdominal representation. *(d)* Some DCN neurons have extensive dendritic trees which would enable them to receive monosynaptic excitatory inputs from afferents terminating about 500 μm from the cell body. The well-documented phenomenon of *in-field* afferent inhibition suggests that such weak inputs to distal dendrites might normally be rendered ineffective as a result of inhibition derived from activity of afferents that innervate the centre of the excitatory receptive field *(e)* Some primary afferent fibres may give rise to sparse terminal arborizations in somatotopically inappropriate regions of the nuclei. The participation of these projections in the *immediate* changes of somatotopy is unknown but is unlikely to be greater that the involvement of inputs to the distal dendrites of some DCN cells (see *(d)* above). *(f)* Changes in somatotopy that occur over long postoperative times may depend on a strengthening of newly effective synaptic inputs and/or the formation of new connections by collateral sprouting. While there is no evidence of long-range sprouting of primary afferent fibres into the deafferented cell-nest region the possibility of short-range sprouting and synaptogenesis by somatotopically inappropriate afferent projections cannot be ruled out.

(5) *Recovery from Sensorimotor Deficits Following Dorsal Column Lesion in Adult Primates.* Dorsal column lesions cause massive deafferentation of neurons in the core region of the DCN. These neurons have small, somatotopically organized, receptive fields, concerned mainly with the processing of inputs from rapidly adapting afferents. In primates, deficits of sensorimotor function have been demonstrated on behavioural tasks specifically designed to test the likely function of such an afferent system. In some cases, there is a time-dependent recovery.

(6) *Plasticity in the DCN of Developing Mammals.* The few studies that have been carried out on the rat DCN add support to hypotheses which have emerged from the findings on the rodent trigeminal system. In neonatal marsupials it has been shown that normal differentiation of the nuclei is dependent on the presence of appropriate limb tissue. In cats with congenital agenesis of the hindlimb, normal differentiation of the DCN proceeds and the somatotopic representation in the mid-dorsal cell-nest region still appears to receive afferents from the distal part of the limb even when those skin regions do not include the foot or the toes. The representation of the missing part is replaced by a representation of the most distal remaining part. This suggests that the topographical relationship between the skin and the central nervous system is established very early in development irrespective of the anatomical identity of the skin. These findings reflect the results of psychophysical observations on humans in whom a complete limb is congenitally reduced to a stump.

6 Plasticity and the Somatosensory Thalamus

6.1 Experimental Strategies and Plasticity in the Ventral Posterior Nuclei of the Thalamus

The ventral posterior nuclei of the thalamus are one of the more inaccessible parts of the somatosensory system. Deep in the brain, they contain compact representations of the body surface and yet they are under strong regulatory forces of afferent inhibition, the somatosensory cortex and the closely associated reticular nucleus of the thalamus (Jones 1986). Only in very recent years have we begun to understand the functional and microstructural organization of the major inputs to this area. Consequently, it is not surprising that few studies that purport to have shown alterations of connections within the somatosensory divisions of the thalamus stand the scrutiny of more recently acquired knowledge of its organization in normal animals. Nevertheless a substantial number of workers have documented physiological and anatomical alterations within the thalamus that suggest that connectivity within this structure might be no more stable than it is at other levels of the somatosensory system.

The approaches that have been taken to the question of plasticity fall into four main groups: First, there have been a number of detailed studies of the microstructure of inputs and outputs to the ventral posterior group in normal animals. Second, many workers have examined the receptive fields and responsiveness of thalamic neurons under a host of anaesthetic and stimulus regimes. Third, some attention has been directed to the effects of blocking peripheral nerves and afferent information ascending in the dorsal columns. Fourth, the effects of lesioning the dorsal column nuclei (DCN) and so depriving the thalamus of a major source of ascending afferent fibres has been examined.

The following chapter does not cover developmental aspects of plasticity in the thalamus for, to our knowledge, this has only been studied within the confines of the vibrissae system of rodents (see Sects. 3.2 and 3.4.2). It provides a sketch of the connectivity and physiology of the somatosensory divisions of the thalamus and critically reviews studies on the plasticity of connections within this region. Our impression is that in view of the recently acquired knowledge on the normal operation and connections of this area renewed efforts to demonstate functional and structural plasticity might be productive.

6.2 Anatomical Organization of Inputs and Outputs of the Ventral Posterior Nuclei

6.2.1 Primate and Cat

Nomenclature, Cortical Connections and Sensory Input. The literature on anatomical organization of the ventral posterior nuclei is large and filled with conflicting interpretations and an ever-changing nomenclature. Current interpretations owe much to recently acquired knowledge of the thalamic projections to particular cytoarchitectonic areas of the cortex and the distribution of particular types of afferent input. The cytoarchitecture of the thalamus continues to be used but some workers appear to regard it as of supportive value in relation to conclusions that are primarily derived from connectional and functional studies.

In contemporary literature there are two views relating to the organization of the ventral posterior nuclei. The first is based on experiments on primates and recognizes that in the coronal plane of the VPL and VPM there is a detailed representation of the skin of the entire body (Fig. 6.1a) (Jones and Friedman 1982; Jones 1983b, 1986). This region is referred to by Jones and Friedman (1982) as the cutaneous core of the VB complex in order to distinguish it from the anterodorsal shell of tissue that contains neurons which respond to deep stimuli (Fig. 6.1b). This anterodorsal shell receives deep inputs from lemniscal fibres and relays them to cortical areas 3a and 2 (Jones and Friedman 1982; Jones 1983a, b). The cutaneous core itself contains a *central core* that receives cutaneous input from lemniscal fibres while the surrounding part of the cutaneous core receive cutaneous inputs both from the medial lemniscus and from other ascending systems. The central core projects only to area 3b while the region around it projects to areas 1 and 3b.

More recently an alternative interpretation has been put forward from studies on both the primate and the cat. In this scheme the VPL and VPM receive cutaneous input that is primarily, though not completely, derived from the DCN and the equivalent subnuclei of the rostral divisions of the trigeminal nuclear complex (Fig. 6.2a, b) (Kaas et al.1984; Dykes et al. 1986). Cells in the VPL and VPM project primarily to areas 1 and 3b although in the primate some projections to the hand representation in area 2 have also been demonstrated (Pons and Kaas 1985). Anterior to this cutaneous representation is a group of cells that respond to muscle spindle afferents and that project to area 3a. In both the cat and the primate these cells are considered to constitute the ventral posterior oralis nucleus (VPO) (Kaas et al. 1984; Dykes et al. 1986). Medial of the VPO is a second group of cells that project to area 2. This group is regarded as the ventral posterior superior nucleus (VPS) and in the primate these cells respond to joint and muscle input (Kaas et al. 1984) while Dykes et al. (1986) have suggested that in the cat the VPS receives input from joint afferents. Finally, at the ventral aspect of the ventral posterior group is a region that receives input from pacinian corpuscles and relays this information to the SII cortex (Fig. 6.2a). This region is regarded as the ventral posterior inferior nucleus (VPI) (Kaas et al. 1984; Pons and Kaas 1985; Herron and Dykes 1986). In addition to these regions, Metherate et al. (1986) have demonstrated in the cat that cells in the lateral division of the posterior group (PO_1) receive input from hair

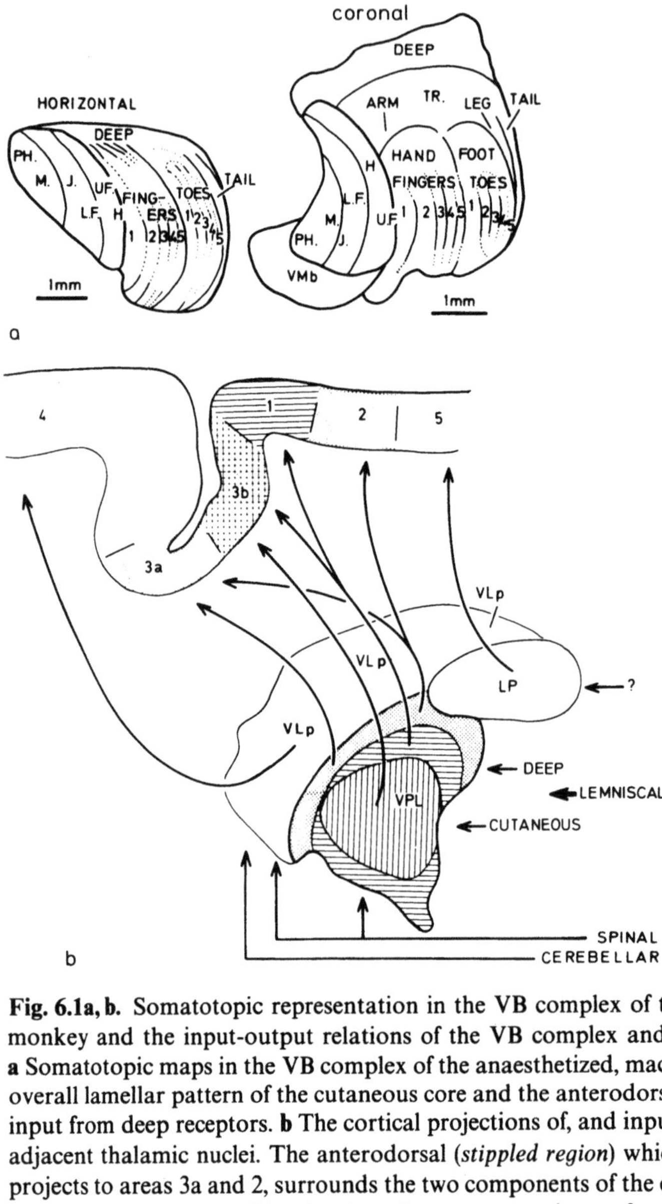

Fig. 6.1a, b. Somatotopic representation in the VB complex of the anaesthetized macaque monkey and the input-output relations of the VB complex and adjacent thalamic nuclei. **a** Somatotopic maps in the VB complex of the anaesthetized, macaque monkey showing the overall lamellar pattern of the cutaneous core and the anterodorsal shell (*deep*) that receives input from deep receptors. **b** The cortical projections of, and inputs to, the VB complex and adjacent thalamic nuclei. The anterodorsal (*stippled region*) which receives deep input and projects to areas 3a and 2, surrounds the two components of the cutaneous core of the VPL. The central core (*vertical hatching*) receives cutaneous inputs from the medial lemniscus and projects to area 3b. The surrounding region (*horizontal hatching*) receives cutaneous inputs from the medial lemniscus and other sources and projects to areas 1 and 3b. *Abbreviations*: **a** *H*, head; *J*, jaws; *LF*, lower face; *M*, mouth; *PH*, pharynx; *UF*, upper face; **b** *VPL*, ventral posterior lateral nucleus; *VLp*, pars postrema of the ventral lateral nucleus; *LP*. lateral posterior nucleus. (**a** Modified from Jones and Friedman 1982, p. 541 and **b** from Jones 1986, p. 361)

follicles via the spinocervical tract (SCT) and the lateral cervical nucleus and convey this information to the SII cortex (Fig. 6.2b). In evaluating these data, it should be remembered that in the cat, and particularly the primate, the VPL and VPM receive projections from the spinothalamic and trigeminothalamic tracts that constitute a source of wide-dynamic-range inputs that are not altogether compatible with the idea of discrete representations of particular modalities either within the ventral posterior nuclei or the SI cortex (see Sect. 6.3).

The above interpretations are clearly not markedly different from one another. The anterodorsal shell described by Jones and Friedman (1982) and Jones (1986) is equivalent to the VPO and the VPS described by Kaas et al. (1984) and Dykes et al. (1986). The cutaneous core of the VB complex described by Jones and Friedman (1982) and Jones (1986) may be thought of as the VPL and VPM of Kaas et al. (1984) and Dykes et al. (1986). The reader should consult the original literature for extensive discussions of possible differences. What is for the moment important is that, in both the cat and the primate, the ventral posterior region appears to contain a number of complete body representations each of which is composed of a different form of somatic input and that in both species each of these representations is relayed to a different cytoarchitectonic area of the SI cortex. The resultant representation of somatic input across the SI cortex is reviewed in Chap. 7 (Figs. 7.1 and 7.14; Sects. 7.2.1 and 7.3.1).

Microstructural Organization of the VPL and VPM. In the primate, injections of small amounts of tracer into restricted parts of cortical areas 1, 2 or 3b labels parasagittally oriented slabs of cells in the VPL and VPM. These slabs extend throughout much of the anteroposterior and dorsoventral extent of the VPL and VPM but they are restricted in the mediolateral axis (Lin et al. 1979; Nelson and Kaas 1981; Jones et al. 1979). They are variously referred to as discs, bands or lamellae, this last term no doubt being derived from the analogous structures that are so clearly seen in the VB complex of the raccoon (Fig. 6.5; Sect. 6.2.2) (Welker and Johnson 1965). If a cortical focus as small as 0.5 to 1.0 mm is injected with HRP, then labelled thalamic cells are found within a longitudinally oriented *rod* of tissue. These rods measure about 500 μm in diameter and extend throughout much of the anteroposterior length of the VB complex. If injections are placed at two cortical foci that are separated along the anteroposterior axis the labelled rods lie dorsoventrally displaced from one another within a single lamella, but if the two nearby injections are separated along the mediolateral axis the labelled rods lie in adjacent lamellae (Jones et al. 1979).

In cats, similar experiments have shown that the thalamocortical cells that project to small regions of areas 3b, 1 and 2 are organized into *clusters* that are about 140–350 μm in diameter but that extend for only about 500 μm in the anteroposterior dimension (Kosar and Hand 1981). It is considered that these clusters are similar to the rods formed by focally projecting thalamocortical cells in primates (Jones 1986).

In the primate, injection of HRP into the medial lemniscus close to the VPL labels the terminals of lemniscal axons (Jones 1983a). The terminals of most of these axons are confined either to the VPL or to the anterodorsal shell (equivalent to the VPO or

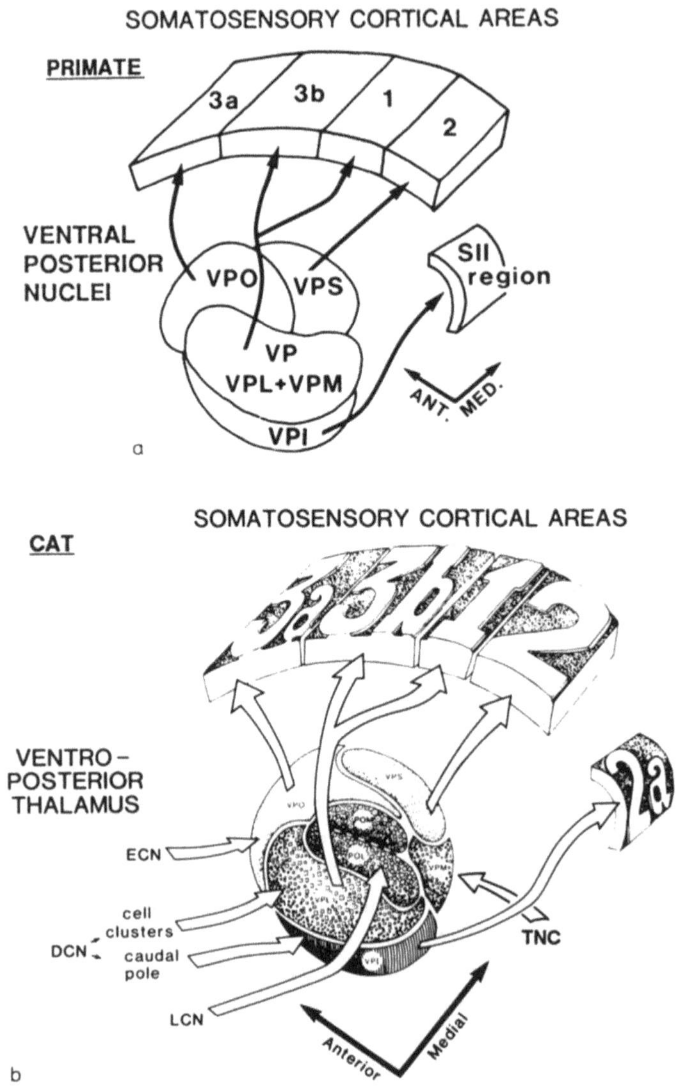

Fig. 6.2 a, b. Hypothesis concerning the modality-specificity of information transfer through the ventral posterior nuclei of the thalamus to specific areas of the somatosensory cortex in the primate and the cat. **a** Primate: VPL and VPM receive cutaneous inputs and project to areas 1 and 3b. VPO receives input from muscle receptors and projects to area 3a and VPS receives input from joint and muscle receptors and projects to area 2. VPI receives inputs from pacinian corpuscles and projects to SII cortex. **b** Cat: Each thalamic nucleus receives a specific class of somatosensory input and conveys this information to a distinct cytoarchitectonic region of the cortex. The cell clusters in the *core* of the DCN receive cutaneous inputs from small areas of skin and project to the VPL while cells in the nucleus principalis of the TNC project to the VPM. Both VPL and VPM project to areas 1 and 3b although the projection of the latter is not shown. The caudal pole of the DCN receives input from pacinian corpuscles and projects to VPI. VPO receives input from muscle receptors and projects to area 3a. VPS receives input from joint afferents and projects to area 2.

VPS, see above). Within the VPL single axons give rise to 2, 3 or 4 branches each of which forms a terminal ramification. These ramifications are contained in parasagittally oriented slabs that do not exceed 300 μm in width and measured 250–500 μm in the dorsoventral axis and 400–1100 μm in the anterioposterior axis. A single ramification embraces 120–200 thalamic cells. While the rods of thalamocortical neurons extend throughout much of the anteroposterior axis of the VPL, the ramifications of single lemniscal axons occupy only one-third to one-half of the anteroposterior dimension (average length of the VPL = 3 mm). It therefore seems that a single lemniscal axon does not activate all cells within a single rod (Jones 1983a).

The terminals of some lemniscal axons are postioned, like the rods of focally projecting thalamocortical cells, at a particular dorsoventral level in the VPL. However, other lemniscal axons give rise to anteroposteriorly directed ramifications at two different dorsoventral levels of the VPL (Fig. 6.3a). Given the topographic organization of the projections of thalamocortical cell rods at different levels within a single lamella, these lemniscal axons would presumably influence foci in the SI cortex that are separated along the anteroposterior axis (Jones et al. 1979; Jones 1983a).

In cats, most single lemniscal axons give rise to only one or two terminal ramifications. When two or more are present they are usually separated along the anteroposterior axis of the VB complex (Rainey and Jones 1983). The terminal ramifications are fan-shaped, extend for 200–400 μm in the anteroposterior axis and are 100–200 μm wide (Fig. 6.3b). Rainey and Jones (1983) have estimated that the terminal ramifications of a single lemniscal axon may provide input to 50–120 neurons but have concluded that a single lemniscal axon does not innervate all cells in a cluster. Intracellular staining of thalamocortical cells in the VB complex of the cat has shown that they have radiate dendritic trees that may be up to 470 μm in diameter (Yen et al. 1985; see also Rainey and Jones 1983). It therefore seems that, at least in the mediolateral axis, the dendritic tree of a single thalamocortical neurons extends well beyond the bounds of a single lemniscal axon.

In both the cat and the primate, it seems possible to conclude that information from spatially separate lemniscal axons is delivered to the vicinity of the somata and dendrites of groups (clusters or vods) of thalamocortical neurons that project to a column of cortical tissue that is 0.5–1.0 mm in diameter (see Figs. 7.13a, b and 7.15a, b Sect. 7.3.1 and 7.3.2). Furthermore it seems that at least some single lemniscal axons might influence cortical foci that are displaced in the anteroposterior axis of the SI cortex. However, this might not greatly influence

◀ **Fig. 6.2 a, b** (*continued*). LCN receives input from hair afferents via the spinocervical tract and projects to PO_1. Note that in these summaries inputs from the spinothalamic tract or the rostral reticular zone of the DCN are not shown. See text for further discussion. *DCN*, dorsal column nuclei; *ECN*, external cuneate nucleus; *TNC*, trigeminal nuclear complex; *LCN*, lateral cervical nucleus; *VPL*, ventral posterior lateral nucleus; *VPM*, ventral posterior medial nucleus; *VPO*, ventral posterior oralis nucleus; *VPS*, ventral posterior superior nucleus; *POl* and *POm*, lateral and medial divisions of the posterior nuclei. (**a** Modified from Kaas et al. 1984, p. 138; **b** modified from Dykes et al. 1986, p. 1537)

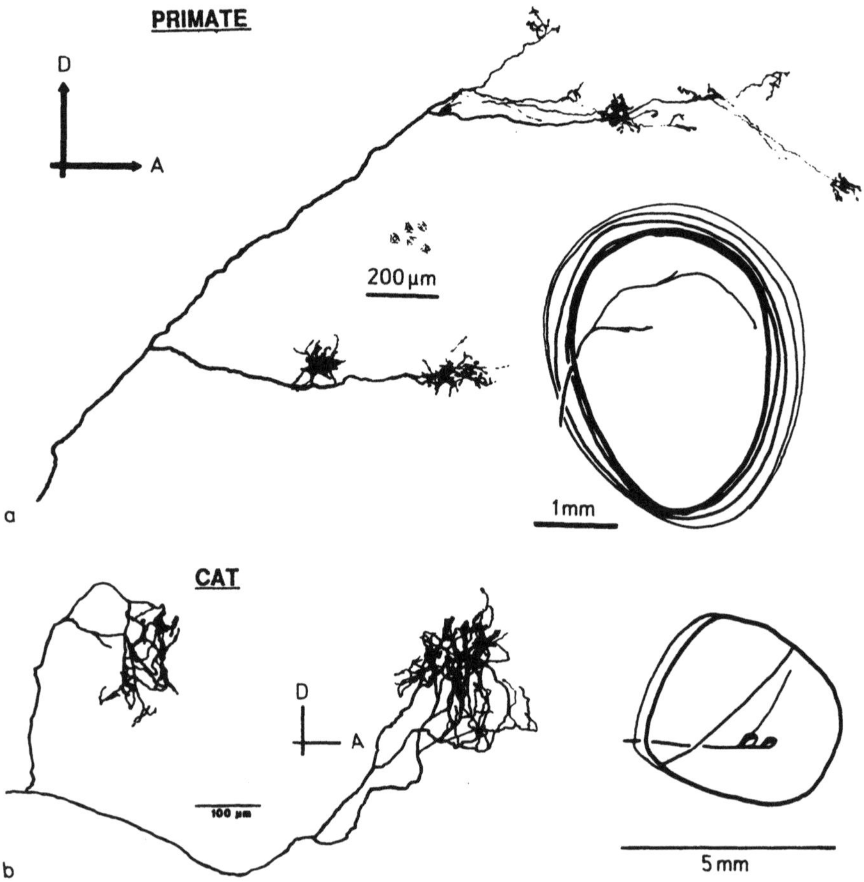

Fig. 6.3 a, b. Terminals of medial lemniscal axons in the VB complex of primate (**a**) and the cat (**b**). **a** Parasagittal reconstruction of an HRP-filled lemniscal axon projecting to two levels of the *central* core of the cutaneous core of the VB complex in the primate. *Inset* shows the extent of the axonal projections in relation to a parasagittal view of the VPL. **b** Parasagittal reconstruction of the HRP-filled lemniscal axon projecting to two levels of the VB complex in the cat. *Inset* on the *right* shows the small size of the arborizations of a lemniscal axon relative to the size of the VB complex in the parasaggital plane. *Abbreviations: D*, dorsal; *A*, anterior. (**a** Modified from Jones 1983a, p. 8; **b** modified from Rainey and Jones 1983, pp. 236 and 241)

somatotopic representation at the cortical level for, in the SI cortex of both cats and primates, the representation of the body is organized into anteroposteriorly oriented, homotopical strips (Figs. 7.1 and 7.14).

In the primate, the injection of tracer into the same parts of the somatotopic representation in areas 1 and 3b labels cells in the same position in the coronal plane of the VPL and VPM (Fig. 6.4) (Nelson and Kaas 1981; Lin et al. 1979). In the owl monkey, Lin et al. (1979) concluded on a numerical basis that some cells must project to both cortical areas. In the macaque, however, the simultaneous injection of two different dyes into areas 1 and 3b has shown that the thalamic projections to

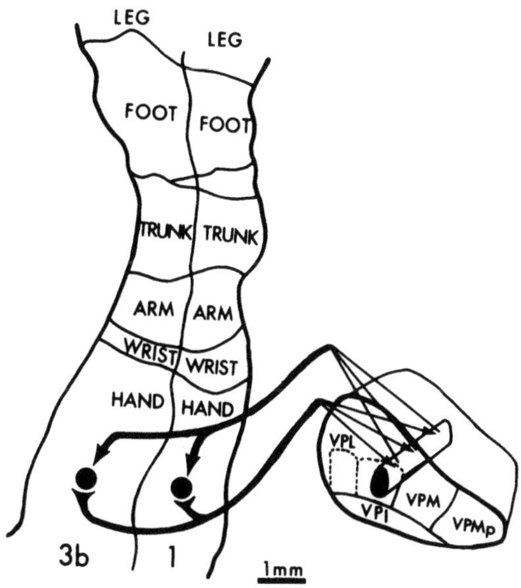

Fig. 6.4. Diagrammatic summary of a modality- and place-specific rod of thalamocortical cells in the owl monkey that receive input from the terminal ramifications of modality- and place-specific lemniscal axons and that relay this information to foci in areas 1 and/or 3b. Note the precisely organized, reciprocal, corticothalamic projections. *VPL*, ventral posterior lateral nucleus; *VPM*, ventral posterior medial nucleus; *VPI*, ventral posterior inferior nucleus. (From Lin et al. 1979, p. 368)

areas 1 and 3b are from separate populations of neurons within the VPL (Jones 1983b). On the basis that area 3b receives input from the central core of the cutaneous core (Jones and Friedman 1982; Jones 1986), while areas 1 and 3b both receive input from the surrounding parts of the cutaneous core, Jones (1983b) has argued that those cells projecting exclusively to area 1 may be responsible for the rapidly adapting responses which in primates are characteristic of the cortical cells in that area. In contrast, thalamocortical cells that project to area 3b receive rapidly adapting or slowly adapting inputs and presumably convey these selectively to the rapidly adapting or slowly adapting subdivisions within area 3b (see Sect. 7.2.2). In the cat, one can regard the VPL and VPM as equivalent to the cutaneous core of Jones and Friedman (1982) and no central core projecting exclusively to area 3b has been identified. In the SI cortex of the cat, inputs from slowly adapting and rapidly adapting receptors contribute to separate, although interdigitated, body representations across areas 1 and 3b (Fig. 7.13 c–e; Sect. 7.3.1) (Stretavan and Dykes 1983) and Dykes et al. (1986) have reported that the VPL and VPM are composed of interdigitating zones that respond to either slowly or rapidly adapting receptors.

While the above studies reveal a precise and easily explained orderliness it is very important to emphasize that they rely almost totally on correlations of neuroanatomical data with the responses of neurons to a restricted stimulus regime, in animals under deep barbiturate anaesthesia. The reader is referred to Sect. 6.3 where

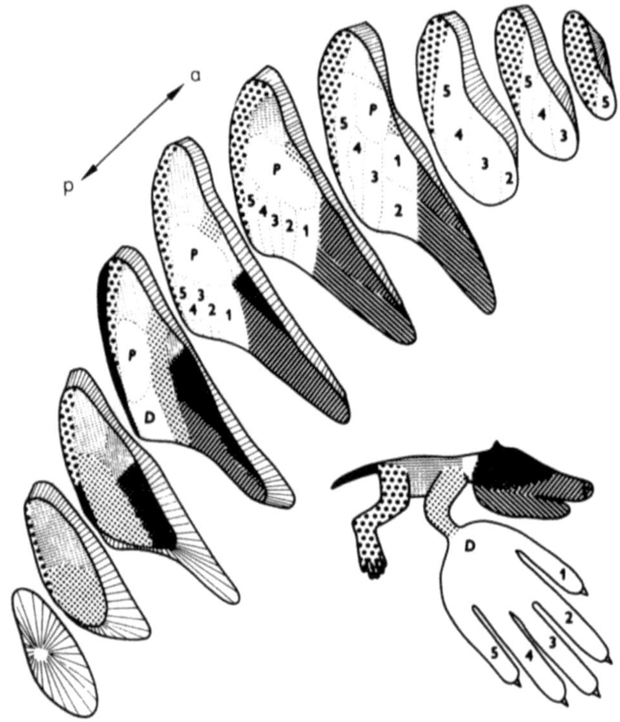

Fig. 6.5. Schematic three-dimensional drawing of the representation of the body surface in the VB complex of the adult raccoon. Note that the 5 cellular lamellae each contain the representation of a digit beneath the representation of the palmar pads (*P*). *Abbreviations*: *D*, dorsal surface of hand; *a*, anterior; *p*, posterior. (Modified from Welker and Johnson 1965, p. 783)

an attempt is made to describe the responses of VB neurons to a wider range of stimuli, in animals that are awake or under a variety of anaesthetics.

6.2.2 Raccoon

In coronal or horizontal sections through the brainstem of the raccoon, four laminae composed of myelinated fibres delineate five cellular lamellae in the central portion of the VB complex (Fig. 7.19d–f) (Welker and Johnson 1965). Welker and Johnson (1965) used single-unit and multiunit recording in pentobarbitone-anaesthetized raccoons and retrograde degeneration following destruction of specific parts of the SI body representation, to show that each of these cellular lamellae constitutes the thalamic representation of tactile input from a single digit of the forepaw (Fig. 6.5). Dorsal and caudal of these lamellae is a region of approximately the same volume which contains cells that respond to stimulation of the glabrous skin of the palm. The representation of all the glabrous skin of the forepaw thus occupies almost 50% of the volume of the VB complex. The face is represented medial to the paw representation and the hindlimb and tail representations are confined to a narrow strip of tissue lateral to the most lateral fibrous lamina (Fig. 6.5) (Welker and Johnson 1965). This most lateral lamina is considered as delineating the VPL from the VPM (Herron 1983).

Since these early studies a number of workers have used retrograde tracing techniques to study the location of thalamic neurons that project to the sensorimotor cortex in the raccoon. Sakai (1982) injected HRP into physiologically identified areas of the primary motor area (MI), an area which he defined as lying along the postcruciate gyrus. He showed thus that the somata of thalamocortical neurons in the ventral lateral nucleus were topographically ordered across the coronal plane of the ventral lateral nucleus in a manner that reflected their projections to the MI representation. From all sites in MI, labelling was also found in the paracentral nucleus and the central lateral nucleus. Injection of HRP into the face and forepaw areas also labelled cells in the ventral medial nucleus and the ventral anterior nucleus, respectively.

Herron (1983) injected HRP into somatotopically identified portions of SI and SII. Injections of HRP into SI labelled cells in the VB complex while injections into SII labelled cells in the VPI and the central lateral, centromedian and parafascicularis nuclei of the intralaminar group. In a similar study, Warren and Pubols (1984) injected HRP into the crowns of single subgyri within the SI forepaw representation. This resulted in labelling of a single crescent-shaped aggregation of thalamic cells that were confined within the appropriate lamella. Large injections into a single subgyrus labelled 80% of the large (18–30 μm) cells in a single lamella. In these studies a few cells were also labelled within the posterior nuclear group.

6.2.3 Rat

In the rat, the VPM is primarily devoted to the representation of the mystacial vibrissae and consequently the organization of that part of the VB complex is dealt with in Chap. 3. Nevertheless some mention will be made here of some anatomical differences between neurons in the VPM and VPL that emphasize overall differences in the organization of these two nuclei. When the VB complex is stained with cresyl violet or with the Golgi technique it can be seen that cells in the VPL are aligned in curvilinear, parasagittal laminae that are roughly parallel to the external medullary lamina (Fig. 6.6a) (McAllister and Wells 1981; Peschanski et al. 1984). In their longest dimension the dendritic trees of both VPL cells and VPM cells measure about 460 μm. Those of VPL cells are elongated along the anteroposterior axis such that they conform to, and enhance, the laminar appearance of the nucleus as seen in the horizontal plane (Fig. 6.6a). A few cells in the VPL resemble the cells in the VPM in that they have radiate dendritic trees. Furthermore, some discoid neurons have long (250–275 μm) medially or laterally projecting dendrites that may cross the dendritic domains of neurons in 3 to 10 adjacent laminae (Fig. 6.6a).

When a small volume of HRP is injected into the forepaw or hindpaw representation of the SI cortex curvilinear arrays of cells are labelled within the VPL (Saporta and Kruger 1977). These arrays occupy most of the anteroposterior extent of the VPL

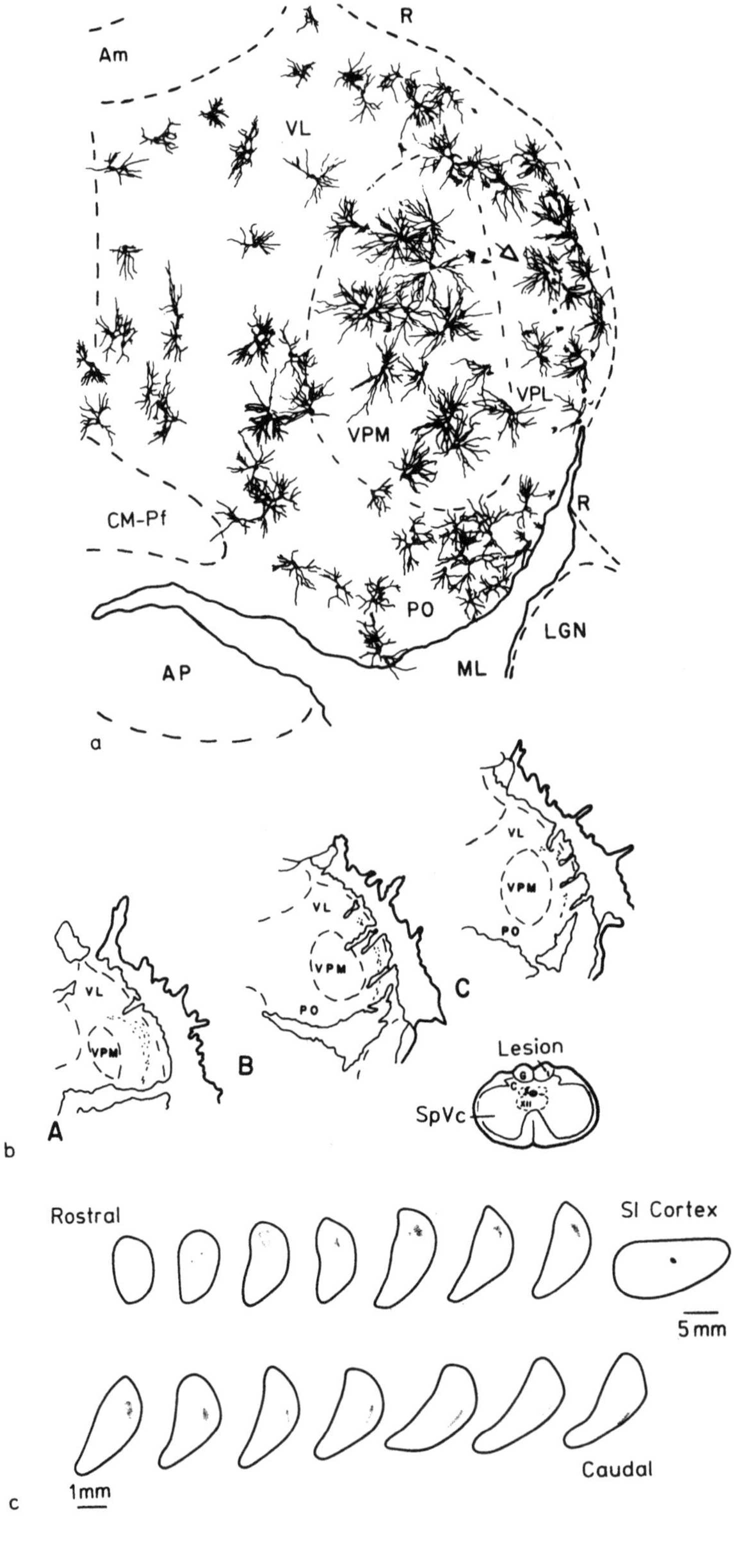
R
Am
VL
VPL
VPM
R
CM-Pf
PO
LGN
AP
ML
a
VL
VPM
PO
C
VL
VPM
PO
VL
VPM
B
Lesion
SpVc
A
b
Rostral
SI Cortex
5 mm
Caudal
c
1 mm

but are confined in the dorsoventral axis. These cell groups thus somewhat resemble the lamellae of thalamocortical cells described in primates and raccoons (Sect. 6.2.1 and 6.2.2). In the rat a single cortical injection results in labelling of *all* cells within a particular array or lamella, supporting the conclusion that there are very few, if any, interneurons within the VB complex of the rat (McAllister and Wells 1981; Houser et al. 1980; Ohara et al. 1983; Jones 1986).

Like the VPL of the primate, the VPL of the rat receives inputs from the spinal cord and the DCN (Lund and Webster 1967a, b; McAllister and Wells 1981; Peschanski et al. 1983). In the rat, however, there are no detailed studies of the projections of single lemniscal axons to the VPL. When a small punctate lesion was placed in the DCN, McAllister and Wells (1981) could detect bands of degeneration that extended throughout the anteroposterior and dorsoventral dimensions of the VPL (Fig. 6.6b). At the widest point such a band occupied up to one-third of the mediolateral extent of the VPL. This degeneration covered a width equivalent to that of 10–15 cell laminae (Fig. 6.6b). Although McAllister and Wells (1981) concluded that the terminations of lemniscal axons were widespread, it would seem that they may well fall within the confines of thalamocortical neurons retrogradely labelled from small areas of the SI cortex (Fig. 6.6c). Further detailed microanatomical work on this system is required before useful comparisons between anatomy and physiology are possible within the VB complex of the rat.

6.3 Responses of Neurons to Cutaneous Stimulation and the Effects of Anaesthetics and Other Drugs

The responses of neurons throughout the ventral posterior group has been reviewed recently by Jones (1986). Consequently, the following section places emphasis on

◀ **Fig. 6.6a–c.** Organization of neurons and connections in the VPL of the rat. **a** Golgi-Cox-stained neurons in the horizontal plane of the VB complex. Note the alignment of fusiform neurons along the lateral margin of the VPL which enhances the laminar appearance of the VPL. The dendritic trees of these VPL neurons should be compared with the radiate dendritic trees of VPM neurons. *Open arrow* indicates a neuron in the medial part of the VPL that has a large group of dendrites extending medially. **b** Drawings of horizontal sections through the VB complex at three different levels showing the distribution of degeneration resulting from a single stab lesion in the DCN (see *inset*). Note that the band of degeneration extends throughout the anteroposterior and much of the dorsoventral axis of VPL and follows the cellular laminae shown in **a**. **c** A series of coronal sections taken at intervals through the VB complex and showing the distribution of thalamocortical neurons, retrogradely-labelled after injection of a small amount of HRP into the SI cortex (see *inset* on *right*). Note that labelled neurons are found throughout the anteroposterior axis and that in the more caudal regions they are organized in lamellae. *Abbreviations; AP*, anterior pretectal area; *R*, reticular nucleus of the thalamus; *CM-Pf*, centromedian-parafascicularis complex; *PO*, posterior thalamic nuclei; *ML*, medial lemniscus; *LGN*, lateral geniculate nucleus; *VL*, ventral nucleus; *VPL*, ventral posterior lateral nucleus; *VPM*, ventral posterior medial nucleus. (**a, b** Modified from McAllister and Wells 1981, pp. 278 and 286; **c** modified from Saporta and Kruger 1977) those issues that seem relevant to the matter of plasticity as manifested by the production of new cutaneous inputs. One such issue is the alteration of neuronal responses by anaesthetics, or other neuroactive agents and another is the nature of inhibitory receptive fields in the VPL and VPM.

Lemniscal Versus Widefield Properties. As emphasized in the above section, neurons in the VPL and VPM are primarily responsible for relaying cutaneous input (including input from claw receptors) to the SI cortex. Although numerous studies have reported activation of neurons in the VB complex by receptors associated with joints and even muscle, it seems possible that these responses were recorded at sites that would now be considered as lying in VPO and VPS (Figs. 6.1 and 6.2; Sect. 6.2.1).

The literature on the responses of thalamic neurons to cutaneous stimuli is greatly complicated by the profound effects of the state of arousal of the animal or the type of anaesthetic or neuromuscular blocking agent used. In their single-unit studies on the VB complex of the monkey, Poggio and Mountcastle (1963) used the term *lemniscal properties* to describe the responses of certain neurons that responded to either cutaneous or joint receptors. This term meant that a neuron had a small, contralateral receptive field and responded to a particular modality. It implied that axons in the medial lemniscus had similar properties and conveyed these faithfully to cells in the VPL and VPM. Poggio and Mountcastle (1963) also defined what they referred to as the *static* and *dynamic* properties of thalamic neurons which had lemniscal characteristics. The receptive field and modality specificity of a neuron constituted its static properties. These properties were considered to be insensitive to the state of the animal (end-tidal CO_2, body temperature, level of consciousness or anaesthesia). The dynamic properties consisted of the temporal elements of the response of a neuron to peripheral stimuli. The dynamic properties were influenced by the state of the animal and, in particular, by anaesthetic agents. While the usefulness of the term "lemniscal" has been questioned (Golovchinsky et al. 1981), it remains, we feel, a historically important criterion that embodies some fundamental aspects of plasticity within the somatosensory portions of the neuraxis.

Since these early experiments, there have been numerous reports of cells in the VB complex that have large and often bilateral cutaneous receptive fields. These cells are often referred to as widefield neurons. In searching for, and defining the characteristics of, these cells, workers have used both electrical stimuli of all four paws and natural, multimodal stimulation over wide areas of the body. Based on the responses of neurons to electrical stimulation of the paws, Jabbur et al. (1972) have proposed a system of classification. This system resembles that used in the SI cortex (Sect. 7.3.3). In the thalamus, cells are classified into one of the following groups: (1) *s*-cells that respond to electrical stimulation of only one paw; (2) *sb*-cells that respond to the ipsi- and contralateral hindpaws or forepaws; (3) *sc*-cells that respond to both ipsilateral or both contralateral paws; (4) *m*-cells that respond to all four paws. A slightly different system has been proposed by Harris (1970) and Berkley (1973) in which *s*-cells are as defined above but *m*-cells include all those cells responding to electrical stimulation of more than one paw *or* to multimodal stimuli *or* to bilateral mechanical stimuli. Harris (1980) has attempted to bridge these systems by referring to the *s*-cells as *sa*-cells and all others (the *m*-cells of Harris (1970) and Berkley (1973)) as ~ *sa*-cell. We will use the system proposed by Harris (1970) and Berkley (1973) but in order to make this section compatible with the terminology established in the SI cortex (Sect. 7.3.3) we will refer to *s*-cells as *sa*-cells and all other cell types as *m*-cells.

Anaesthetics and Neuron Responses. The most conspicuous feature of the studies utilizing a combination of electrical and natural stimulation to define the *m*-cell and *sa*-cell has been the elucidation of profound effects of neuroactive agents on the convergence of input to the thalamus. In this context we might consider convergence as being the convergence of inputs from different receptor types in the same skin area (wide-dynamic-range neurons), as well as from the same receptor type over a large area of skin (widefield neurons). Clearly, the results from these experiments are at variance with the view, expressed in Sect. 6.2.2, of the VPL and VPM containing a neat (if complex) somatotopically organized array of cells that respond to particular types of cutaneous receptors and that transfer this information to specific cortical areas.

In unparalysed, awake cats Baker (1971) has shown that many VB cells have stocking-like tactile receptive fields that cover the entire paw and extend onto the trunk. These receptive fields were substantially larger than those seen under barbiturate anaesthesia (Poggio and Mountcastle 1960; Nakahama et al. 1966). Surprisingly, in awake cats, receptive fields did not change with the behavioural state of the animal or (in 4 out of 5 cells) in response to small doses of barbiturate (Baker 1971). Stocking-like receptive fields have been reported in cats anaesthetized with chloralose (Harris 1970). In the unanaesthetized, paralysed monkey Loe et al. (1977) reported that 96% of cells in the ventral posterior nuclei had lemniscal properties, thus supporting the much earlier observations of Poggio and Mountcastle (1963) in the VPL of the same preparation. The latter study has been often cited as evidence that, in the normal monkey, most VB neurons have lemniscal properties, but Berkley (1973) has suggested that the use of the competitive nicotinic blocker gallamine triethiode might be responsible for the smaller receptive fields reported by Poggio and Mountcastle (1963). Perhaps the findings of Pollin and Albe-Fessard (1979) in the unanaesthetized and unparalysed monkey that only 86% of cells in the ventral posterior nuclei had lemniscal properties add some weight to Berkley's (1973) suggestion. Finally, in cats paralysed with gallamine and anaesthetized with halothane (0.5–1.0%) and N_2O, Golovchinsky et al. (1981) reported that only 11% of VB neurons had widefield properties and that responses to more than one type of mechanoreceptor was very rare (see also Nakahama et al. 1966). These authors do, however, point out that 8.5% of the neurons they could isolate were unresponsive to natural stimulation and they suggest that these might well be cells responsive to noxious thermal stimuli or thalamic afferents with wide-dynamic-range responses.

The effects, on a small sample of VB neurons, of transference of animals from 0.5% halothane and N_2O/O_2 to 60 mg/kg of chloralose has been studied in rats (Guilbaud et al. 1981). In these animals 15 neurons responded to noxious cutaneous stimuli, 6 to innocuous and 4 to both types of stimulation. Chloralose abolished all responses to noxious stimuli and elicited an enlargement of the area of the body over which innocuous stimuli elicited activity. In addition, it caused 12 of the 15 neurons that initially responded selectively to noxious stimuli to develop tap, brush or light touch responses. Similarly, Gordon and Manson (1967) have noted that in the barbiturate-anesthetized cat, small additional doses of barbiturate caused cells initially responsive to light tactile and noxious thermal stimuli to lose their response to heat. Both these experiments are interesting not only for their implications to plasticity but in particular because of their relevance to future studies of the cortical and thalamic representation of nociceptive input.

In unparalysed cats that were very lightly anaesthetized with barbiturate, Berkley (1973) has reported that 29% of VB cells had *m*-characteristics. Similarly, Harris (1970, 1980) and Jabbur et al. (1972) found that 29% of VB cells in the chloralose anaesthetized cat were *m*-cells. In a study specifically designed to compare the effects of pentobarbitone with chloralose, Harris (1980) reported that under deep barbiturate anaesthesia less VB cells responded to electrical stimulation of the periphery and that only 18% of these were *m*-cells. Haimann et al. (1978) and Harris (1980) discuss the relative effects of chloralose and barbiturates on the responses of cells in the VB complex, suggesting that differences in the effects of the two anaesthetics on the influence of corticofugal fibres on VB neurons might underlie the differences in the responses of VB neurons to peripheral stimulation. One can certainly conclude that the enhancement of GABAergic effects by pentobarbitone reduces the convergence of peripheral inputs onto neurons in the VB complex, although these effects seem to be less pronounced than those observed under similar circumstances in the SI cortex (Sect. 7.3.3).

In relation to the effects of anaesthetics on the responses of cells in the VPL of cats and monkeys, it is interesting to note that bilateral tactile receptive fields have only been reported in experiments where electrical and natural stimuli have been used (Harris 1970, 1980; Jabbur et al. 1972; Berkley 1973). In contrast, such receptive fields are very rare in experiments where, in the presence or absence of anaesthetics, only natural tactile stimuli have been applied (Poggio and Mountcastle 1960; 1963; Nakahama et al. 1966; Gordon and Manson 1967; Loe et al. 1977; Pollin and Albe-Fessard 1979; Baker 1971). Perhaps prolonged electrical stimulation of the periphery alters the effectiveness of the inputs from the ipsilateral side of the body to cells in the VB complex. If so, then this could be regarded as in itself an interesting form of plasticity that may bear some relationship to the effects of usage on the central representation of the body surface (see Sects. 3.3.5, 7.2.9 and 7.3.6).

For the purposes of assessing the existence of plasticity at the levels of the thalamus and the cortex it is important to consider whether the above variations in the size of the receptive fields of thalamic neurons reflect mechanisms in the thalamus itself and whether it is likely that the neurons showing these changes project to the SI cortex. It would indeed seem that some thalamocortical neurons qualify as *m*-cells. The rat, for example, lacks interneurons in the VB complex (Sect. 6.2.3) and yet under urethane anaesthesia Angel and Clark (1975) have shown that 24 of 998 cells responded to electrical stimulation of more than one appendage. Furthermore, in the cat, Shimazu et al. (1965) have shown that at least some VB cells with *m*-characteristics could be antidromically activated from the pericruciate cortex.

The question of *m*-cells in the DCN has been considered by Berkley (1973). She points out that such cells have been described in the rostral portion of these nuclei. Subsequently it has been shown that not only do some cells in the rostral regions project to the VB complex but that, compared to those in the core region of the DCN, cells in the rostral region are particularly sensitive to anaesthetics (Sect. 5.2). The possibility thus remains that in the normal animal *sa*- and *m*-cells in the VB complex simply reflect lemniscal inputs even when those inputs are not truly *lemniscal* in character. An alternative explanation for the origin of the *m*-cell characteristics is the well-known projections of the spinothalamic tract to the VB complex of the rat, primate and cat (Jones 1986). These inputs could well account for the non-lemniscal properties of some neurons within the VB complex and, in fact, Poggio and Mountcastle (1963) expressed surprise at the absence of non-lemniscal responses in the VB complex of the unanaesthetized monkey.

Inhibitory Receptive Fields. Although inhibition of resting discharge or excitatory cutaneous inputs by deep stimuli over wide areas of the body has been occasionally mentioned (Nakahama et al. 1966), only inhibition arising from skin close to, or within, the excitatory receptive field has been carefully studied. In unanaesthetized, paralysed monkeys Poggio and Mountcastle (1963) found that only 5% of cells in the VPL were inhibited by skin surrounding the excitatory receptive field. In the awake, unparalysed cat, Baker (1981) reported surround inhibition in less than 15% of cells and in the barbiturate-anaesthetized cat, Gordon and Manson (1967) reported surround inhibition in an average of 10% of cells but noted that this varied between animals from 0% to 35%. In animals anaesthetized with chloralose, Gordon and Manson (1967) reported surround inhibition in 47% of cells. In the unanaesthetized cat paralysed with gallamine triethiode, Janig et al. (1979) found surround inhibition in 92% of a sample of identified thalamocortical neurons. They attributed this high percentage to the fact that previous workers used inhibition of spontaneous discharge to detect an inhibitory receptive field, whereas they used inhibition of a response evoked from the excitatory receptive field.

In the monkey, Poggio and Mountcastle (1963) applied pairs of electrical stimuli to the functional centre of the excitatory receptive field. They observed a 75%–80% inhibition of the response to the second stimulus even when the interval between the first and second stimulus was in the order of 20–45 ms. In the presence of barbiturates, this inhibitory period was prolonged such that there was only a 50% recovery of the response to the second stimulus after 70 ms. Since then, similar cases of *in-field* inhibition have been described using both electrical and natural stimuli in the urethane-anaesthetized rat (Angel and Clarke 1975), the unanaesthetized, paralysed cat (Fig. 6.7a, b) (Janig et al. 1979) and the barbiturate-anaesthetized, paralysed cat (Iwamura et al. 1972; Iwamura and Inubushi 1974; Tsumoto and Nakamura 1974). In all these studies inhibition has been shown to be maximum from the functional centre of the excitatory receptive field such that the spatial distribution of inhibition approximately resembled that of excitation. Recently, infield inhibition in the cat and the rat has been shown to be GABA ergic (Salt, 1989). Several authors, including Poggio and Mountcastle (1973), have pointed out that surround inhibition is a special case of this form of organization in which the

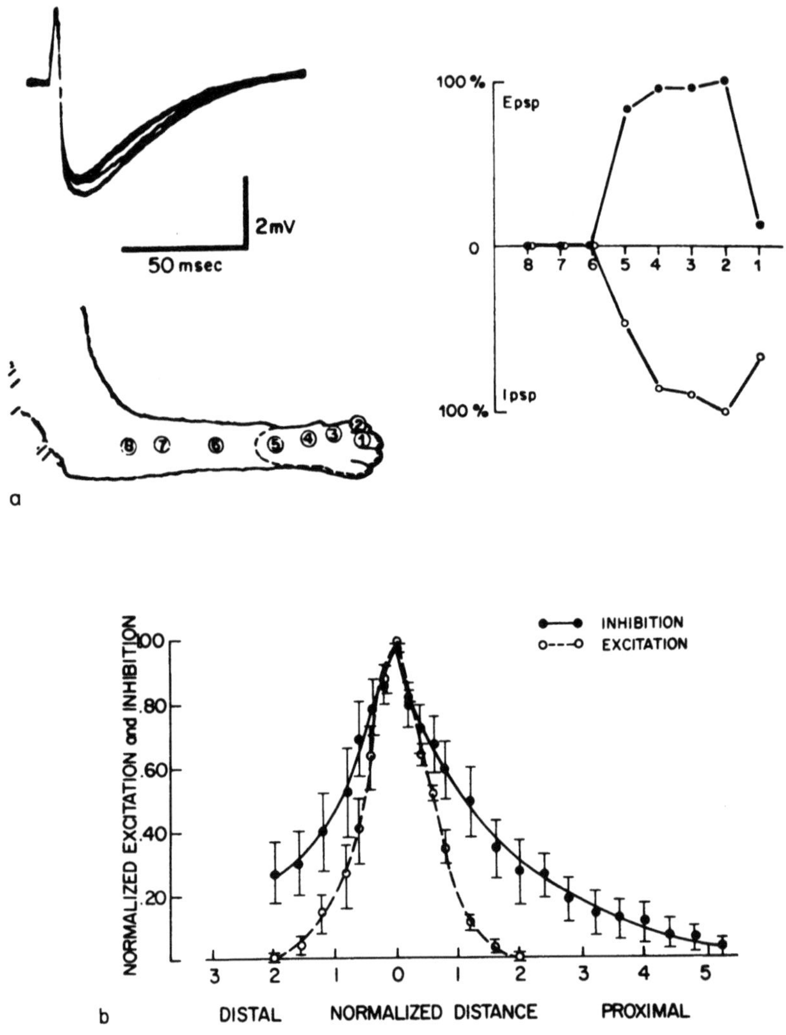

Fig. 6.7 a, b. In-field inhibition in neurons in the VB complex of the barbiturate-anaesthetized (**a**) and unanaesthetized (**b**) cat. **a** *Top left*: Intracellular recording of an EPSP-IPSP complex in a VB neuron following a single electrical stimulus to skin within the excitatory receptive field. *Right*: Graph of relative size of the EPSP and the IPSP of the complex when electrical stimulation is applied at the skin sites indicated in the diagram of the forepaw (*lower left*). Note the matching of the size of the EPSP with the size of the IPSP across the receptive field (*dotted line* on diagram of forepaw). **b** The response of a thalamocortical neuron in the VB complex to short, supramaximal puffs of air across the excitatory receptive field (*open circles*) and the percent inhibition (*filled circles*) of the response to a supramaximal, test puff applied to the centre of the excitatory receptive field by a supramaximal, conditioning puff applied at various distances from the receptive field centre about 20 ms before the central, test puff. For normalization purposes the units of distance shown on the *abscissa* represent places where excitation from a single puff has fallen to 20% of its maximum value. (**a** From Iwamura et al. 1972, p. 189; **b** Janig et al. 1979, p. 1458)

inhibitory envelope is wider than the excitatory envelope. In particular, Iwamura et al. (1972) have suggested that in these cases there is a fringe of pure inhibition around the excitatory receptive field whereas in other cases, where the inhibitory envelope is narrower than the excitatory, one might expect there to be a fringe of pure excitation. In the thalamus the strength of in-field inhibition seems to be related to the part of the body representation being studied. Thus Tsumoto and Nakamura (1974) have reported that in-field inhibition is greater the closer the excitatory receptive field is to the distal tips of the limbs and that it is weak or absent in cells with receptive fields on the trunk. Similarly, Iwamura and Inubushi (1974) have claimed that for cells with distal receptive fields the inhibitory envelope may be wider than the excitatory receptive field. Particularly where electrical stimuli are used, one wonders whether these differences are not a reflection of the lower density of sensory innervation of the proximal skin.

Finally, it is important to emphasize that phenomena similar to in-field inhibition have been seen in the spinal cord, DCN and the SI cortex (Fig. 8.5; Sects. 4.2.5, 5.2.3 and 7.3.1). It is therefore difficult to imagine that this property of the somatosensory system is not in some way involved in the alterations of somatotopy that have been observed following localized deafferentation or block of afferent input (Sect. 8.5.1).

6.4 Somatotopic Organization of the VPL and VPM

Not surprisingly, the most detailed somatotopic maps of the VB complex have been derived from deeply anaesthetized animals and in these cases the choice of anaesthetic has generally been those substances that produce a very high incidence of cells with lemniscal properties. In all mammals studied, the entire body is represented across the coronal plane of the VB complex. Within the VPL the representation of the trunk generally lies dorsal or dorsolateral to that of the limbs while the tail lies along the extreme lateral boundary. The face and head are represented medially within the VPM (Figs. 3.5e, 6.1a, 6.5 and 6.8) (Mountcastle and Henneman 1949; Welker and Johnson 1965; Cabral and Johnson 1971; Angel and Clarke 1975; Jones and Friedman 1982; Kaas et al. 1984). As in the SI cortex (Chaps. 3 and 7) the proportion of tissue devoted to each skin surface is related to the importance of that area in the life of the animal. Thus in the raccoon (Fig. 6.5) much of the VB complex is allocated to the representation of the forepaw while in the rat inputs from the mystacial vibrissae dominate much of the VB complex (Fig. 3.5e).

A second important characteristic of the somatotopic organization of the VPL and VPM is the representation of the skin in place- and modality-specific columns that are oriented along the anteroposterior axis of these nuclei (Figs. 6.1a, 6.5 and 6.8). In the monkey, Jones and Friedman (1982) and Jones et al. (1982) have shown that sequences of units, encountered over 500–800 μm of anteroposteriorly directed electrode tracks, often had identified receptive field characteristics in terms of both place and modality. They have suggested that this phenomena reflects the anteroposteriorly directed rods of thalamocortical cells that project to a small cortical focus and receive from lemniscal fibres that arise from a place- and modality-specific cluster of cells in the core region of the DCN (Sect. 6.2.1 and Chap. 5).

The somatotopic representation of the rat forepaw warrants some special attention as this animal was used in an oft-cited study of plasticity in the VPL (Sect. 6.5) (Wall and Egger 1971). The general representation of the entire body of the rat is shown in Fig. 3.5e. From recordings in 15 rats, Wall and Egger (1971) concluded that the limbs of this animal were represented such that the distal extremities activated neurons ventral in the VPL. In a detailed study of the forelimb representation in 304 rats, Angel and Clarke (1975) recorded from many neurons with receptive fields restricted to parts of a single digit. These neurons were located along the rostral, dorsomedial part of the VPL close to the ventrolateral border of the VPM while caudally in this part of the VPL they found representation of the palmar and wrist pads and the vestigal thumb of the forepaw (Fig. 6.8). Angel and Clarke (1975) concluded that the representation of the paw is a partially closed fist and they pointed out that this is a mirror image of the representation found in the rat cuneate nucleus in an unpublished study (Berridge 1973). In the SI cortex, subsequent work has shown that many neurons in layer IV have receptive fields that are restricted to a single digit (Chapin and Lin 1984) and some suggestion of a closed-fist representation has been reported by Dawson and Killackey (1987) (Fig. 7.24). In conclusion, despite the laminar structure of the VPL in the rat (Sect. 6.2.1), as yet there does not seem to be any evidence for the modality- and place-specific, anteroposteriorly oriented columns seen in the primate.

6.5 Effects of Alteration of Input on Somatotopic Organization

While numerous studies have shown that neuroactive agents affect the responses of VP neurons, few studies have adequately demonstrated changes of responsiveness or receptive field size or position following alteration of thalamic input. The representation of the mystacial vibrissae in the VPM of the rodent will no doubt continue to be exploited to study such changes during early developmental phases and recently that system has been used to provide some interesting insights into plasticity in the thalamus of the adult rat (Sect. 3.4.2). The following section deals with attempts to measure the effects of altered input on cells in the VPL of rats, cats and monkeys. In the thalamus of adult mammals it remains important to ask whether connections are indeed labile and to try to distinguish between mechanisms that alter the efficacy of existing synapses and mechanisms that involve the generation of new connections. There is little data on either of these issues but what there is suggests that connections in the thalamus are plastic.

6.5.1 Reversible Blockade of Afferents and the Immediate Expression of New Inputs

Two major studies exist in which the *immediate* expression of new inputs have been described. In the earliest, Nakahama et al. (1966) studied the responses of cells in the VB complex of cats before and after block of their excitatory receptive field by subcutaneous injection of 0.5% procaine. These animals were paralysed with

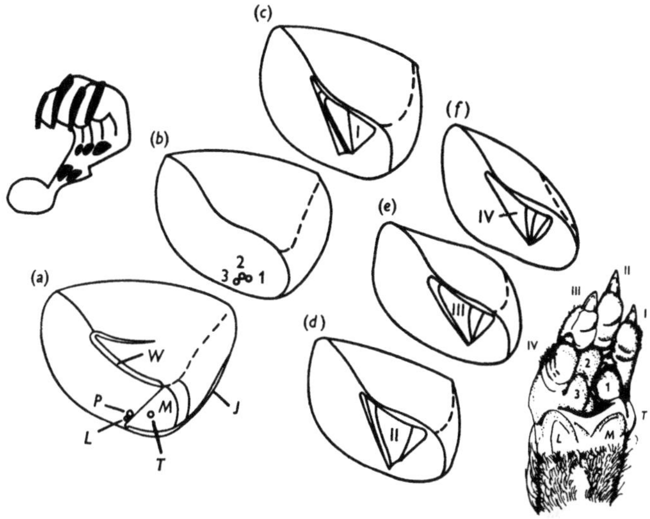

Fig. 6.8. Somatotopic representation of the forepaw in the VPL of the urethane-anaesthetized rat. The parts of the paw are shown on the *right* and relative positions of the representation of the parts of the paw within the VPL are shown on a set of successively more anterior coronal sections (*a–f*) through the VB complex. A 3-dimensional diagram is shown on the *left* illustrating that the forepaw is represented as a closed fist that is stretched along the length of the VPL. *Abbreviations: I, II, III* and *IV*, finger digits; *T*, vestigial thumb; *1, 2* and *3*, palmar pads; *M*, medial wrist pad; *L*, lateral wrist pad; *W*, wrist; *P*, palm. (From Angel and Clarke 1975, pp. 406 and 407)

gallamine and held under light barbiturate anaesthesia. The injection of procaine caused the disappearance of responses from the previous receptive field. In 16 out of 59 cells this was accompanied by the appearance of a new receptive field on skin proximal to, but abutting, the skin of the original receptive field (Fig. 6.9). With the reversal of anaesthetic block the original receptive field reappeared, but not until some hours later did the new receptive field disappear. Although it is not well documented in their paper, Nakahama et al. (1966) imply that the new cutaneous receptive fields were *not* fashioned from the type of receptor that originally excited the neuron. In considering these results one is reminded of the powerful effects of in-field inhibition that might be held at a tonic level by ongoing activity in afferents supplying the excitatory receptive field. Thus within the VB complex, Jones et al. (1986) have shown regional decreases in the levels of the activity-related mitochondrial enzyme, cytochrome oxidase (CO) following peripheral nerve section. An anaesthetic block of the original receptive field would therefore remove much ongoing input and the attendant tonic inhibition to a small thalamic region thereby enabling previously sub-threshold excitatory inputs from proximal skin to activate the neurons in this region (Sect. 6.3). In their discussion, Nakahama et al. (1966) regarded the persistence of the new receptive fields as evidence against any theories involving a release from inhibition. However, it remains quite possible that the reappearance of the normal excitatory inputs from the original receptive field might occur at levels of local anaesthesia that still effectively block the expression of in-field

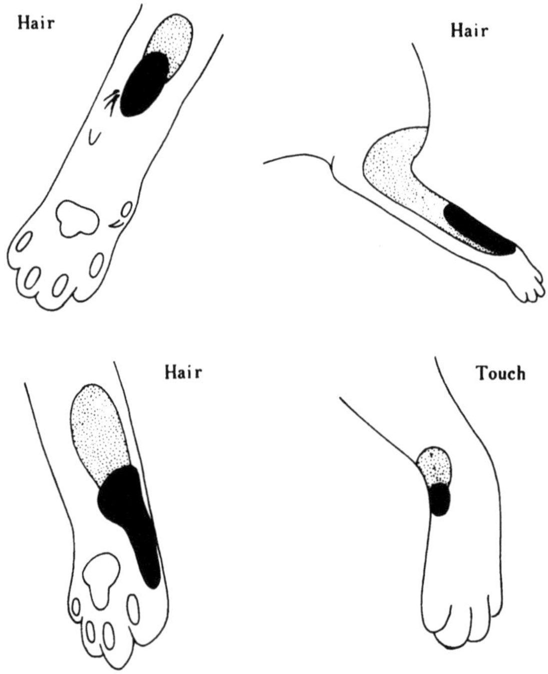

Fig. 6.9. Excitatory receptive fields of neurons in the VB complex of the cat under light barbiturate anaesthesia, before (*black areas*) and after (*stippled areas*) anaesthetic block of the original excitatory receptive field. Note that after anaesthetic block a new receptive field appears on proximal skin that abuts the original receptive field. The modalities shown refer to those that excited the neuron before anaesthetic block and do not necessarily relate to those effective after block. (From Nakahama et al. 1966, p. 190)

inhibition. The question of shifts in the effective modality remains more difficult to explain. While the data in these experiments are rather scant it seems to us this approach may be worthy of further investigation.

In another study Fadiga et al. (1978) applied reversible cold block to the DCN and examined the receptive fields of neurons in the VPL of the cat. Out of 55 cells that were studied, 15 remained responsive to peripheral stimuli during cold block at the DCN. However, changes in the receptive field were seen in only 5 of these cells. Unfortunately, the small sample size together with the absence of information concerning the extent of the block makes useful interpretation of these results impossible.

6.5.2 Chronic Lesion of Afferent Pathways and Sprouting of Thalamic Afferents

Electrophysiological Studies. The effect on VPL neurons of chronic lesion of the fasiculus gracilis and Morin's bundle (roughly equivalent to the SCT in the dorsolateral fasciculus) between T3 and T5 has been tested in awake monkeys (Pollin and Albe-Fessard 1979). Unfortuntely, data were presented on only two animals each with lesion of the contralateral Morin's bundle and the contra- and ipsilateral fasiculus gracilis. In these monkeys histological localization of recording sites showed that within what can be considered to be the hindlimb representation, responses to forelimb stimulation were recorded in only a single electrode track through the lateral parts of the VPL. Stimulation of the forelimb elicited activity at nine sites within this single track. In normal monkeys, only hindlimb responses could be recorded in this region although forelimb responses were recorded in tracks that were only 200–300 μm medial. As the delay between cord lesion and thalamic recording was 2 weeks, the authors have suggested that those thalamic afferents that respond to forelimb stimulation might have sprouted axons into the deafferented hindlimb representation thereby eliciting an expansion of the forelimb representation across the VPL. It is of course possible that similar observations might have been made had animals been studied at postoperative times that were sufficiently short to preclude the sprouting hypothesis. An alternative explanation of these results is that the single electrode track through the deafferented hindlimb representation picked up evoked activity of cells that were actually located in the not-so-distant (200–300 μm away) forelimb representation. The paucity of data in this study relegates most discussion of it to the realm of theory.

Wall and Egger (1971) have studied the effects of the destruction of the nucleus gracilis on the representation of the forelimb in the VPL of urethane-anaesthetized rats. Unfortunately the receptive fields and recording sites are not shown in their paper. The somatotopy of the forelimb and hindlimb representations are described by the presentation of a partial figurine illustrated in the coronal plane of the VB complex. As mentioned in Sect. 6.4 the representation of the forepaw and wrist in Wall and Egger's (1971) paper is different from that described in the subsequent extensive and well-documented study of Angel and Clarke (1975). Wall and Egger (1971) made recordings throughout the VPL in a grid of tracks in which the intertrack spacing was 200–250 μm along the anteroposterior and mediolateral axes. Recordings were made in normal rats and in rats at various times after destruction of the contralateral nucleus gracilis. Again in these operated animals conclusions are presented as a partial figurine drawn across the coronal plane of the VB complex at an unspecified anteroposterior level. As no data are given on the size or shape of receptive fields or the precise location of recording points, one is left to guess how the figurine boundaries were defined. The conclusions drawn by the authors were that 3 weeks after destruction of the nucleus gracilis the lateral edge of the arm and the paw representation had expanded 200–400 μm (between 1 and 2 intertrack spacings) into the hindlimb representation. A small sample of rats (11) was studied over the first postoperative week and from these experiments Wall and Egger (1971) concluded that spread of the forelimb representation began at 3 days. Given the intertrack spacing this spread must have been assessed in 200–250 μm steps to a distance of 200–400 μm. Again it would seem that an adequate explanation for these results is the ease with which recordings from distant neurons may be made when an electrode is inserted into a deafferented region (see Sect. 8.2.3). The picture which emerges is one of apparent somatotopic reorganization in which innervated skin that is normally represented adjacent to the denervated skin appears to occupy an expanded representation that lies partially within the deprived region of the central nervous tissue. This is precisely the result illustrated in Fig. 3 of Wall and Egger (1971). As their paper deals only with effects of lesions to the nucleus gracilis it should be stated that there is certainly no evidence for their conclusion that: "Mapping of the receptive fields of cells in the thalamus and cortex after section of the dorsal columns in the rat reveals evidence of functional reorganization in the central nervous system".

Anatomical Studies. When cells in the VB complex of the adult rat are destroyed with kainic acid a proliferation of unmyelinated axonal processes is observed within the affected area. It has been suggested that these may be growth-cone-like structures the presence of which reveals the ability of mature thalamic afferents to sprout new axonal processes (Peschanski and Besson 1986). One might therefore ask if this process can be elicited by less traumatic methods.

The effects of sectioning the dorsal columns in adult rats on the numbers of synapses in the rat VPL has been studied by Ganchrow and Bernstein (1981). In this work they assessed the number of boutons on the soma and on a 5 μm and 10 μm length of the proximal dendrite of cells that were within 100 μm of the lateral boundary of the VPL, 1 to 120 days after bilateral lesion of the fasiculus gracilis at T12. They concluded that there was a significant decrease in the number of boutons along a 5 μm segment of the proximal dendrite from the first 3 days to later survival times. Unfortunately, many apparent changes were not statistically significant. For instance, no significant difference was found between one-day survivors and animals that survived for 7 or 60 days, or between two-day survivors and animals that survived for 7, 14, 45, 60 or 90 days. Furthermore, bouton numbers in unoperated animals were not significantly different from animals that survived dorsal column lesion for 1, 2, 3, 7, 14, 60 or 90 days. It is thus difficult to make any conclusions from these experiments regarding the effects of dorsal column lesions on the synaptology of the VPL.

While it seems that there is no reliable evidence that in the adult rat the removal of the nucleus gracilis elicits changes in the somatotopic organization of the VPL the reader should not entirely despair for there is some good evidence that lesions to the DCN do in fact cause anatomical changes within the VPL. In a well-documented study, Tripp and Wells (1978) and Wells and Tripp (1987) removed most of the contralateral nucleus gracilis and nucleus cuneatus and assessed the effects of these lesions on the numbers of three morphologically distinct types of terminals; the LR, SR and F terminals. The SR terminals are small, contain round vesicles and are thought to be those of corticothalamic fibres. The F terminals contain flattened vesicles and, in the rat, are thought to belong to cells in the reticular nucleus of the thalamus. The remaining 3% of the terminals in the VPL are the LR type. These are large and contain round vesicles. Approximately 75% of the LR terminals originate from neurons in the DCN while 25% originate from spinal neurons (Wells and Tripp 1987).

Using morphometric techniques, Wells and Tripp (1987) assessed the numbers of the three types of terminals at postsurvival times ranging from 12 h to 50 days. Over this time-frame they found no change in the numbers of SR and F terminals but over the

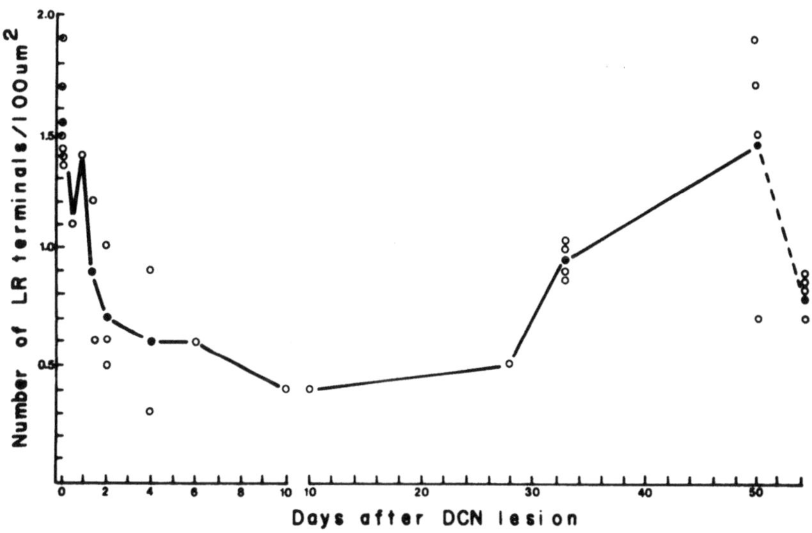

Fig. 6.10. Time-course of changes in the number of large, round-vesicle terminals (*LR*) following lesions to the DCN in adult rats. The number of LR terminals/100 μm^2 that showed no signs of degeneration are plotted against time after the lesion. Note the initial decrease followed by a relatively long period (50 days) before synaptogenesis restores the number to normal levels. At this time a second lesion to the DCN elicits a second decrease to a level similar to that caused by the first lesion. See text for further explanation. (From Wells and Tripp 1987, p. 470)

first 2 days the numbers of LR terminals decreased markedly to about 50% of the normal complement (Fig. 6.10). Synaptogenesis began after 30 days and by 50 days the number of LR terminals was normal. In some animals the remnants of the contralateral DCN were subjected to a second lesion 46 days after the first. Ultrastructural examination of these animals 4 days after the second lesion showed that the number of LR terminals had returned to the level seen shortly after the initial lesion (i.e. 50%) (Fig. 6.10). From these results Wells and Tripp (1987) concluded that both spinal afferents and afferents from DCN must contribute to the new population of LR terminals in the VPL. In light of this interesting work it would be extremely important to compare the terminal morphology of single, physiologically identified lemniscal and spinal afferents in the VPL of normal animals with the terminal morphology in animals that have been subjected to lesion of one DCN.

Finally, a recent approach to correlating the anatomy of connections with function in the VB complex of the primate is worthy of mention in relation to future experiments on plasticity. In this study Jones et al. (1986) showed that 2 weeks after sectioning peripheral nerves in the monkey there is a reduction in the histochemically detectable levels of the mitochondrial enzyme CO within the VPL. Reductions in the level of CO are considered to indicate a reduction of neural activity. Sectioning one nerve resulted in several rod-like regions that contained reduced CO activity. These extended throughout the axis in a manner that reflects the terminations of groups of DCN cells (Sect. 5.3.1), the cell rods and somatotopic organization of the VPL (Sect. 6.2 and 6.4). These results suggest that sectioning a peripheral nerve reduces the activity in the terminals of thalamic afferents that normally respond to this nerve. It would be interesting to use this approach in combination with electrophysiological recordings to test the responses of cells within the regions of low CO activity to peripheral stimulation.

6.6 Major Conclusions

Most studies in which attempts have been made to demonstrate plasticity within the VPL have been initiated in times when we had only a scanty knowledge of the physiology, pharmacology and connectivity of the normal thalamus. Perhaps this cavalier approach has been exaggerated by a prevalent view that the thalamus is simply a relay station through which information must pass to reach the cerebral cortex. As we learn more of the state-dependent responses of thalamic neurons and the physiology, pharmacology and anatomy of the pathways that influence transmission through the thalamus we move towards the view of the thalamus as a station concerned not only with relaying information but with performing certain integrative functions prior to passing messages to particular parts of the cerebral cortex (Jones 1986). It is against this background that we must design and interpret experiments relating to plasticity.

The following statements may be made concerning plasticity of somatosensory connections within the ventral posterior nuclei of the thalamus:

(1) *Input-Output Connections in the Normal Thalamus.* Axons from the medial lemniscus terminate in restricted regions within the VB complex that are comparable in size with groups of thalamocortical cells that project to columns in the SI cortex that are about 0.5–1.0 mm in diameter. However, some evidence exists for wider-ranging lemniscal terminations and the dendrites of some thalamic cells clearly extend beyond the major focus of the terminations of a group of place-specific lemniscal axons. We know little about the effectiveness or distribution of the terminals of spinal afferents to the VB complex although we may be sure that many of these afferents carry information that is spatially and qualitatively different from that carried by lemniscal afferents.

(2) *Lemniscal Properties and the Effects of Anaesthetics.* In the cat there seems to be good evidence that what are called *lemniscal properties* are actually the properties of cells in the *core* of the DCN. Thus although cells in the rostral reticular zone of the DCN have non-lemniscal properties some do project to the VB complex via the medial lemniscus. The receptive fields of these cells are particularly sensitive to anaesthetics (Chap. 5). In the unanaesthetized primate, the anaesthetized rat and unanaesthetized and anaesthetized (chloralose or light barbiturate) cat an appreciable percentage of VB cells have non-lemniscal properties. These non-lemniscal properties may arise from inputs from the rostral portion of the DCN and inputs from the spinothalamic tracts. These properties are particularly sensitive to barbiturates, suggesting a potential

importance of GABAergic mechanisms in the control of the receptive fields of thalamic neurons and emphasizing the existence of two classes of excitatory inputs that may be discerned on the basis of their sensitivity to GABAergic inhibition and to the behavioural state of the animal or simply their strength (Sect. 8.5).

(3) *Nerve Blocks and Lesions of the Dorsal Columns.* There is no acceptable evidence that lesion to the dorsal columns elicits changes in receptive field size, somatotopic organization or synaptology within the VB complex. There is, however, some evidence that temporary anaesthetic block of the excitatory field causes the development of an excitatory input from skin proximal to, but abutting, that which formed the original receptive field. Perhaps these observations reflect the fact that the maximal inhibition of a thalamic neuron comes from the centre of the excitatory receptive field. Block of the excitatory receptive could thereby release a small group of VB neurons from tonic inhibition caused by ongoing levels of afferent input, thereby rendering effective sub-threshold excitatory inputs from the skin adjacent to the excitatory receptive field.

(4) *Lesions to the Dorsal Column Nuclei.* There is no acceptable physiological evidence that lesions to the DCN elicit anything other than the loss of the input from these nuclei to cells within the VPL. There is, however, one study in the adult rat where it has been shown that partial destruction of the DCN elicits, within the VPL, sprouting of new terminals by spinal afferents *and* surviving DCN afferents. Furthermore, as reviewed in Sect. 3.4.2, there is good evidence that, in adult rats, chronic lesion of the principal nucleus of the trigeminal nuclear complex results in the expansion of the receptive fields of many neurons throughout the VPM.

7 Plasticity and the Somatosensory Cerebral Cortex

7.1 Experimental Strategies and Cortical Plasticity

If the representation of the vibrissae (Chap. 3) is placed aside, there are six basic types of experiments that are related to the study of plasticity in the somatosensory cortex of adult and developing mammals. The first approach has resulted from the use of anaesthetic agents as part of the standard experimental protocol for recording in the central nervous system of mammals and it involves studies on the effects of general anaesthetics, transmitter agonists and antagonists on the receptive fields of cortical neurons. The second approach has been to map the representation of the skin supplied by a particular nerve and then to crush the nerve, or section and repair it, before allowing it to regenerate. Using the blood vessel patterns on the cortical surface as a guide, it is then possible to repeatedly map the receptive fields of neurons at specific points across the temporarily deprived patch of cortex and to test whether the reinnervated skin surface is represented in a normal or disorganized manner. The third approach is identical to the second except that the nerve is ligated to prevent regeneration and the deprived cortex is mapped to determine whether other intact nerves have established an influence. In approaches 2 and 3 the cortical territory in which the skin supplied by the nerve is normally represented may be mapped at various times (hours, weeks, months) after nerve damage. A variation on approach 3 involves lesioning or applying transient blocking to spinal nerves, dorsal roots or particular parts of the spinal cord. In the fourth approach attempts have been made to alter the frequency of natural stimulation of a particular area of skin in order to test whether the cortical representation of this area of skin increases. In the fifth approach a lesion is placed in a defined part of the cortical representation. Some time later either the cortex around the lesion is remapped to see whether the skin previously represented within the lesioned area of cortex is newly represented in the surrounding cortex, or behavioural tests are made to detect persistent deficits in the performance of tasks that utilize related somatosensory input. The sixth and final approach involves the use of cytotoxic agents to grossly alter the development of the cortex and other brain regions. The effects of this procedure may then be assessed using anatomical, behavioural and electrophysiological techniques.

As the following sections indicate, there have been many demonstrations of plasticity at the level of the somatosensory cortex. It could, in fact, be argued that perhaps more research should have been devoted to seeking out the underlying mechanisms. In our view, however, the detail with which plasticity has been documented in the SI cortex is not without benefit, for taken together, the results of the many studies do strongly suggest a finite number of underlying mechanisms. Already some workers have embarked on experiments designed to test the involvement of these mechanisms and even their preliminary findings are illuminating in relation to both the normal and compensatory functioning of the cerebrum.

7.2 Plasticity in the Cortex of Adult and Developing Primates

7.2.1 Multiple Representations

In 1972 Paul et al. (1972b) made multiunit microelectrode recordings in many tracks throughout the hand representation of Brodmann's areas 3b and 1 of the postcentral gyrus of the anaesthetized rhesus monkey, *Macaca mulatta.* They found that much of the skin of the hand and wrist was represented in each of these areas. This work was followed by a large series of papers in which the somatotopic organization of the SI cortex was derived from single-unit and multiunit recordings in anaesthetized simian and prosimian primates (Kaas et al. 1979; Merzenich et al. 1978; Nelson et al. 1980; Merzenich et al. 1983a; Carlson and Welt 1980; Carlson and Fitzpatrick 1982; Sur et al. 1980; Sur et al. 1982; Felleman et al. 1983a; Kaas 1983; Pons et al. 1985a). These studies demonstrated that the simian, but not the prosimian, cytoarchitectonic areas 3b, 1 and 2, and probably also 3a, each contain a complete representation of the body surface (Fig. 7.1b) (Pons et al. 1985a). Similar results have also been obtained from the unanaesthetized macaque (Fig. 7.1a) (McKenna et al. 1983). The existence of these multiple somatotopic maps together with the differences in the stimuli required to excite cells in the different cytoarchitectonic areas led Merzenich et al. (1978) to propose that in primates the SI region be called the "parietal somatosensory strip" and that area 3b should be considered the homologue of SI in subprimate species. They further proposed the terms "SI proper" for area 3b and "posterior cutaneous field" for area 1 (see Fig. 16 in Merzenich et al. 1978; also Sur et al. 1978; Kass 1983). In this review we will retain the old terminology in which SI refers collectively to areas 3a, 3b, 1 and 2.

7.2.2 Thalamic Input and Intracortical Connectivity

Thalamic Input. The relay of specific forms of somatosensory information to the SI cortex is dealt with in Sect. 6.3. Nevertheless, areas 1 and 3b deserve some special attention because they are the regions that have been exclusively used in studies of deafferentation-induced changes in somatotopic organization. It is generally agreed that neurons in area 1 of the primate receive information from slowly adapting receptors while those in area 3b respond to either slowly or rapidly adapting receptors (Paul 1972b; Kaas et al. 1979; Sur et al. 1981; 1984). In the digit representation in area 3b the sites responding to rapidly adapting receptors are grouped into anterposteriorly oriented bands that are surrounded by similar bands in which cells respond to slowly adapting receptors (Sur et al. 1981). This form of organization thus bears some resemblance to that seen in the SI cortex of the cat (Sect. 7.3.1). In neither the primate nor the cat is the thalamic origin of this organization clear (Sect. 6.2.1). However, in the primate, Dykes et al. (1981) have

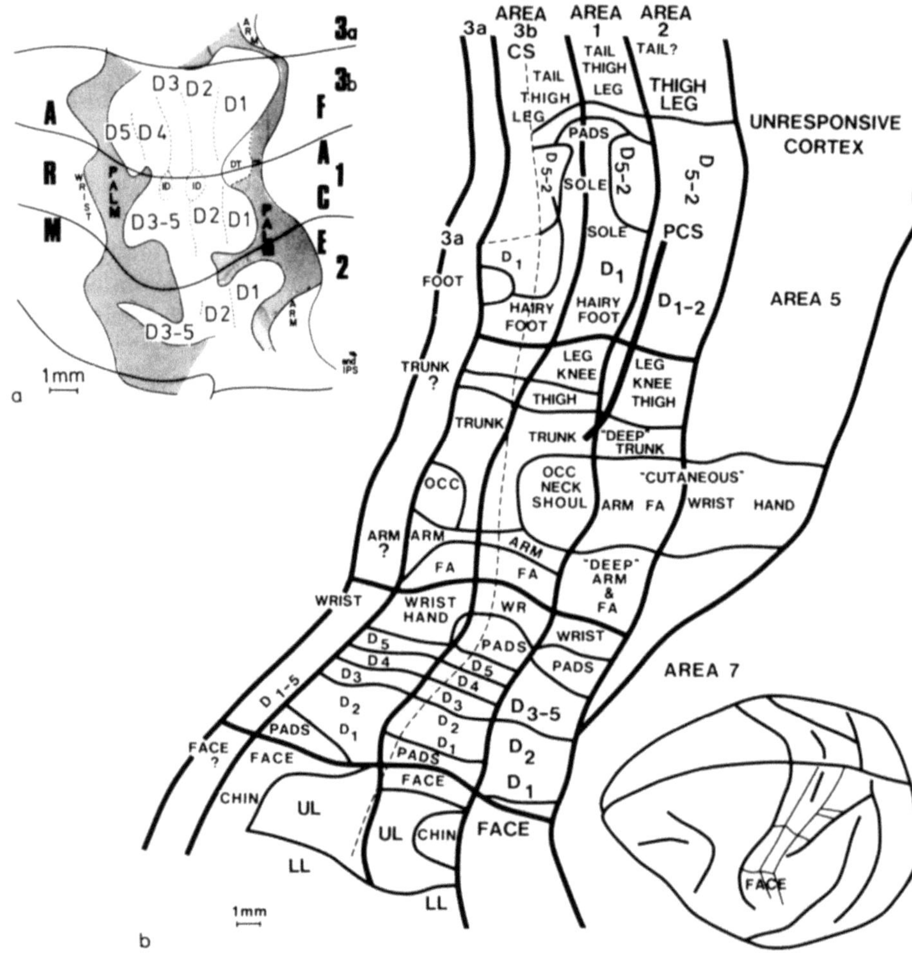

Fig. 7.1 a, b. Somatotopic maps in the first somatosensory area of the unanaesthetized (**a**) and the anaesthetized (**b**) macaque. Note the similarity of the representations in both states. *D1* to *D5*, digit representation; *1D*, interdigital pad zone; *DT*, distal thenar eminence. *CS*, central sulcus; *IPS*, intraparietal sulcus; *PCS*, postcentral sulcus; *UL*, upper lip; *LL*, lower lip; *FA*, forearm. (**a** Modified from McKenna et al. 1982, p. 307; **b** from Pons et al. 1985a, p. 464)

provided evidence for the segregation of neurons in the VPL into rapidly adapting and slowly adapting groups, thereby providing evidence that there is, in the VPL, two representations of the body with respect to each of these inputs.

Intracortical Connectivity. Small injections of tracers into areas 3a, 3b, 1, 2, 5, 4 and 6 have been used to study the intracortical connections in monkeys (Jones et al. 1978). Two types of connections have been found: (1) non-specific projections from all areas to layers I, III, V and VI which show no regard for direction or cytoarchitectonic boundaries, and (2) axons which enter the white matter and re-enter specific cytoarchitectonic areas to terminate in layers I and IV. Via the latter group of connections, area 3b projects to areas 2 and 1 while area 1 is reciprocally connected to 3a and 2 and to 4. Areas 3a, 4, 1, 2 and 5, but not area 3b, project to area 6 which itself projects to areas 3a, 4 and 5. In terms of somatotopy the authors state that on the assumption that one area projects to homotypical parts of other areas, then the results imply separate representations in 3b, 1–2, 5, 3a, 4 and 6 (Jones et al. 1978).

7.2.3 Responses of Cortical Neurons to Natural Stimulation

The connections between thalamic neurons and areas 3a, 3b, 1 and 2 together with the specific intracortical connections are quite compatible with the stimulus-response properties of neurons in these areas in both anaesthetized and unanaesthetized monkeys. Difficulties, however, often arise in comparing different experiments owing to a lack of precise data concerning the localization of cells to particular cortical layers, the standardization of the stimuli used and the position of a given track with respect to the overall somatotopic organization of the surrounding cortex. The general picture which emerges for cells in the "middle" cortical layers (III and IV) is as follows. Most cells in area 3b have simple homogeneous excitatory cutaneous receptive fields (Sur 1980; Iwamura et al. 1980; Hyvarinen and Poranen 1978a, b; McKenna et al. 1982; Iwamura et al. 1985). In anaesthetized monkeys, cells with slowly adapting or rapidly adapting responses have been shown to be spatially subdivided into 200–600 μm-wide bands which cross area 3b along the antero-posterior axis (Sur et al. 1981). In contrast, cells in area 1 of anaesthetized monkeys show mainly rapidly adapting responses to cutaneous stimuli (Sect. 7.2.2) (Paul et al. 1972b; Sur 1980). The receptive fields of these cells are not homogeneous, many having centre-surround organization with separate on and off zones (Sur 1980). Although data in unanaesthetized monkeys vary somewhat there seems to be a general agreement that there are more cells in area 1 that have large receptive fields than are found in area 3b (Hyvarinen and Poranen 1978b; McKenna et al. 1982) and this has led Iwamura et al. (1985) to conclude that there is no simple somatotopic representation of fingers in area 1.

Powell and Mountcastle (1959) showed that in anaesthetized monkeys neurons in areas 3a and 2 differed from those in areas 3b and 1 in that the former were driven by stimulation of deep receptors. In particular, neurons of area 2 responded to joint movement. Subsequent studies in both anaesthetized and unanaesthetized monkeys have supported these observations. Within area 2 of both preparations there is evidence that the contribution of cutaneous versus deep receptors to the responses of a cortical neuron depends somewhat on its position in the overall somatotopic representation (McKenna et al. 1982; Pons et al. 1985a, b). Thus many cells within the palm and arm representation of area 2 receive deep input while within the digit representaion of area 2 almost all cells receive input of cutaneous origin – an observation which is doubtless related to the projection of the VPL to the hand representation in area 2 (Pons and Kaas 1985). This relationship between receptive field position and submodality was less conspicuous for areas 1 and 3a and was not at all apparent for area 3b where cutaneous input dominates.

The increasing complexity of the receptive fields of cortical neurons from those in area 3b, through area 1, to those in area 2 has led Hyvarinen and Poranen (1978b) to suggest that these areas are hierarchically organized (see also Jones 1967). This view has been supported by similar studies of Iwamura et al. (1985) who have referred to area 2 as more "associative" in function than area 3b. Finally, Costanzo and Gardner (1980) have suggested that cells in 3b extract quantitative data relating to the characteristics of a stimulus at a particular skin locus whereas the cells of area 1 and 2 extract the more complex features (e.g. velocity detection, edge detection) of a stimulus that relate to the temporal pattern of stimulation across areas of the receptor surface. The view, that the processing of somatic information occurs along the anterior to posterior axis of the postcentral gyrus, is supported by the demonstrations that (1) area 3b projects to area 1 and area 2, while area 1 is also reciprocally connected to areas 2 and 3b (Jones et al. 1978), (2) the influence of deep receptors is relayed selectively to areas 3a and 2 from the anterodorsal shell of the VPL (Fig. 6.1a, b) (Sect. 6.2.1) (Jones and Friedman 1982), and (3) the exclusively cutaneous input to the central core of the VPL is relayed selectively to cortical areas 3b and 1 with the former receiving the more intense projection (Fig. 6.1b) (Jones and Friedman 1982; Jones et al 1982; Jones 1983b).

7.2.4 Somatotopic Representation of the Hand in Areas 3b and 1

The most complete somatotopic maps of the SI cortex have been made in anaesthetized monkeys. The reasons for this are clearly (a) the difficulty, in an awake animal, of making the large number (often between 400 and 1000) of tracks needed to accurately map the representation of the body, and (b) the fact that receptive fields in these anaesthetized animals are more easily localizable to a discrete area of skin (see above). In studies of plasticity the hand representation of areas 3b and 1 have been used exclusively because the skin of the hand, particularly that of the digits, is represented over a sufficiently large cortical area to permit high-resolution mapping of the component skin surfaces – a feature which also permits the most accurate detection of induced or spontaneous changes in somatotopy (Sur et al. 1980). The accessibility of areas 3b and 1 in the squirrel and the owl monkey have made these animals the species of choice. Consequently, the hand representation has been mapped in enough animals to permit comparisons to be made between individuals of the same species. Such comparisons have shown that while, as in the raccoon (Fig. 7.20; Sect. 7.4.1) and the rat (Sect. 7.5.2), the internal topographic relationships in the maps were similar, the areas devoted to specific skin surfaces and the shapes of these areas were quite variable (Fig. 7.2) (Merzenich et al. 1987). In general, however, the glabrous skin of the hand pads and digits is represented in clearly defined, topographically ordered fashion within easily identified areas. In contrast, the hairy skin of the dorsal surface of the hand is represented in small islands of cortical tissue between the representations of associated glabrous skin (Fig. 7.2c–g) (Merzenich et al. 1983a; Merzenich et al. 1978). The representation of the glabrous skin of each digit occupies about 1 to 2 mm^2 of cortical territory, digit 1 being represented most laterally while digit 5 is represented most medially. The representation of the digits

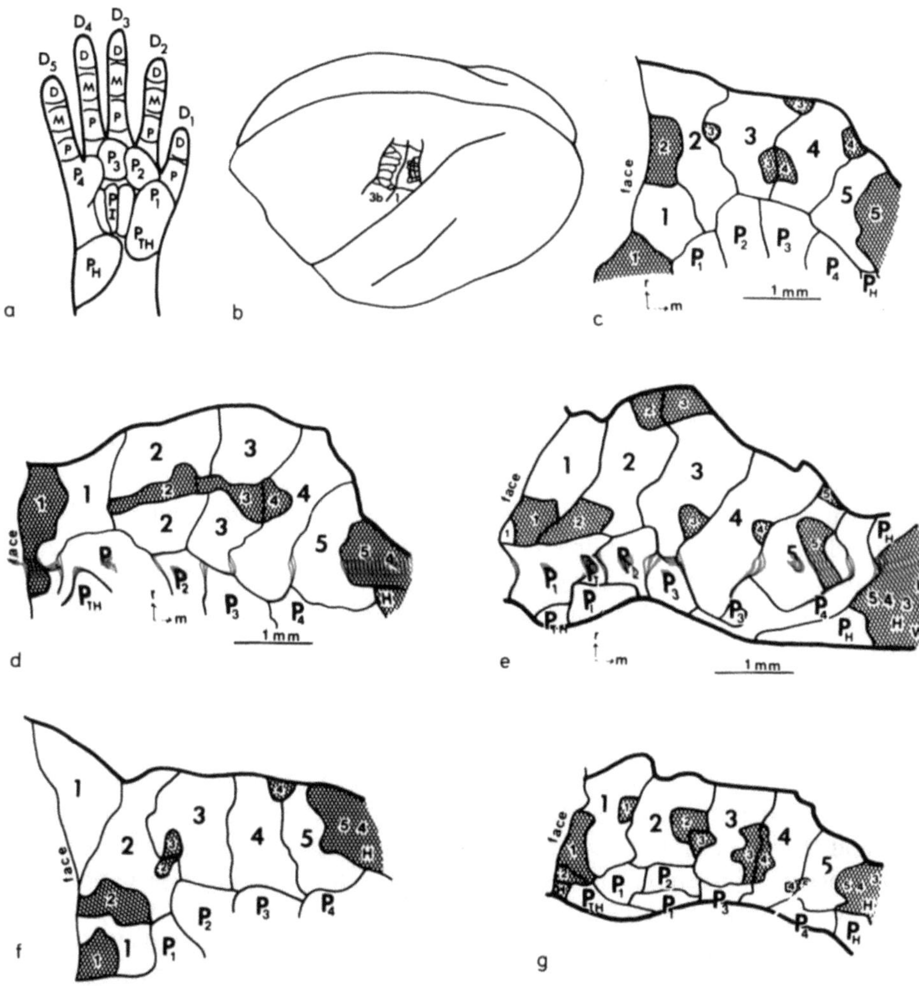

Fig. 7.2 a–g. Somatotopic representation of the skin of the hand in area 3b of five anaesthetized, owl monkeys. **a** Divisions of the skin of the hand. **b** The hand representation in areas 1 and 3b. **c–g** The inter-animal variation in the representation of the glabrous and hairy (*shaded region*) skin of various parts of the hand. It has been proposed that such variation might result from inter-animal differences in the use of various parts of the skin surface. *D1* to *D5*, digit representation; *P1* to *P4*, palmar pads; P_{TH} and P_H, thenar and hypothenar eminences; P_I, insular pads; *P*, *M* and *D*, proximal, middle and distal phalangeal sufaces; *H*, hand dorsum; *W*, wrist; *F*, face. (**a, b** From Merzenich et al. 1983a, p. 35; **c–g** from Merzenich et al. 1987, p. 285)

in area 3b is such that the digit tips are located rostrally along the boundary between areas 3a and 3b whereas in area 1 the digit tips are found along either the caudal or rostral boundary depending on the species of monkey being studied (Fig. 7.2a) (Sur et al. 1978, 1982; Kaas 1983).

7.2.5 *Anatomy and Innervation of the Monkey Hand*

The spatial terms of reference used in describing the hand of the monkey are those pertaining to the hand of man as it is described in the *anatomical position.* Thus the thumb, or digit 1 (D1), lies on the lateral aspect of the hand and the little finger or digit 5 (D5) lies on the medial aspect. The two surfaces of the hand will be referred to as the palmar surface and the dorsal surface. In both the owl monkey and the squirrel monkey the glabrous skin extends over the palmar surface of each digit and over the four pads (numbered lateral to medial as P1 to P4) at the base of the digits, the three insular pads (P_I) in the centre of the palm and the lateral thenar (P_{TH}) and medial hypothenar (P_H) eminences (Fig. 7.2a). The hairy skin covers the dorsal surface of the hand. The median nerve innervates exclusively all the glabrous skin on D1, D2, P1, P2 and P_{TH}, the lateral portion of D3, P3 and P_I. The radial nerve innervates an identically shaped area of the hairy skin which overlies the lateral half of the dorsal surface of the hand. The ulnar nerve innervates both the glabrous and hairy skin which covers the medial half from a line dividing longitudinally D3 on the dorsal surface to a similar line on the palmar surface (Fig. 7.3) (Merzenich et al. 1983). In the macaque monkey (and man) the lateral boundary of the ulnar nerve territory is along a line which bisects D4 on the dorsal and palmar surface of the hand (Hartman and Straus 1961; Paul et al. 1972a). It is important to mention that with the exception of a reference to the unpublished observations of Wall and Kaas (Wall et al. 1985), these boundaries have only been established by gross dissection of nerves and by establishing those regions of skin where, following transection of the median nerve, light tactile stimuli fail to excite cortical neurons. On the basis of these recordings in anaesthetized monkeys, these regions are referred to as "anaesthetic skin" and they are presumed to be regions which are totally denervated following median nerve transection. On the basis of the same criteria these regions are considered to remain totally denervated following the ligation of the proximal stumps of the severed nerves (Merzenich et al. 1983; Merzenich et al. 1983). It is notable that in fact there is no direct evidence that ligation of a severed nerve totally prevents regeneration (Chap. 2).

7.2.6 *Anaesthetics and the Representation of the Hand*

In unanaesthetized monkeys, cells in area 2 have much larger receptive fields than are found in anaesthetized preparations (Ducan et al. 1982; Pons et al. 1985a; Hyvarinen and Poranen 1978b; Iwamura et al. 1985). Nevertheless, the somatotopic map of the hand in area 2, as well as those in areas 3b and 1 are very similar in each preparation (compare Fig. 7.1a with Fig. 7.1b), and this has led Pons et al. (1985a) to comment that somatotopic maps in the anaesthetized monkeys probably "reflect robust, but not necessarily all, inputs".

In primates, all detailed mapping studies of the SI cortex have been carried out using ketamine or pentobarbitone anaesthesia. Both these anaesthetics require repeated supplementation and it is therefore a reasonable assumption that throughout an experiment the level of anaesthetic would vary considerably. If such fluctuations in

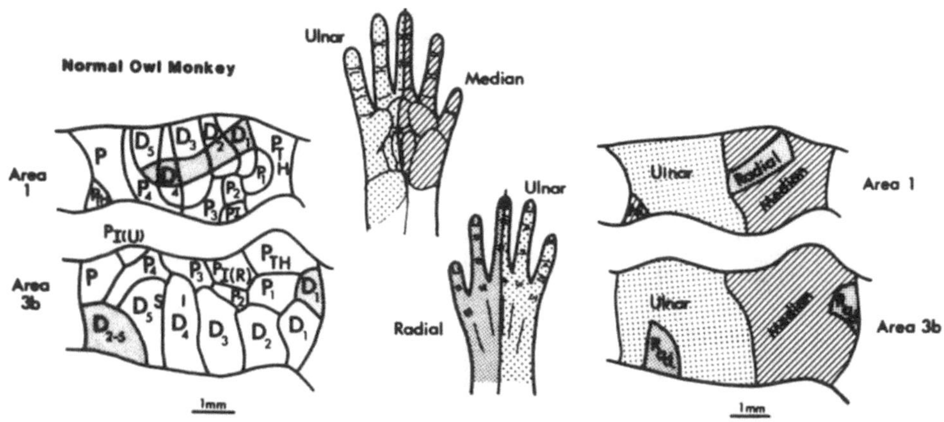

Fig. 7.3. Representation in areas 1 and 3b of the owl monkey of the skin areas of the hand that are innervated by the ulnar, median and radial nerves (*right*). These should be compared with the representations of the hairy (*stippled*) and glabrous skin of the hand (*left*). *Abbreviations* as for Fig. 7.2, except $P_{I(R)}$ and $P_{I(U)}$ refer to the representations of the parts of the insular pads that are innervated by the radial and ulnar nerves, respectively. (From Merzenich et al. 1983a, p. 36)

anaesthetic level were to produce alterations of the receptive fields of cortical neurons then the results of experiments in which a particular part of a representation is mapped on a number of occasions could be affected.

The effects of ketamine on the responses of SI neurons have been tested in initially unanaesthetized macaque monkeys (Ducan et al. 1982). This study showed that doses in excess of 7 mg/kg (far less than the 30–55 mg/kg normally used in mapping experiments) altered the frequency and pattern of spontaneous action potentials and depressed the responses of cortical neurons to mechanical stimulation within that area of skin which previously constituted their receptive field. The latter effect elevated the threshold at the edge of the receptive field so far that in some cases the "minimal receptive field" was drastically reduced in size (Fig. 7.4a). In some neurons this reduction did not occur symmetrically around the geometric centre of the pre-drug receptive field and in these cases the centre of the post-drug receptive field was thus at a different locus from the pre-drug centre. In a second study, McKenna et al. (1982) have tested the effects of nitrous oxide, methohexital, ketamine (15–23 mg/kg) and pentobarbital (2 mg/kg) on a small group of SI neurons of unanaesthetized monkeys. All these agents decreased receptive field size (Fig. 7.5).

The reduction of receptive field size by ketamine led Ducan et al. (1982) to comment that the use of this anaesthetic might have important effects on the position of some representational boundaries. In particular, they mentioned the boundaries between the representations of the hairy and glabrous skin of the hand. As the position of these boundaries are important to the conclusions of a number of studies that report map changes induced by nerve damage, it is necessary to evaluate the comment of Duncan et al. (1982) critically. There are two difficulties in accepting *a priori* their cautionary statements. First, in neither of the above studies were the

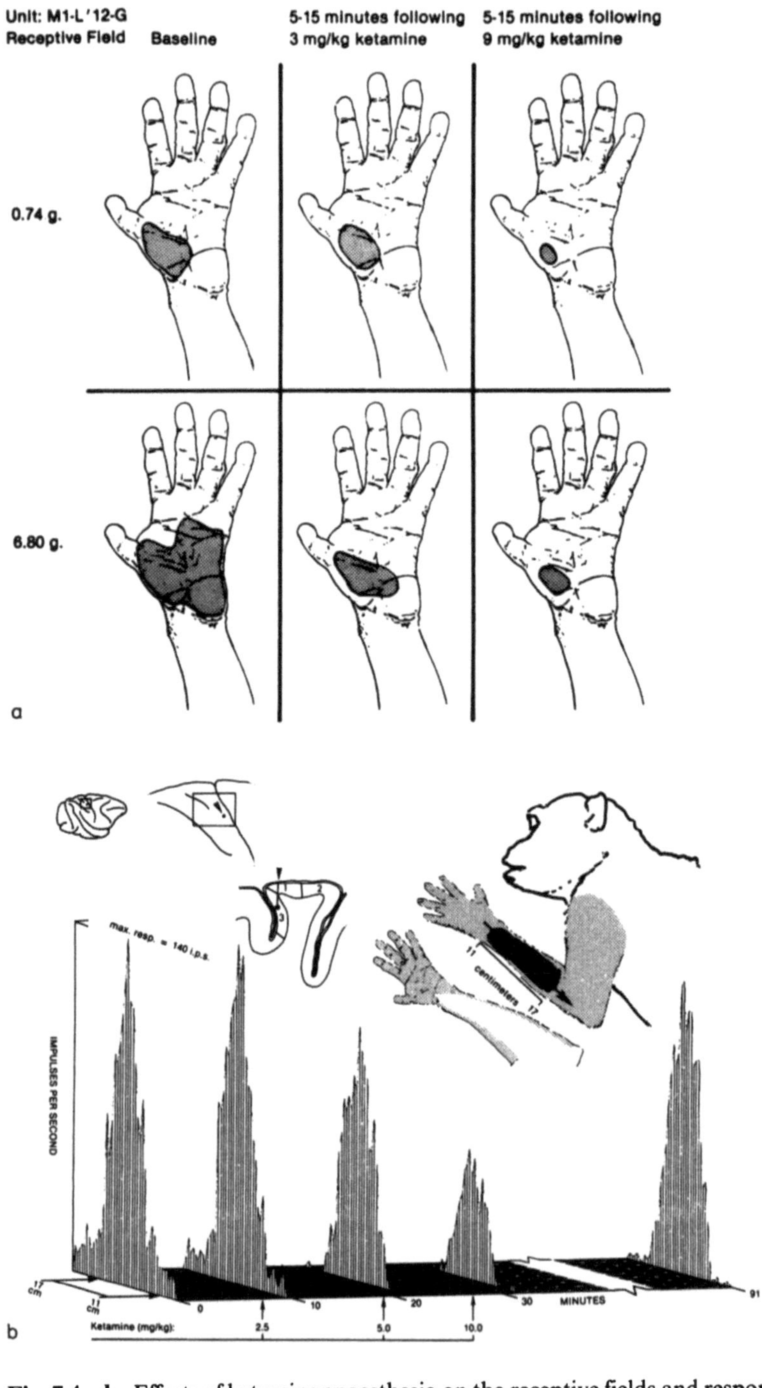

Fig. 7.4 a, b. Effects of ketamine anaesthesia on the receptive fields and responses of SI cortical neurons of the macaque. **a** Minimal (*top row*) and maximal receptive fields of a neuron in the unanaesthetized (*left*) macaque and following the two different doses of ketamine. Note the shrinkage of the minimal and maximal relative fields.

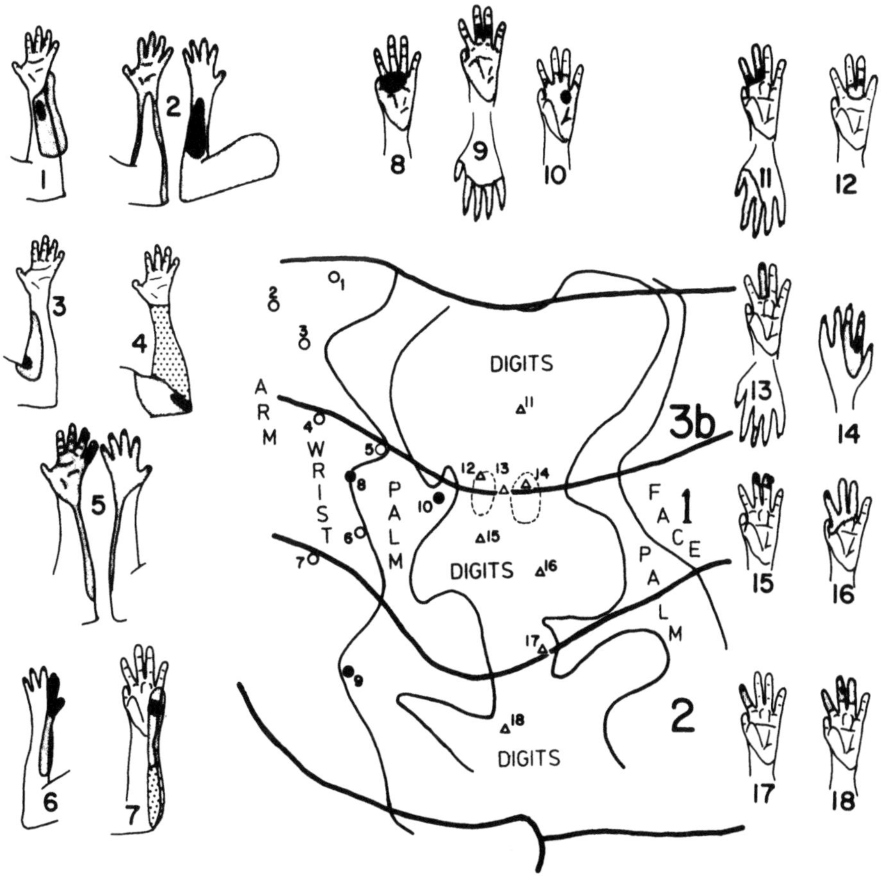

Fig. 7.5. Effects of anaesthetics on the receptive fields of single neurons at selected locations within specific cytoarchitectonic subdivisions of the forelimb representation in SI cortex of the macaque. Pre-drug receptive fields are indicated by *grey stippling*, post-drug receptive fields by *solid black* and the pre-drug inhibitory receptive fields are indicated by *zones of low-density stippling*. Recording sites were located as follows; neurons 8, 15 and 18 in layer III; neurons 2, 4, 5, 7, 9, 11, 13, 14, 16 and 17 in layer IV; neurons 1, 3, 10 and 12 in layer V. The anaesthetics used were; ketamine, 15–23 mg/kg (neurons 2,5,6,7,8,9,18); methohexital 2.5 mg/kg (neurons 3, 4, 10, 11, 12, 13, 16, 17); pentobarbital, 2 mg/kg (neuron 1); nitrous oxide, 75%, in oxygen for 5 min (neuron 14). Note that only four neurons (11, 12, 13 and 14) lie in the hand representation of area 3b (see text for further discussion). (Modified from McKenna et al. 1982, p. 310)

◀ **Fig. 7.4 a, b** (*continued*). The cytoarchitectonic area containing the recording locus and the route of ketamine introduction were not reported. **b** Post-stimulus time histograms showing the response of a rapidly adapting unit in area 3b to 25 identical brush strokes moved distal to proximal across the receptive field centre, in the unanaesthetized animal and after the intraveneous introduction of ketamine. Note the decrease in response magnitude and in the length of the effective stimulus path with the level of ketamine. Recovery from these effects was complete after 91 min. (From Duncan et al. 1982, pp. 683 and 691) effects of ketamine (or pentobarbitone) adequately tested on neurons within the hand representation in area 3b. McKenna et al. (1982) looked at the effects of methohexital and N_2O on only four such neurons, three of which were very close to the 3b–1 boundary (Fig. 7.5). Furthermore, none of these neurons had a pre- or post-drug receptive field on the distal or middle phalanx of a digit. Duncan et al. (1982) concluded that ketamine reduces the size of the receptive fields of neurons in area 3b but, with an exception (Fig. 7.4b) where moving stimuli were used, their figures show pronounced effects only on neurons recorded in areas 1 and 2 or in unspecified parts of SI. As mentioned above, in unanaesthetized monkeys, cells in areas 1 and 2 have much larger and more complex receptive fields than cells in areas 1 and 2 of anaesthetized monkeys. Thus, the effects illustrated by Duncan et al. (1982) in these areas are, perhaps, not surprising. Second, in the study of Duncan et al. (1982) ketamine was sometimes given intravenously and sometimes intramuscularly. These different routes could be expected to result in significantly different concentration/time profiles within the CNS and are not strictly comparable with mapping experiments where supplementary ketamine was given regularly via a specific route. In conclusion, it does seem as though anaesthetics have an important effect on the size of the receptive fields of cells in areas 1 and 2 but that there is little evidence for such effects on neurons within the finger representation of area 3b.

The effects of anaesthetics on neurons in the finger representation of area 3b have recently been explored in two owl monkeys (Stryker et al. 1987). Many electrode tracks were made at specific cortical sites across the hand representation in one monkey which was paralysed and artificially respired with a mixture of N_2O and O_2. This animal was then anaesthetized with pentobarbitone (22 mg/kg i.v. 5 mg/kg/h suppl., i.v.) and, in the absence of N_2O and the paralysing agent, the hand representation was remapped by making tracks close to those made in the first map (Fig. 7.6a, b). In the second monkey a line of tracks was made across the digit representation initially in the N_2O/O_2-paralysed state, the ketamine-anaesthetized, N_2O free, unparalysed state (50 mg/kg i.m. suppl. 5 mg/kg/0.5 h), and again following a second induction dose of ketamine (50 mg/kg i.m., suppl. 10 mg/kg/0.5 h) (Fig. 7.6c). No statistically significant change in receptive field size was detected between monkeys in the N_2O/O_2-paralysed, pentobarbitone- (Fig. 7.6b) or ketamine-anaesthetized states. Furthermore, within the limits of track placement there seemed to be little evidence for shifts in the centres of receptive fields between the various states (Fig. 7.6c). It therefore seems that in the middle cortical layers, within the hand and digit representation of area 3b in the monkey, the receptive fields of at least some cells may be particularly insensitive to ketamine and pentobarbitone anaesthesia. Consequently, under these anaesthetics, the somatotopic representation of the body in the responses of some layer III and IV neurons is resistant to variation in the level of anaesthesia. It is presumed in this study that maps derived in the presence of N_2O and a paralysing agent will resemble closely maps derived from awake, alert monkeys. However, it should be emphasized that when confronted with anaesthetic agents, the receptive fields of neurons in area 1 or the arm representation of area 3b are not so robust as those of many neurons in the hand and digit representation of area 3b (Duncan et al. 1982; McKenna et al. 1982).

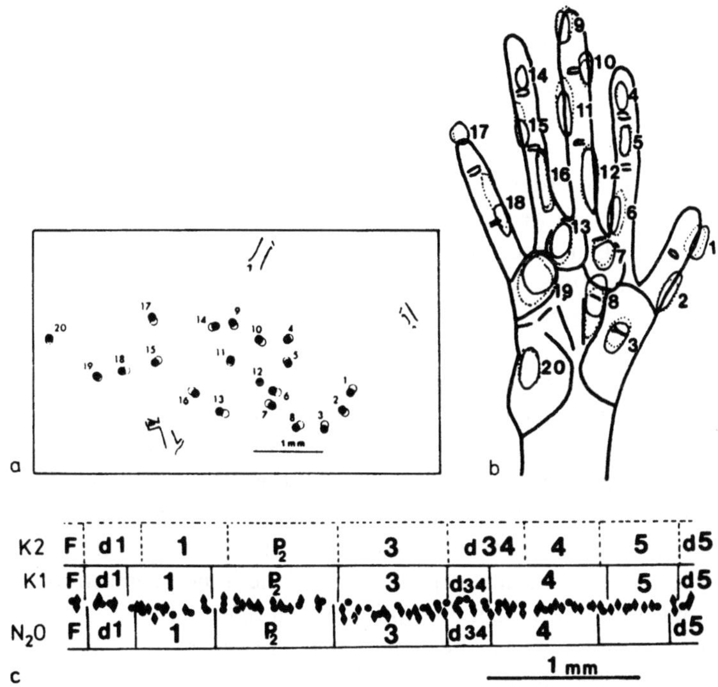

Fig. 7.6 a–c. Effects of anaesthetics on the receptive fields of neurons in the hand representation in area 3b of SI cortex of the owl monkey. **a, b** Comparison of the receptive fields from nearby recording sites in an owl monkey anaesthetized with nitrous oxide (*filled circles*, **a**) and in the same animal anaesthetized with sodium pentobarbitone (*open circles*, **a**). The number next to a recording site in **a** indicates which of the numbered receptive fields in **b** was represented at that site. Note that the receptive fields defined under nitrous oxide (75% in oxygen) (*solid lines*, **b**) are very similar to those defined at a nearby site under pentobarbitone (22 mg/kg, plus 5 mg/kg/h) (*dotted lines*, **b**). **c** Sites of three closely overlapping rows of microelectrode penetrations across the hand representation in area 3b showing the region of the hand represented along the row. One row (*filled circles*) was performed in an owl monkey anaesthetized first with nitrous oxide (75% in oxygen), the second row (*filled diamonds*) was performed under a normal level of ketamine anaesthesia (50 mg/kg i.m. plus 5 mg/kg/30 min) and the third row (*open diamonds*) was performed after the addition of a second dose of ketamine (50 mg/kg i.m.) and increased supplementary doses (10 mg/kg/30 min). The numbers in **c** indicate the digit on which receptive fields at any site were located and *d* indicates places where receptive fields were on the dorsal hairy skin of a digit. Similarly, *F* indicates the face representation and *P2* the representation of the second palmar (or interdigital) pad. Note that even at the very high dose of ketamine there is little change in boundaries between the representation of particular skin areas. (From Stryker et al. 1987, pp. 299, 200 and 300)

7.2.7 Injury and Subsequent Regeneration of Peripheral Nerves

In humans, peripheral nerve injury is followed either by a persistent inability to correctly localize cutaneous stimuli or by recovery of normal sensory function (Sect. 2.3). These different outcomes appear to relate to whether nerves are transected or crushed. Such observations have raised the question of whether regeneration of peripheral connections after nerve section, but not nerve crush, results in a disruption of cortical somatotopy.

Effects of Section, Repair and Subsequent Regeneration of Nerves in Adult Primates. In order to determine the effects of nerve injury on cortical somatotopy, areas 1 and 3b were mapped in anaesthetized macaque monkeys 6.5 to 10 months after the ulnar, median and radial nerves had been sectioned and repaired (Paul et al. 1972b). In these animals the normal somatotopic organization was disrupted (Fig. 7.7) and there was also some loss of modality-specific inputs to individual cortical columns (Mountcastle, 1957). In addition, a substantial number of tracks contained units with multiple receptive fields (Fig. 7.7a, b). At least some of these tracks contained clusters of units with identical multiple receptive fields, suggesting, on the basis of point-to-point connections between the thalamus and the cortex, that thalamic neurons or neurons lower along the neuraxis might also have multiple receptive fields. In general, these studies supported the contention that the psychophysical phenomenon of false localization in man results from a disruption of orderly topographical organization of the hand representation across the primary somatosensory area, a disruption which reflects the disorderly regrowth of severed primary afferents to the denervated area of skin (Fig. 8.4) (Sects. 2.5 and 8.4.3) (Trotter and Davies 1909; Penfield and Boldrey 1937; Ford and Woodhall 1938; Penfield and Ramussen 1950; Wall and Kaas 1985).

Recently, the effects on the hand representation in areas 3b and 1 of transection, repair and subsequent regeneration of the median nerve have been tested in three owl monkeys 76 to 322 days after insult (Fig. 7.8a, c & d) (Wall et al. 1986b). As previously reported in the macaque monkey (Paul et al. 1972b) many (19% in area 1 and 65% in area 3b) neurons within the cortical representation of the regenerated nerve developed multiple receptive fields. In area 3b, up to five separate patches of skin made up these multiple receptive fields while in area 1 more than five components were sometimes found. At 65% of the recording sites in area 1, and 25% in area 3b, cells had a single receptive field which often overlapped with the components of the multiple receptive fields of other cells. Both the single receptive fields, and the component receptive fields of the multiple receptive fields were smaller in area than the cortical receptive fields found on that skin area in normal animals – an observation which suggests that both may be formed by a few or even one regenerated primary afferent fibres. Finally, at 3% to 4% of sites in areas 1 and 3b "pacinian-like" responses, such as are not seen in area 3b of normal owl monkeys, were recorded. In agreement with previous studies in the macaque monkey (Paul et al. 1972a), the authors reported that throughout any track made perpendicular to the cortical layers at a point where multiple receptive fields were recorded, all recorded units were excited by a "relatively constant" combination of component receptive fields. This emphasizes further that disruption of the normal system occurs at a level of the neuraxis below that responsible for the cortical columns as described by Mountcastle (1957) (see also Jones et al. 1986).

Section of the median nerve results in an expansion of the representation of the hairy skin into much but not all of the median nerve representation (Figs. 7.8, 7.10b and

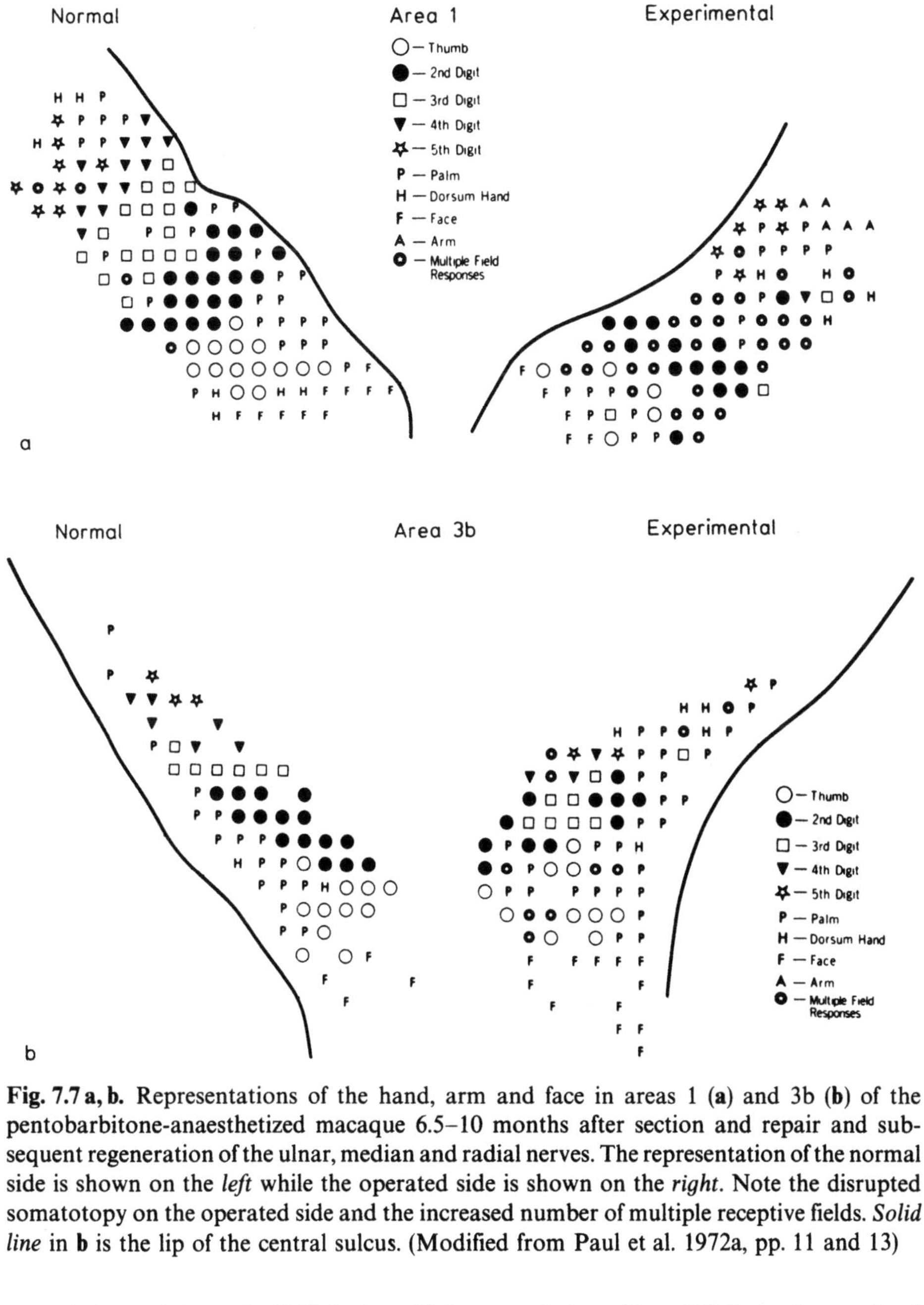

Fig. 7.7 a, b. Representations of the hand, arm and face in areas 1 (**a**) and 3b (**b**) of the pentobarbitone-anaesthetized macaque 6.5–10 months after section and repair and subsequent regeneration of the ulnar, median and radial nerves. The representation of the normal side is shown on the *left* while the operated side is shown on the *right*. Note the disrupted somatotopy on the operated side and the increased number of multiple receptive fields. *Solid line* in **b** is the lip of the central sulcus. (Modified from Paul et al. 1972a, pp. 11 and 13)

7.11) (Merzenich et al. 1983a). As will be seen below (Sect. 7.2.8), in the squirrel monkey, much of this expansion occurs immediately and may be maintained by prevention of median nerve regeneration. If, however, the median nerve is repaired and allowed to regenerate, the reinnervated skin is represented in a topographically disorganized manner. In the owl monkey the representation of the reinnervated skin was still abnormal 322 days after median nerve repair (Fig. 7.8a–d). Thus, whereas

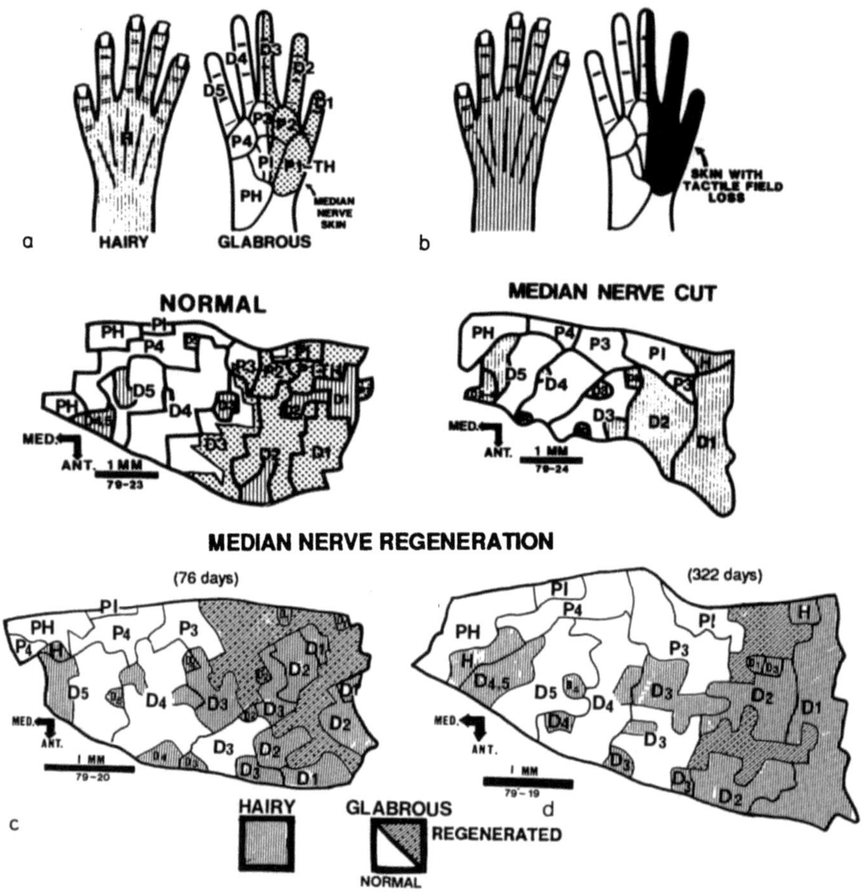

Fig. 7.8 a–d. Representation of the hand in area 3b of ketamine-anaesthetized owl monkeys following section, repair and subsequent regeneration of the median nerve. **a** Representation of the anatomical parts of the hand as well as the regions supplied by the median (*stippling*), radial (*lined*) and ulnar nerves. **b** Representation of the hand in a different owl monkey some time after section and ligation of the median nerve. In the deprived median nerve representation (lateral) there is an enlarged representation of hairy skin of digits 1 and 2 – skin that is supplied by the radial nerve. **c, d** Representation of the hand in two monkeys 76 and 322 days after section, repair and subsequent regeneration of the median nerve. There is a clear representation of the skin that has been reinnervated by the median nerve (*shaded stippling*). Note that in none of these maps is there a region of cortex that is unresponsive to cutaneous input and note also the actual dimensions of particular skin areas. Abbreviations as in Fig. 7.2. (Modified from Wall et al. 1986b, pp. 219 and 221)

the glabrous skin supplied by the median nerve usually occupies an almost continuous block in the lateral half of the hand representation, after repair and regeneration it was represented in one or more relatively small patch- or strip-like areas. This confinement of the median nerve representation is accompanied by an expansion of the representation of the hairy skin on the dorsal surface of D1, D2 and, to some extent, D3 (Fig. 7c, d) (Wall et al. 1986b). On the assumption that cortical territory, even in adulthood, appears to be allocated on the basis of numbers of functionally active primary afferents (Sects. 3.3.3, 3.4 and 7.3.5), these findings are perhaps not surprising. Using minimum regeneration rates of peripheral nerve of 1 mm/day (Sunderland 1978) it would take 60–70 days for the median nerve in the monkey to reinnervate the digits. Up to this length of time is thus available for the establishment of the expanded cortical representation of surviving afferent fibres. In addition, it is likely that not all median nerve fibres survive transection (Sect. 2.2) and so even after regeneration it is possible that the median nerve contains fewer functional afferent fibres to be represented at the cortical level.

Examination of the detailed somatotopy within the representation of the repaired and regenerated median nerve revealed that reinnervated skin regions were represented at inappropriate locations (Wall et al. 1986b). However, some patches of reinnervated skin were represented in a topographically ordered manner. While it is possible that this reflects mechanisms within the CNS that function to restore topographic representation of the body surface, it is just as likely that some severed afferents that originally had neighbouring receptive fields preferentially reinnervate neighbouring patches of denervated skin. Thus the degree of topographic orderliness in the emergent map might depend on the precision of the procedure of nerve repair (Sect. 2.5).

Effects of Neonatal Nerve Section, Repair and Subsequent Regeneration. Finally the effects of section and repair of the median nerve in the neonatal (7-day-old) marmoset monkey have recently been described in an abstract (Wall et al. 1986a). One year after nerve injury, area 3b was mapped under ketamine anaesthesia. The representation of the regenerated median nerve was found to be in a normal position with respect to the representation of the ulnar nerve but the normal orderly topographic representation of the skin within the peripheral territory of the median nerve was disrupted. Furthermore, neurons at some cortical sites had abnormally large receptive fields or multiple receptive fields or were unresponsive to cutaneous stimuli. These observations are remniscent of those made following section and repair of this nerve in the adult animal (Wall et al. 1986b). Like the report that the effects on cortical maps of nerve section and ligation in adult and neonatal monkeys are similar (Sect. 7.2.8), these results are surprising in view of a preliminary report that the projections of thalamocortical neurons are more widespread in macaque monkeys of less than 6 months of age than those in older monkeys (Cheema et al. 1986). However, experiments on the rodent vibrissae system (Sect. 3.3.2) have indicated that the periphery is critical to development of cortical maps only at a much earlier stage of embryonic development than that present in the newborn primate.

Effects of Nerve Crush and Subsequent Regeneration. The effects of crushing the median nerve on the somatotopic organization of neurons in the hand representation of area 3b have been studied in five anaesthetized owl monkeys 21 to 568 days after nerve crush (Wall et al. 1983). Data presented show that after 142 days there was good recovery of the normal representation (Fig. 7.9a, b). Tracks made at a

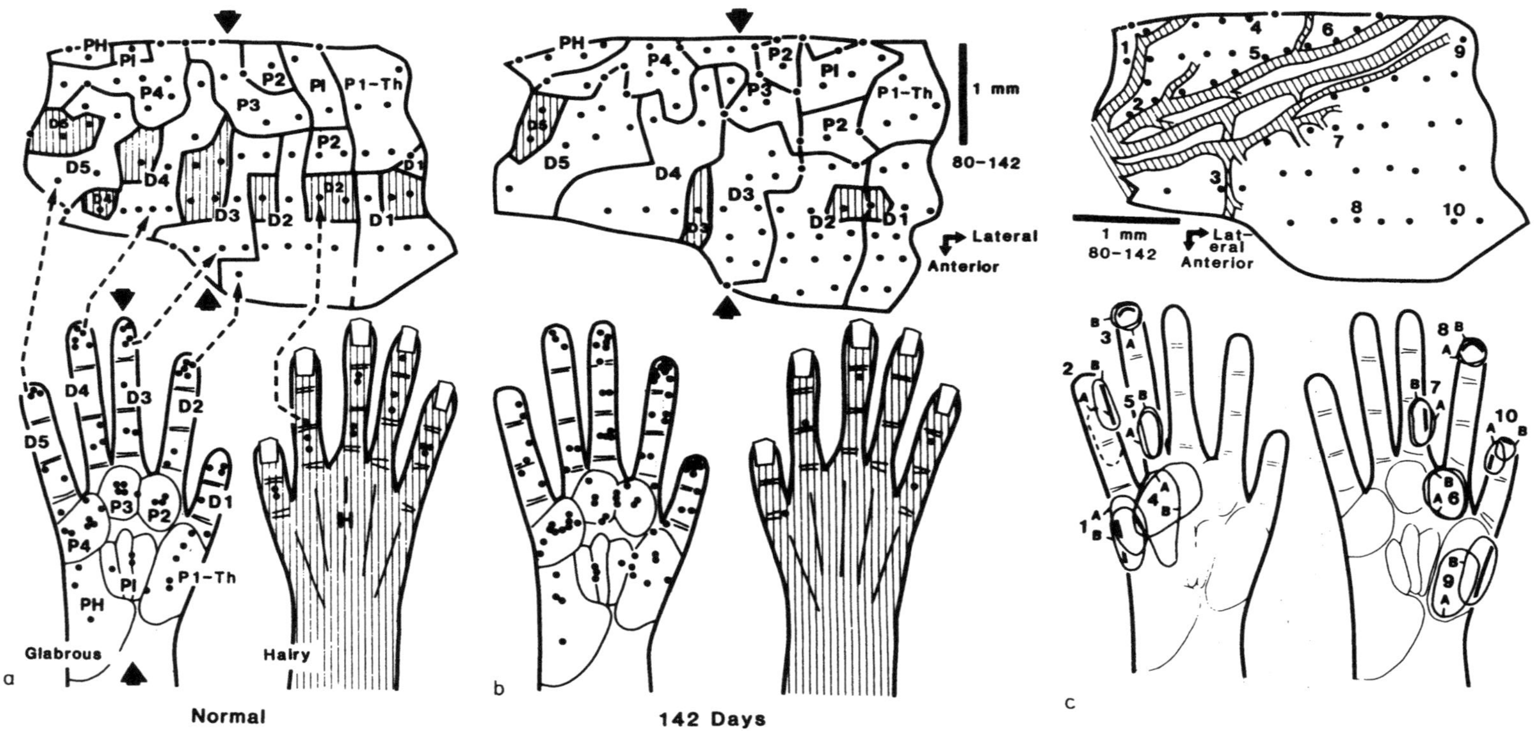

Fig. 7.9 a–c. Representation of the hand in area 3b of the ketamine-anaesthetized owl monkey 142 days after crush and subsequent regeneration of the median nerve. **a** Representation of the hand, recording sites (*dots*) and receptive field centres (*dots*) in a normal individual. *Arrows* indicate the boundaries between the peripheral territory and the central representation of the median and ulnar nerves. **b** Representation of the hand, recording sites (*dots*) and receptive field centres (*dots*) in the same individual 142 days after median nerve crush. Note the recovery of a clear somatotopic representation of the skin within the median nerve territory. **c** Comparison of receptive fields recorded at particular cortical loci (see corresponding numbers) in a normal monkey (*A*) and at approximately the same loci in that individual 142 days after nerve crush (*B*). Note the similarity of the normal and post-regeneration receptive fields. (Modified from Wall et al. 1983, pp. 771 and 772)

particular locus (estimated position error = 50–300 μm), before, and 142 days after, nerve crush yielded almost identical receptive fields (Fig. 7.9c). At this survival time it is most likely that the restoration of somatotopy results from peripheral rather than central phenomena for, as mentioned elsewhere, (Sect. 2.5) it has been shown that the original afferent innervation pattern of skin is restored after nerve crush but not after nerve transection.

At shorter post-crush survival times some evidence was found for a contribution of central mechanisms to the existing representation. Thus in a map made only 32 days after nerve crush two small areas (total size approximately 1 mm^2) of cortex in the centre of the median nerve representation were unresponsive to tactile stimuli (Wall et al. 1983). Around these areas the hairy skin of the hand occupied an expanded representation much of which had presumably been "unmasked" by nerve crush. The fact that after regeneration of a crushed nerve such expanded representations of normally innervated skin are reduced to normal size whereas after repair and regeneration of a sectioned nerve they remain large (Wall et al. 1986b) perhaps reflects both the greater topographically specific regeneration of afferent fibres following nerve crush compared with transection and repair, and the greater survival of afferent fibres following nerve crush. The expansion and subsequent contraction of the representation of the hairy skin following, respectively, crush and regeneration of the median nerve, can only be explained by alterations in central circuitry presumably in response to nerve injury and regeneration (Wall and Kaas 1987).

Long-Term Effects of Cross Anastomoses of Nerves. In an attempt to discover the long-term effects on cortical somatotopy of forcing a sectioned peripheral nerve to reinnervate foreign skin, Wall and Kaas (1986) cut the ulnar and median nerves of two owl monkeys and reconnected the central stump of the ulnar nerve to the distal stump of the median nerve. To prevent regeneration of the median nerve, the central stump was embedded into muscles. Specifically, the question being addressed was whether, after a long period of time, the skin normally innervated by the median nerve would be represented in the medially positioned, ulnar nerve representation or in the laterally positioned, median nerve representation. The former would be expected on the basis of the new peripheral connections of the ulnar nerve but should the latter be the case, then massive reorganization of central connections would be implicated. Maps in area 3b, 2.6 to 2.9 years postoperatively, showed that cells in the lateral median nerve representation responded only to the hairy skin within the innervation territory of the radial nerve. Medially within the ulnar nerve representation cells had receptive fields on skin normally innervated by the median nerve – skin which dissection showed was now innervated by the reconnected ulnar nerve. As in the other nerve section and repair experiments, described above, there was no topographically ordered representation of reinnervated skin. While all these experiments provide a basis for persistent mislocalization of tactile stimuli following nerve section and repair the authors point out that observations in humans suggest that sensory mislocalization may be corrected by training and go on to suggest that perhaps central areas other than 3b may have the neural machinery necessary to extract useful spatial information from a disrupted periphery (Wall and Kaas 1986). It seems likely that mechanisms of directed attention which have been shown to influence activity in the postcentral gyrus of monkeys and man (Hyvarinen et al. 1980; Roland 1981) play a vital role in the psychophysics of sensory retraining.

7.2.8 Section and Ligation of Peripheral Nerves

The effects on the hand representation in areas 3b and 1 of sectioning and ligating the median nerve of owl and squirrel monkeys have been studied under ketamine anaesthesia (33 mg/kg i.m. suppl. 1–4 mg/h) immediately after section and at times up to 273 days (Merzenich et al. 1983a, b).

Immediate Effects of Median Nerve Section. In one owl monkey area 3b was mapped immediately before and a few hours after median nerve section. Neurons within the cortical representation of the median nerve were found to be generally unresponsive to cutaneous stimuli (Fig. 7.10a) (Merzenich 1983b). Similar recordings in areas 3b and 1 of three squirrel monkeys showed that within a few hours of sectioning the median nerve much but not all of the median nerve representation was unresponsive to cutaneous stimulation anywhere on the hand. However, in the squirrel monkey some parts of the median nerve representation contained neurons that *were* clearly responsive to cutaneous stimulation. The tracks in which many of these neurons were encountered were generally grouped in cortical regions adjacent to the original representations of the radial nerve. Furthermore the skin to which these neurons responded was that normally supplied by the radial nerve. In area 3b these new "unmasked" receptive fields were on the hairy skin of the dorsum of D1, D2 and D3 of the same part of the digit as that which bore the glabrous receptive fields of neurons recorded at that site prior to median nerve section. In area 1 the new receptive fields were on the dorsum of D1 and D2. It thus seems that in the squirrel monkey the immediate expression of new inputs is restricted to inputs from the skin bordering the denervated skin. That is to say, denervation reveals repressed connections between a point on the skin and a point in the cortex that is very close to the normal representation of that point. What is meant by "very close" is critical to any assessment of plasticity within this system. Figure 7.10b shows that in area 3b the hairy skin of D2 is represented up to about 600 μm from the edge of its cortical representation prior to nerve section. The apparent expansion of the ulnar nerve representations within areas 1 and 3b involves substantially smaller distances.

Time-Course of Changes Following Nerve Section. The time-course of changes in a single individual following nerve section and ligation has been studied. In one squirrel monkey the hand representations in areas 1 and 3b were mapped five times; immediately before, immediately after and 11, 22 and 144 days after median nerve section. In area 3b large zones in which neurons were unresponsive to cutaneous stimuli were found immediately after, and 11 days after nerve section but not after 22 and 144 days (Fig. 7.11). The reduction in size of these "silent" zones occurred as a

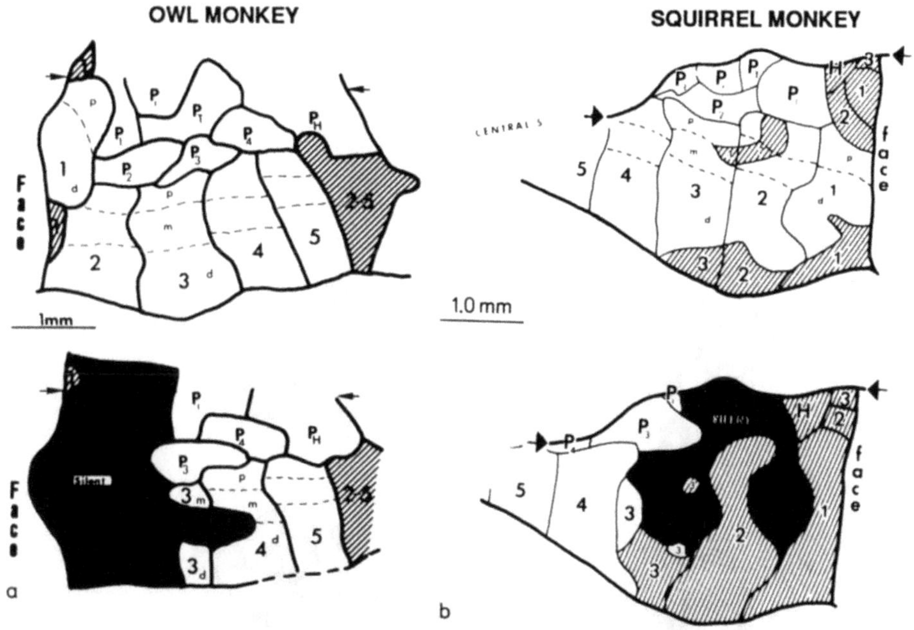

Fig. 7.10 a, b. Representation of the hand in area 3b before (*top*) and immediately after (*bottom*) section of the median nerve. **a** After nerve section in the ketamine-anaesthetized owl monkey, neurons throughout the deprived median nerve representation are unresponsive (*black*) to cutaneous stimulation. **b** After nerve section in the ketamine-anaesthetized squirrel monkey, neurons throughout a large portion of the deprived median nerve representation *immediately* become responsive to skin innervated by the radial nerve (*hatching*) but neurons in an equally large area of the deprived region become unresponsive (*black*) to cutaneous stimulation. *H*, dorsal surface of the hand; others as in Fig. 7.2. (Modified from Merzenich et al. 1983b, pp. 644 and 645)

result of an ever-increasing expansion of the representation of the hairy skin and that glabrous skin (D3, P_I and P3) which borders the median nerve territory and is innervated by the ulnar nerve. In this animal, the time-course of reorganization seen in area 1 resembled that seen in area 3b. However, changes in area 1 differed from those in area 3b with respect to the source of new inputs to the region of the former median nerve representation. Thus even 144 days after median nerve section the overall size of the representation of hairy skin in area 1 was little different from that seen in normal animals and most of the median nerve representation responded to glabrous skin (P3, P_I and P_H) that was innervated by the ulnar nerve. Particularly marked in area 1 was the *de novo* establishment of a second representation of the skin of the hypothenar pad (P_H) about 2 mm from its original, and maintained representation medial of the D1 representation (Fig. 7.11) (Merzenich et al. 1983b). Finally, although there is little evidence for the appearance of new representations immediately after median nerve section in the owl monkey, Merzenich et al. (1983b) claim that a map made 30 days after nerve section showed changes resembling those seen at 22 days in the squirrel monkey.

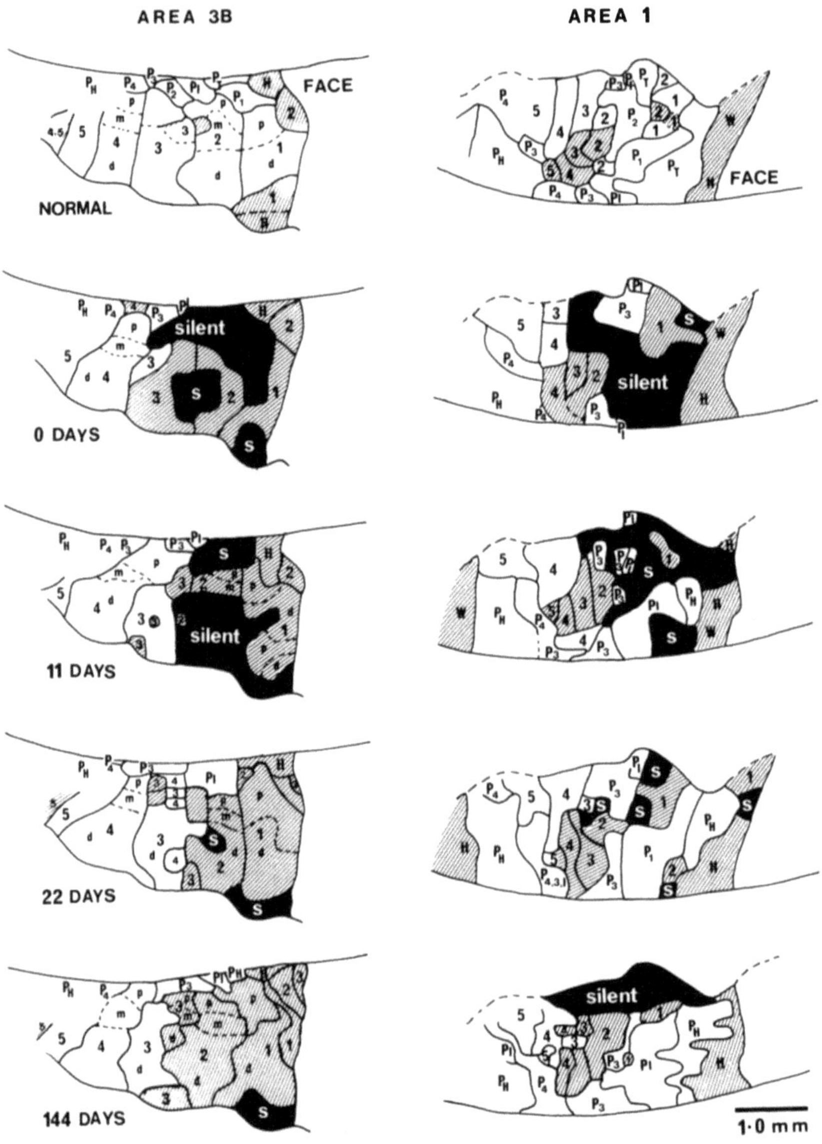

Fig. 7.11. Somatotopic organization of the hand representation in areas 1 and 3b of the ketamine-anaesthetized squirrel monkey, before and at various times after section and ligation of the median nerve. Note the progressive expansion of the representation of hairy skin (*hatching*) of the dorsal surface of the hand (*H*), wrist (*W*) and the digits and the progressive reduction in the areas that are unresponsive to peripheral stimulation (*black*). Other abbreviations as in Fig. 7.2. (Modified from Merzenich et al. 1983b, p. 648)

Within area 3b of the squirrel monkey, 0–11 days after median nerve section, neurons with new receptive fields on the dorsum of D1 and D2 were responsive only to stimulation of regions proximal to the nail beds. Over the subsequent 133 days the representation of these digits expanded and a topographically ordered representation of the nail beds developed in the appropriate region of the new representation. Over this period the receptive fields on the dorsum of D1 and D2 decreased in size by a factor of three. The emergent picture is thus one in which the once-confined representation of the skin on the dorsum of the digits within area 3b expands into a coarse representation of the hairy skin which is subsequently refined in a way which would best serve accurate localization of stimuli (Merzenich et al. 1983b).

Effects of Nerve Section on the Boundaries Between the Representations of Intact Nerves. The repetitive mapping of the entire hand representation in areas 1 and 3b after median nerve transection shows that not only do changes occur within the median nerve representation but that they also occur across the boundaries between the representations of intact nerves (Fig. 7.11) (Merzenich et al. 1983b). Merzenich reports that 144 days after median nerve section the representation of the face (a trigeminal projection) is shifted "hundreds" of microns laterally by the development of new (spinal) inputs from chest and neck (Fig. 14 from Merzenich et al. 1983b). It is important to note that in the monkey quantitative changes in the position of the boundaries that are illustrated within the ulnar nerve representation are small (maximum few 100 μm at a few points) and that in the absence of information on the exact position of individual electrode tracks and recording loci, the accuracy of these boundaries is difficult to evaluate. This is especially so if laminar difference in the size of the receptive fields of cortical neurons is considered (Chapin 1986). Even if further experiments confirm these preliminary observations it would still be difficult to accept the contention of Merzenich et al. (1983b) that such changes are related to median nerve section. First, it is possible that these changes would have occurred anyway. To our knowledge, repeated maps from normal individual animals have not been published. Repeated maps have been made in monkeys before and after crush of the median nerve and finger amputation but in available published illustrations the track densities are insufficient to permit accurate assessment of the somatotopic boundaries within the ulnar nerve representation (Wall et al. 1983; Merzenich et al. 1984). Second, as pointed out above, in the macaque the receptive fields of neurons in area 1 are very sensitive to ketamine – a feature which could affect the representational boundaries between experiments or over the course of a long experiment. Third, the resistance to ketamine of the receptive fields of neurons in area 3b has only been established in the owl monkey (Stryker et al. 1987).

Effects of Nerve Section in Neonatal Primates. Recently a comparison of the effects of median nerve section and ligation in the adult and the neonatal (7- to 14-day-old) marmoset monkey have been reported in an abstract (Huerta et al. 1986). Adult animals from both groups were mapped under ketamine anaesthesia at an unspecified survival time. In both groups, cells within the median nerve representation became responsive to ulnar and radial nerve inputs. Like those experiments involving median nerve repair in neonatal and adult monkeys (Sect. 7.2.6) (Wall et al. 1986) no differences were reported between adult and neonatal operated animals. The authors conclude that the central circuits responsible for alteration in cortical somatotopy are developed to "some degree" by postnatal week 1–2.

Effects of Digit Amputation in Adult Primates. All the above nerve ligation experiments involved the median nerve and spared the ulnar and radial nerves. Normal innervation of the dorsal surface of the hand was thus maintained following median nerve section and it was this dorsal skin which most clearly established an excitatory influence over the representation of the deafferented skin. The question of what map changes would occur if both the hairy and glabrous skin were deafferented has been tested by the amputation of one or two fingers in the owl monkey (Merzenich et al. 1984). Data were derived from four monkeys in which area 3b was mapped under pentobarbitone anaesthesia (dose not given), 1 to 8.5 months after amputation of both D2 and D3, or D2 alone. One monkey was mapped before and 62 days after amputation of D3 and another was mapped 2 months and again 8.5 months after amputation of D2 and D3. Severed digital nerves were ligated to prevent regeneration.

Remapping area 3b 62 days after the amputation of D3 showed that the cells within the D3 representation had become responsive to glabrous skin on the tips of D2 and D4 and on the palmar pads P2 and P3 at the base of D3 (Fig. 7.12a–c). In a second animal mapped 71 days after amputation of D3 a large (approximately 1 mm^2) silent zone was observed. The effects of removing D2 and D3 were assessed in 2 monkeys mapped 64 or 76 days after amputation. Maps of the hand representation showed what appears to be enlarged representations of D1 and D4 and the stump of D3 (Fig. 7.12d). In one case the representation of P2 also appeared to be enlarged while in the other the stump of D2 appeared to have an enlarged representation. A second map made in the latter monkey 6.5 months later revealed a further expansion of the representation of the stump of D2 into an abnormally placed part of the D4 representation and a contraction of the silent zone. Although no pre-amputation map was made of these two-digit amputees, comparison of post-amputation maps (Fig. 7.12d) (Merzenich et al. 1984) with a map in a normal animal does suggest that the expanded representation is into areas once dominated by inputs from the missing fingers.

Across regions that were confirmed (1 monkey) or, on the basis of map topography, suspected (3 monkeys) to have adopted new cutaneous inputs, the skin appeared to be represented in a topographically ordered manner. There was, however, a bias for these new representations to involve skin on the distal parts of the digits or skin near, or on the stump of excised fingers (Fig. 7.12c). Thus although the receptive fields in the regions were reported to be smaller than normal, their confinement to a relatively small region of the digit resulted in considerable receptive field overlap (Merzenich et al. 1984). It is important to note that once again the maximum shift in the representation of skin with intact innervation was between 500 and 700 μm.

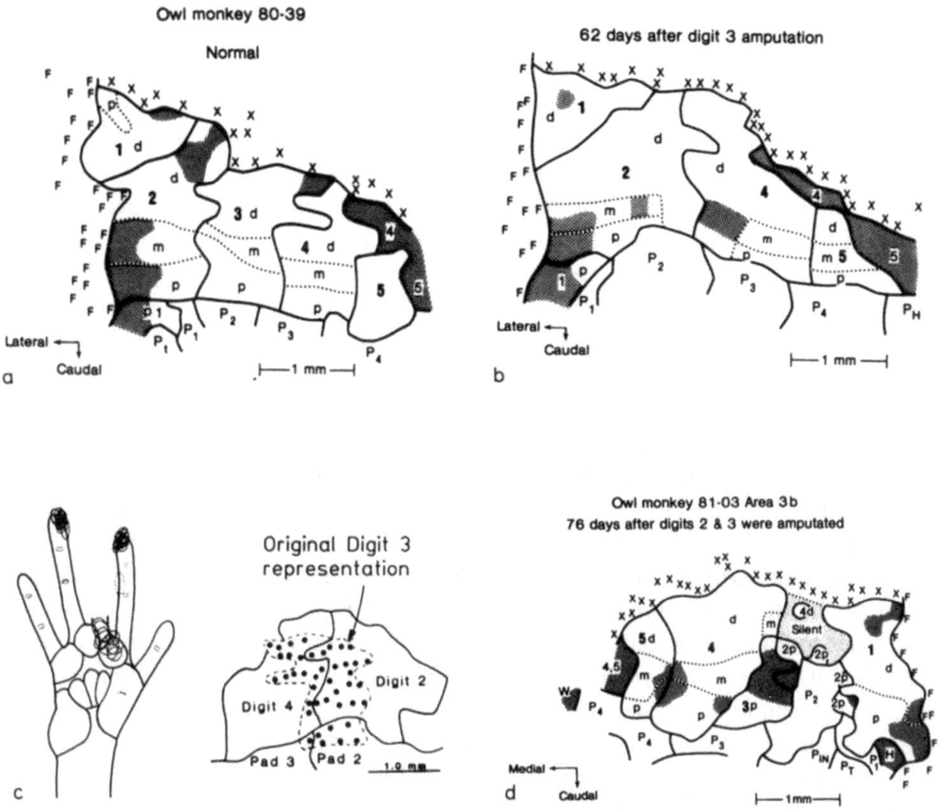

Fig. 7.12 a–d. Effects of amputating one or two digits on the somatotopic organization of the hand representation in area 3b of the pentobarbitone-anaesthetized owl monkey. **a** Hand representation before amputation. **b** Hand representation in the same monkey 62 days after the amputation of digit 3. Note that the representation of the glabrous skin of digits 2 and 4 has expanded to occupy all of the region that once responded to inputs from digit 3. **c** Receptive fields of neurons within the area of cortex that prior to amputation of digit 3 contained the digit 3 representation. **d** Hand representation 76 days after amputation of digits 2 and 3. Note the large representations of digits 1 and 4 but the persistence of an area of cortex that is unresponsive to cutaneous stimulation (*light stippling*). In **a**, **b** and **c** *heavy stippling* indicates the representation of hairy skin on the hand (*H*), wrist (*W*) or digits. *F* indicates tracks in which neurons responded to cutaneous input from the face. *X* indicates tracks where neurons did not respond to cutaneous inputs. Other abbreviations as in Fig. 7.2. (Modified from Merzenich et al. 1984, pp. 594–596)

7.2.9 Effects of Repeated Stimulation on Cortical Representations

The effect of repeated stimulation of an area of skin on the size of the cortical representation of that area has been postulated to explain the inter-animal variation seen in normal individual monkeys (Fig. 7.2) (Merzenich et al. 1987). Similar variability has been reported in the SI cortex of the raccoon (Fig. 7.20, Sect. 7.4.1) (Welker and Seidenstein 1959) and the rat (Sect. 7.5.1) (Wall and Cusick 1984).

Chapter 3 (Sect. 3.3.5) reviews similar experiments carried out in the rodent vibrissa system. In the primate three types of experiments have been conducted. In the first, an innervated patch of skin was removed from one finger and grafted onto another finger. Six to 12 months later area 3b was mapped and the skin patch was found to be represented twice – once in the somatotopically appropriate part of the representation of the recipient finger and once in the representation of the donor finger. In both regions multiunit responses were recorded to stimulation of skin areas which spanned the scar line around the patch. Block of the other nerves in the recipient finger indicated that these changes "could not be accounted for by peripheral axonal sprouting across the scar line" (Clark et al. 1986). In the second experiment a longitudinal incision was made along the medial (ulnar) surface of digit 3 and the lateral surface of digit 4 in 3 owl monkeys. The digits were then fused by suturing together the palmar and dorsal skin flaps. After the wound had healed maps of the representation of digits 3 and 4 of the syndactylus hand were assessed by making microelectrode recordings in area 3b under barbiturate anaesthesia. Immediately after mapping the digits were surgically separated and the border region between their representations was searched for points at which responses could be evoked from skin on both digits (2 monkeys only). The report shows data obtained from one monkey. In the syndactylus condition 41 recording sites were found in which responses could be evoked by stimulation of skin that spanned the scar line between digits 3 and 4. These sites were grouped within a strip of cortex that would normally contain the boundary of the representations of digits 3 and 4 and the authors argue that this number of sites is far greater than the occurrence of sites activated by skin on both digits 3 and 4 in the normal monkey. Immediately after separation of the digits, activity at 21 sites could be evoked from skin patches on both digits (Clark et al. 1988). For two reasons it is difficult to accept the authors' implications that in the boundary between the representations of digits 3 and 4 single cortical neurons adopted input from both fingers as a consequence of syndactyly. First, the medial and lateral surfaces of the digits have smaller representations than the palmar surfaces. Thus in the typical mapping experiments which were used to provide control data for the syndactylus condition, relatively few tracks would fall in the area likely to contain neurons (or groups of neurons) responsive to skin on both digits. In contrast, during examination of the effects of syndactyly, tracking would be focused on this border area. Second, no information is given on whether responses recorded were from single cells but the terminology adopted suggests that in fact multiunit responses were accepted. If multiunit recordings were made in the border region then it cannot be concluded that any single neurons developed inputs from adjacent digits.

These preliminary experiments have been used to argue that the temporal coincidence of stimulation of two skin areas is all that is required to establish *de novo* convergent inputs to single cortical neurons. On the basis of available data these conclusions cannot be accepted. However, these matters are of sufficient theoretical interest to warrant further experimentation in which greater effort is made to isolate single units at localized and regularly-spaced recording sites across the digit representation of a larger sample of animals.

In the third experiment, operant conditioning has been used to test for use-dependent alterations in the details of the hand representation in area 3b. In this experiment owl monkeys were rewarded for reaching out and touching, with the tips of two digits, an abrasive surface (Jenkins et al. 1984, 1990). The digit representations in area 3b were examined under barbiturate anaesthesia in a total of 5 monkeys. In 2 of these, the digit representations were assessed prior to training and again after 75 to 285 days of training. In the other 3 monkeys a training period of 52 to 580 days preceded the first assessment of the digit representation and a second map was made 31 to 80 days after cessation of training. The results obtained are complicated by the variation in the protocol used for each monkey (see Table 1 of Jenkins et al. 1990) but the authors reach the following interesting conclusions on the effects of repeated use elicited by training: (1) The representation of the stimulated skin expands; (2) the receptive fields recorded within the expanded zones are unusually small; (3) the representations of the cutaneous input from all digits spreads rostrally up to about 1000 μm, into territory which in the absence of training was functionally defined as area 3a on the basis of it being unresponsive to cutaneous input; (4) there was a lateral shift in the border between the digit and face representations of up to about 800 μm. This interesting experiment should be replicated a number of times before the conclusions are accepted. In unanaesthetized monkeys it would be most interesting to evaluate the influence of directed-attention mechanisms which, even in the absence of stimulation, appear to alter activity at specific loci across the postcentral gyrus and which may well tonically alter representation boundaries (Hyvarinen et al. 1980; Roland 1981). Furthermore, the random delivery of stimuli to a point on the body surface would perhaps enable investigators to test more directly the effects of repeated input on cortical representations in the absence of influences of reward or reinforcement which have been implicated in changes in somatic sensation (see Chap. 1 and Sect. 8.6).

7.2.10 Cortical Damage

The effects on somatotopic organization, of very small lesions (approximately 1 mm^2) within areas 3b and 1 of the owl monkey have been tested and reported in an abstract (Jenkins et al. 1982). From this preliminary report it would seem that 1.5–4 months after such a lesion the area of skin that was represented within the lesioned area becomes newly represented in a nearby area. The representation of inputs that were represented in the nearby area shift to regions further from the lesion site. The overall picture is thus one where the original thalamocortical inputs to the lesioned area are redistributed within intact cortex such that their topographical organization is preserved. Whether this is accompanied by an anatomical relocalization of the terminals of thalamocortical neurons or whether it is achieved via intracortical mechanisms or pre-existing ineffective terminations of thalamocortical fibres (Sect. 7.3.2) is unknown. However, these results from adult monkeys are reminiscent of the effects of lesions placed within the vibrissa representation of neonatal rodents (Fig. 3.20; Sect. 3.3.6) (Ito and Seo 1983).

Most of the examples of plasticity described in this review deal with somatotopic maps and involve relatively tiny areas of cerebral cortex. In contrast to these, extensive damage to SI in infant primates has provided an intriguing demonstration of plasticity which deserves further detailed exploration. In the adult macaque monkey, removal of the hand representation in SI causes severe tactile deficits (Semmes and Porter 1972; Randolf and Semmes 1974; Semmes et al. 1974; Carlson 1981). Removal of areas 1 and 2 produces deficits in texture and size-shape discriminations, respectively. Removal of area 3b interferes with all forms of tactile discrimination – a finding in keeping with the view that areas 3b, 1 and 2 are hierarchically ordered networks which extract ever more complex information from somatosensory activity flowing posteriorly across the postcentral gyrus (Sect. 7.2.2). Lesion of the second somatosensory area (SII) in the adult macaque also produces deficits in texture and size-shape tasks (Murray and Mishkin 1984). If, however, lesion to part or all of SI is performed in an infant monkey then there ensues full recovery of the ability to perform size and texture discriminations (Carlson 1984a, b) (see also Sects. 7.3.7 and 7.5.5).

Recently the dependence of this amazing recovery on the integrity of SII has been tested (Burton et al. 1987). In the infant, destruction of SII produced only a delayed acquisition of texture discrimination but destruction of SII with SI had "devastating" consequences for this form of learning. The authors suggest that the recovery of size discrimination following damage to SI and SII possibly reflects compensatory changes in areas 3a, 5 and 4 but they conclude that the recovery of texture discrimination depends on the integrity of SI or SII in the damaged hemisphere. As to the mechanism(s) underlying the compensatory changes, they point out that the age range of the infant monkeys covered the period of maximum synaptic exuberance in the cerebral cortex (Rakic et al. 1986). It seems to us that these behavioural studies are of vital importance in understanding the functional significance of cortical areas, but that to assess the relevance of the data to plasticity, they will have to be combined with anatomical and physiological investigations.

7.3 Plasticity in the Cortex of Adult and Developing Cats

7.3.1 Somatotopic Organization, Cytoarchitectonics and Neuronal Responses

Columnar Organization of Receptive Fields. In 1957 Mountcastle reported that neurons in the SI cortex of the barbiturate-anaesthetized cat responded to hair movement, skin pressure or deformation of deep tissues. In any electrode track perpendicular to the cortical surface, cells encountered responded to only one of these types of stimuli and had receptive fields approximately on the same area of skin. The fact that receptive fields of cells in a given track were both place-and modality-specific formed the basis for regarding them as being grouped within a functional column, the diameter of which was estimated to be about 500 μm (Mountcastle 1957). Similar observations have since been made in the unanaesthetized cat (McKenna et al. 1981) and Favorov et al. (1987) have provided evidence that the cortex is subdivided into columns or *segregates* of 350–400 μm in diameter in which the receptive fields of all neurons encompass a small common patch of skin. For any segregate the common patch of skin is known as the *receptive field centre of the segregate* and it is quite separate from the receptive field centre of adjacent

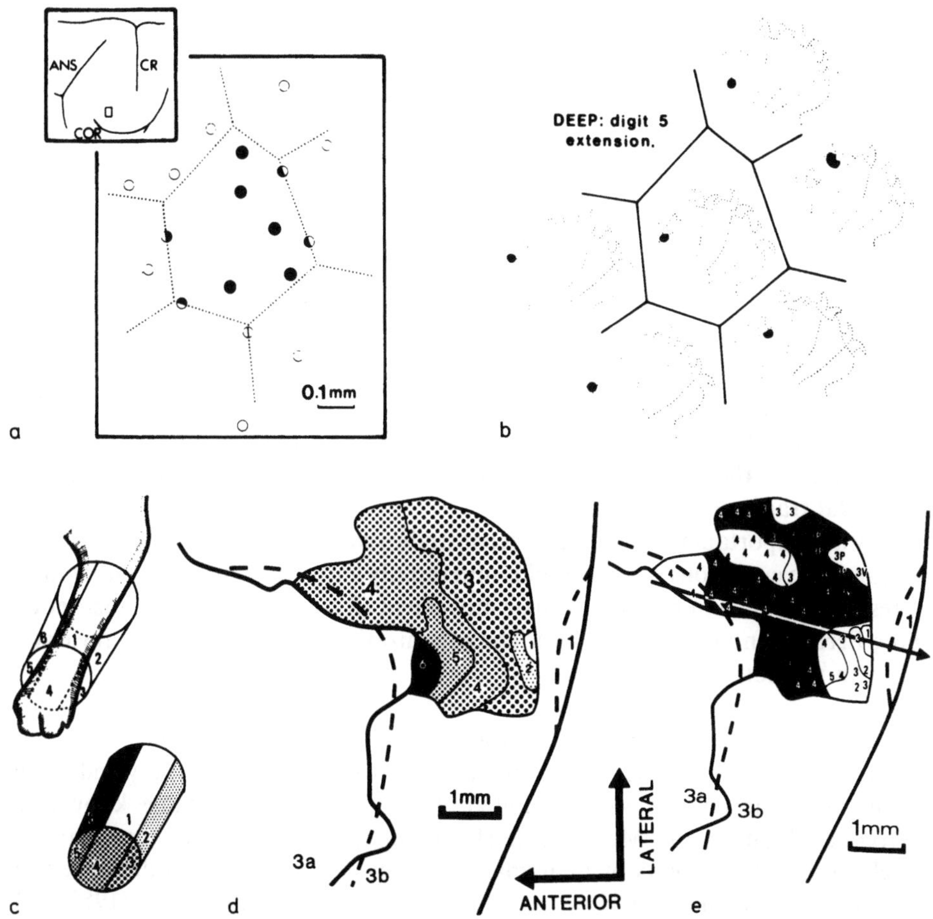

Fig. 7.13 a–e. The spatial representation of place and modality across the first somatosensory area of the cortex of the cat. **a** The size and shape of the cortical segregates in the anaesthetized cat. The area studied is enclosed in the rectangle (see *insert*). The *circles show the* location of penetrations. In the five penetrations (*filled circles*) all neurons encountered had receptive fields that shared a common skin area – the receptive field centre of this segregrate. Four other penetrations (*half-filled circles*) crossed the border between this segregate and the adjacent one such that the receptive fields of neurons in one part of the track consistently encompassed the receptive field centre of one segregate while in the other they consistently encompassed the receptive field centre of the adjacent segregate. **b** Shows the receptive field centres of six segregates. Note that adjacent segregates may have receptive field centres on adjacent digits. **c, d** The representation in area 3b of the pentobarbitone-anaesthetized cat of longitudinal strips of skin covering the forearm and wrist. **e** The representation of slowly (*white area*) and rapidly adapting (*black area*) responses across the same cortical region shown in **d**. The *numbers* refere to the strips of forearm skin shown in **a**. Note that the area is organized into meandering bands of slowly and rapidly adapting regions that are superimposed on the somatotopic representation such that any point on the body surface is represented, as nearly as possible, in adjacent slowly and rapidly adapting regions. *V* indicates that the receptive field at that site included the carpal vibrissae while *P* indicates the inclusion of the glabrous skin of the carpal pad. *ANS*, ansate sulcus; *COR*, coronal sulcus; *CR*, cruciate sulcus. (**a, b** Modified from Favorov et al. 1987, p. 6610; **c–e** from Sretavan and Dykes 1983, pp. 393 and 394)

segregates (Fig. 7.13a, b). As the injection of HRP into a very small (0.5–1.0 mm) area of the SI cortex labels one small group of cells in the VB complex, it would seem that the concept of segregrates may be applicable to individuals clusters (or in the primate, rods) of thalamocortical cells in the VB complex (Sect. 6.2).

The modality specificity of cortical columns has been challenged by workers who have studied the receptive fields of neurons in cats anaesthetized with alpha-chloralose (Morse et al. 1965; Doetsch and Towe 1976; see also Sect. 7.3.3). Their results suggest that only when the actions of GABA are sufficiently potentiated (as in the presence of barbiturates) is there a precise area-to-area cortical representation of the skin surface (Sects. 3.2, 7.2.6 and 7.3.3). Furthermore, within the posterior sigmoid gyrus (roughly equivalent to SI) of the unanaesthetized cat, Brooks et al. (1961b) reported that 40% of the pyramidal tract cells, but none of the non-pyramidal tract cells, had labile receptive fields. However, subsequent evidence suggests that the majority of the observations of labile receptive fields are attributable to the pyramidal tract stimulation used to antidromically identify pyramidal tract neurons (Mann 1979).

Finally it is important to note that in recent studies workers have subdivided cortical neurons with cutaneous receptive fields into rapidly and slowly adapting rather than as responding to skin pressure or hair movement (Rasmusson 1979; Bennett et al. 1980; Ferrington and Rowe 1980b). The organization of the SI cortex is reviewed below in relation to these criteria.

Inhibitory Receptive Fields. Mountcastle (1957) reported that when areas of skin around a receptive field were stimulated, the cortical cell under study and those within the same column were inhibited. This phenomenon became known as surround inhibition and Mountcastle (1957) suggested that it was a mechanism which increased the ability of an animal to perceive as separate, stimuli applied to two nearby sites on the skin. However, other studies by Mountcastle et al. (1957) and Baker et al. (1971) have shown that only relatively few (5%–15%) cortical neurons show the property of surround inhibition (Laskin and Spencer 1979b). Recent, as well as earlier work on SI cortical neurons in unanaesthetized cats has shown that when inhibitory receptive fields are present they occur on separate patches of skin which often do not even abut the excitatory receptive field (Brooks et al. 1961a, b; Iwamura and Tanaka 1978a). Furthermore, when, in unanaesthetized, paralysed (gallamine triethiodide) cats, small groups of hairs within the receptive field of a cortical neuron were stimulated sequentially, the most powerful inhibition of the response to the second, or test, stimulus was obtained by applying the first, or conditioning, stimulus to the centre of the excitatory receptive field. The time-course of this *in-field* inhibition was such that it reached a maximum 5–15 ms after the conditioning stimulus (Laskin and Spencer 1979a, b). A functionally identical phenomenon has been shown at lower levels of the neuraxis (Sects. 4.2.5, 5.2.3, 6.3 and 8.5.1). At all levels it would seem that spontaneous or incidental afferent activity might well suppress weak synaptic inputs so that they do not contribute to the receptive field of a central neuron. The removal of afferent input from an excitatory receptive field (i.e. as in partial deafferentation) might thus permit the expression of these weak inputs and a consequential shift in somatotopic organization (Sects. 7.3.2 and 8.5.1).

Cytoarchitectonics, Effective Stimulus and Somatotopy. According to Hassler et al. (1964) four cytoarchitectonically distinct areas are found in the region between the cruciate sulcus and the ansate sulcus in the cat – the area known to contain SI (Kaas 1983). In rostral-to-caudal sequence these areas have been called 3a, 3b, 1 and 2, each occupying a relatively thin mediolaterally orientated strip of post-cruciate cortex. Early work on the response of cortical neurons in anaesthetized cats showed that neurons responsive to deep stimuli were more common in the rostral regions of SI (Mountcastle 1957; Levitt and Levitt 1968a) although in general the distribution of functional columns was described as being a mosaic. Using tangentially orientated, caudal-to-rostral, electrode tracks through the middle layers of SI in barbiturate-anaesthetized cats, Rasmusson et al. (1979) and Dykes et al. (1980) showed that the effective stimulus of cortical neurons switched from cutaneous to deep when the electrode tip was within about 150 μm of the border between area 3b and 3a. In addition, largely by inference from previous work on primates (Sect. 7.2.3) Rasmusson et al. (1979) and Dykes et al. (1980) suggested that neurons in area 2 in the cat might be responsive principally to input from joint receptors (Sect. 6.2).

In unanaesthetized cats, Iwamura and Tanaka (1978b) reported that neurons in the ansate region (probably areas 1 or 2) may respond to (a) deep stimuli, (b) cutaneous stimuli that involved large skin areas, (c) complex cutaneous stimuli or (d) stimuli which activate deep or cutaneous receptors. From these observations they have suggested that information may be passed through area 3b to areas 1 and 2 as part of hierarchically ordered processing network. In contrast, McKenna et al. (1981) categorized the effective inputs to neurons in various somatotopically and cytoarchitectonically defined regions of SI in unanaesthetized cats as *skin* or *deep*. They did, however, report neurons in 3a and 2 which received convergent input from both deep (in the case of 3a, proximal and axial muscles) and cutaneous afferents. They found little evidence for an effect of receptive field position (e.g. proximal versus distal on a limb) on the submodality represented in a particular area, with the exceptions being the anterior part of 3b and area 3a. In these areas many neurons with receptive fields on the distal part of the limb responded to cutaneous inputs while most of those with proximal receptive fields responded to deep input (Table I of McKenna et al. 1981). In the unanaesthetized cat there were clear functional differences between different parts of the cytoarchitectonic areas of SI. For example, the anterior part of area 1 and the posterior part of area 3b both contained a high percentage of neurons responding to cutaneous input. Similarly, when deep versus cutaneous (including claw) inputs were used as indicators of function in the *unanaesthetized* cat, no functional boundary corresponding to the cytoarchitectonic boundary between 3a and 3b could be found (Table I of McKenna et al. 1981) and yet using deep versus cutaneous stimuli in the *anaesthetized* cat, Rasmusson et al. (1979) and Dykes et al. (1980) were able to show the clear functional designation of the 3a–3b boundary.

The Question of Multiple Representations of the Body. The complication concerning the relationships between cytoarchitectonics and function across the SI cortex in the cat has led to considerable controversy in relation to the question of multiple

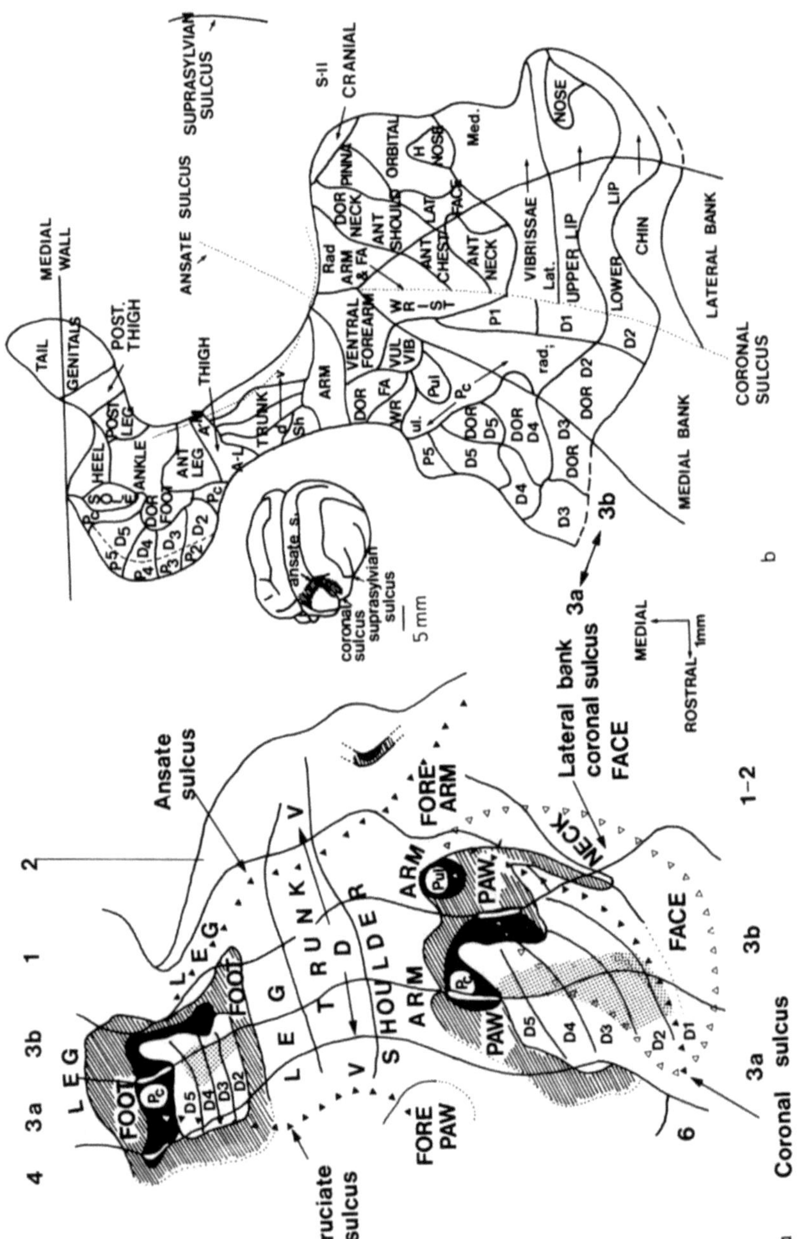

Fig. 7.14 a, b. The representation of the entire body surface in the first somatosensory area of the unanaesthetized (**a**) and ketamine- or pentobarbitone-anaesthetized cat (**b**). In both maps the coronal sulcus is laid open to illustrate the representations occupying the medial and lateral banks. **a** Neurons in areas 3a, 3b, 1 and 2 responded to cutaneous inputs (see text) and thus these areas contributed to the representation of the entire body surface (cf. Fig. 7.1a). No attempt is made to delineate the representation of the thigh from that of the leg and the representation of the digital pads is not shown.

versus single representation of the body surface. First, in unanaesthetized cats, the occurrence of cells with large receptive fields which encompass the smaller receptive fields of cells at a different location was used as a basis for suggesting multiple representation of the forepaw (Iwamura and Tanaka 1978a, b). Second, on the basis of segregation of submodalities in various regions of SI of the anaesthetized cat and previous work on monkeys (Sect. 7.2.1), Rasmusson et al. (1979) and Dykes et al. (1980) proposed that as many as five representations might exist across area 3a, rostral 3b, and caudal 3b, 1 and 2 (Fig. 7.6; Dykes et al. 1980). Subsequently, however, in a more detailed study of the forelimb representation in areas 3b and 1, Sretavan and Dykes (1983) have reported that the same body part is represented in adjacent parts of the slowly adapting and rapidly adapting regions. Thus one might regard these regions as containing meandering bands which contain either rapidly adapting or slowly adapting neurons but which do not disturb the underlying cortical somatotopy (Fig. 7.13c–e) (Sretavan and Dykes 1983). Third, in unanaesthetized cats, McKenna et al. (1981) proposed that there exists one continuous representation of the body across areas 3a, 3b, 1 and 2 throughout which neurons in particular regions are excited by afferents subserving different submodalities (deep or cutaneous) (Fig. 7.14a) (McKenna et al. 1981). It is interesting to note that, ignoring for the moment the cytoarchitectonic areas, this view is not unlike that adopted by Sretavan and Dykes (1983) in that the spatial representation of both afferent modality and receptive field position coexist as best they can across SI. Finally, Felleman et al. (1983b) have made extensive somatotopic maps of the SI cortex in cats deeply anaesthetized with either ketamine or pentobarbitone. Using cutaneous stimuli and multiunit recordings in between 300 and 500 tracks they found that the entire body surface was represented within areas 3b, 1 and "much of area 2" (Fig. 7.14b). In these cats, neurons within 3a responded only to taps or body movements which presumably activated deep receptors. It is not clear from their accounts whether neurons in the caudal part of area 2 responded to deep input. In response to the previous work of McKenna et al. (1981), Felleman et al. (1983b) have pointed to (a) the traditional electrophysiological definition of SI as that area responsive to tactile stimulation in barbiturate-aesthetized animals, and (b) the co-localization of a single representation with the extent of koniocortex (cortex with a thick inner granular, layer IV – characteristic of area 3b) in rats and primates (Kaas 1983). In relation to the latter point they emphasize that, in the cat, the borders between 3b, 1 and 2 are difficult to define anatomically and imply that areas defined as 1 and 2 by Hassler and Muhs-Clements (1964) might actually contain koniocortex such as formerly characterized area 3b.

◀ **Fig. 7.14a, b** (*continued*). *Stippling* shows the representation of the claws. **b** Although the boundaries between cytoarchitectonic areas are not shown, it was reported that in the presence of anaesthetics cells in 3a did not respond to cutaneous inputs and that the body surface is represented across areas 3b and 1 and the rostral part of area 2. *D*, dorsal; *V*, ventral; *Sh*, *shoulder*; *H*, *nose*, hairy skin on nose; *WR*, wrist, *D1* to *D5*, digits 1 to 5; *P1* to *P5*, toe pads 1 to 5; *Pc*, central pads; *Pul*, ulnar pad. (**a** Modified from McKenna et al. 1981, p. 667; **b** modified from Felleman et al. 1983b, p. 1651)

While the preceding observations and arguments are persuasive in their simplicity, it seems to us that, in relation to cortical plasticity, their primary importance is to underscore the profound effects of anaesthetics on neurons in the SI cortex of the cat (Sect. 7.3.3). For the purpose of studying the effects on cortical circuitry of imposed alterations of input it is pleasing to find that in both anaesthetized and unanaesthetized preparations there exist more or less stereotyped somatotopic maps which manifest particular sets of interconnections within the somatosensory system (compare Fig. 7.14a with Fig. 7.14b). In the anaesthetized cat, as in the primate (Sect. 7.2.4), somatotopic maps in the cortex may reflect those connections which best survive the effects of the particular anaesthetic agents. The specific effects of anaesthetics on the responses of single neurons in the cat cortex is considered below in Sect. 7.3.3.

7.3.2 Thalamic Input and Ineffective Thalamocortical Connections

Cortical Terminations of Thalamic Neurons. A number of approaches have been devised to study the distribution and microstructure of thalamocortical neurons that project to unidentified and identified areas of the SI cortex. In the cat, the primary focus of thalamocortical projections from the VB complex is layers IV and IIIb although terminals were also reported in layer I (Jones and Powell 1969; Hand and Morison 1970). More recently, Kosar and Hand (1981) have placed electrolytic microlesions of 30–175 μm in diameter in the medial portion of the VPL (VPLm) and described the resultant degeneration in the forelimb representation of SI and SII cortex. When viewed in coronal sections through SI, degeneration in layers IV and IIIb was contained within one or more discrete patches which ranged from 80–500 μm in width. Degeneration in layer I was very slight, but when observed it seemed to overlie that in layers IV and IIIb. Reconstruction of the distribution of degeneration in a plane tangential to the surface of the cortex showed that a microlesion in the VPLm produced bands of degeneration within layers IV and IIIb of areas 3b and 1–2. These bands extended up to 2.5–3.0 mm along the rostrocaudal axis, often branching and then rejoining to produce a sinuous "zebralike" pattern, several such patterns resulting from a single microlesion (Kosar and Hand 1981). No difference was observed in the density of degeneration of bands produced in areas 3b and 1–2 by a single microlesion. Furthermore, bands sometimes crossed the boundary between these two areas, suggesting that sites on either side of the 3b–1 boundary receive a common thalamic input.

The functional correlate of the bands in area 3b and rostral area 1–2 of the cat is unknown. They somewhat resemble the distribution of slowly adapting and rapidly adapting responses in areas 3b and 1 reported by Sretavan and Dykes (1983) and the existence, in the VB complex of the cat, of place- and modality-specific cell clusters is strongly inferred by the demonstration that small aggregations of cells in the DCN give rise to terminal arborizations approximately the size of the clusters of thalamocortical cells which are retrogradely labelled from areas 3b and 1–2 (Sect. 6.2) (Hand and Van Winkle 1977; Berkley and Hand 1978; Jones 1986). The

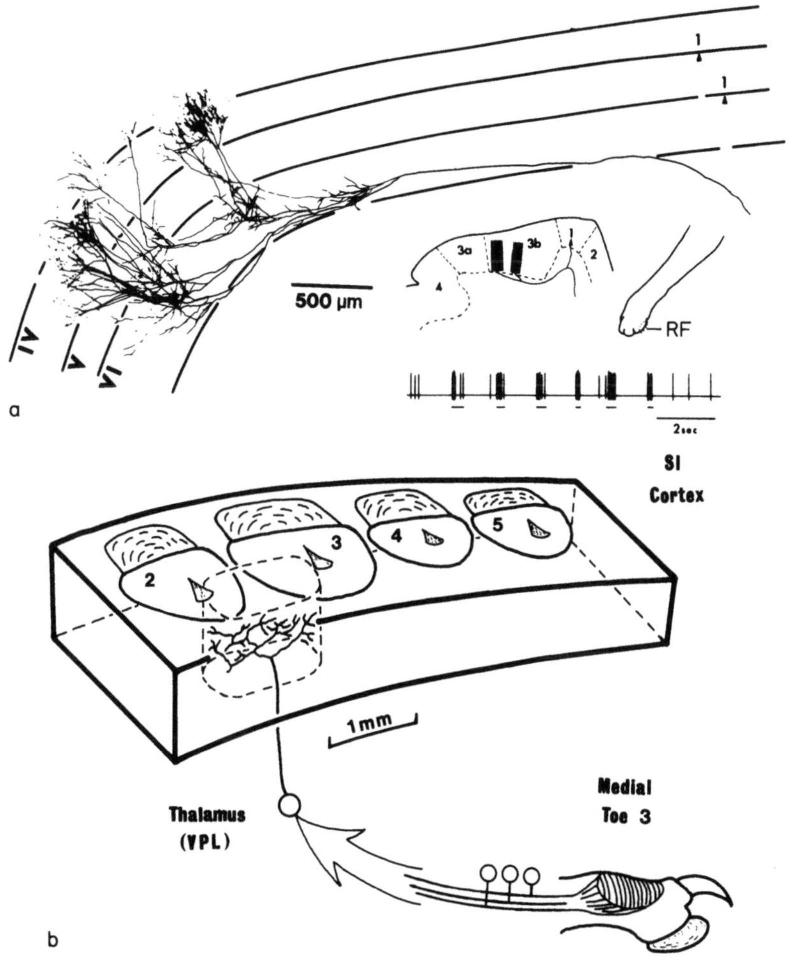

Fig. 7.15 a, b. Intracortical spread of the terminals of single thalamocortical neurons. **a** The terminations, in area 3b of the cat cortex, of a single thalamocortical axon that has been injected with HRP. Under pentobarbitone anaesthesia this neuron responded to the movement of hairs on the lateral surface of 5th digit of the forelimb. Within the cortex it gave rise to two "bushes" of terminals in each of layers IIIb, IV and VI. *Arrows* indicate the boundary between areas 1 and 3b near which the main axon gives rise to a sparsely branched collateral that enters the cortical grey matter about 2.5 mm away from the main arborization. The main arborization measured 2080 μm in the mediolateral axis and 1250 μm in the anterioposterior axis. These dimensions should be compared with the size of the cortical representations of the entire 5th digit in the unanaesthetized and the anaesthetized cat (see Fig. 7.14a, b). **b** Diagrammatic summary of the results of intracortical microstimulation experiments in which single thalamocortical neurons were antidromically activated from particular parts of the somatotopic representation in the SI cortex of the barbiturate-anaesthetized cat. In the case shown, the thalamocortical neuron had a receptive field on toe 3 and could be antidromically activated from the somatotopically appropriate part of the toe 3 representation *as well as* the representation of toe. *Abbreviations: RF*, receptive field; *VPL*, ventral lateral posterior nucleus. (**a** Modified from Landry and Deschenes 1981, pp. 363 and 362; **b** modified from Snow and Wilson 1989, p. 300)

function of these bands might be partially elucidated by results of the intra-axonal staining of single thalamocortical axons in the cat (Landry and Deschenes 1981; Landry et al. 1982, 1987). In these studies thalamocortical axons were injected with HRP in the subcortical white matter of barbiturate-anaesthetized cats, following the identification of their receptive fields and verification that they could be orthodromically activated from the VPL and driven transynaptically from the medial lemniscus. Single, stained thalamocortical fibres gave rise to "bushes" of terminals in layers IV, IIIb and VI (Fig. 7.15a) and the upper portion of layer V. Some fibres also sent a few collaterals to layers I and II (Landry et al. 1987). The arborizations of stained axons tended to be elongated along the mediolateral axis of the cortex (mediolateral extent = 720–2080 μm; cf. rostrocaudal extent = 390–1250 μm, Landry and Deschenes, 1981) and the overall size of an arborization was not related to the position of the receptive field of the parent axon. Of 17 stained axons 10 had both a main and a secondary arborization that were separated by up to 300 μm of uninvaded cortex. Six axons responded to joint movement or deep stimuli and had their main arborization in area 3a or 3b whereas the remainder responded to light touch or hair movement and projected to area 3b or in one case to area 1. When a secondary arborization was present it was usually in a different cytoarchitectonic area (Fig. 7.15a). One rapidly adapting cutaneous thalamocortical fibre was also shown to project to area 2 (Landry et al. 1987).

In commenting on the existence of two bushes of terminals on some thalamocortical neurons, Landry and Deschenes (1981) and Landry et al. (1982) draw attention to the cortical segregration of rapidly adapting and slowly adapting responses in the cat reported by Dykes et al. (1980) (see also Sretevan and Dykes 1983). They argue that in the forearm and forepaw representations these functional regions tend to become rostrocaudally orientated bands. Thus, for example, two adjacent bands of rapidly adapting cortical neurons might well receive input from the two bushes of a single rapidly adapting thalamocortical fibre. On the basis of the suggestion that within the VB complex there is interdigitation of the slowly adapting and rapidly adapting regions (Dykes et al. 1986), it is possible that two bushes of single thalamocortical neurons are responsible for the two or more bands of degeneration seen after microlesions within the VB complex (Kosar and Hand 1981) and that the boundaries of these bands do not cross the boundaries between the slowly adapting and rapidly adapting regions within SI. Thus, while the rapidly adapting and slowly adapting bands within SI may be made up of inputs from rapidly adapting and slowly adapting regions within the VB complex, any small group of modality- and place-specific thalamocortical cells might provide input to only a small part of a rapidly adapting or slowly adapting band. Confirmation of these matters must await studies in which detailed mapping of cortical neurons and intra-axonal staining of single thalamocortical neurons is performed within individual animals.

Ineffective Terminations of Thalamocortical Neurons. To date there exists only one attempt to correlate the somatotopic organization of SI with the projection patterns of single thalamocortical fibres. In that study, recordings were made in the VPL from single thalamocortical cells with receptive fields on the toe of the barbiturate-anaesthetized cat. A second microelectrode was then used to record from cortical neurons or to apply small (0–10 μA, 300 μs) cathodal stimulating pulses to points throughout the cortical depth. By this process attempts were made to antidromically activate thalamocortical cells from somatotopically identified areas within SI (Snow et al. 1988). Currents which antidromically activated thalamocortical neurons were as low as 0.5 μA and typically less than 2.0 μA.

The dimensions of the area and the depth within the cortex from which thalamocortial neurons could be activated agreed with the anatomical descriptions of Landry and Deschenes (1981) although no evidence of two terminal bushes was obtained. More interestingly, however, the results showed that some thalamocortial neurons with receptive fields on a single toe projected to points in the middle cortical layers that were up to 800 μm away from tracks in which cortical neurons had receptive fields encompassing the receptive field of the thalamocortical neuron under study. In some instances a thalamocortical neuron with a receptive field on a particular toe could be antidromically excited from well within that region of cortex which contained the representation of an adjacent toe (Fig. 7.15b) (Snow et al. 1988; Snow and Wilson 1989). The somatotopically inappropriate afferent terminals in the SI cortex and those in the dorsal horn of the spinal cord (Sects. 4.3.3 and 8.3.2) (Meyers and Snow 1984; Meyers et al. 1984; Snow and Meyers 1985; Snow and Wilson 1989) must be ineffective in firing neurons; at least under the circumstances normally used for mapping receptive fields. Their presence invites the hypothesis that they have much to do with the reorganization of somatotopy seen after peripheral nerve damage and possibly also the effects of anaesthetic agents on the receptive fields of neurons in SI. Landry et al. (1987) have shown that a single thalamocortical axon gives rise to a high concentration of boutons at the centre of its arborization. Perhaps the more peripheral parts of the arborization provide a weak input to cells in a somatotopically inappropriate region that may be held at subthreshold levels by in-field inhibition (Fig. 8.5; Sects. 7.2.3 and 8.5.2).

7.3.3 Effects of Anaesthetics and Other Drugs

Anaesthetics and the Responses of Cortical Neurons. In a unique series of experiments, reviewed by Mann (1979), Towe's group studied the responses of neurons in the forelimb and hindlimb representations of SI cortex to electrical stimulation of all four paws (Morse et al. 1965; Doetsch and Towe 1976). In these studies cats were lightly anaesthetized with alpha-chloralose (50 mg/kg, i.p.) and paralysed with decamethonium bromide. Most neurons in SI cortex (area 3b only) fell into one of two categories: (1) *sa*-cells; those responding only to stimulation of the contralateral paw appropriate to the recording site (hindlimb or forelimb) and (2) *m*-cells; those responding to stimulation of each of the four paws. In the hindlimb representation 62% and 36% of units were classified as *sa*-cells and *m*-cells, respectively. In comparison, within the representation of the arm and the shoulder, 70% of units were *sa*-cells and 30% were *m*-cells. No *m*-cells were found in the forepaw representation, an area devoted to that part of the body most often used by the cat for tactile searching. Mann (1979) emphasized that the receptive fields of *sa*-cells are stable under a wide variety of conditions which include variations in anaesthesia (chloralose, barbiturate, unanaesthetized), the presence or absence of a paralysing agent, the conscious state of the animal and the topical application of the glycine antagonist, strychnine, or the GABA antagonist, bicuculline. In contrast, the responsiveness of *m*-cells is markedly reduced by barbiturates such that in the chloralose-anaesthetized cat they are immediately silenced by as little as 5 mg/kg of pentobarbitone (Towe et al. 1976; Harding et al. 1979; Mann 1979). Projection cells with the widefield properties of *m*-cells have been identified in the DCN and the VB complex (Sects. 5.4.4 and 6.3) suggesting that in the SI cortex at least some widefield neurons are merely reflecting the characteristics of neurons lower in the neuraxis.

In reviewing their results from 583 neurons in areas 3a, 3b and 4γ of barbiturate-anaesthetized cats and 4176 neurons from these areas in cats anaesthetized with chloralose, Harding et al. (1979) concluded that neither preparation yields cell responses similar to those seen in the unanaesthetized cat. This is, however, not surprising as it was previously concluded by Baker et al. (1971) and Towe et al. (1976) that the excitability of *m*-cells seems to depend on the level of arousal of an animal and Towe et al. (1976) have proposed that during normal behaviour *widefield* neurons may not be found unless the animal is searching an unfamiliar or potentially dangerous environment. In relation to the subject of plasticity these studies provide an abundance of evidence that the functional pathways between the periphery and the cerebral cortex are labile and that this property is somewhat at the mercy of anaesthetic agents. They further suggest that in the free and awake animal the receptive fields of cortical neurons may be regulated as a consequence of both the "behavioural state" and environmental cues or, as we have suggested in relation to studies on primates, directed attention (Sects. 7.2.7, 7.2.9 and 8.6).

Iontophoretic Application of Drugs and the Responses of Cortical Neurons. The systemic application of a substance has several disadvantages, not the least of which is that the agent may, and probably does, act at all levels of the neuraxis. The use of the microiontophoretic administration of substances from a multibarrel pipette electrode has permitted the study of those effects that are more likely to be a consequence of intracortical circuitry. From literature cited in the preceding paragraphs it is clear that the inhibitory neurotransmitter GABA is intimately involved with the mechanisms underlying the somatotopic organization of the cortex. In addition, however, early microiontophoretic studies established beyond any doubt that ACh plays an important role in regulating the responses of cortical neurons (Krnjevic and Phillis 1963a–c). In the following paragraphs the effects of the iontophoretic application of ACh, GABA and their antagonists on cortical neurons will be reviewed.

Acetylcholine. In cats anaesthetized with N_2O and methoxyflurane, Krnjevic and Phillis (1963a–c) showed that, particularly in the primary sensory areas, ACh produced excitation of some cortical neurons and greatly potentiated the responses of others to exogenously applied glutamate. Similar effects were found in a small number of rabbits and monkeys. In contrast to the rapid excitation produced when

ACh was applied to Renshaw cells in the spinal cord, the facilitatory effect of ACh on cortical neurons had a slow time-course. While the majority of ACh receptors on Renshaw cells are nicotinic, the receptors responsible for the prolonged excitatory action of ACh on cortical neurons are muscarinic (Krnjevic and Phillis 1963c; Krnjevic 1974).

The biophysical mechanism underlying the effects of ACh on cortical neurons has been investigated by Krnjevic et al. (1971) who found that ACh produced an increase in membrane resistance and a slowing of the repolarization phase of the somatic action potential. Consequently they suggested that the slow excitatory effects resulted from a reduction in the resting K^+conductance and in the delayed K^+ conductance during the action potential (see also: Cole and Nicoll 1983; Cole and Nicoll 1984; Richardson and DeLong 1988). They pointed out that neither of these effects will lead to excitation of a cortical neuron but that both might explain the "marked enhancement of other inputs" and the "particularly striking prolongation of evoked discharges" (Krnjevic et al. 1971). Finally, these authors stated that "a discrete or diffuse liberation of ACh may greatly change both the general responsiveness and the pattern of firing of small or large groups of cortical neurones". They thus established the role of ACh as an important modulator of excitatory transmission in cortical pathways.

In these early studies attention was focused on the excitatory effects of ACh. However, Randic et al. (1964) pointed out that superficial neurons were sometimes clearly inhibited by ACh. Subsequently, ACh has been shown to produce a decrease in K^+ conductance and rapid excitation of GABAergic interneurons which in turn elicit short-latency inhibition of pyramidal cells in slices of the cingulate gyrus of the guinea-pig (McCormice and Prince 1985). In addition, ACh produced an increase in the input resistance and slow excitation of the same pyramidal cells. Both responses were shown to be mediated by muscarinic receptors.

Hicks et al. (1985) reported greater difficulty in assigning cells to rapidly adapting or slowly adapting subclasses in the SI cortex of cats under urethane anaesthesia than in cats anaesthetized with pentobarbitone. This they suggested might result from a prolongation of rapidly adapting responses brought about by an enhancement of the life of ACh that resulted from the anticholinesterase activity of urethane (Machne and Unna 1963). These observations together with the pronounced effects of ACh on the responses of cortical neurons to visual stimuli (Sillito and Kemp 1983), led Metherate et al. (1987) to investigate the effects of iontophoretically applied ACh on the response of SI neurons, in cats anaesthetized with pentobarbitone, to tactile stimulation and to iontophoretically applied glutamate. While only 21% (13/61) of the neurons responded to ACh alone, in the presence of ACh 75% (66/88) displayed altered responses to glutamate or stimulation of the receptive field. In 47% (29/62) of the latter group responses remained potentiated from 8 min to over 1 h. A total of 107 cells were studied in order to assess the effects of ACh on the responses to somatic stimuli and the responses of 53 (50%) of these were found to be altered. When a controlled mechanical stimulator was used to quantitatively assess the responses of 30 cells it was found that, in the absence of glutamate, ACh often produced no effect or only short-term (< 5 minutes) changes in the responsiveness to somatic stimuli. More interesting was the fact that 68%

(70/107) of cells had no receptive field. In 64% (45/70) of these cells, low doses of glutamate revealed previously subthreshold inputs from the skin which disappeared within seconds on cessation of glutamate application. Of these cells, 47 were tested with ACh alone and 14 (30%) "displayed long-term changes" in their responses to somatic stimuli – presumably developing receptive fields which persisted for more that 5 min after the application of ACh. In order to quantify the effects on neuronal responses, measurements were made on cells which had a receptive field before the application of ACh. In 11 such cells displaying short-term effects (< 5 minutes), ACh produced a 296% increase over control responses while in 6 cells showing long-term effects (> 5 minutes) the responses reached 357% within 2–5 min.

On the basis of these observations Metherate et al. (1987) suggested that ACh may contribute to the reorganization of somatotopy seen in the SI cortex following peripheral nerve injury (Sects. 7.2.8, 7.3.5 and 7.4.4). In arguing for this involvement they point to the evidence for projections of thalamocortical neurons in the SI cortex of the cat (Sect. 7.3.2) suggesting that an increase in ACh levels might increase the presumably subthreshold influence of these synapses to threshold levels. Thus it could be proposed (Metherate et al. 1987) that peripheral nerve section may give rise to an increase in the levels of ACh or in the numbers of ACh receptors (or a reduction in acetylcholinesterase) within the deprived cortex.

Recently, the possibility that ACh released from the cortical terminations of cholinergic neurons in the basal forebrain modulates the responses of SI neurons to stimulation of the receptive field has been explored (Rasmusson and Dykes 1988). In this work, conditioning stimuli were applied near the rostral pole of the globus pallidus, a region containing cholinergic neurons which project to many cortical regions including the somatosensory areas. The somata of the activated neurons were probably located in the nucleus basalis of Meynert which has been implicated in the cholinergic regulation of arousal and the aquisition of information during learning (Richardson and DeLong 1988). While forebrain stimulation alone did not enhance the responses to subsequent stimulation of the skin, pairing of forebrain stimuli with cutaneous stimuli resulted in an increase in the cutaneous evoked potential for over 90 min (3 out of 4 cats, anaesthetized with 2.5% halothane). Atropine did not block this enhancement unless it was applied before the conditioning-test regime, suggesting that enhancement relies on the binding of ACh to muscarinic receptors but that once this has taken place further ACh binding is unnecessary for the maintenance of increased levels of excitability. The implication of this work is that cholinergic neurons in the forebrain have, in the presence of cutaneous stimulation, the capacity to regulate the responsiveness of neurons in SI cortex. While this work is extremely interesting in that it relates to natural mechanisms that control receptive fields in SI cortex, it remains unclear as to whether it bears any relationship to the denervation-induced shifts in the centres of receptive fields of cortical neurons (Sects. 7.2.7, 7.2.8, 7.3.4, 7.3.5, 7.4.4, 7.5.2, 8.5 and 8.6).

GABA. Early pharmacological studies on the inhibition of cortical neurons in the cat following direct and indirect activation of the cortex suggested that GABA rather than glycine was the inhibitory transmitter in the cortex (Krnjevic et al. 1966;

Kelly and Krnjevic 1969). A number of immunocytochemical studies have shown that a large number of neurons in the sensory-motor cortex contain the GABA-synthesizing enzyme glutamic acid decarboxylase (GAD) (Hendry et al. 1983; Hendry and Jones 1981; Houser 1983). Combined retrograde labelling of thalamocortical cells together with immunocytochemical localization of GAD, has been used to show that although GABAergic neurons exist in the VPL of the cat these neurons do not project to the SI cortex (Spreafico et al. 1983). At present it seems most likely that the transmitter in thalamocortical neurons projecting to SI is glutamate (Jones 1986).

The possibility that cortical neurons receive subthreshold excitatory inputs from skin areas outside their receptive fields suggests that (a) increasing the excitability of cortical cells or (b) decreasing the level of pre- or postsynaptic inhibition might increase the size of the receptive fields of cortical neurons by raising the subthreshold inputs to threshold levels. In the cat cortex the first investigation of this hypothesis was that of Batnev et al. (1982). This study involved 37 neurons in the forelimb representation of cats anaesthetized with pentobarbitone (40 mg/kg i.p.). Glutamate and the non-competitive GABA antagonist, picrotoxin, were applied iontophoretically to single cortical cells. The skin within and around the receptive field was stimulated electrically and the responses observed before and after drug application. While no increase in the size of the receptive fields was seen during glutamate application, "relatively short" (90 nA, 3–6 min) application of picrotoxin markedly increased the area of skin from which cortical neurons could be driven. In a subsequent series of studies, Hicks and Dykes (1983) and Dykes et al. (1984) applied glutamate, GABA and the competitive antagonist of the $GABA_A$ receptor, bicuculline, to neurons within the slowly adapting and rapidly adapting regions of the SI cortex of cats anaesthetized with pentobarbitone (35 mg/kg i.p.) or urethane. Although the authors report that in urethane-anaesthetized cats neurons showed more spontaneous activity, and were thus harder to type in relation to the slowly adapting/rapidly adapting criteria, the responses to glutamate, GABA and bicuculline were not different from those of neurons in cats anaesthetized with pentobarbitone. It must be remembered that in the presence of barbiturate the actions of the resting levels of GABA are enhanced and that this action is borne out by the fact that fewer cells have large receptive fields when studies are made using barbiturate anaesthesia. Similarly one might expect urethane to enhance GABAergic inhibition by its prolongation of ACh lifetime and the consequent augmentation of excitation of cortical GABAergic interneurons (McCormice and Prince 1985). Thus some caution should be used in relation to the significance of expansions of receptive fields by GABA antagonists in animals under urethane or barbiturate anaesthesia (Sect. 7.5.4) as these might simply be produced by a reversal of an effect of the anaesthetic in use.

A total of 209 neurons were examined in the barbiturate-anaesthetized cats, 123 (59%) of which had receptive fields (Dykes et al. 1984). Glutamate was applied to 173 neurons, GABA to 125 and bicuculline to 152. Cells were activated by natural stimulation of the skin, electrical stimulation of the thalamus or by iontophoretic ejection of glutamate from one barrel of the multibarrel electrode. Recording sites were marked and the depth and cytoarchitectonic area determined by histological examination. No changes in the modality or submodality of cortical cells were seen with glutamate, GABA or bicuculline. Thus rapidly adapting cells never developed slowly adapting responses and neurons responding to cutaneous or deep receptors retained their receptor specificity. These observations suggest that at any point in the SI cortex, the modality or submodality that is represented is characteristic of the modality or submodality of all neurons giving rise to excitatory afferent terminations at that point. It would be interested to test this proposal by applying bicuculline or glutamate to neurons within a few hundred microns of the border between the slowly adapting and rapidly adapting bands in the SI cortex (Fig. 7.13b).

The application of low doses of glutamate lowered the threshold for activation by cutaneous inputs and increased the number of spikes resulting from both cutaneous and thalamic stimulation but even higher doses of glutamate produced no increase in the receptive field size of cells. Curiously, some cells in the middle layers of slowly adapting cortex were relatively insensitive to glutamate suggesting perhaps that this excitatory transmitter is utilized less in these regions. Most interesting, however, was the finding that in both rapidly adapting and slowly adapting cortex, glutamate or bicuculline application revealed distinct cutaneous receptive fields in some cells which were initially unresponsive to somatic stimuli (Fig. 7.16a, b).

In all cortical regions bicuculline blocked the poststimulus inhibitory period that normally follows thalamic stimulation. However, although the application of bicuculline increased the size of the receptive fields of most cells in the rapidly adapting regions (mean increase for 38 cells = X 6.5; SE = 0.73), it rarely affected the receptive field size of cells in the slowly adapting region (compare Fig. 7.16a with Fig. 7.16b) (Dykes et al. 1984). In the rapidly adapting regions increases in the receptive fields were greatest for cells between 600 and 1100 μm from the cortical surface – a region corresponding to a major termination site of thalamocortical axons within layers IIIb and IV (Sect. 7.3.2).

The application of bicuculline increased the effectiveness of inputs all across the pre-drug receptive field. Furthermore, Dykes et al. (1984) reported that increasing concentrations of GABA progressively increased the threshold of cells to natural, cutaneous stimulation and to electrical stimulation of the thalamus without altering the size of the receptive field. In a more recent paper Kaneko and Hicks (1988) have reported that while GABA does reduce the excitability of a neuron from the centre of its receptive field it does, in fact, also reduce the size of the receptive field. In addition, these workers compared the effects of GABA with those of the $GABA_B$ agonist baclofen and presented some evidence that, unlike GABA, baclofen more effectively inhibited inputs from skin that was originally on the borders of the receptive field. It can therefore be tentatively proposed that the $GABA_B$ receptor might selectively mediate receptive field size while the $GABA_A$ receptor might influence receptive field size in the rapidly adapting regions by exerting a general reduction of excitability (Kaneko and Hicks 1988).

That drug-induced increases in the size of the receptive fields of cortical neurons were not always concentric around the pre-drug receptive field (Fig. 7.16a) perhaps adds weight to the concept that the SI cortex is organized into segregates (Fig. 7.13a; Sect. 7.3.1) (Favorov et al. 1988). Thus we might envisage a cell in a segregate as

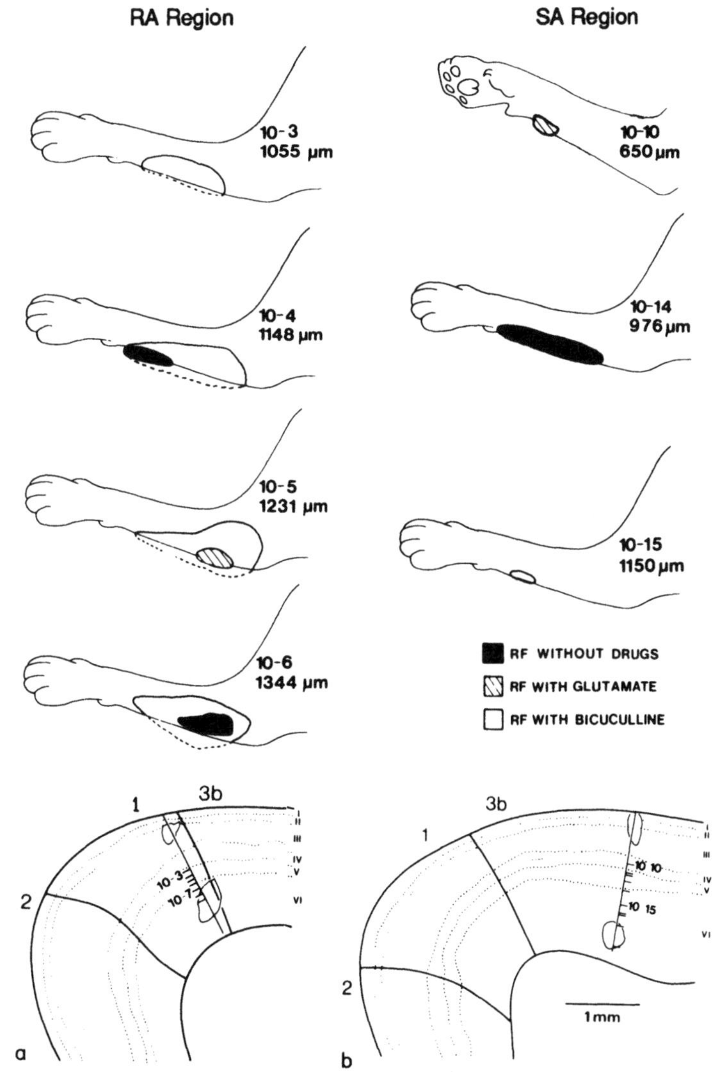

Fig. 7.16 a, b. The effects of iontophoretically applied glutamate and the $GABA_A$ antagonist bicuculline on the receptive fields of neurons in (**a**) rapidly adapting (*RA*) and (**b**) slowly adapting (*SA*) regions of the SI cortex in the pentobarbitone-anaesthetized cat. The presence of a *hatched area* or an *unshaded area* not encompassing a *filled-in area* indicates that no receptive field could be located without drugs. See text for further discussion. (Modified from Dykes et al. 1984, pp. 1074 and 1075)

receiving an above-threshold input from the skin forming the *receptive field centre of the segregate* and skin abutting one side of this area. The subthreshold inputs to this cell might thus arise totally from skin abutting the opposite side of the *receptive field centre of the segregate*. Raising these inputs to threshold levels would thus cause an asymmetrical expansion of the receptive field similar to that which is often observed in the presence of a GABA antagonist or glutamate (Dykes et al. 1984; Kaneko and Hicks 1988).

Dykes et al. (1984) note that the extent of the cortical arbors of rapidly adapting and slowly adapting thalamocortical neurons are the same (Landry and Deschenes 1981) and suggest that perhaps the size of the receptive field of cells within slowly adapting regions of the SI cortex may be controlled by other inhibitory transmitter candidates such as taurine, β-alanine or glycine. To test this, Tremblay et al. (1986) applied the glycine antagonist strychnine to cortical neurons in rapidly adapting and slowly adapting regions. This antagonist only produced increases in the receptive field size of some cells (11/27) in rapidly adapting regions. It was, however, more potent in rapidly adapting regions in revealing receptive fields in cells which were initially unresponsive to somatic stimuli and in decreasing the threshold of cells to cutaneous stimuli. Out of β-alanine, glycine and taurine, β-alanine was most effective in increasing the threshold to somatic stimuli although the authors concluded that none of these putative inhibitory transmitters decreased the size of receptive fields in cortical neurons.

Several conclusions may be drawn from these studies of the cat SI cortex. First, in anaesthetized animals cortical neurons receive many subthreshold excitatory inputs from the skin around, or nearby, their normal receptive field. Second, many cells which do not have a detectable receptive field ("silent cells") receive subthreshold inputs from skin represented in the receptive fields of nearby cortical neurons (Dykes and Lamour 1988). Third, it is clear that neuroactive factors such as ACh or GABA can alter the efficacy of inputs to cortical neurons. Fourth, the effects of operative procedures, such as denervation or amputation, or altered states of attention, all of which clearly alter somatic sensation, could well act totally through the pharmacological regulation of the cortical connections of VB neurons, rendering unnecessary the proposal that structural alterations might occur in this important link to the adult cortex.

7.3.4 Cordotomy and Section of Ascending Tracts

Normal Adult Cats. The effects on the responses of neurons in the cat SI cortex of lesions of ascending tracts in the spinal cord have been examined in two studies by Levitt and Levitt (1968a and b). Cats were anaesthetized with sodium thiopental and paralysed with gallamine triethiodide. In 20 normal cats these workers sampled a total of 588 units all of which were located in three 2.5 mm strips (rostral to caudal order – A, B and C) which ran across the postcruciate cortex parallel to the cruciate sulcus. These strips approximately overlay cytoarchitectonic areas 3, 1 (plus part of 2) and 5, as illustrated by Hassler and Muh-Clements (1964). No driveable units were found outside these strips. Neither histological reconstruction of the recording sites nor verification of the cytoarchitectonic areas were performed. Of the units, 63% were influenced by what might be crudely defined as cutaneous stimulation of the contralateral hindlimb, a further 17% being driven by deep input. Particularly prominent were responses to hair movement which were seen in 36%, 52%, and 31% (overall 42%) of units in strips A, B and C. All excitable units responded with a short latency to peripheral stimulation and showed clear modality and place specificity. A clear somatotopic organization was recognized which included a disproportionate area devoted to the representation of the toes.

Effects of Lesioning Various Funiculi in Adult Cats. Given this background, Levitt and Levitt (1968b) placed acute or chronic lesions in the mid-thoracic spinal cord of 16 cats prior to studying the responses of neurons in the region encompassed by strips A and B (a 5 mm wide strip posterior and parallel to the cruciate sulcus) of the postcruciate cortex. The responses of cortical neurons were studied, again under sodium thiopental anaesthesia. In three cats acute section of the lateral funiculus and ventral funiculus contralateral to the recording site (two cases) and bilaterally (one case) did not alter the percentage of cells responding to particular types of stimulation. In contrast, acute section of the dorsal columns (bilaterally) produced a reduction in the number of cortical neurons excited by light mechanical stimuli from 88% to 48%. In particular, responses to hair movement were found less frequently (19% as compared to 44%) although the authors suggest that this might have been a consequence of trauma to the contralateral spinocervical tract at the site of the lesion. Also reported was an increased number of neurons responding only to tap. In one cat, in which the dorsal columns were cut bilaterally and the dorsolateral funiculus (a region containing axons of the spinocervical tract) was also damaged contralateral to the recording site, no units in the digit area could be driven by mechanical stimulation. However, in this preparation five units responded to "strong mechanical stimuli", one to forceful flexion of the digits, three to intense pinch of the skin of the digits of both hindpaws and one to intense cutaneous pinch of both hindlimbs as well as both forelimbs and the trunk. In control animals no neurons were encountered which responded to, or showed increased responses to intense mechanical stimulation. Similarly, in a further six cats in which the dorsal columns and the dorsolateral funiculus were sectioned contralateral to the recording site, not one of the 334 units responded to light mechanical stimuli. In contrast, 19% (64) were excited by noxious peripheral stimuli such as pinching, striking the limb, kneading deep tissues and forceful rotation of the joints. The responses to these presumably noxious forms of stimulation occurred with a long latency (100–200 ms) suggesting that they ascended the spinal cord along axons with a very low conduction velocity. Cortical recording sites were localized either within 750 μm of the surface (61%) or between the depths of 1350 and 1650 μm (39%)-sites which are probably above or below the granular layer. The responses to noxious stimulation were abolished by small, additional, systemic doses of sodium thiopental or the paralysing agent and nicotinic antagonist, gallamine triethiode, suggesting that the effectiveness of noxious inputs in these may depend on both GABAergic and cholinergic mechanisms (Sects. 6.3, 7.33 and 8.5.1, 2). In normal animals responses to noxious stimuli were never seen (Levitt and Levitt 1968a, b).

The somatotopic organization of these nociceptive responses depended on whether the ventrolateral quadrant contralateral to the recording site was intact. When it was, 10 of the 55 units which responded to "strong mechanical" stimulation were so affected from three or four limbs, the strongest response nevertheless resulting from stimulation of the contralateral hindlimb (Levitt and Levitt 1968b). In two cats in which the ventrolateral quadrant together with the dorsolateral funiculus and dorsal columns were sectioned contralateral to the recording site (contralateral hemisection), noxious stimuli were only effective when applied to the contralateral hindlimb and somatotopic organization was preserved.

In conclusion, it would seem that activity ascending in the contralateral dorsolateral funiculus and dorsal columns is capable of inhibiting nociceptive input to the SI cortex (layers I, II, V, VI and possibly III) from the contralateral ventrolateral quadrant of the spinal cord. Furthermore, it appears that GABAergic and cholinergic-nicotinic systems might be involved in this inhibition. Whether inhibition is achieved via descending systems to the thalamus, DCN or spinal cord, or via inhibitory processes within relay stations below or at the level of the cortex or within the cortex itself is unknown. However, these observations may be important for research into the cortical representation of pain.

Lesions Sparing the Dorsal Columns. In a later study aimed at determining the function of the dorsal columns, Frommer et al. (1977) transected the cat spinal cord at T7, sparing or partially sparing the dorsal columns. Electrophysiological observations were made in cats paralysed with gallamine triethiodide and anaesthetized with barbiturates. Immediately (within 7 h) after this operation, normal-sized, evoked potentials were recorded in SI cortex in response to electrical stimulation of the contralateral hindpaw but 2 days after the lesion evoked potentials were either attenuated, or in many parts of SI, undetectable. Recordings from 338 neurons before the lesion and 36 neurons after the lesion showed that the percentages of neurons responding to various receptor types (hair, claw, joint, touch, deep, skin movement) were relatively similar in the two conditions. Approximately 75% of neurons recorded before (253/338) and also immediately after (28/36) the lesion were driven by the periphery. In contrast, only 27% of units recorded 10 days after the lesion were excited by peripheral stimulation, an observation which, together with the effects of chronic lesion on cortical evoked potentials, the authors attribute to post-lesion trauma of the dorsal columns. Most significant, however, is their report that within the hindlimb representation of these chronically lesioned cats 38% (15 units) of the units were driven by stimulation of the forelimb. Of these, 13 responded exclusively to forelimb stimulation while two responded to both forelimb and hindlimb stimulation. Nine of these units required only tactile stimulation. Such units were never encountered in normal or acutely sectioned cats.

Spinal Transection in Young Kittens. Between the time of birth and postnatal day 2, the distribution of thalamocortical terminals throughout the SI cortex of the cat first takes on an appearance similar to that seen in the adult (Wise et al. 1977; Sect. 7.3.2). Furthermore, in newborn kittens (6–24 h old) anaesthetized with pentobarbitone (18 mg/kg) the somatotopic organization of SI cortex has been described as being identical to that of adult cats (Rubel 1971). In this developmental context, the effects of chronically sectioning the spinal cord at T12 have been tested (McKinley et al. 1987). Lesions were placed in twelve 2-week-old, and six 6-week-old kittens which were maintained for 4 months before examination. Under ketamine anaesthesia the SI cortex was then thoroughly mapped. Unfortunately no information is given concerning how many animals survived and were mapped and no illustrations of the receptive fields in various regions are shown. The general projection of various areas of the body surface across SI is illustrated for one normal cat and for one cat from each experimental group (Fig. 7.17a–c). In animals lesioned at 6 weeks, 40% of sites

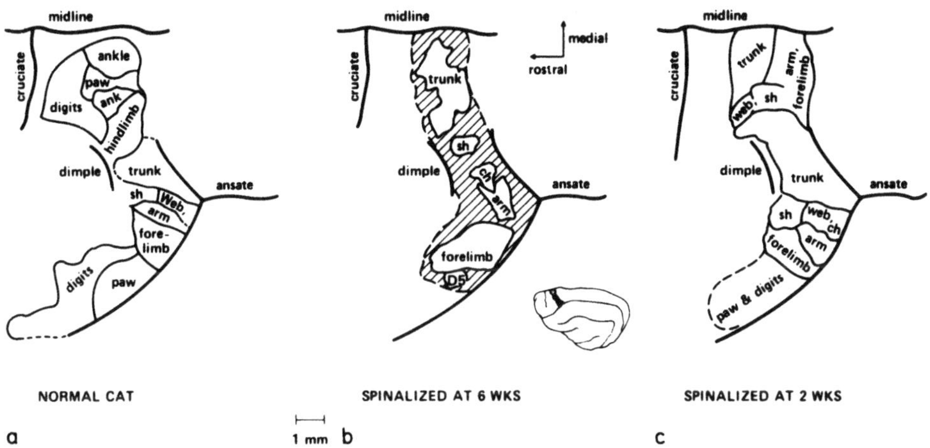

Fig. 7.17 a–c. Effects of spinal cord section at T12 in young kittens on the body representation in the SI cortex of the 120-day-old cat. Note that if section is performed at 2 weeks of age (**c**) a second representation of the forelimb and trunk develops in that region that would normally contain the hindlimb representation (**a**). In contrast, if section is performed at 6 weeks of age (**b**) no double representation develops but instead the parts of the body supplied by the spinal cord rostral of the lesion are represented in islands of responsive cortex that are surrounded by cortex that is unresponsive to peripheral stimulation (*hatching*). (From McKinley et al. 1987, p. 317)

sampled within the area of the normal body representation contained neurons which were unresponsive to cutaneous stimulation. Areas of the body which were supplied by that part of the cord rostral of the lesion (e.g., trunk, shoulders, chest, arm, etc.) were represented in islands of responsive cortex that were surrounded by unresponsive regions presumed to be SI cortex (Fig. 7.17b). Furthermore, the position of these representations was often different from that seen in normal animals (compare Fig. 7.17a with Fig. 7.17b). Thus while the forelimb representation was found in its normal position with respect to the ansate sulcus and the postcruciate dimple, the shoulder and trunk representations were found in medially displaced islands of responsive cortex that were in regions normally occupied by the trunk representation and the representation of the ankle and the hindpaw (Fig. 7.16b). In cats spinalized at 2 weeks the SI cortex lateral of the postcruciate dimple contained a relatively normal representation of the upper half of the body. In addition, however, there was a second and discontinuous representation of the trunk, shoulder and the forearm within the 2–3 mm of cortex medial to the normal trunk representation (Fig. 7.16c) (McKinley et al. 1987). Very recently this work has been extended by the investigation of the effects of chronic cordotomy in adult cats (McKinley and Smith 1990). In these animals only scattered sites throughout the deprived cortical region responded to tactile stimulation of the trunk and most cells within this region were unresponsive to low-threshold cutaneous input.

Some progress in establishing the origin of this new representation has been made by injecting different retrograde tracers into the two separate forelimb representations of adult cats that had sustained cord section at 2 weeks of age (McKinley and

Kruger 1988). These experiments showed that the thalamocortical neurons projecting to the normal, forelimb representation were located only in the medial part of VPL whereas a separate and more laterally located population of thalamocortical cells projected to the novel, medially-placed, forelimb representation. This important experiment shows that the effects of neonatal cordotomy cannot be explained by alterations in the connections between the VB complex and the SI cortex.

From their observations McKinley and colleagues suggest that in older animals (e.g. 6 weeks) there may be an upper limit to the amount of cortex which may be activated by a given body part. This idea gains some support from the results of experiments on the time-course of changes following cordotomy in 6-week-old kittens (McKinley and Swyter 1986). In these experiments animals were mapped immediately after lesion and after 3, 6 and 9 weeks survival. No trunk representation was found in the hindlimb region immediately before or after lesion. However, after 3 weeks the deafferented hindlimb cortex was fully occupied by inputs from the trunk but no topographical order was present. Between 6 and 9 weeks a dramatic reduction in the size of this new trunk representation was apparent such that only a "central core" of topographically organized trunk representation was found within the original hindlimb area. Unfortunately, no information is available on the somatotopic organization at the level of the DCN or the thalamus in these animals. Given the maturity of the thalamocortical connections in neonatal kittens and the thesis that the topographical order of projections is established early in development (Sects. 3.2 and 7.5.3) it would seem that further analyses of the basis of these changes might be useful in elucidating entities that are malleable in the early postnatal life of the cat.

7.3.5 Blockage of Primary Afferent Input in Specific Dorsal Roots

This and the following section (Sect. 7.3.6) contains a review of several experiments aimed at studying the effects of altering sensory input to the spinal cord of young and adult cats on the receptive fields of SI cortical neurons. These are described separately as the methodologies differ.

Reversible Block of Spinal Nerves and Somatotopic Organization. The reversibility of epidural anaesthetic block of conduction along the spinal nerves has been exploited by Metzler and Marks (1979) in a careful study designed to show the effects of transiently depriving the SI cortex of selected sensory inputs. Recordings were made using a chronic recording chamber from four cats in 14 recording sessions over a 3-month period. During recording sessions animals were paralysed with gallamine triethiode and artifically respired with a 2:1 mixture of N_2O:O_2. The body representation was determined for each of the four animals and a total of 213 units were studied. Of these units 68% responded to cutaneous (hair or light touch) input while 22% responded to joint movement or pressure on the bones or fascia and were thus classified as "deep". The remaining 10% were spontaneously active but did not respond to peripheral stimuli. Prior to epidural blocks, units in the hindlimb representation were isolated between the depths of 750 and 1350 μm. Input from the

L4 and L5 spinal nerves was then blocked by the epidural administration of local anaesthetic. Blocks took 10–15 min to appear and lasted 2–4 h. A total of 46 units were studied while blocks were in effect and 21 of these were held until the block disappeared, as assessed by the reestablishment of the pre-block receptive field. The general effects of the blocks were to abolish input from the upper part of the leg, the area represented in the L4 and L5 dermatomes of the cat (Brown and Koerber 1978). During blocks 26 units (56.5%) failed to respond to mechanical stimulation of the periphery. For 7 units (10.9%) that had pre-block receptive fields encompassing the upper leg and the tail or abdomen, the latter portion of the receptive field disappeared (Fig. 7.18a, b). In the remaining 13 units (28.3%) receptive fields were initially found proximal to the knee, within the L4 and L5 dermatome, but during the block these units developed new receptive fields on the foot or toes (5 units) or on the abdomen (7 units) (Fig. 7.18c, d). The skin within these new receptive fields was not effective in activating the units before blocks or after at which time these units were again responsive to stimulation of the skin of the upper leg.

During the block the responses of those units that developed new receptive fields varied in three ways from the normal responses of SI cortical neurons. First, these cells often habituated to stimuli applied at more than 1/s. Second, modality specificity was not always preserved. Thus some units which before the block responded only to cutaneous inputs responded to deep input during the block. Third, receptive field size was not specifically related to place. Thus, the new receptive fields on the foot could be either extremely small (4 mm^2) or stocking-like.

In their concluding remarks, Metzler and Marks (1979) suggest that widespread ineffective afferent connections in the intact animal could be the basis of their observations. They also, however, point to a most important conclusion which may be drawn from experiments utilizing reversible procedures "that modifications in synaptic connectivity are not always a prerequisite for functional plasticity" but that a "change in [the efficacy of] available inputs may be sufficient". No doubt they had in mind changes in synaptic efficacy such as those that may be elicited by the application of GABA antagonists or ACh (Sect. 7.3.3). With the aid of the detailed somatotopic maps illustrated in Figs. 7.14a, b it is now possible to see that the new receptive fields they reported were on skin (Fig. 7.18) that was probably represented only 500–1500 μm from the recording site—a distance that is not radically outside the reported displacement of the somatotopically inappropriate terminations of thalamocortical neurons from their somatotopically appropriate region (Figs. 7.15a, b, Sect. 7.3.2) (Landry and Deschenes 1981; Snow et al. 1988).

Dorsal Root Section, Receptor Representation and Somatotopic Organization. The effects on the receptive fields of SI neurons of chronically blocking input from a large number of dorsal roots have been studied by Franck (1980). It is important to emphasize that this study differs from that of Metzler and Marks (1979) in that available inputs were blocked permanently by sectioning dorsal roots, a procedure which may have profound affects on the synaptology of the spinal cord (Chap. 4). Cats were anaesthetized with ketamine (10 mg/kg i.m.) and a 2:1 mixture of N_2O:O_2 and paralysed with gallamine triethiodide. The medial portion of SI (mainly the

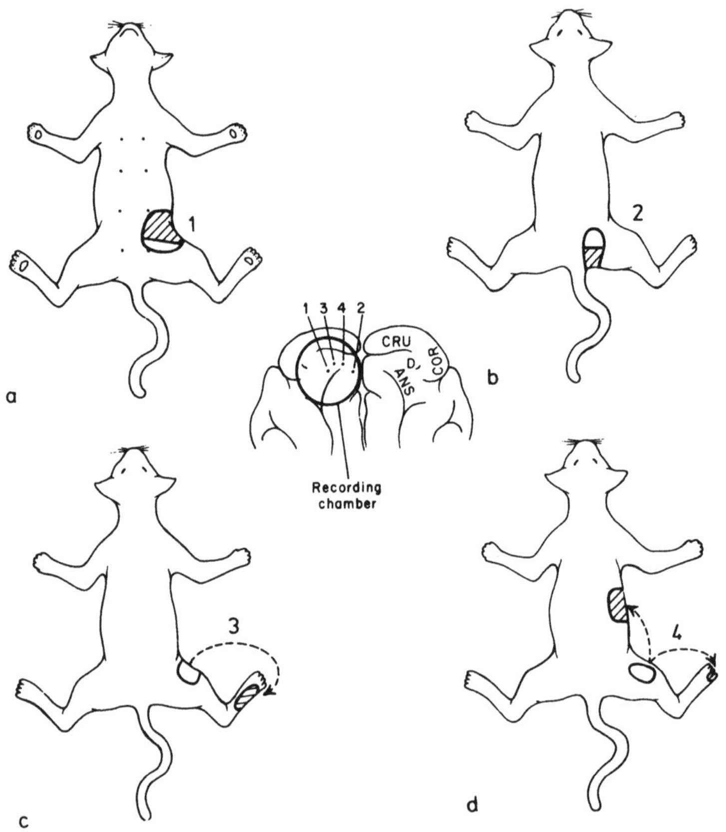

Fig. 7.18 a–d. The receptive fields of cortical neurons before, during and after epidural anaesthetic block of the L4 and L5 dorsal roots in cats anaesthetized with nitrous oxide. The *solid lines* surround pre- and post-block receptive fields and the *hatched areas* indicate the receptive fields during block. In **a** and **b** the post-block receptive field (*hatched region*) is formed by shrinkage of the pre-block receptive field. In **c** the pre-block receptive field completely disappeared and a new receptive field was established on the lateral foot. In **d** the pre-block receptive field disappeared completely and two new receptive fields appeared, one on the thigh and one on the toes. The receptive fields shifts involved should be referred to Fig. 7.14 in order to give the reader an appreciation of the real size of the transcortical shifts in effective input that underlie these results. *CRU*, cruciate sulcus; *ANS*, ansate sulcus; *COR*, coronal sulcus; *D*, post-cruciate dimple. (Modified from Metzler and Marks 1979, p. 380)

hindlimb representation) was mapped in five normal cats, a total of 144 units being sampled in 53 penetrations made throughout nine recording sessions. In all, 93% (134/144) of units responded to either hair (61 units, 46%), touch (28 units, 21%), deep, non-noxious pressure (24 units, 18%), and joint manipulation (21 units, 16%). The remaining 7% of units exhibited "unusual" properties. In three cats all dorsal roots caudal to L3, except L7, were sectioned unilaterally. After 8–55 days, 169 units in the medial portion of SI were studied in 59 penetrations made throughout nine recording sessions. Compared with the control studies there was no change in the proportion of unresponsive units. This is in marked contrast to the effects of epidural block of the input from the L4 and L5 roots where the percentage of unresponsive units increased from 10% to 50% during block (Metzler and Marks 1979). Following dorsal root section only 67% (113) of the units sampled responded to specific stimuli, there being a marked reduction in the percentage of units responding to hair (27%) and light touch (13%) compared with the control group (Franck 1980). Concurrently, there was a marked increase in the percentage of units responding to deep, non-noxious pressure (42% compared with 18% in the control group). Similarly, during epidural block there was an increase in the percentage of neurons responding to deep touch and pressure (Metzler and Marks 1979). Thus, taking account of Franck's (1980) study together with that of Metzler and Marks (1979), it would seem that blockage of input at the level of the dorsal roots causes a transient reduction in the numbers of cortical neurons responding to peripheral stimulation but a permanent increase in the numbers of neurons responding specifically to deep pressure at the expense of inputs from hair, light touch and joint receptors.

Following transection of all the lumbosacral dorsal roots other than L7 most (68%) SI neurons were responsive to joint manipulation and deep pressure while the remainder were driven by a combination of deep and cutaneous stimuli (Franck 1980). Some of the latter group had multiple or bilateral receptive fields or stocking-like receptive fields like those observed during epidural block (Metzler and Marks 1979).

The most dramatic effects of dorsal root section were those on the somatotopic organization of SI. In the experiments of Franck (1980) the L7 dorsal root which supplies much of the distal hindlimb was left intact. Dermatomal maps in the cat have shown that skin within the L7 dermatome is also supplied by the L6 and S1 dorsal roots but that the L7 dermatome is restricted to the foot and is centred on the toes (Brown and Koeber 1978). Within 8–55 days of sectioning all dorsal roots caudal to L3 except L7, the representation of the toes and the foot contracted to 40% of the control size. This was not accompanied, even transiently, by the development of unresponsive regions around the contracting hindlimb representation. Instead, neurons around the representation of the contralateral hindlimb were responsive to stimulation of skin on the abdomen, thorax and proximal forelimb, areas of the body that supply afferents to those dorsal roots rostral to L3. Unfortunately, from the data provided it is not possible to determine the immediacy of the emergence of these new receptive fields, but Franck (1980) infers that the contraction of the hindlimb (or L7) representation occurred progressively between 8–55 days. Thus the establishment of the new receptive fields is probably a progressive phenomena that is related to the contraction of the representation of distal skin. An immediate emergence of new receptive fields would be expected on the basis of the results of Metzler and Marks (1979). However, the effects of spinal cord section in the kitten (McKinley et al. 1987) and the adult cat (Frommer et al. 1977) suggest that the emergence of new inputs is a process which proceeds slowly following the abolition of input to a large area of cortex. Perhaps the simplest explanation of these apparently conflicting results lies in the suggestion that following the removal of its normal afferent input, a cortical neuron may *immediately* develop a sensitivity to inputs from that innervated skin that is represented in nearby (within 1000 μm) areas of the SI cortex. After more time (such as that which might be required for the contraction of the L7 representation) innervated skin gradually exerts an influence over a wider area of "available" cortex. In Franck's (1980) experiment sectioning the L6 and S1 roots that supply at least some afferents to the L7 dermatome would reduce the number of afferents innervating the edge of this dermatome. The slow contraction of the representation of the L7 dermatome perhaps indicates the survival into adulthood of a slow competitive mechanism that governs the allocation of cortical territory on the basis of the number of primary afferent fibres with an influence over a partially deprived cortical region (cf. Sect. 3.4). It would be interesting to know the nature of the psychophysical experience elicited by stimulation in the L5 dermatome of man in a situation in which all lumbosacral dorsal roots were cut or blocked with the exception of the L5 dorsal root – a situation in which only a small isolated island of cortex retains an input.

7.3.6 Damage to Peripheral Nerves and Effects of Usage on Cortical Representation

The effects of peripheral nerve section on the receptive fields of SI neurons in the cat have been assessed by Kalaska and Pomeranz (1979). In this study the forepaw was completely denervated by sectioning, ligating and removing the distal stumps of the radial, median and the palmar and dorsal cutaneous branches of the ulnar nerves in fourteen 9- to 12-week-old kittens and eight adult cats. The responses of neurons in the SI cortex were studied under pentobarbitone or Dial anaesthesia, 8–10 weeks (6–7 months in two adult denervated cats) after surgery. Potentials evoked in the paw area by electrical stimulation of the medial cutaneous nerve of the forearm and the ulnar nerve were also examined. A control group of 12 unoperated and 2 sham-operated, litter-mate kittens was also studied. In all cases the experimenter was unaware of whether any given animal was normal, sham-operated or denervated. Recording sites were located histologically.

The identity of the cortical areas containing neurons of the forepaw and the forearm representation was determined in control cats and related to the pattern of sulci and gyri, the authors claiming that this method of determining these representations was sufficiently accurate for the purposes of studying the effects of chronic paw denervation. These regions of SI are referred to as "paw cortex" and "forearm cortex" and in Fig. 1 of the paper by Kalaska and Pomeranz (1979) these regions are illustrated as being about 2–3 mm across and being separated at their closest point by about 3 mm of cortex (compare Fig. 7.14a with Fig. 7.14b) (Felleman et al. 1983b; McKenna et al. 1981).

In control kittens (aged 17–22 weeks), 90% of all cells (107/119) within the area referred to as paw cortex, had receptive fields on the paw, while in forearm cortex 76% of cells (53/70) had receptive fields on the forearm (Kalaska and Pomeranz 1979). No cells in paw cortex had receptive fields on the forearm and no cells in the forearm cortex had receptive fields on the paw. In denervated kittens 52% of cells (67/128) in paw cortex had forearm receptive fields and 19% had no receptive fields compared with 6% in controls. Clear paw receptive fields were found in 3% of cells, confirming that there was some reinnervation of the denervated skin. No change in neuron responses was found in the forearm cortex of cats denervated as adults or kittens. The receptive fields of cells in the denervated paw cortex were typical of the receptive fields in forearm cortex of normal cats. Furthermore, in any track, modality specificity was present and units with multiple receptive fields were not observed. Kalaska and Pomeranz (1979) also note that while some cells were exquisitely sensitive many required more intense natural stimuli and some showed habituation.

In denervated adults only 7% of cells (8/121) in paw cortex had receptive fields on the forearm and 37% of cells in this region failed to respond to any peripheral stimuli. Two adult denervated cats were allowed to survive for 6–7 months before cortical mapping. No greater number of neurons with forearm receptive fields were found in the paw cortex of these animals than in adult denervated cats mapped 8–10 weeks after nerve section.

It is clear from this study that chronic section of a peripheral nerve either in a young kitten or an adult cat elicits reorganization of the receptive fields of neurons within the SI representation of that nerve (Kalaska and Pomeranz 1979). In addition, the greater number of unresponsive cells in the deafferented region of adult denervated cats compared with the number in adults denervated as kittens (39% compared with 19%), together with the smaller number of forearm receptive fields in paw cortex of adult denervated cats (7% compared with 52% in kitten denervates) suggests that the CNS of the adult is less prone to functional reorganization than the kitten. In contrast to results reported elsewhere for the rodent and the primate (Sects. 7.2.8, 7.5.2) peripheral nerve section in kittens or adult cats did not produce a somatotopically organized representation of the forearm within paw cortex. Indeed Kalaska and Pomeranz (1979) make the important observation that in four out of seven electrode tracks that were histologically verified as being perpendicular to the cortical surface, the receptive fields of even nearby units were scattered "and often non-overlapping". As they point out, these preliminary observations suggest that the columnar organization of paw cortex is disrupted by peripheral nerve lesion and so emphasize the potential plasticity in the neural circuitry which underlies this clear and widely accepted feature of cortical organization. In the cat and probably other mammals, place specificity involves a finer level of organization than modality specificity (compare Fig. 7.13a, b with Fig. 7.13c, d). As in the raccoon (Sect. 7.4.4) denervation of the forepaw in the cat removes the input from a relatively large area of cortex. The development of new inputs to such large areas of cortex possibly involves mechanisms that are different from those described in the monkey and the rat where new inputs were somatotopically organized but confined to areas of deprived cortex nearby (within 500–600 μm of) normal cortex. Perhaps in the case of paw denervation in the cat the cortical changes reflect alterations at lower levels of the neuraxis. In relation to this, Kalaska and Pomeranz (1979) have reported similar changes in the DCN after identical lesions in very young kittens (Sect. 5.3.4).

There exists one preliminary report on a set of experiments in which the receptive fields of cortical neurons were measured before, during and after increased sensory input (Recanzone et al. 1986). These studies were performed on an unstated number of adult cats anaesthetized with pentobarbitone. The ulnar nerve in the distal forelimb was stimulated at levels that "massively excited large afferents" at 20 Hz for

6–8 h. Multiunit recordings were made at particular sites in the forelimb and hindlimb representation with floating iridium microelectrodes. The "multiunit receptive fields" at these sites were examined prior to stimulation and every hour during the stimulation period. In all cats, receptive field sizes had increased after 2 h of stimulation and after 6–8 h the receptive fields had increased in size on average 3.2 times. Presumably this figure refers to the increase of receptive fields in the forelimb representation but it is stated that similar changes in receptive field size were noted elsewhere, "for example the hindpaw" representation. These workers also report movement of the receptive fields from one digit to another over the stimulation period—a shift equivalent to a transcortical distance of about 400–1000 μm in the digit representation of the cat (Figs. 7.13a, b and 7.14a, b) (Sect. 7.3.1).
Without more data and the appropriate controls (identical experiments in which the stimulator was not turned on) it is difficult to interpret these very preliminary results. Even with these controls it would surely be difficult to relate the effects of 432 000–576 000 sychronous volleys in the larger afferent fibres of the ulnar nerve, delivered over 6–8 h in an anaesthetized preparation, to changes that might occur during natural utilization of a portion of the skin surface.

7.3.7 Cortical Damage

Behavioural Effects. The recovery of the ability to discriminate roughness following lesion of SI and SII in newborn (P5) and adult cats has been tested 6 months after lesion. Lesions were performed by aspiration and were "generously large", including parts of the visual, auditory and motor cortex (Benjamin and Thompson 1959). The ability of animals to learn to discriminate between different grades of roughness was then tested. The test included four levels of difficulty which were, in order of increasing difficulty: (a) very rough versus smooth (smooth cardboard), (b) various degree of roughness (grades of emery paper) versus smooth, (c) very rough versus least rough, and (d) various degrees of roughness versus standard grades of roughness. At 6 months of age unoperated kittens performed identically to normal adults and operated kittens performed identically on all but the most difficult task (task (d)). In contrast, all but one of the cats operated on as an adult, failed to learn to discriminate roughness in even the easiest task (task (a)). The authors concluded that the "degree of behavioural impairment resulting from ablation of somatic cortex is a function of age at the time of operation". Unfortunately, tests were not made to compare the overall decrement of learning ability following such large lesions in neonatal and adult animals. As detailed in Sect. 7.3.1 there now exist for cats highly detailed somatotopic maps (Rubel 1971; McKenna et al. 1981; Felleman et al. 1983b) which would make it possible to compare the neurological effects of lesions in specific and restricted parts of the body representation (e.g., the SI forepaw representation) in adult and neonatal animals. The use of such focal lesions should facilitate the establishment of controls that isolate the effects of the lesion on the ability to learn any somatosensory discrimination task from the loss of the ability to perceive a particular stimulus on a restricted area of the body surface following restricted lesions to the cortical representation of that part. With these controls it should be possible to establish the relationship between the developmental stage of an animal and the ability of the CNS to compensate for focal lesions of the first and second somatosensory areas (see also Sects. 7.2.10 and 7.5.5).

7.4 Plasticity in the Cortex of Adult and Infant Raccoons

7.4.1 Somatotopic Organization and Cytoarchitectonics

The locations of the visual, auditory, motor and first and second somatosensory areas across the cerebral cortex of the raccoon were first studied by Welker and Seidenstein (1959). Surface evoked potentials were recorded from the somatosensory cortex of raccoons anaesthetized with pentobarbitone. Attention was focused on the representation of the forepaw within SI. Tactile stimuli were applied using an electromechanical device or handheld probes. Using these methods Welker and Seidenstein (1959) reported that SI of the raccoon covers an area of 1000–1100 mm^2 compared with 150–200 mm^2 in the cat (compared with 90 mm^2 measured from the representation of the body across areas 3b, 1 and 2 in the cat, see Fig. 2 in Felleman et al. 1983b, Fig. 7.14b). Approximately 50% to 60% (about 500 mm^2) of SI in the raccoon is devoted to the representation of the palmar surface of the forepaw (Fig. 7.19a) and in any individual the representation of the glabrous skin of the palmar pads and digits is related to surface features of the cortex. It is therefore possible to identify particular parts of the somatotopic map of the paw by visual inspection of the surface of the cortex. As this situation has made the raccoon a most useful animal for studying the effects of digit amputation on the somatotopic organization of the forepaw representation (Sect. 7.4.4), some description of the relationship between the representation of the forepaw and the surface anatomy of SI is appropriate (Fig. 7.19). The representation of the forepaw is laid out around the three arms of the triradiate sulcus so that it is divided into three segments (Fig. 7.19b, c). The somatotopic map of the forepaw is such that the most lateral of these segments contains the representation of digit 1 posterior to that of digit 2. The most rostral segment contains the representation of digit 3 lateral to that of digit 4 and the most caudal segment contains the representation of digit 5 medial to the representation of the pads of the palm. Within each segment, the representations of the digits 1 to 4 or digit 5 and the palmar pads are found on separate "subgyri" which are often anatomically delimited by sulcal spurs radiating from the triradiate sulcus (Kelahan et al. 1981; Warren and Pubols 1984). Laterally the forepaw representation is separated from the face representation by the coronal sulcus, the rostral end of which curves medially to form a boundary between the lateral part of the forepaw representation and the postcruciate (or posterior cruciate) gyrus. This medially directed branch is sometimes referred to as the lateral central sulcus (Johnson et al. 1982). Medially the forepaw representation is separated from the foot representation by the postcruciate (or interbrachial) sulcus which projects a branch laterally to form a boundary between the medial part of the forepaw representation and the postcruciate gyrus. This branch is sometimes referred to as the medial central sulcus (Johnson et al. 1982). Often the lateral and medial central sulci appear to fuse and

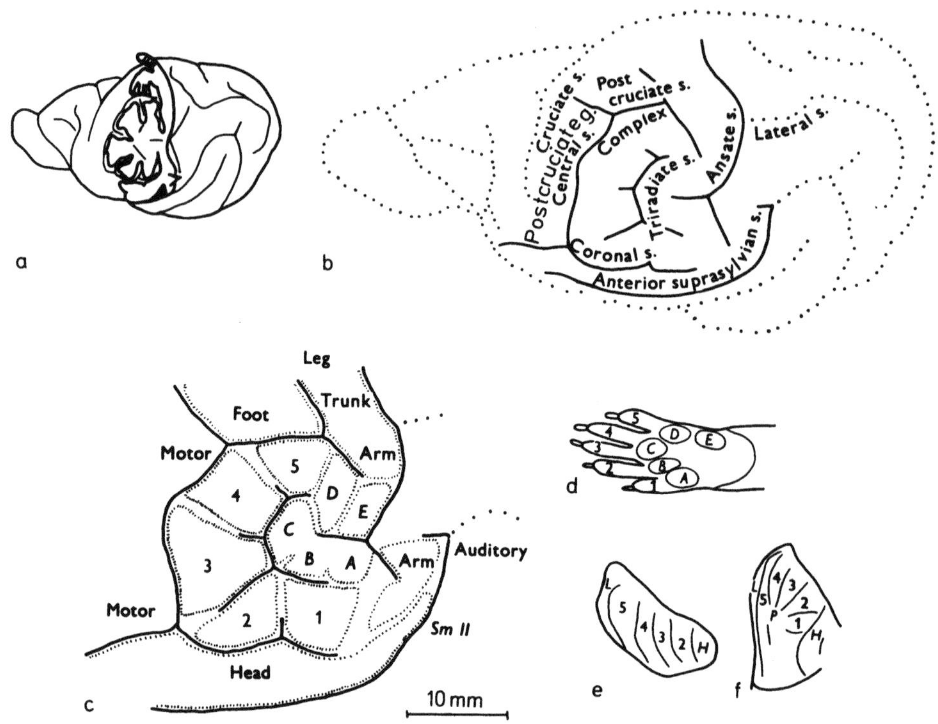

Fig. 7.19 a–f. Somatotopic representation of the forepaw across the SI cortex and the VB complex of the pentobarbitone-anaesthetized raccoon. **a** Representation of the entire body surface across the SI cortex. **b** Anatomy of the cortex containing and surrounding the paw representation. **c, d** Details of the representation of the glabrous skin of the digits and the palmar pads around the triradiate sulcus and the function of surrounding cortical areas. **e, f** Schematic diagrams of the representation of the digits (*1* to *5*), pads (*P*), head (*H*) and hindlimb (*L*) in coronal and horizontal planes of the VB complex, respectively. *Scale* refers only to the approximate dimensions of **b** and **c**. (**a** Modified from Herron 1983, p. 249; **b–f** modified from Welker and Johnson 1965, p. 249)

when this happens that may be referred to as simply the central sulcus (Fig. 7.19b) (Welker and Seidenstein 1959). Caudally the paw representation is separated from "silent" cortex and auditory cortex by the ansate and anterior suprasylvian sulci, respectively. It is interesting to note that Welker and Seidenstein (1959) and Rasmusson (1982) have documented some considerable inter-animal variation in the dimensions of the representations of particular digits (Fig. 7.20), a finding which has since been reported in the rat (Sect. 7.5.2) (Wall and Cusick 1984) and in the primate (Fig. 7.2) where it has been interpreted as being related to inter-animal variation in the utilization of particular regions of the surface of the hand (Merzenich et al. 1987; Sects. 7.2.4, 9).

Surface potential (Welker and Seidenstein 1959) and single-unit (Rasmusson 1982) recordings across the gyral or subgyral crowns have revealed that the glabrous skin of single digits is represented in a topographically ordered manner. In contrast, the hairy skin of the digits is represented in relatively small regions in the walls of the

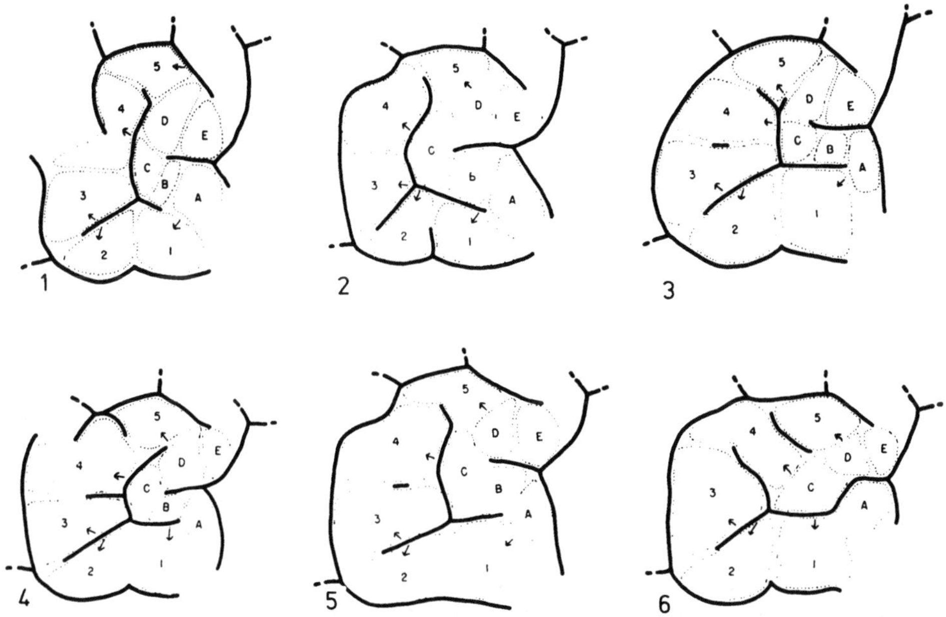

Fig. 7.20. Variation in the size and shape of the representation of the glabrous skin of the digits and the palmar pads between different adult raccoons anaesthetized with pentobarbitone. The numbering of the pads and the digits corresponds to that shown in Fig. 7.19d. It is possible that this variation results from individual differences in the frequency of use of particular skin surfaces and in this regard this figure should be compared with Fig. 7.2 (Modified from Welker and Seidenstein 1959, p. 481)

triradiate sulcus or in sulcal spurs. The hairy skin of the dorsal surface of the hand is located in two small areas at the junction of the palm representation and the representation of the lateral and medial surface of the forearm.

Anterior of the central sulci, Welker and Seidenstein (1959) reported a sharp rise in the intensity of stimuli required to activate neurons. Recently, Johnson et al. (1982) have used microelectrodes to explore the tissue around the central sulci, the interbrachial (or postcruciate) sulcus and in the postcruciate gyrus (Fig. 7.19b) of raccoons anaesthetized with urethane, chloralose or methoxyflurane and N_2O. They confirmed the representations of the glabrous and hairy skin described by Welker and Seidenstein (1959) and described, in addition, a "heterogeneous zone" most of which lay rostral and medial to the forepaw representation. The responses of neurons in the heterogeneous zone vary. Caudally this zone abuts the representation of the digits and interdigitates with the representation of the dorsal hairy skin. Within these most caudal parts of the heterogeneous zone, stimulation of the dorsal surface of a number of digits produced responses in multiunit recordings. However, when single units were isolated they never responded to stimulation of the dorsal surface of more than one digit. More rostrally and medially in the heterogeneous zone, cells responded to stimulation of the dorsal surface of the digits, the claws and occasionally the glabrous surfaces. Further rostrally and medially, within the banks of the central sulci and the interbrachial sulcus, cells responded to the hairy and glabrous surfaces of the digits, palm and the skin of the forearm. In the anterior wall of the central sulci cells responded to muscle stretch, joint movement and sometimes also cutaneous stimuli applied to the hand, arm and shoulder regions. Johnson et al. (1982) have called this region the "muscle projection zone". Rostral to this zone, on the postcruciate gyrus, neurons were responsive to stimulation of the proximal hand or distal forearm but difficulty was experienced in deciding whether muscle, cutaneous or both receptor types were involved.

It is difficult to subdivide the sensory-motor region of the raccoon cortex into those cytoarchitectonic areas (4, 3a, 3b, 1 and 2) which have been clearly described in cats and monkeys. Johnson et al. (1982) point out that the "cell free stripe" which is pronounced in layer V of the cortex between the central sulci and the ansate sulcus, is indicative of area 3b. Both Johnson et al. (1982) and Sakai (1982) report that the cortex on the posterior bank of the cruciate sulcus lacks a distinct layer IV but has a well-developed layer V containing large pyramidal cells. In other mammals these features are indicative of area 4 and motor cortex (Hassler and Muhs-Clement, 1964). In the raccoon, cells in this region were unresponsive to mechanical stimulation (Johnson et al. 1982). Between this region and the rostral banks of the central sulci there is a progressive decrease in the number of large pyramidal cells without a concomitant increase in the size of layer IV. Johnson et al. (1982) consider that the caudal boundary of area 4 lies close to the posterior lip of the cruciate sulcus. Thus by what they refer to as the "remainder approach", they suggest that the cortex between this boundary and the rostral boundary of area 3b (including the heterogeneous and muscle projection zones) constitutes cytoarchitectonic area 3a and that this region contains a reversed representation (digits caudal) of the forelimb and upper body. In contrast, Sakai (1982) considers that the rostral boundary of area 3a lies along the anterior lips of the central sulci. It is notable that Welker and Seidenstein (1959) and Hardin et al. (1968) considered the entire postcruciate gyrus to be part of motor cortex and that the latter group of workers have presented evidence that the motor representation of the forepaw digits extends caudally to the fundus of the central sulci.

In summary, there exists across area 3b of the raccoon cortex a single representation of the body surface. The exact position of the boundary between area 3a and area 4 is controversial but, caudally, area 3a probably constitutes at least the heterogeneous zone of Johnson et al. (1982). It is important for both the studies of thalamocortical connections (Sect. 7.4.3) and the effects of digit amputation on cortical somatotopy (Sect. 7.4.4) to emphasize that the caudal region of the heterogeneous zone interdigitates with the representations of the glabrous skin of individual digits within area 3b. Within the raccoon, as in rat (Sect. 7.5.1), it thus seems reasonable to consider that SI is indeed equivalent to cytoarchitectonic area 3b.

7.4.2 Neuronal Responses in SI Cortex and the Effects of Anaesthetics

Pubols and Leroy (1977) studied the responses of cortical neurons within SI of 21 adult raccoons anaesthetized with methoxyflurane. Of 128 cells, 110 responded to light touch of the glabrous surface of the forepaw. The remaining 18 cells responded to joint or claw manipulation, deep pressure or stimulation of the hairy surface of the hand. The recording loci of the latter cells tended to be in the wall of sulci while those cells responding to light touch were located on the gyral crowns. The response of 51 cells in the gyral crowns to punctate stimuli (glass rods, von Frey hairs) or stimulation with edges that had a width of 0.3 mm and lengths of 5.0, 7.5, 10.0 or 20.0 mm, were tested. Of these, 21 cells responded to edge stimuli and were called "linear" units, while 30 responded to punctate stimuli and were referred to as "round-field" units. The response of linear units to punctate stimuli such as von Frey hairs was often unreliable or absent. Instead, linear units responded to indentation of the skin with an edge set at a particular orientation. On the palm the preferred orientations of a sample of 9 linear units appeared to be random but on the digits the 8 units studied all had preferred orientations that lined up with the long axis of the digit. Insufficient data are available to determine whether there exists orientation-selective columns of cells or whether preferred orientation is represented in an orderly topographical manner (Pubols and Leroy 1977). Finally, Pubols and Warren (1985) have reported four linear units in the VB complex suggesting that in the raccoon the physiological mechanisms underlying edge detection might lie below the level of the cortex.

Kelahan et al. (1981) studied the responses of cells and cell clusters in 10 normal raccoons anaesthetized with pentobarbitone. Using von Frey hairs, they determined the "threshold" receptive field of cells. In addition, they used a calibrated nylon fibre to apply punctate stimuli with a force of 15 g to the skin and called the receptive fields so mapped "suprathreshold" receptive fields. Threshold receptive fields on glabrous skin tended to be ovoid, had a mean area of 13 mm^2, and were found on the palmar surface of one segment of a digit or on one pad. In contrast, the suprathreshold receptive fields had a mean area of 101 mm^2 and often covered the entire palmar surface of a digit, sometimes extending onto the adjacent palmar pad. Units encountered at different depths in tracks perpendicular to the cortical surface had concentric or overlapping threshold and suprathreshold receptive fields (Fig. 7.21a,b) (Kelahan et al. 1981).

In a similar study Rasmusson (1982) examined the responses of cells in 148 penetrations in the forepaw area of raccoons maintained areflexive under deep pentobarbitone anaesthesia. In 24 of these penetrations, 15 of which were located near the arms of the triradiate sulcus, cells responded only to "high" intensity stimuli. In the remaining 121 tracks units responded to stimuli of "low" and "moderate" intensity. Rasmusson (1982) specifically noted the absence of suprathreshold input from wide regions of the hand although he reports that a probe 2–3 times the diameter of a von Frey hair was often required to activate cortical units. The threshold receptive field of neurons in the representation of the glabrous surface of a digit were usually small and Rasmusson (1982) suggested that the large suprathreshold receptive fields reported by other workers resulted from the spread of mechanical stimulus from distant sites, to skin within the threshold receptive field. While this point requires further experimentation, the report of split suprathreshold receptive fields at 4% of cortical sites studied in normal animals (Kelahan et al. 1981) and the discovery of cells with large receptive fields in the VB complex and the cuneate nucleus (Pubols and Warren 1985) suggest that the large suprathreshold

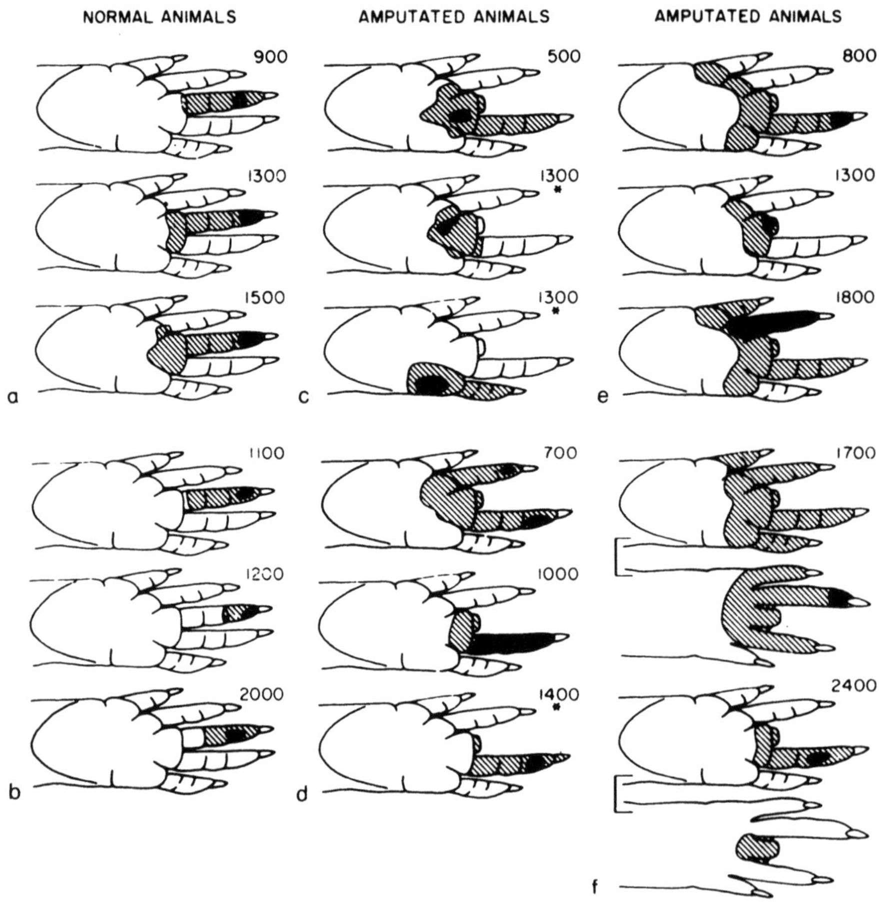

Fig. 7.21 a–f. Threshold and suprathreshold multiunit and single-unit receptive fields recorded in single tracks within the digit 3 subgyrus of normal adult raccoons (**a**, **b**) and adult raccoons following amputation of the 3rd digit at 4 weeks of age (**c**) and 8 weeks of age (**d–f**). Raccoons were anaesthetized with pentobarbitone. **a**, **b** In a single track in the normal animal both the threshold receptive fields (*black*) and the suprathreshold (15 g; *hatched*) receptive fields recorded at different depths overlap. Amputation of digit 3 in the young raccoon disrupts the normal columnar organization of both the threshold and suprathreshold receptive fields within the digit 3 subgyrus (**c**–**f**). *Asterisks* indicate sites where a single unit was recorded. The receptive field shifts involved should be compared with Fig. 7.19c in order to give the reader an appreciation of the real size of the transcortical shifts in effective input that must underlie these results. (Modified from Kelahan et al. 1981, p. 157)

receptive fields reported by Kelahan et al. (1981) and Kelahan and Doetsch (1984) may not result from stimulus spread.

Johnson et al. (1982) reported no differences between animals anaesthetized with a mixture of Dial and urethane (144:463 mg/kg), chloralose (70 mg/kg) and methoxyflurane plus N_2O with respect to the location, extent and type of receptive fields in the forepaw representation, the heterogeneous zone or the muscle zone. The only effect of anaesthesia they noted was an increase in spontaneous activity with lower anaesthetic levels. In contrast, Kelahan and Doetsch (1984) have reported that in animals anaesthetized with N_2O and thiopental the mean size of the threshold receptive fields of cells in the palmar representation is 9.2 mm^2 compared with 13.4 mm^2 in animals anaesthetized with pentobarbitone. In conclusion, it thus seems that pentobarbitone and other anaesthetics may have a differential effect on the receptive fields of some neurons within area 3b of the raccoon cortex. As will be shown below, similar effects of anaesthetics are also apparent in studies of the responses of cortical neurons following removal of their normal peripheral input (Sect. 7.4.4).

7.4.3 Ineffective Afferent Connections

In normal adult raccoons the largest surface potentials evoked by electrical stimulation of the tip of one digit are recorded within a relatively small centrally placed focus on the appropriate subgyrus (Carson et al. 1981). However, potentials of more than 50% of the size of these *on-focus* potentials can be recorded across a region of cortex which extends near to, or even to within, the focus of the adjacent digit. This discovery, together with the observation that when a single digit is removed cells in the associated subgyrus become responsive to surrounding skin (Kelahan et al. 1981; Rasmusson 1982; Sect. 7.4.4), has raised the question of whether thalamocortical neurons in the VB complex of the normal raccoon give rise to ineffective terminations in the somatotopically inappropriate parts of the SI body representation.

With this problem in mind, Rasmusson and Nance (1986) injected different fluorescent dyes into adjacent subgyri and then ascertained the distribution of retrogradely labelled cells in the VB complex. In agreement with Warren and Pubols (1984) they found that the cells projecting to a single subgyrus were located within a single crescent-shaped lamella. Labelled cells were also found in the VPI, VPO and the intralaminar group. When different dyes were injected into adjacent subgyri, cells labelled with one dye were found in one lamella while cells labelled with the other were found in the adjacent lamella. Only when dye from the two cortical injection sites overlapped were neurons found labelled with both dyes.

In summary, these findings suggest that in normal raccoons all thalamocortical cells within a single lamella project to the somatotopically appropriate subgyrus. However, two cautionary comments should be made. First, the capricious nature of the retrograde tracing method casts some doubt on conclusions drawn from negative results. Second, if somatotopically inappropriate projections do exist between the thalamus and the cortex, since by definition they must be relatively ineffective, they may not take up detectable amounts of tracer from the extracellular space.

7.4.4 Effects of Amputation on Cortical Somatotopy

The unique relationship between surface structures and the representation of the glabrous surface of the forepaw of the raccoon (Sect. 7.4.1) has made this animal a useful experimental model for studying the effects, on cortical somatotopy, of digit amputation at both the neonatal and adult stages. In the first of these studies the 3rd digit was removed from 10 raccoons at 2, 4, 6 or 8 weeks of age (Kelahan et al. 1981; Carson et al. 1981). Nine to 12 months later the digit and pad representations in SI were mapped under pentobarbitone anaesthesia. Initially, averages were made of surface potentials evoked by electrical stimulation of the distal glabrous skin of digits 2 and 4 and the stump of digit 3. Following this, the responses of single units and clusters of units to tactile stimulation of the forepaw were recorded. The receptive fields of control units were mapped using threshold tactile stimuli and also using a calibrated 15 g nylon probe to apply suprathreshold stimuli. Data were compared with similar recordings made in 10 normal adult animals.

Evoked-Potential Studies. As mentioned in Sect. 7.4.3, in normal animals, electrical stimulation of the tip of one digit produces large evoked potentials even within the central part of the representation of an adjacent digit (Carson et al. 1981). The ligation of digit 3 ("functional amputation"), just prior to recording, completely abolished all responses to electrical stimulation of digit 3 but did not alter the amplitude (or presumably the spatial distribution) of the surface potentials evoked in the digit 3 representation by electrical stimulation of digits 2 or 4. In contrast, in adult raccons from which digit 3 had been amputated at 4–8 weeks of age, evoked potentials of more than 50% of the maximum amplitude could be recorded over a much larger area of cortex. The results of these experiments led Kelahan et al. (1981) and Carson et al. (1981) to conclude that: (a) in the cortex of normal adults there must be some spatial overlap between inputs from adjacent digits which is only reflected in the responses of cortical neurons when synchronous volleys are evoked in the afferent fibres supplying a digit, and (b) the underlying pathway and mechanism responsible might have something to do with the altered evoked potential maps seen in the amputated group. In our view, however, it remains to be shown that these subthreshold pathways in normal adults are responsible for the enlargement of the areas affected by similarly synchronous volleys following the removal of a digit in infant raccoons.

Microelectrode Studies. In the second part of their experiments, Kelahan et al. (1981) made microelectrode recordings from 32 single units and 129 cell clusters in the digit 3 subgyrus of amputated animals. No silent or unresponsive regions were found within SI but neither was there any somatotopic organization in the deprived region. This disruption of somatotopy was not restricted to the horizontal dimension of the cortex but was also observable within single electrode tracks. Thus, within a single track, the position of both threshold and suprathreshold receptive fields, based on activity recorded from clusters of units and single units, varied with depth in a nonsystematic manner (Fig. 7.21c–f) (Kelahan et al. 1981). The deprived subgyrus thus lacks the columnar organization of receptive fields that typifies normal SI cortex in the raccoon (Fig. 7.21a, b) (Kelahan et al. 1981) and other mammals (Fig. 7.13a, b).

In terms of the receptive fields recorded within the deprived digit 3 subgyrus, Kelahan et al. (1981) reported three major differences from the receptive fields seen in area 3b of normal animals. First, there was an increase in the size of both threshold (13–25 mm^2) and suprathreshold (101–353 mm^2) receptive fields in amputated animals (Fig. 7.21). Second, multiple, threshold receptive fields were recorded at 7% of the cortical sites whereas none was recorded in normal adult raccoons (Fig. 7.21d). Third, effective inputs from the dorsal surface of the digits and the paw were found in 59% of sites that in normal adults would receive inputs only from glabrous skin (Fig. 7.21f) (Kelahan et al. 1981). The first two of these three differences might arise from the acceptance of receptive field data that are based primarily (129/161 cases) on multiunit recording with deprived cortex that no longer shows a columnar organization of receptive fields. Indeed the receptive fields of two isolated units in deprived cortex do not look any larger than those seen in normal animals (Fig. 7.21c, d). However, the third difference is reminiscent of the enhanced representation of the dorsal surface of the hand seen in primates following partial deafferentation of the hand (Sect. 7.2.8).

Kelahan et al. (1981) and Carson et al. (1981) claimed that in both neonatal and adult, barbiturate-anaesthetized raccoons, immediately after amputation of digit 3, cells in the digit 3 subgyrus were unresponsive to peripheral stimuli. Similar findings were reported following amputation of digit 5 in adult raccoons (Rasmusson 1982). Subsequently, however, Rasmusson and Turnbull (1983) studied the effects of amputation of one digit (the 4th or 5th) in three groups of three adult raccoons. In each group two animals were anaesthetized with chloralose and one was anaesthetized with pentobarbitone. Multiunit recordings were made in the cortices of one group 2–14 h after amputation, in another after 1–2 weeks, and in the third 8–20 weeks postoperatively. In 61% (38/62) of penetrations made immediately after amputation (2–14 h), neurons were encountered that fired in response to the cessation of tapping or in which tapping produced a suppression of spontaneous activity. These inhibitory responses were found with both anaesthetics. In contrast, those penetrations in which no neurons were found to be excited by peripheral stimulation were all anaesthetized with barbiturate. One to 2 weeks after amputation, 33% (14/42) of penetrations had inhibitory responses and after 8–20 weeks no penetrations exhibited purely inhibitory responses. One to 2 weeks postoperatively, excitatory responses were seen in chloralose-anaesthetized animals. After 8–20 weeks all penetrations yielded excitatory responses to peripheral stimulation. From these data it was concluded that: (a) surviving inputs initially have inhibitory actions upon cells in deprived cortex of the raccoon but that with time these inputs develop excitatory effects, and (b) immediately after digit amputation the inhibitory actions of surviving inputs are mediated by GABAergic mechanisms which as a consequence of pentobarbitone anaesthesia have an increased effectiveness (Rasmusson and Turnbull 1983). It is interesting to note with reference to (b) that Kelahan and Doetsch (1984) report preliminary observations that systemic administration of the GABA antagonist, picrotoxin, produces a marked augmentation of both on-focus and off-focus evoked potentials in the forepaw representation of normal raccoons, as well as changes in the size and position of the receptive fields of single neurons.

The effects of amputation of a digit from adult animals on single neurons in the deprived subgyrus has also been studied by Rasmusson (1982) and Kelahan and Doetsch (1984). Rasmusson's (1982) experiments were carried out using pentobarbitone anaesthesia, 2, 8 and 16 weeks after amputation. Kelahan and Doetsch (1984) used pentobarbitone (35 mg/kg i.p.) in one group of animals to study the deprived cortex immediately (<1 day), and 2 and 36 weeks after amputation. A second group were studied under N_2O supplemented with thiopental sodium (30 mg/kg i.p.) immediately, and 1, 2 and 4 weeks after amputation. Both these groups of animals were paralysed with tubocurarine hydrochloride (0.2 mg/kg/h i.v.) (Kelahan and Doetsch 1984).

Rasmusson (1982) reported that 2 weeks after removal of the 5th digit, cells in the appropriate subgyrus were "almost totally" unresponsive to sensory input. By 8 and 16 weeks, cells at many cortical sites responded to high intensity stimulation although by 16 weeks there were fewer unresponsive sites and more sites that were responsive to low intensity stimuli (Fig. 7.22a). At 8 weeks, receptive fields covering large parts of the palm and the digits were present at 50% (16/31) of sites while at 16 weeks this percentage had fallen to 30% (18/53) with a concurrent increase in the proportion of small receptive fields and sites excited by low intensity stimuli (Fig. 7.22b–e). Although receptive fields in the deprived digit 5 subgyrus tended to be on digit 4 or around the stump of digit 5, no topographical organization of these inputs was apparent. From the time-course of these changes Rasmusson (1982) concluded that extensive anatomical changes could have occurred within the cortex or at subcortical levels.

In marked contrast to the findings of Rasmusson (1982) and Rasmusson and Turnbull (1983), Kelahan and Doetsch (1984) reported that less than 1 day after digit amputation in adult raccoons, many neurons in the deprived subgyrus were responsive to peripheral stimulation and even to low intensity, cutaneous stimuli (Fig. 7.23). In addition, Kelahan and Doetsch (1984) stated that: "(1) The percentage of isolated neurons that responded to cutaneous stimulation increased after 1 day. (2) The ratio of neuronal sensitivity to skin touch versus claw tap increased greatly after 1 day. (3) The stimulus thresholds increased sharply after 1 day, remained high for a period of at least 3 weeks and declined by 36–52 weeks. (4) The proportion of receptive fields located on digits 2 and 4 versus digit 3 stump decreased dramatically after 1 day, was low for at least 3 weeks and increased by 36–52 weeks. (5) The ratio of receptive fields located on the glabrous skin versus the hairy skin or claws declined after 1 day, remained relatively low for a period of at least 3 weeks and rose by 36–52 weeks. (6) The receptive field sizes decreased greatly after 1 day, were small for at least 3 weeks and increased by 36–52 weeks" (see Figs. 2–6 in Kelahan and Doetsch 1984). (7) No topographic organization was apparent in the deprived subgyrus and in fact the receptive fields of cortical neurons encountered within any track varied in a non-systematic manner with depth (Fig. 7.23a–c).

From a comparison of the barbiturate and the N_2O-anaesthetized groups Kelahan and Doetsch (1984) concluded that N_2O-anaesthetized raccoons had "(1) proportionately more neurons that were responsive to cutaneous stimulation of "novel"

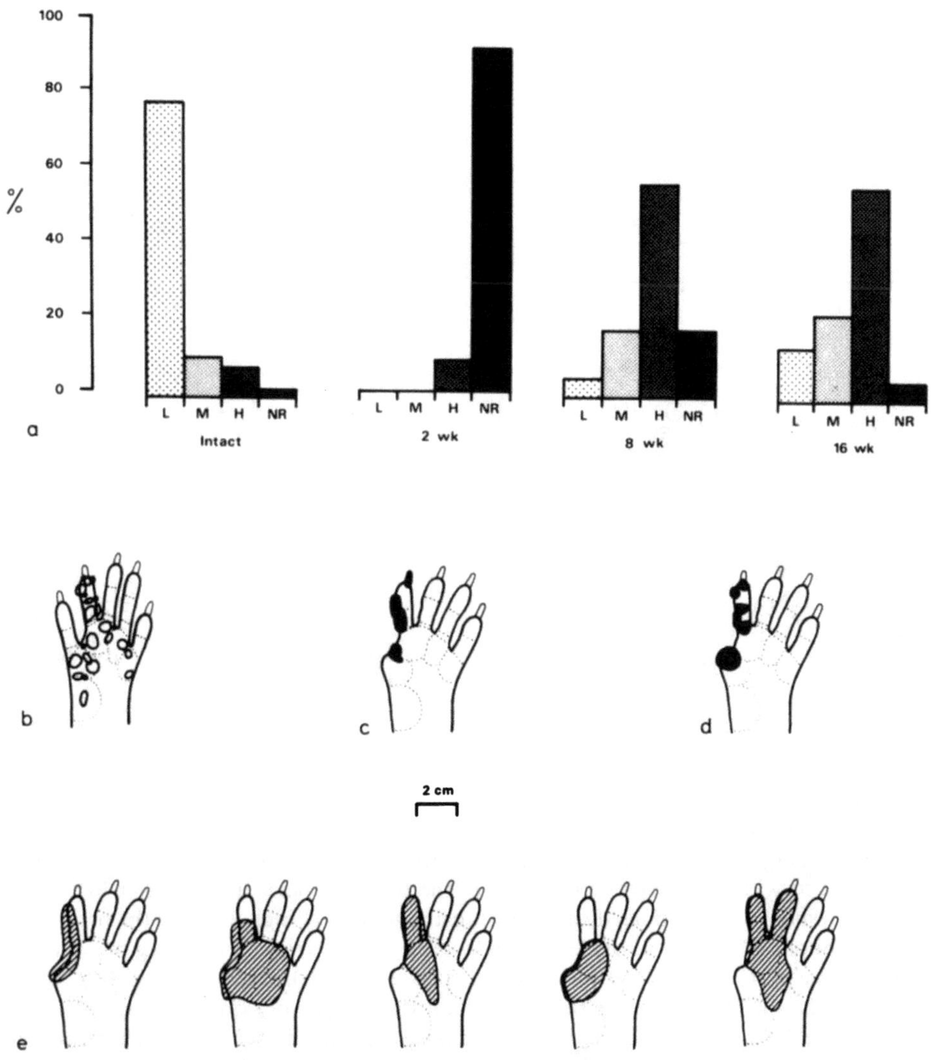

Fig. 7.22 a–e. Progressive effects of amputation of the 5th digit from adult animals on the inputs and receptive fields of neurons in the digit 5 subgyrus in the pentobarbitone-anaesthetized raccoon. **a** The proportion of units in the digit 5 subgyrus that required a low (*L*), moderate (*M*) or high (*H*) intensity cutaneous stimulus to elicit a response or that were unresponsive (*NR*) to peripheral stimulation, in intact raccoons ($n = 52$), and in raccoons 2 ($n = 39$), 8 ($n = 38$) and 16 ($n = 56$) weeks after digit amputation. Note the lack of responsive neurons 2 weeks after amputation and the progressive increase in responsive units at subsequent postoperative times. **b** Typical receptive fields obtained in intact animals from the digit 4 subgyrus and from within the palm representation. **c, d** Typical "small" receptive fields obtained from the digit 5 subgyrus 8 (**c**) and 16 (**d**) weeks after the amputation of digit 5. **e** Examples of "large" receptive fields obtained from the digit 5 subgyrus 16 weeks after amputation of digit 5. In general these neurons had high thresholds to cutaneous stimulation. Note that the "small" receptive fields that develop in the deprived subgyrus (**c, d**) are similar in size to those normally found in cortex surrounding this subgyrus (**b**). (From Rasmusson 1982, pp. 320 and 321)

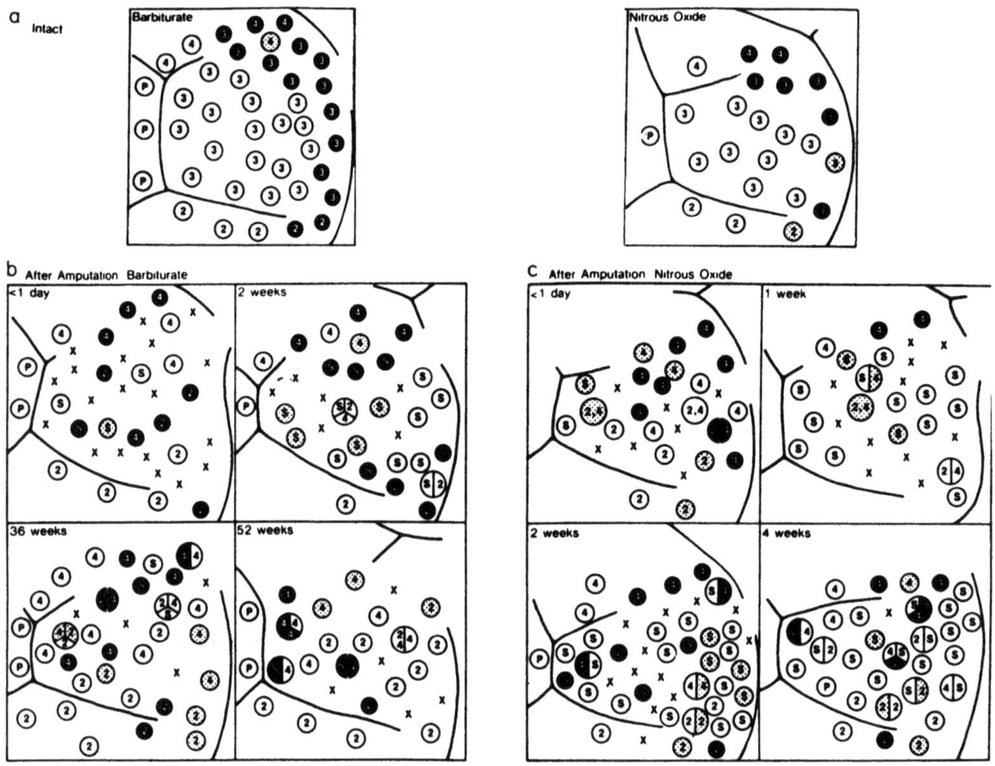

Fig. 7.23 a–c. The receptive fields of neurons in and around the digit 3 subgyrus of intact adult raccoons and at various times after amputation of the 3rd digit. **a** The receptive fields in the digit 3 subgyrus in an animal anaesthetized with pentobarbitone (*left*) and in an animal anaesthetized with nitrous oxide (*right*). **b, c** The receptive fields at particular loci across the digit 3 subgyrus at various times following amputation of digit 3 and in the presence of pentobarbitone (**b**) or nitrous oxide (**c**). *Circles* represent the tracks containing neurons that were responsive to the skin on one intact digit (see *numbers*), stump of digit 3 (*S*) or the palmar pads (*P*). The *shading in the circles* indicate the type of skin encompassed by the receptive field; *black*, glabrous; *open*, hairy; *stippled*, both. The *subdivided circles* indicate tracks where different receptive fields were encountered at different depths, the receptive field of the first unit encountered being indicated at the *top left* and the receptive fields of deeper neurons being indicated in a clockwise order. Note the early expression of digit 4 and digit 2 inputs across the deprived digit 3 subgyrus when nitrus oxide is used and barbiturates are absent. Note also that after digit amputation a number of tracks contain neurons with spatially separate receptive fields. (From Kelahan and Doetsch 1984, p. 64)

skin regions…, (2) relatively more cells that were sensitive to skin touch than to claw tap…, (3) proportionately more receptive fields that were located on digits 2 and 4 than on the digit 3 stump…, (4) relatively more receptive fields that involved the glabrous skin than the hairy skin or claws…, (5) more receptive fields that were larger than normal", and (6) stimulus thresholds that were in the same range as those of barbiturate-anaesthetized animals (compare Fig. 7.23b with Fig. 7.23c). These differential effects of pentobarbitone and N_2O, again emphasize the role of GABA in the suppression of preexisting subthreshold inputs from skin outside the receptive fields of cortical neurons (Sect. 8.5.1).

Kelahan and Doetsch (1984) have also compared the effects of digit amputation in adults with those of digit amputation in early postnatal life (Kelahan et al. 1981) and concluded that with the exception of neurons in the latter group, which developed larger receptive fields, the effects on both groups were similar. The mean gestation period of the raccoon is 63 days, a figure similar to the domestic cat. It might therefore be expected that, as in the newborn cat (Rubel 1971), the SI cortex of the raccoon would be somatotopically organized at birth. However, such arguments do not always hold, for paw denervation in young kittens has been shown to have less severe consequences in terms of a larger number of responsive neurons, than denervation carried out in the adult cat (Sect. 7.3.6) (Kalaska and Pomeranz 1979).

Deafferentation and Sprouting. Although surface potentials evoked from the electrical stimulation of the tip of one digit may be recorded on the subgyrus appropriate to an adjacent digit (Carson et al. 1981), retrograde tracing studies have produced no evidence for somatotopically inappropriate projections to the digit representation from thalamocortical cells within the VB complex (Sect. 7.4.3) (Pubols and Warren 1984; Rasmusson and Nance 1986). Given that some workers have found that the major effects of digit amputation develop over a period of up to 16 weeks (Rasmusson 1982; Rasmusson and Turnbull 1983), it is possible that the sprouting of the axons of thalamocortical neurons with intact inputs into the subgyri related to an amputated digit might account for the physiological effects of amputation. To test this, Rasmusson and Nance (1986) amputated digit 5 in six raccoons and 4 months later injected one fluorescent dye into the representation of the 4th digit and another into the reorganized subgyrus of the 5th digit. In the three animals in which there was no dye overlap at the injection site only one showed a few (6, 1.5%) double-labelled neurons in the VB complex. It is therefore possible to conclude that, within the constraints of the retrograde labelling technique (Sect. 7.4.3) (Snow and Wilson 1989), there was little evidence that the growth of new collaterals from thalamocortical neurons had occurred, that might account for the physiological effects of digit amputation (see Sect. 7.2.9).

Large Somatotopic Shifts. A bizarre case of the effects of limb amputation has been documented by Rasmusson et al. (1985). In this study the cortex was mapped in a raccoon that some months prior to capture had suffered amputation of its forearm. The area of the cortex so deprived of input was estimated to be 500 mm^2. Within the forepaw representation only 2/32 sites were unresponsive to peripheral stimuli while 18/32 sites responded to the forelimb stump. The remaining 12 sites responded to glabrous skin of the hindlimb and also, in 11 cases, the forelimb stump. The most lateral of these 12 sites was estimated to be 23–28 mm from the normal hindlimb representation. No sites responded to inputs from the face, supporting the suggestion of Wall and Kaas (1985) that reorganization occurs only between the representations of adjacent skin areas. In the deprived forepaw area, thresholds were high, receptive fields were large, responses tended to habituate and no somatotopic organization was apparent. Clearly, more information about the physiology and the microanatomy of connections at various levels of the somatosensory system of the normal raccoon is required in order to interpret adequately the results of digit or forelimb amputation. The large transcortical distances over which reorganization appears to occur in this species makes it an important experimental model for the investigation of the mechanisms underlying plasticity and the telescoping of phantom limbs (Figs. 1.2, 8.2b and 8.6 and Sect. 8.6).

7.5 Plasticity in the Cortex of Adult and Developing Rodents

7.5.1 Somatotopic Organization and Cytoarchitectonics

In rats there is a single representation of the body across the SI cortex with only the forepaw being represented twice within two neighbouring regions (Welker 1971, 1976b; Kaas 1983; Chapin and Lin 1984). Furthermore, in tangential sections through layer IV, the entire body representation may be revealed by (a) Nissl staining of the densely packed granule cells (Fig. 7.24c, d) (Welker 1976b; Chapin and Lin 1984), (b) SDH labelling of the thalamocortical projections from the VB complex within this granular region (Fig. 7.24a, b) (Wise and Jones 1978; Dawson and Killackey 1987), and (c) anterograde labelling of the commissural projections from the agranular cortex surrounding SI on the contralateral hemisphere to the agranular cortex surrounding SI in the ipsilateral hemisphere (Kaas 1983; Jones 1986; Wise and Jones 1976, 1978; Ivy et al. 1979).
A peculiarity of the rat is an overlap between the sensory and motor maps in the hindlimb representation – a condition that correlates with an overlap between a well-developed granule cell layer and the presence of giant pyramidal cells in this region (Hall and Lindholm 1974; Donoghue et al. 1979; Jones and Porter 1980). Nevertheless, the cortex containing the representation of the body is highly granular and is not otherwise divisible cytoarchitectonically (Fig. 7.24c, d) (Welker 1976b; Welker et al. 1984). Thus Jones (1986) has suggested that the area of representation of the body is probably comparable with area 3b in primates whereas the surrounding less granular areas are equivalent to areas 1 and 2 or possibly area 3a (Chapin and Lin 1984; Welker et al. 1984). Cells within the granular regions project to cells in the surrounding agranular cortex suggesting that, as has been proposed in primates, there may be some hierarchical processing of incoming somatosensory information. This contention is supported by observations that in unanaesthetized rats cells in the agranular cortex respond differentially to complex cutaneous stimuli or deep stimuli which excite joint or muscle afferents. Like the neurons in areas 1 and 2 in the primate, these cells are more sensitive to anaesthetics than cells in the granular region. However, unlike cells in area 3b of the owl monkey (Stryker et al. 1987), the receptive fields of cells in granular cortex of the rat *are* reduced in size by anaesthetics. This reduction in receptive field size does not alter the position of receptive field centres and so, as in area 3b of the primate, somatotopic organization across granular cortex is not changed by anaesthetics (Chapin and Lin 1984). In barbiturate-anaesthetized rats, cells in the surrounding cortex are not responsive to light cutaneous stimuli (Welker 1976b; Welker et al. 1984).

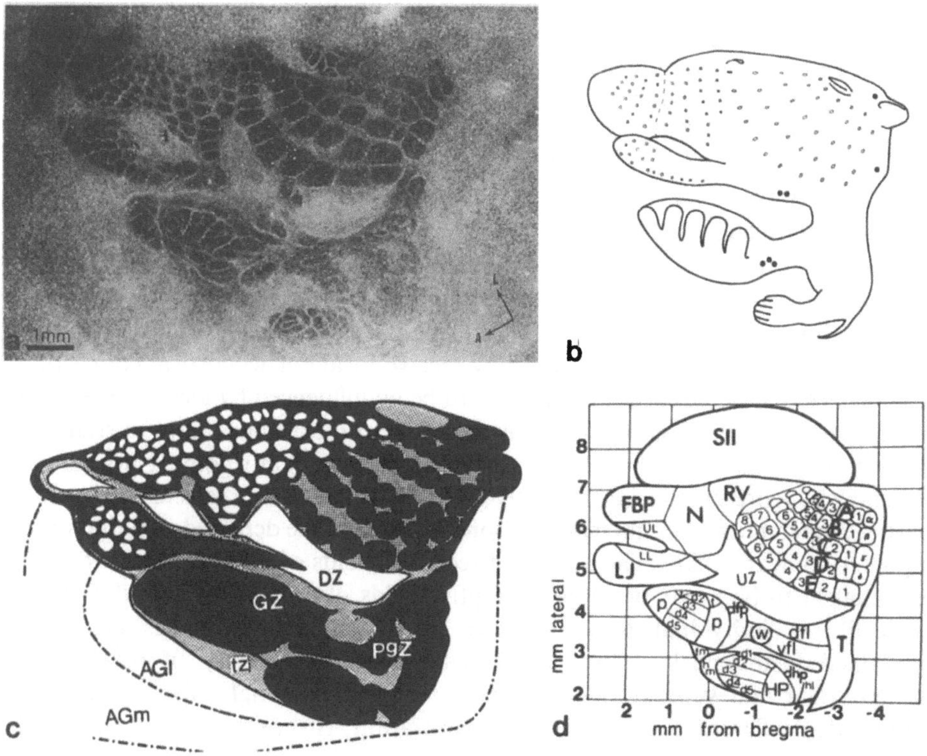

Fig. 7.24 a–d. Somatotopic representation of the entire body in the SI cortex of the rat as seen with three different methods. **a** Tangential section through layer IV that has been reacted for succinic dehydrogenase (SDH). The *dark areas* show the distribution of the terminals of thalamocortical neurons that respond to cutaneous stimulation. **b** A caricature of the body surface of the rat based on the distribution of SHD staining shown in **a**. **c** Schematic drawing taken from a 75 μm, thionin-stained, tangential section through layer IV. Note the regions of high granule cell density or granule zones (*GZ*; *black*), the perigranular zones (*pgZ*; *stippled*), the transition zone (*tz*) just rostral to SI, the dysgranular zones in the centre of SI (*DZ*) and the medial and lateral frontal agranular areas (*AGm* and *AGl*) rostral of SI. *AGl* contains most of the primary motor cortex. **d** Representation of the body surface in the anaesthetized (pentobarbitone or halothane) rat derived from single-unit recordings. *Other abbreviations*; *SII*, second somatosensory area; *T*, trunk; *hl*, hindlimb; *HP*, hindpaw; *dhp*, dorsal hindpaw; *d1* to *d5*, digits; *vfl*, ventral forelimb; *dfl*, dorsal forelimb; *w*, wrist; *dfp*, dorsal forepaw; *p*, palm; *t*, thumb; *A* to *E*, *1* to *8*, *alpha*, *beta*, *gamma* and *delta*, mystacial vibrissae; *RV*, rostral small vibrissae; *N*, nose, *FBP*, frontobuccal pads; *UL*, upper lip; *LL*, lower lip; *LJ*, lower jaw; *UZ*, unresponsive cortex. (**a**, **b** From Dawson and Killackey 1987, p. 249; **c**, **d** from Chapin and Lin 1984, pp. 202 and 204)

In the SI granular cortex of the rat, laminar differences in the size of the receptive fields of neurons have been reported (Simon 1978; Lamour et al. 1983) and this has subsequently been quantitatively studied in rats under light barbiturate and halothane anaesthesia (Chapin 1986). Although Chapin (1986) states that these experiments support the columnar organization of the SI cortex proposed by

Mountcastle (1957), at different depths the variation in receptive field size does not always appear to be concentric around a particular patch of skin. Chapin's (1986) data fit better a model in which the SI cortex of the rat is divided into columns in which all cells share input from a common patch of skin that excites, exclusively, cells within that column. This organization is identical to the organization of the SI cortex of the cat into *segregates* that has been proposed by Favorov et al. (1987) (Fig. 7.13a, b) (Sect. 7.3.1).

Quantatively, Chapin's (1986) work has shown that the receptive fields of cells in the supragranular layer are up to twice the size of those of cells in the granular layer while the receptive fields of cells in the infragranular layer were up to three times as large as those of granular layer cells. To explain this, Chapin (1986) has suggested that cells outside the granular layer may receive local corticocortical connections from adjacent areas of cortex, in addition to being influenced by local thalamocortical inputs. Clearly, these observations are of the utmost importance to studies of deafferentation-induced plasticity, for they suggest that, at any point, a large area of skin may be represented in the receptive fields of at least some cells. Consequently, partial deafferentation might leave neurons well within the deprived area that retain inputs from nearby skin. The truncated receptive fields of these neurons would be easily taken to be newly developed receptive fields of cells in the granular layer – especially as individual recording sites in reorganization studies are rarely confirmed histologically (Sect. 7.5.2). The greater sensitivity of cells in the infragranular and supragranular layers to anaesthetics might also explain why immediate deafferentation-induced "reorganization" is suppressed by barbiturates (Sect. 7.6).

7.5.2 Section and Ligation of Peripheral Nerves in the Adult

Normal Somatotopic Representation of the Hindpaw. In rats anaesthetized with ketamine (50–75 mg/kg i.m.) the hindpaw is represented within an area of cortex that is approximately 1.0 mm^2 and is bounded on three sides by cortex which is unresponsive to cutaneous stimuli (Fig. 7.25a) (Wall and Cusick 1984). The hairy skin on the dorsal surface and the ankle is represented mainly along the posterior and medial margins of the hindpaw representation while the representation of the glabrous skin occupies the central and rostral two-thirds. Moving medial to lateral across the paw representation results in a lateral to medial shift in the receptive fields of cortical neurons (Fig. 7.25b). Recordings from the sciatic nerve show that it innervates the skin on the ventral, the lateral and the lateral part of the dorsal surface of the foot leaving the saphenous nerve to innervate skin on the medial surface and the medial part of the dorsal surface (Fig. 7.25d) (Wall and Cusick 1984). Thus neurons responsive to saphenous nerve afferents are normally found along the lateral boundary of the paw representation where they occupy the anterior part of the representation of the hairy skin (compare Fig. 7.25b with Fig. 7.25c). Only 15% (0.14 mm^2) of the total area of the paw representation is occupied by saphenous inputs. In normal rats the shape of the saphenous nerve representation varies considerably between individuals (Wall and Cusick 1984), a situation similar to that

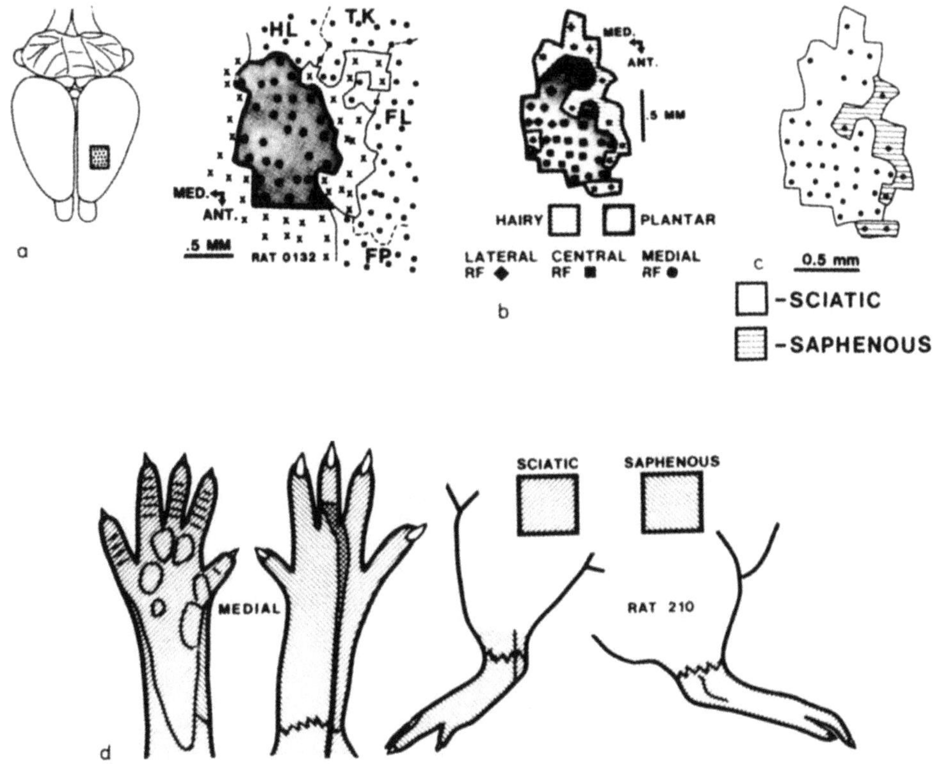

Fig. 7.25 a–d. Representation of the hindpaw skin and the sciatic and saphenous nerves in the SI cortex of the ketamine-anaesthetized rat. **a** shows the location of the hindpaw representation (*rectangle*) on the brain surface and the relationship of the hindpaw representation (*shaded area*) to the representation of the hindlimb (*HL*), forelimb (*FL*), trunk (*TR*) and the forepaw (*FP*). *Filled circles* show the position of tracks containing responsive neurons whereas *X*s show tracks in which neurons were unresponsive to cutaneous stimulation. **b** Representation of hairy skin (*shading*) and glabrous skin (*hatching*) within the hindpaw area. **c** Representation within the hindpaw area of the skin supplied by the sciatic and saphenous nerves in the same animal illustrated in **b. d** Skin areas supplied by the sciatic and saphenous nerves. Note that there is little overlap and that as suggested by its cortical representation (**c**), the saphenous nerve supplies only hairy skin. (From Wall and Cusick 1984, pp. 1502–6)

seen also in the hand representation of the primate (Fig. 7.2; Sect. 7.2.4) and the raccoon (Fig. 7.20; Sect. 7.4.1).

Reorganization of Somatotopy. The effect on the hindpaw representation of sectioning and ligating the sciatic nerve has been tested 1–156 days postoperatively, in adult rats anaesthetized with ketamine. The cortex was not mapped before nerve section but maps in lesioned rats were compared with those made in normal rats (Wall and Cusick 1984). The results showed that only 1 or 2 days after nerve section the mean area of the saphenous nerve representation (0.33 mm^2 after 1

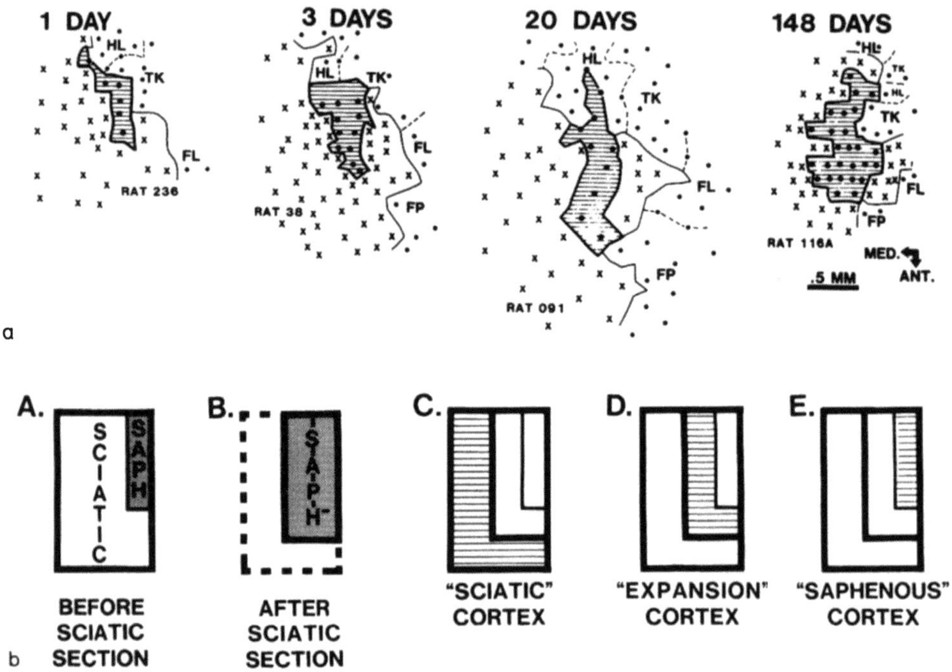

Fig. 7.26a, b. Progressive effects of sciatic nerve section on the representation of the hindpaw in the SI cortex of ketamine-anaesthetized, adult rats. **a** Hindpaw representation in individual rats studied 1–148 days after sciatic nerve section. *Filled circles* show the position of tracks containing responsive neurons whereas *X*s show tracks in which neurons were unresponsive to cutaneous stimulation. **b** Schematic summary of the results showing *A* normal representations of the sciatic and saphenous nerves; *B* the expanded saphenous nerve representation some days after sciatic nerve section; *C* that part of the sciatic representation that *does not* become responsive to saphenous nerve inputs ("sciatic" cortex; *hatching*); *D* that part of the cortex that *does* becomes responsive to saphenous nerve inputs ("expansion" cortex; *hatching*); *E* that part of the cortex which Wall and Cusick (1984) suggest receives only saphenous nerve input (*hatching*). Abbreviations as in Fig. 7.25. (From Wall and Cusick 1984, pp. 1507 and 1511)

day) appeared larger than normal. This change was statistically significant and occurred at the expense of the sciatic nerve territory. At these early survival times, cells in the remainder of the sciatic nerve territory (0.7 mm^2) were unresponsive to cutaneous stimuli (Fig. 7.26d). In animals studied 20–157 days postoperatively, the mean size of the saphenous nerve representation had grown to 0.47 mm^2 (Fig. 7.26a). This was significantly different from the representations of both the saphenous nerve and the whole hindpaw in normal animals. Within the cortical territory newly responsive to saphenous nerve input, a coarse but topographically ordered representation of innervated skin was apparent in which distal skin was represented anterior to the representation of proximal skin. Finally, representational boundaries between other parts of the body or hindlimb were not detectably altered by sciatic nerve section.

From these experiments Wall and Cusick (1984) concluded that in adult rats peripheral input was a necessary requirement for the maintenance of normal somatotopic organization of the cortex. While within 1 or 2 days some deprived cortex often becomes responsive to nearby innervated skin, they suggest that the maximum spatial extent over which this process of reorganization of inputs may occur "depends partly on how inputs from injured and intact nerves distribute to cortex". In quantitative terms, their figures suggest that the maximum linear distance over which there were changes in representation of saphenous nerve afferents was about 600 μm. Perhaps thalamocortical neurons receiving inputs from skin along the edge of the saphenous nerve territory have ineffective terminals within a strip of the sciatic nerve representation that abuts the normal saphenous nerve representation (Sect. 7.3.2). If sciatic nerve section were to increase the effectiveness of these terminals then the saphenous nerve representation would enlarge and the newly activated strip of cortex would contain an orderly anterior-to-posterior representation of a strip of skin supplied by the saphenous nerve and bordering the sciatic nerve territory.

Recently Dykes and Lamour (1988) have supported the view that, in the rat, reorganization of cortical somatotopy following nerve section represents the response of normal cortical circuitry to removal of input. The sciatic nerve was sectioned 14–24 days before recordings were made from a total of 302 cells within the hindlimb representation of 13 urethane-anaesthetized rats. Cells were sampled from all cortical layers and compared with a sample of cells previously gathered from normal rats. Although they do not show the position of their recording tracks in relation to the recorded receptive fields, they state that stimulation of the skin innervated by the saphenous nerve did excite neurons within the sciatic nerve representation. Furthermore, they note that the receptive fields within this representation were concentric throughout a tract made perpendicular to the cortical surface (Mountcastle 1957). Most interesting, however, were their observations that: (a) an equal number of cells had cutaneous receptive fields in lesioned and control animals, (b) more cells (45% compared with 32%) in lesioned animals were spontaneously active than found in control rats, (c) after deafferentation cells with large and small receptive fields were found in all cortical layers whereas in normal cortex only the cells in middle cortical layers had large receptive fields (compare with Chapin, 1986), and (d) some neurons had very large receptive fields. Dykes and Lamour (1988) suggest that these observations reveal that deafferented cortex is not depressed but that it is in a state of "elevated excitability" which coexists with an "autoregulatory process" that functions to maintain relatively constant the number of responding neurons before and after nerve section.

7.5.3 Effects of Perinatal Nerve Section or Limb Amputation

Reorganization of Somatotopic Organization. The effects on the hindpaw representation in adult rats of sectioning the sciatic nerve in the first day of life (P0) have been tested (Wall and Cusick 1986). Neonatal section of the sciatic nerve led to a stunted development of the hindpaw which consequently had a reduced skin area. From

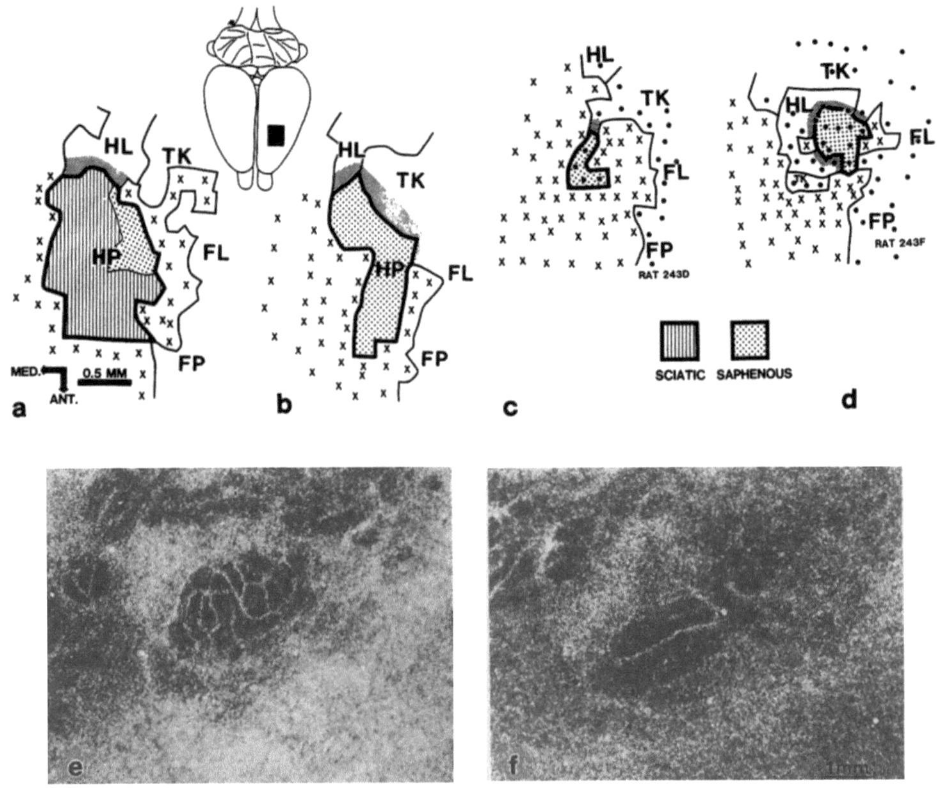

Fig. 7.27 a–f. Effects of sectioning the sciatic nerve on the day of birth on the hindpaw (*HP*) representation of rats. **a, b** Representation of the sciatic and saphenous nerves and somatotopic organization of surrounding cortex in normal adult rats (**a**) and in adult rats some days after transection of the sciatic nerve. **c, d** Representation of the hindlimb and somatotopic organization of surrounding cortex in two adult rats following section and ligation of the sciatic nerve on the day of birth. Note that in both cases the hindpaw (saphenous) representation is about the same size as in normal rats (cf. **c** and **d** with **a**). Note also that in one case (**d**) cortex normally activated by sciatic nerve inputs has become responsive to hindlimb inputs (*HL*). **e, f** Succinic dehydrogenase staining in the hindpaw representation of a normal 7-day-old rat (**e**) and a 7-day-old rat following section of the sciatic nerve on the day of birth (**f**). Note the disruption of the pattern of staining without a gross reduction in the size of the total area containing the terminals of thalamocortical neurons (compare with Fig. 7.24a). Abbreviations and symbols as in Fig. 7.25. **a–d** From Wall and Cusick 1986, p. 1130; **e, f** modified from Dawson and Killackey (1987, p. 254)

recordings in both the saphenous nerve and the SI cortex these workers claimed that neonatal section of the sciatic nerve did not alter the skin territory in which low-threshold mechanical stimulation activates saphenous afferents (Sect. 7.5.2). There was, however, a permanent loss of low-threshold input from skin normally supplied by the sciatic nerve. Curiously, the area of the cortical representation of the saphenous nerve in these neonatally denervated adult rats was not significantly different from that in normal rats (Fig. 7.27). In relation to these results it is notable that sprouting of low-threshold saphenous afferents, after neonatal sciatic nerve section, has been reported by other workers (see Sect. 2.3.2).

In 5 of the 11 neonatally denervated rats studied, cells in the sciatic nerve representation were unresponsive to cutaneous stimuli (Fig. 7.27c). However, in 6 neonatally denervated rats, cells in a small part (12%) of what would normally have been the sciatic representation did respond to stimulation of skin on the medial hindlimb and adjacent pelvic-abdominal trunk region (Fig. 7.27d). That this abnormal representation was not the result of the sprouting of fibres from the sciatic nerve was confirmed in 2 neonatally denervated rats in which receptive fields were found medial and rostral to the saphenous nerve representation even though the sciatic nerve had been resectioned before mapping. Thus it appears that only in some rats is the cortical territory normally dominated by sciatic nerve inputs taken over by surviving inputs. Indeed, in neonatally denervated animals the area of the hindpaw (saphenous) representation is only 28% of that of the sciatic and saphenous representation in normal rats while following sciatic nerve section in the adult, the saphenous representation is about 50% (Wall and Cusick 1984, 1986). Finally, the most striking result of this study is the effect sciatic nerve section has on the representation of skin innervated by nerves other than the saphenous. Thus in some rats the skin of the trunk was found represented between the saphenous nerve representation and that of the hindlimb (Fig. 7.27c) (Wall and Cusick 1986) and at times skin of the trunk was found represented in two separate locations separated by the hindlimb and saphenous nerve representations (Fig. 7.27d) (Wall and Cusick 1986). It is difficult to see why the saphenous nerve representation is unresponsive to neonatal denervation of adjacent skin and yet other peripheral inputs appear to exploit the cortex that would normally be dominated by sciatic inputs. Perhaps the cortical topography of the terminals of those thalamocortical neurons that receive inputs from the paw is fixed at birth, while the terminals of thalamocortical neurons serving the rest of the body are sufficiently labile to exploit nearby cortex that is deprived of normal physiological input. A similar mechanism at a lower level of the neuraxis (for example the DCN) would, of course, lead to the same result.

Recently, Dawson and Killackey (1987) have shown that between P 0 and P 4 or 5 section of the sciatic nerve, denervation of the forelimb, amputation of the forelimb or amputation of the forepaw digits alters the patterns of thalamocortical terminal clusters in layer IV of the appropriate part of the body representation. Both the level of SDH activity indicating the general region of termination of VB neurons (Fig. 7.24a, b) and the size of the representation as assessed by SDH activity were unchanged and only the pattern of clustering of their terminals within the region was altered (Fig. 7.27e, f). Unfortunately, neither the somatotopic organization of the effected region nor the location of the cells of origin of the disrupted terminals across the body representation in the VB complex is known. These data are therefore not yet of use in understanding the anatomical basis for the physiological effects of neonatal sciatic nerve section on adult rats described by Wall and Cusick (1986).

In an abstract, Dawson and Killackey (1986) have reported large alterations in the SDH patterns in SI cortex of the rat following cautery of the forelimb on embryonic day 16 (G16, birth = G22). This procedure resulted in a 100% expansion of the hindlimb representation as assessed by SDH labelling of VB terminals at P14. However, even in these animals, labelling in the forelimb regions was still present although disrupted in pattern. It is, of course, possible than in this study the effects of damage to the entire forelimb, including nerves carrying deep input which might project to the agranular cortex around the hindlimb representation (Welker et al. 1984), might have more to do with the expansion of the hindlimb representation than the attendant deafferentation of forelimb skin. A more complete account of this work has now appeared (Killackey and Dawson, 1989) (see also Sect. 5.6.1).

7.5.4 Pharmacological Mechanisms Underlying Somatotopic Reorganization

Acetylcholine and Glutamate. Lamour and Dykes (1988) have sought the biological identities of *elevated excitability* and the *autoregulatory process* by the iontophoretic application of glutamate and acetylcholine (ACh) to neurons within the cortical area that is newly responsive to the saphenous nerve following sciatic nerve section. Both normal and operated rats were anaesthetized with urethane which, via its anticholinesterase activity, could be considered as enhancing the effects of ACh and so of those inhibitory influences that are mediated by ACh receptors (McCormice and Prince 1985) (Sect. 7.3.3). Compared with SI neurons in normal rats, fewer of these neurons had receptive fields that could be uncovered by glutamate but more neurons responded with an increase in receptive field size. From this, the authors concluded that not all subliminal inputs are uncovered following nerve section but that after this procedure less tonic inhibition was operative in the cortex. Implicit in this suggestion is the idea that, in intact animals, tonic inhibition suppresses the expression of inputs from skin near the normal receptive field without altering the size of the receptive field. This phenomenon is discussed further in Sect. 8.5.1 in relation to in-field inhibition and its potential role in deafferentation-induced reorganization.

Nerve section reduced the effectiveness of ACh in revealing new somatic inputs and in enhancing neuronal responses. This was particularly apparent in the infragranular layers. In addition, after nerve section ACh produced inhibition of more cells (presumably by an increased excitability of inhibitory neurons) than in normal rats. Together with evidence that normally ACh can bring about long-term changes in the excitability of cortical neurons to afferent input (Sect. 7.3.3) (Lamour et al. 1982; Metherate et al. 1987), these findings have led Lamour and Dykes (1988) to suggest that the *immediate* appearance of new receptive fields following deafferentation reflects a loss of normal intracortical inhibition while the subsequent more gradual changes are controlled by changes in ACh receptor distribution or the selective release of ACh.

Finally, some recent evidence suggests that the neuropeptides, vasoactive intestinal polypeptide and somatostatin are capable of modulating the effects of GABA and ACh on SI cortical neurons in the rat (Waterhouse et al. 1986). These observations, taken together with the known modulatory actions of ACh and the interactions of ACh and GABAergic interneurons (Sect. 7.3.3) (Krnjevic et al. 1971; McCormice and Prince 1985), suggest that the list of substances with the potential to modify somatotopic organization and reorganization might be large.

Norepinephrine. In the mammalian cortex norepinephrine (noradrenaline) is released exclusively from the terminals of neurons located in the locus coeruleus (Foote and Morrison 1987). The effects of norepinephrine on neurons in the SI cortex of the rat have been investigated by Waterhouse and Woodward (1980). Microiontophoretic application of norepinephrine was found to enhance both the excitatory and inhibitory responses generated by afferent input. In 78% of cells tested low doses of norepinephrine suppressed background discharge more than stimulus-bound excitation, such that there was a twofold increase in the signal: noise ratio. In 82% of cells tested norepinephrine augmented the postexcitatory suppression of activity that is a characteristic of in-field inhibition (Sects. 7.3.1, 7.3.3 and 8.5.1). These modulatory effects were observed on cells located throughout the vertical extent of the cortex.

In an attempt to ascertain the pharmacological basis of some of these actions Waterhouse et al. (1980) studied the effects of iontophoretically applied norepinephrine on the responses of neurons to iontophoretically applied GABA and ACh. Again, norepinephrine suppressed the spontaneous activity more than the excitatory responses of neurons to applied ACh, such that the signal:noise ratio of the ACh effect was enhanced in 86% of cells. In 13 out of 35 cells the ACh responses were enhanced over normal levels. In 94% of cells the inhibitory effects of GABA were potentiated. Taken together, these results suggest that norepinephrine may actually modulate the effectiveness of ACh which in turn may enhance the responses of cortical neurons to glutamate and to natural stimulation of their receptive fields. Furthermore, they suggest that norepinephrine may enhance the effects of tonic inhibition as well as postexcitatory inhibition so as to increase the spatial and temporal resolution of the cortical effects of a peripheral stimulus. These effects are discussed elsewhere, in relation to systems that regulate the responses of cortical neurons in intact animals and following deafferentation (Sects. 3.4.1, 8.5 and 8.6).

7.5.5 Cortical Damage

Cytotoxic Agents, Development and Somatotopy. The use of the cytotoxic agent methylazoxymethanol acetate to alter radically the development of the rat cortex has yielded results which are of particular value in relation to the effects of prenatal and neonatal cortical damage (see also Sect. 3.3.6). In these studies Yurkewicz et al. (1984) injected pregnant rats on G 14 or 15 with the drug thereby killing those neural precursor cells which were synthesizing DNA at the time of injection. Following similar treatment with this agent, the behaviour of progeny was superficially normal although closer examination revealed learning deficits (Haddad et al. 1969). As adults, the cerebral cortices of these rats were greatly reduced in area (microencephaly) and cortical layers II to IV had failed to develop (Yurkewicz et al. 1984). In particular, the cortex of the frontal and occipital poles was reduced and in relation to this the thalamus was found to lack the mediodorsal nucleus and large parts of the three anterior nuclei and the dorsal lateral geniculate and medial geniculate nuclei. The ventral nuclei were also reduced in size and there was no distinct boundary between the VPM and VPL. In normal rats thalamocortical neurons project in a bistratified manner to (a) layer IV and the deep part of layer III, and (b) the junction of layer V and VI (Wise and Jones 1978). In the microencephalic rats the terminals of thalamocortical neurons were also distributed in a bistratified manner to the superficial part of layer V (where they synapse with the apical dendrites of pyramidal neurons) and in a thinner band to the junction of layers V and VI. Yurkewicz et al. (1984) pointed out that the maintenance of the bistratified pattern of thalamocortical terminals in the absence of layers II, III and IV suggests that it is a property intrinsic to the thalamocortical axons rather than depending on the distribution of target cells (Sect. 3.3.6) (Caviness et al. 1976).

Topographic relations between SI cortex and the thalamus were still apparent in microencephalic rats, a finding which suggests that the broad topographic relations between the thalamus and the cortex either form very early in development (G 15 to G 17 in mouse, see Table 1, Sect. 3.2) or that this process occurs independently of the microstructure of the cortical target. The preservation of thalamocortical topography is paralleled by a preservation of an orderly, though partial somatotopic representation of the body across the SI cortex (Fig. 7.28). Within this representation the receptive fields of SI neurons were normal in size and on this basis Yurkewicz et al. (1984) speculated that "individual thalamocortical axons or bundles of such axons specifying a place on the body surface, during innervation of the cortex, establish synaptic domains of constant size within the target cortical area". If this were the case then it would clearly be impossible for the entire body to be represented within the abnormally small cortical area which in microencephalic rats constitutes the SI cortex. In fact the SI representations in these animals are indeed incomplete such that those body regions, the hindlimb and upper face and head, normally represented on the edge of the somatotopic map, are unrepresented (compare Fig. 7.28 with Fig. 7.24d). In terms of neuronal specificity it would seem that only a certain cortical area is able to receive and sustain the axonal sprouts of cells in the VB complex. The highly modified laminar structure of the SI cortex in these microencephalic rats further argues that it is the area rather than the cell type that is essential for the maintenance of thalamocortical connections. Yurkewicz et al. (1984) suggest that if insufficient space is available across the SI cortex then those thalamocortical axons (presumably those receiving inputs from the hindlimb or upper face and head) on the edge of the topographically organized projection die and that this process accounts for both the reduced size of the ventral thalamic nuclei and the peripheral deletions in the somatotopic representation across SI. These results are important for they highlight the significance of the topography of ingrowing axons and the specificity of cortical areas, as well as microscopic features such as the laminar distribution of thalamocortical terminals in relation to the distribution or presence of the correct target cells during embryonic development.

Behavioural Effects of Cortical Lesions. Bilateral damage to SI, SII or both in the adult rat results in permanent somatosensory deficits (Finger et al. 1972; Finger and Reyes 1975; Finger and Simons 1976). This has led researchers to ask whether permanent deficits also result from cortical damage in the neonatal rat. Some such experiments have already been discussed in dealing with the rodent vibrissa system (Sect. 3.3.6) and what follows will thus deal with only that part of the somatosensory system devoted to spinal afferents.

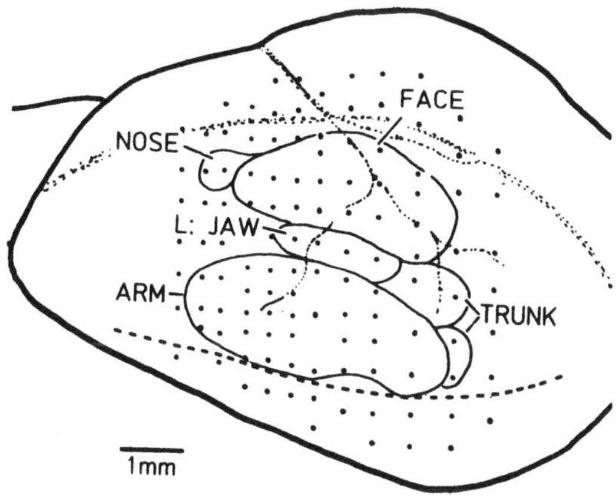

Fig. 7.28. Somatotopic organization of the SI cortex of a ketamine-anaesthetized, microencephalic, adult rat. Microencephaly was induced by injection of a cytotoxic agent on the 14th or 15th day of gestation. *Dots* indicate the position of the microelectrode tracks that were made at approximately 0.5 mm intervals. Note the absence of a representation of the hindlimb at the medial edge of the map and the upper face and head at the lateral edge. This figure should be compared with Fig. 7.24b. (Modified from Yurkewicz et al. 1984, p. 312)

At birth, crude topographically organized thalamocortical projections exist but the immaturity of the rat cortex at this time precludes the cytoarchitectonic manifestations of somatotopy that are an outstanding feature of the cortex of the adult rodent (Fig. 7.24; Sects. 3.2 and 7.5.1) (Dawson and Killackey 1985; Chapin and Lin 1984). The effects of bilateral lesions of SI and SII on P 1 have been tested in 37 adult rats (Finger et al. 1978). These rats were enucleated when they reached the age of 90 days and 2 weeks later they were tested in a T maze designed to determine their ability to learn a graded series of 5 different two-choice tactile discriminations. A group of neurological tests were also used to assess the motor as well as the sensorimotor capabilities of the operated animals. Finally, animals were anaesthetized with sodium pentobarbital and recordings were made in and around lesioned cortex.

Lesions in these animals often encompassed a large percentage of SI (70%–100%) and also probably parts of SII. Electrophysiological recordings showed that when parts of SI were spared they usually involved the representation of the face and to a lesser extent the forepaw and not suprisingly it was these particular animals which showed in behavioural tests the highest ability in tactile discrimination tasks. Indeed, rats with near total (> 90%) ablations of SI failed to complete the series of tactile discrimination tasks. Furthermore, the number of sensorimotor deficits revealed by the neurological tests was related to the tactile learning scores of the rats. Thus, sparing of a part of the SI cortex permits tactile acuity to develop for that part of the body represented in the spared portion.

By matching the size of the lesions, the tactile discrimination ability of neonatally lesioned rats could be compared with that of a group of adult-lesioned rats. When this was done, no significant difference was found between the groups (Finger et al. 1978) and it can thus be concluded that, in the rat, compensation for extensive damage to the somatosensory cortex is not more or less pronounced when it is inflicted neonatally (see also Sects. 7.2.10 and 7.3.7).

It seems to us that the subject of the effects of damage to the somatosensory cortex of the rodent at various ages deserves further study. With the availability of very precise somatotopic maps and related cytoarchitectonic and histochemical profiles (Lin and Chapin 1984; Dawson and Killackey 1987) more precise lesions and lesion assessment should be possible. In particular, lesions to specific parts of the SI and SII body representations in both adult and neonatal rats might provide some interesting comparisons. It may be that, as in those studies on the effects of cortical lesions placed in the vibrissa representation of neonatal rats (Sect. 3.3.6), the ability of pieces of healthy cortex to adopt those inputs that normally supply the damaged cortical area might depend on the linear distance between the healthy and damaged cortex (Dawson and Killackey 1985; 1987; Lin and Chapin 1984; Wall and Cusick 1984; 1986; Ito and Seo 1983; Seo and Ito 1987). Based on studies in various mammals (Chaps. 3 and 7) of denervation-induced reorganization of somatotopy, a maximum estimate of this distance would be 600–700 μm.

7.6 Major Conclusions

For technical reasons alone, studies of plasticity in the cerebral cortex are more difficult to interpret than similar studies conducted at other levels of the neuraxis. Any part of the SI cortex receives a variety of inputs and contains a number of cell types that cover a range of sizes. Although these cell types have been studied in detail (Jones 1975; Peters and Jones 1984a, b) neither their structure nor function is discussed in the many studies of plasticity that have been reviewed in this chapter. Certain types of cells are located in particular layers and the depth of a given layer might vary considerably across a relatively small area of cortex. In a typical experiment involving many electrode tracks it is often not possible for experiments to determine even the precise location of their recording sites or for that matter to know the proximity of these sites to the cells from which recordings are being obtained. The dendrites of a cortical cell extend up to 200–300 μm from the soma providing an opportunity for distant synaptic input that may be added to the widespread terminations of some thalamocortical afferents (Sect. 7.3.2) in the formulation of subthreshold inputs from areas of skin beyond the receptive field of the neuron. Finally, many cortical neurons have intracortical axons that extend beyond the extent of their dendritic trees, again providing pathways that might contribute significantly to transcortical shifts in the representation of the skin – a phenomenon that has been implicated in the laminar variations in the receptive fields of cortical neurons in the rat (Sect. 7.5.1).

In cortex that has been deprived of its normal input it must certainly be very easy to monitor the activity of cells distant from the electrode and so perhaps to incorrectly conclude that there has been a redistribution of input across the cortex. Consequently, the acceptance of multiunit recordings by many researchers would seem a dangerous concession to expediency. Again, this problem becomes acute if one accepts that there are large laminar variations in the size of the receptive fields of neurons in the SI cortex (Sect. 7.5.1).

Compounding these difficulties are the profound effects of anaesthetics on cortical neurons. These agents may alter the responsiveness of cells at all levels of the neuraxis or alter the size of the receptive fields of cortical neurons or even render many cortical neurons completely unresponsive to peripheral input. In addition, it would seem reasonable to suspect that in the cortex anaesthetics may well have different effects on different cell types.

To some extent these difficulties could be overcome if it were possible to study a particular type of cell under a truly constant anaesthetic regime. In this regard the pyramidal cells present an interesting possibility. They are large, they have cutaneous receptive fields and they may be identified antidromically and perhaps even individually recognized by their antidromic latency. While care would have to be taken that the procedures of identification did not themselves elicit alterations in receptive fields (Sect. 7.3.1) studying these cells could do much to establish more accurate boundaries of spatial reorganization and the precise responses of SI neurons to deprivation of input. The compromise of such an approach is, of course, that the results will not pertain to other types of cortical cells that may be less involved with fashioning cortical outflow.

Keeping in mind these qualifications and difficulties the following statements may be made from studies of plasticity in the mammalian somatosensory cortex:

(1) Deprivation of Small Areas of Cortex in Adults: (a) Immediate Effects of Nerve Section. Immediately after cutting a peripheral nerve, intact afferents that supply skin adjacent to the denervated skin develop an excitatory influence over many cortical cells in the deprived cortex that abuts the normal representation of these afferents. The maximum width of this reorganized area is about 600 μm and some somatotopic organization is preserved (Species tested: rats and primates). These immediate effects appear to be sensitive to barbiturates, implying that release from GABAergic inhibition underlies the observed reorganization. *(b) Immediate Effects of Reversible Block of Dorsal Roots.* Some, but not all, neurons in the deprived area immediately develop new receptive fields (Species tested: cat). These cells adopt their original receptive field on removal of the block. From available detailed somatotopic maps it would seem that the skin making up these new receptive fields might normally be represented at a distance that is compatible with the transcortical spread of the arbors of single thalamocortical afferents. *(c) Long-Term Effects of Nerve Section.* In the weeks following peripheral nerve section in adult rats and primates, the influence of intact afferent fibres spreads across the deprived area of cortex to a distance of up to 700 or 800 μm from their original (preoperative) representation. The mechanism for this extension of influence is unknown but it may well explain the progressive disappearance of phantom limb sensations – a process known clinically as telescoping.

(2) Chronic Deprivation of Large Areas of Cortex: (a) Forepaw Denervation and Spinal Section in Adult Cat and Digit Amputation in Adult Raccoon. New receptive fields from skin around the denervated region occur throughout the deprived

area of cortex but no clear somatotopic organization is apparent (Species tested: cat and raccoon). Subcortical mechanisms are implicated, for in contrast to the effects of deprivation on small areas of cortex (see *1(a)* and *(b)* above), there is no columnar organization of receptive fields in the reorganized region. *(b) Spinal Section in Young Kittens.* A second representation of the upper body develops in the deprived region apparently as a result of subcortical alterations in connectivity.

(3) Nerve Crush and Regeneration: The responsiveness of deprived cortex to intact peripheral inputs after nerve crush is reversed following regeneration of a nerve and normal somatotopic organization is restored (species tested: primates).

(4) Nerve Section, Repair and Regeneration: The invasion of deprived cortex by intact peripheral inputs is only partially reversed following repair and regeneration and a disorganized somatotopy is apparent within the representation of the regenerated nerve. The degree of restoration of a normal somatotopic organization seems to depend on the precision of nerve repair (species tested: primates).

(5) Use of a Skin Surface: Variation between individuals in the representation of a skin surface is often observed (species tested: rat, raccoon and primates). Some evidence exists that the size might increase with use but long-term effects of directed attention have not been adequately ruled out (species tested: primates) (Sect. 8.6).

(6) Thalamocortical Projections: In both anaesthetized and unanaesthetized animals the terminations of single thalamocortical neurons occupy a considerably larger area than the representation of the skin areas contained in their receptive fields (species tested: cat). Consequently, many parts of the arborizations of thalamocortical neurons must be ineffective in exciting cortical neurons. The regulation of the effectiveness of these terminations could form the basis for immediate and long-term development of new receptive fields following the removal of normal input to small areas (1 or 2 mm^2) of cortex (see *(1)* above). However, reorganization of receptive fields at lower levels of the neuroaxis cannot be ruled out (Sect. 8.2.4).

(7) Cortical Pharmacology: Anaesthetics and specific transmitter substances and their agonists and antagonists alter the size and shape of the receptive fields of cortical neurons. Thus in the intact animal the receptive fields of cortical neurons reflect a balance of inhibitory and excitatory mechanisms operating on a set of inputs from much wider areas of skin (see *(1)* and *(6)* above). There is in fact some evidence that cortical neurons are under tonic GABAergic inhibition and that this is reduced following nerve section (Sect. 8.5.1). This matter, however, requires further investigation using unanaesthetized animals.

(8) Cortical Modulators: Acetylcholine acts as a modulator of cortical activity by increasing the responsiveness of SI neurons to peripheral stimuli and to the probable transmitter of thalamocortical neurons, glutamate. It has been proposed that changes in ACh receptor distribution may underlie the long-term changes of effective input that have been observed following chronic removal of input. The effectiveness of ACh on cortical neurons is increased by norepinephrine as are the inhibitory effects of GABA. Applied norepinephrine enhances the signal to noise ratio possibly by enhancing the cholinergic facilitation of excitation and

the tonic GABAergic suppression of spontaneous discharge. The systems implicated in ACh and norepinephrine release are the nucleus basalis of Meynert and the locus coeruleus, respectively. These systems are discussed further in Sects. 8.5 and 8.6.

(9) Transcortical Distances, Development of Functional Connections and Axonal Sprouting: There is no evidence that anatomical changes in the cortex underlie the development by cortical neurons of new receptive fields following peripheral nerve section. Most observed changes can be explained by the physiological modulation of transmission from the widespread pre-existing arborizations of thalamocortical neurons (see *(6)* above) but particularly where changes occur over large transcortical distances (e.g. following digit amputation in the raccoon), axonal sprouting may be involved (Sect. 8.4).

(10) Reduction of Cortical Territory During Development: When cytotoxic agents are used to reduce the overall size of the cortex, the resultant somatotopic representation is truncated by the loss of the representation of those body parts that form the edges of the representation in normal animals. This result suggests that only a certain area of cortex can sustain afferents from the VB complex and that the topographically organized terminals of these developing afferents complete for this type of cortex.

8 Concluding Remarks

8.1 Plasticity During Development

8.1.1 Disruption of a Growing System and the Influence of the Periphery

The responses of the developing nervous system to peripherally or centrally placed lesions reflect the disruption of a number of finely tuned but ever-changing processes. Lesions may redirect or even prevent these processes. Unlike plasticity in the mature nervous system, which serves a previously established and successful behavioural repertoire, plasticity in the developing nervous system serves the rules of developmental biology and as often as not may thus have serious consequences for survival. The topographic organization of the rodent somatosensory system, and in particular that part serving the mystacial vibrissae, provides an excellent system in which to observe, in action, the rules of developmental neurobiology along the mammalian neuraxis (Chap. 3). Like the nervous system as a whole, neurons at the lower levels of the somatosensory system differentiate before those at higher levels. Following neuronal differentiation, axonal growth establishes broad topographical relations between various levels and precise connectivity occurs at a later time.

The interconnections between levels follows a similar peripheral to central sequence. Thus the connections of primary afferent somata to the trigeminal nuclear complex or the spinal cord are the first to form, while those between the thalamus and the cortex are the last. In neonatal mammals (cats and rats) the survival of sensory cells is jeopardized by severing their connections with the periphery, suggesting the availability there of some life-sustaining factor (Chaps. 2 and 3). In the mature animal the skin that is distal on the limbs or on specialized somatosensory structures such as the whisker pad of rodents is more densely innervated and it is tempting to suggest that during development it is more liberally endowed with the survival factor. In fact, afferents that innervate such skin areas perhaps have smaller receptive fields because high concentration of survival factor in these regions negates the necessity to form widespread peripheral branches. The distal to proximal gradient in receptive field size seems to be independent of particular peripheral structures since in the gracile nucleus of cats with congenital agenesis of a hindfoot, the leg stump was represented in receptive fields that were smaller than those typically found on that part of the limb – an observation that suggests that the receptive fields of primary afferents innervating this region might also be smaller (Fig. 5.23; Sect. 5.6.1).

The influence of peripheral structures does not stop at the site of termination of the primary afferent fibres. In fact, it appears to be carried throughout the neuraxis such that, if peripheral nerve section precedes a critical period, it is sufficient to prevent correct development at the thalamic and cortical levels. It is not even necessary to expose the system to an event as traumatic as peripheral nerve damage. Thus, in the vibrissae system there is a correlation between the size of the anatomical manifestation of the cortical representation of a single vibrissa (a barrel) and the number of primary afferents innervating the vibrissa. If this number falls below a threshold level, a corresponding barrel will not develop (Sect. 3.3.3). Again, it would seem that some factor in the skin is necessary for the development of normal central connections and that the allocation of central nervous tissue is determined by its availability.

Sensory deprivation or hyperstimulation during early development are also effective in determining the allocation of central tissue to a receptive structure or skin area (Fig. 3.21; Sect. 3.3.5). Perhaps hyperstimulation mobilizes more of the factor from somatic cells in or around the receptor or nerve endings, thereby facilitating the allocation of central space. It is easy to appreciate how such a mechanism could account for the enlarged representation of important sensory areas that have denser afferent innervation and also how this could be finely tuned by early, stimulus-evoking behaviour.

These observations suggest that the general topographic relations within the somatosensory system might be achieved by crudely directed axonal growth between nearby structures but that the emergence of a true somatotopic map, that reflects the regional sensory biases of the skin, depends upon the level and/or availability of an unknown chemical factor at different locations on the body surface.

8.1.2 The Influence of Afferent Axons and the Target Tissue

Again it is the vibrissae system of rodents that best illustrates the way in which these factors influence development of the somatosensory system. While in the previous section we considered that the survival of afferents, and the allocation of central tissue, depended upon the peripheral availability of some unknown factor, closer scrutiny of the effects of lesioning the vibrissae, or the infraorbital nerve, suggest that particular skin areas are in some way specifically labelled. This is particularly apparent because of an affinity for one another of the terminals of primary afferent, trigeminothalamic or corticothalamic cells that represent a single (or principal) vibrissa (Sect. 3.2). Similarly, when the infraorbital nerve is cut in the neonate (Sect. 3.3.1) or a lesion is placed in the PMBSF (Sect. 3.3.6), a second order affinity between the terminals of neurons representing the vibrissae of a single row becomes apparent. This intra-row affinity is manifested in many aspects of the system, including the responses of single cortical cells in the adult rodent to movement of several vibrissae (Sect. 3.1) and the embryonic development and innervation of the vibrissae on the snout (Figs. 3.8 and 3.9; Sect. 3.2).

Despite the above points it would be incorrect to believe that the periphery is entirely responsible for the organization of the central parts of the somatosensory system. In fact several entities can be shown to have considerable independence from the periphery. First, in the rat, neonatal thalamotomy does not disrupt the development of cortical layers or the invasion of the deprived layers by commissural neurons (Wise and Jones 1978). Second, thalamocortical terminals appear to distribute in the depth of the cortex in accordance with the distribution of granule cells but in the absence of granule cells they still adopt a bistratified distribution somewhat characteristic of normal thalamocortical terminals (Caviness et al. 1976; Yurkewicz et al. 1984). Third, only a certain amount of cortex is allocated for somatosensory projections and reduction of the overall size of the cortex (by chemically-induced microencephaly or neonatal lesion) produces a truncated somato-sensory representation of the entire body and/or a condensation of the representation of *all* the mystacial vibrissae to a smaller area (Fig. 3.20 and 7.28; Sects. 3.3.6 and 7.5.5). The factors involved in any of these phenomena may be complex or simple but they are certainly completely unknown.

8.2 Evaluation of Experimentally Induced Plasticity in Adult Animals

When the response of an organism to a particular stimulus changes, we must conclude that there has been either a structural alteration in the connections between neurons or that the strength of connections has been varied. It is therefore as easy to accept that the brain has the property of plasticity as it is to appreciate the importance of behavioural adaptability to survival of higher life forms. In addition to these lines of reasoning, neuroscientists are daily exposed to a mass of data that show that our perceptual and motor functions may be tonically modified by a wide variety of procedures, and that connections between nerve cells may vary both structurally and functionally. In the somatosensory system these preoccupations have provided an important source of motivation in the search for models of plasticity that might be useful for the elucidation of underlying mechanisms. Unfortunately, it would seem that they have also been responsible for the publication of a number of studies (particularly on the spinal cord, DCN and the thalamus) which contain conclusions that cannot be substantiated by the data presented. While these conclusions may match our intuitive acceptance of the existence of brain plasticity, they are often misleading to those who wish to understand the functional role of plasticity in a particular part of the nervous system, as well as to those who wish estabilish, by experiment, the underlying mechanisms. In the elucidation of plasticity within the somatosensory system, the general principle that the *specific* conclusions of a scientific paper should stand on the data presented seems often to have been sacrificed for the authority of contemporary beliefs. Fortunately, there exist also some carefully planned and well-executed studies which already are stimulating scientists to consider more carefully the role of plasticity at various points along the neuraxis and enabling them to identify some of the underlying mechanisms. In general, however, we are left with the feeling that less confusion would exist if, in the past, scientists had attempted to meticulously disprove the null hypothesis that *plasticity is not a property of the somatosensory system.* In this section we will attempt to emphasize some basic areas that we believe require careful consideration in the evaluation of experimentally induced plasticity.

8.2.1 Plasticity in the Peripheral Nervous System

Experimental studies of the effects of peripheral lesions on central somatosensory neurons have tended to play down any possible role of changes in the peripheral sensory apparatus (receptors and afferent fibres) in the changes observed centrally. This criticism does not apply, of course, to studies designed specifically to investigate the possible role of central and peripheral mechanisms responsible for abnormal pain sensations (particularly hyperalgesia and chronic pain states) that result from peripheral nerve lesions, but this is an issue that we have regarded as being largely outside the scope of this review.

Following a peripheral nerve injury a large number of processes within the peripheral nervous system are set in motion and many of these have a bearing on functional connectivity of central somatosensory neurons. Changes include collateral sprouting of intact afferents within denervated tissue, regeneration of cut or crushed axons, and the degeneration and death of some axotomized sensory neurons. The implications, for plasticity, of these well-documented effects of nerve injury have been given appropriate recognition in most studies. However, there are also aspects of peripheral plasticity which are poorly understood and which may influence the interpretation of experiments in an unpredictable way.

The influence of collateral reinnervation on the receptive fields of central neurons may or may not have been underestimated, depending upon the distances over which it can occur in different species, the types of afferents capable of sprouting, and the nature of the stimulus used to determine somatotopic organization. It is now widely accepted that there is little or no functional collateral sprouting by large-diameter myelinated cutaneous afferent fibres (at least those originating in spinal ganglia) into denervated regions of skin in adult mammals. However, there is rapid collateral reinnervation by A∂ and C nociceptive afferents (Sect. 2.3.1). In rats and cats a small but significant proportion of A∂ and C fibres are connected to low-threshold cutaneous mechanoreceptors (C low-threshold mechanoreceptors and D-hair afferents), and there is really very little direct evidence to show that these do not undergo collateral sprouting along with high-threshold afferents of the same fibre groups (Sect. 2.3.1). The distances over which collateral sprouting occurs are known in only a few species although both Nixon et al. (1984) and Kinnman (1987) have shown that in the dorsal trunk skin of the rat axons can sprout across dermatomal boundaries (Sect. 2.3.1). Nevertheless, even a limited degree of collateral reinnervation could substantially affect the outcome of experiments in which multiunit receptive fields are used to map central reorganization in the cat and rat.

Problems of interpretation due to the possible regeneration of injured axons from the dorsal root ganglion into foreign nerves and thereby into inappropriate regions of skin have been raised in relation to the effects of peripheral transection of trigeminal nerve branches in neonatal rats on the somatotopic organization of neurons in the trigeminal system (Sect. 3.3.1). Clearly these problems would not be revealed in electrophysiological studies of single afferents in regenerated or intact nerves. Thus, it is important to eliminate the possibility that aberrant regeneration has occurred before concluding that alterations in central connections are responsible for the reorganization of the receptive fields of central neurons.

A fundamentally new perspective on the anatomy of the peripheral sensory neuron has relevance to the involvement of peripheral nerve regeneration in the altered central representation of the skin. For many years the primary afferent neuron in the sensory ganglion has been (and largely still is) considered to be a pseudo-unipolar cell whose stem process divides into two parts: a single axon which enters the spinal cord via the dorsal root and a single axon which enters a peripheral nerve to innervate peripheral receptors in either skin, or muscle, joints, viscera, and so on. Both central and peripheral axons are known to branch repeatedly in the central nervous system and the peripheral tissue, respectively, thereby producing large numbers of central arborizations and, in the case of cutaneous myelinated afferents, often innervating a number of peripheral receptors in the skin. Recent electron-microscope studies, mainly by Coggeshall and his colleagues (Coggeshall 1986), have indicated that this is an oversimplified view, and that sensory afferent axons in peripheral nerves and dorsal roots in fact outnumber the cell bodies of sensory neurons by about 2 to 1. Even more disturbing for the interpretation of central plasticity after peripheral nerve injury is the electrophysiological evidence that the peripheral process of a single dorsal root ganglion cell may branch while still in major peripheral nerve trunks (Langford and Coggeshall 1981; Pierau et al. 1982; McMahon and Wall 1987) and that these cells may even support axons that run in different peripheral nerves. At the same time, they possess only a single receptive field, suggesting that there may be branches of sensory axons in peripheral nerves that do not have a sensory function. Careful and detailed electrophysiological studies of single dorsal root ganglion cells and their axons in normal and severed peripheral nerves are required in order to assess the possible contribution of these factors to those central changes that we currently refer to as plasticity. The implications of the presence of possible multiple *central* projections of sensory neurons for somatotopic organization and the stability of that organization are also obvious.

Whether the existence of collateral sprouting in the deafferented dorsal horn is finally proven or not, the effect of Wallerian degeneration of the central arborizations of primary afferent fibres on anatomical plasticity of intact neurons which share common target neurons is easily understood in terms of the competition for available synaptic space on central neurons which has been so convincingly demonstrated in the adult rat septal nuclei (Raisman 1977). The same intuitively satisfying hypothesis cannot be formulated for the effect of the slow retrograde cell death on the central reorganization that follows peripheral nerve section. All of the central or peripheral changes consequent to nerve injury (somatotopic reorganization at all levels, changes of neuropeptide and FRAP levels in the dorsal horn and collateral reinnervation of the periphery) occur over a period of days to weeks, whereas cell death appears to be a much more gradual process. Nevertheless, each primary afferent makes a number of synaptic connections with some dorsal horn neurons, and the gradual removal of these connections could lead to collateral sprouting either by preterminal axons that innervate the same neuron, or by terminal-free somatotopically inappropriate collaterals (Meyers and Snow 1984; Snow and Wilson 1989, 1990) (Sect. 4.5.5). Furthermore, it has been suggested that relatively slowly changing somatosensory phenomena such as the gradual shrinking of phantom limbs (Fig. 1.2) and disappearance of dysaesthesias associated with nerve injury, could be related to a continuous slow loss of sensory neurons (Devor et al. 1985). Accordingly, the much greater loss of sensory neurons that occurs after neonatal nerve injury (Sect. 2.2.2) could account for reports that phantom limb sensations occur relatively rarely following amputation in human infants (Poeck 1964; Simmel 1966). However, following injury to the stump (for references, see Melzack 1990), or during Yoga-like exercises called "autogenic training" (Schultz 1956, cited by Poeck 1964), a faded or telescoped phantom may revert to its original vividness or size. This remarkable resurgence would indicate that the changes of phantoms with time are not due to irreversible neuronal death. Indeed, such observations strongly suggest that physiological suppression of abnormal activity in deprived regions of the somatosensory system may underlie these changes (see also Sects. 8.4.1 and 8.6).

8.2.2 Somatotopic Organization in Intact Animals as a Baseline for Assessing Altered Connections

Somatotopic maps have been frequently used for elucidating alterations in central connectivity following either peripheral or central lesions. As our knowledge of these maps increases we have come to understand that the representation of the receptor sheets of the body in the central nervous system is not simple but that particular modalities are often projected onto regions that interdigitate with one another in a complex, and as yet poorly understood, manner. In order to accurately detect and measure experimentally induced alterations in connections it is imperative that the details of the somatotopic organization are thoroughly established in the intact animal. It is astounding to find that establishing this complicated baseline has often been relegated to a minor preamble in papers principally devoted to detailing its plasticity.

Superimposed on these somatotopic representations are the effects of anaesthetics which act upon all levels of the neuraxis. Notwithstanding one report in the primate (Sect. 7.3.3), there is an abundance of evidence that anaesthetics act upon central neurons so as to alter the size of the receptive fields and their responsiveness to particular receptor types (Sects. 3.1, 4.5.1, 5.2.4, 6.3, 7.2.6, 7.3.3, 7.4.2 and 7.5.1). If part of a somatotopic representation is deprived of its afferent input then the new receptive fields that are observed under anaesthetics might reflect the influence of the agent on an abnormally functioning, unbalanced system, rather than any change which could be exclusively attributable to deafferentation. The same argument could be applied to the alterations following centrally placed lesions. In a complicated circuit, removal of an input is bound to have an effect on what remains. That effect may be influenced by parameters that determine overall excitability differently from those properties that are part of the normal intact system. In neurophysiological studies of experimentally induced plasticity we often measure that effect in a highly altered state – that of deep anaesthesia.

Evidence from the DCN and the SI cortex suggests that there is considerable variation between the somatotopic representations seen in different individuals of the same species (Sects. 5.2.5, 7.2.4, 7.4.1 and 7.5.2). Such variation is sometimes apparent in maps made within the same individual at two different times but unfortunately these cases are all derived from experiments where some closely associated part of the peripheral or central nervous system has been lesioned. The origin of the inter-animal variations in somatotopic representations is unknown but it is possible that they result from differences in the levels of anaesthesia or natural mechanisms that regulate the central representation of the body (Sects. 8.5, 6) or an interaction of both these factors.

8.2.3 Somatotopic Artifacts in Regions Deprived of Their Normal Input

In intact animals somatotopic maps were traditionally derived to reveal the manner in which various body regions are represented in central nervous tissue. Assuming that, in neurophysiological studies, spikes over a certain size are routinely accepted, then a relatively large number of recording sites across the representation will reveal a more or less accurate picture of the central representation of the body surface. What will not necessarily emerge are accurate measures of (a) receptive field size, and (b) an accurate assessment of the central representation of a small skin area. These problems are exaggerated by the use of large electrodes and the acceptance of multiunit recordings. They are minimized by the use of fine electrodes and the selection of those single units that produce relatively large spikes.

Alterations in somatotopic representations are detected by recording the responses of central neurons to stimulation of a small area on the body surface. It is generally presumed that the recording site corresponds to the location of the neurons generating the activity. Clearly, this assumption is invalid and the question remains as to how accurately the recording site represents the position of the responsive neurons. To put it another way, how far away from the electrode does a neuron have to be before it is *not* monitored by the experimenter? The answer to this question depends partially on the size of the electrode tip and the extracellular spread of the electrical fields of the action potentials of the responsive cells in that part of the central nervous system. It also depends on the size of the action potentials the experimenter accepts as being representative of a nearby cell.

In regions acutely deprived of their normal afferent inflow, it is particularly easy to monitor responses that normally would not be easily seen amongst the larger spikes of the now unresponsive cells close to the electrode tip. These responses may be from neurons that lie within the nearby representation of skin that retains an intact innervation (and borders on the denervated skin). Alternatively, the neurons may lie in nearby parts of the central nervous tissue where cells might normally have receptive fields that span the boundary between denervated and normal skin (e.g. deeper cortical layers, Sect. 7.5.1) or in nearby parts where cells receive inputs from sensitive receptors in deep tissues (such as muscle spindle afferents) that can be easily activated by stimuli that are often used routinely to crudely localize cutaneous receptive fields during neurophysiological mapping experiments (Sect. 4.4.3). Misidentification of denervation-induced plasticity as a result of both causes is apparent in a number of published accounts. These problems would have been minimized if workers had carefully determined the functional organization in the intact animal *before* embarking on experiments on the effects of lesions and if, in both intact and lesioned animals, they had taken greater care to confirm histologically all recording sites, to present data on the receptive fields of single neurons with well-isolated action potentials at *each* site and to use more precise forms of natural stimulation.

A second approach to attempting to elucidate denervation-induced plasticity is what we might call the *number-of-points method.* In comparing chronically and acutely denervated animals this method is less concerned with the receptive fields of individual neurons and more concerned with the number of recording sites within a fixed grid where responses to the periphery can be recorded. In comparing acutely or chronically denervated animals with normal animals the number of sites responsive to particular regions of the periphery (e.g. foot, leg, etc.) are ascertained. The number-of-points method is less time-consuming for it places less emphasis on individual receptive fields. Unfortunately, it cannot produce results that are less equivocal than those utilizing individual receptive fields and one cannot help questioning whether its practice encourages the careful definition of sensory inputs that is necessary to localize accurately a receptive field to a particular body region. The most common failing of this method occurs when the spacing of the grid points is less than the distance over which one might record the activity of a central neuron. In such cases it can be assumed that an increase in the number of active sites indicates only the recording of the responses of distant neurons in an unresponsive, deprived region of the central nervous system. When chronic deprivation involves removal of the afferent supply to the nucleus at an early stage of development, the apparent reorganization may be further exacerbated by atrophy of the nucleus.

As spatial changes in somatotopy are so often used as an indicator of plasticity it would seem wise to choose part of the somatotopic representation in which a relatively large area of tissue is devoted to skin which is to be denervated. The huge digit representations in the raccoon cortex are an extreme example of a good preparation, in which there is minimal possibility of recording, within the deprived region, the responses of distant neurons which normally receive inputs from intact nerves. In a number of published accounts on denervation-induced plasticity, scrutiny of the normal somatotopic representation shows that the skin within the newly developed receptive fields could easily have been originally represented quite close to the recording site. One cannot help being sceptical that the responding cell was distant, not from the original representation of its receptive field but from the tip of the electrode.

Many of the above technical problems disappear if antidromically identified projection neurons are chosen for studying plasticity. Using such cells has four advantages: (1) they probably represent a more precisely definable, physiological subpopulation; (2) repeated tracking can be used to maximize the spikes and so clearly isolate the responses of a projection cell from those of surrounding neurons; (3) projection neurons usually lie within a particular subdivision of the responsive tissue (e.g. a particular cortical layer); (4) with the help of manipulator coordinates or surface blood vessel patterns, the same projection neuron (identified by its position and its characteristic antidromic latency) can be monitored at different stages throughout an experiment or even in separate recording sessions. There are two major disadvantages of using these cells. The first is that repeated application of the stimulus required for antidromic identification might elicit, via axon collaterals or some other pathway, long-lasting changes in the receptive fields of the projection neurons (Sects. 4.4.3 and 7.3.1). The second is that plasticity in the connections to projection neurons might not be totally illustrative of the mechanisms that influence connections to local circuit neurons.

8.2.4 Plasticity and the Level of the Neuraxis

As will be discussed in Sect. 8.3 there is an abundance of evidence to suggest that ineffective connections exist at many levels of the somatosensory system. Similarly, at all levels of the somatosensory system there is a strong likelihood that what is considered to be the normal somatotopy is, at least partially, a construction of tonic inhibitory processes that render certain afferent connections ineffective (Sect. 8.5.1). It is therefore useful to ask if denervation-induced plasticity in SI cortex is merely a reflection of a reorganized somatotopic representation at a lower level of the neuraxis? Certainly, there exist some striking similarities in the reorganization seen at different levels. For example, following the amputation of a single digit in the monkey, cells in the deprived representation of the SI cortex become responsive to skin on the adjacent digits while, following section of the nerves to a single digit in the cat, cells in the spinal representation of the digit become responsive to skin on the adjacent, innervated digits (Figs. 4.21 and 7.12a–c). From similarities such as this it might be argued that the *preferred* direction of the somatotopic change is always across the representation of the digits and an attempt could be made to attach some adaptive value to this pattern of reorganization, as against, for example, alterations along the axis of the limb.

Notwithstanding the above similarities in the form of somatotopic reorganization there are two important pieces of information that suggest that changes in the cortex do not simply reflect changes in the spinal cord or other subcortical levels of the nervous system. The first involves the time-course of the changes. In the spinal cord reorganization of the digit representation following denervation of a digit takes place slowly over a period of 1–2 months. In contrast, in the SI cortex of both rats and primates, partial deafferentation results in the immediate expression of inputs from nearby skin. Thus the immediate changes at the level of the cortex cannot be attributed to changes at the spinal level, although the immediate expression of new inputs has been reported in the DCN and the VB complex. The second piece of information concerns a comparison of the effects of peripheral nerve section with those produced by lesions at various levels of the central nervous system. Lesions of ascending spinal tracts have been shown to elicit some reorganization in the DCN and the SI cortex (Sects. 5.7 and 7.6). Similarly, lesions to the principal sensory nucleus of the TNC have been shown to produce alterations in the receptive fields of neurons in the VPM of the thalamus and synaptogenesis in the VPL has been shown to follow lesions of the DCN (Sects. 3.4 and 6.5.2). While there are no studies showing reorganization in SI cortex after restricted thalamic lesions, there is a brief report that damage to a small area of the hand representation in the primate

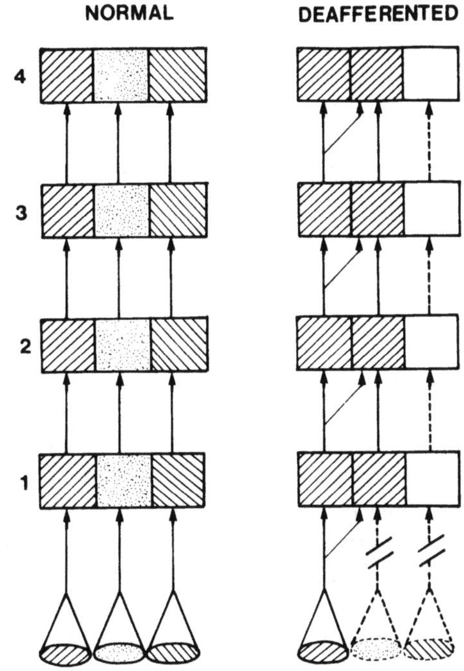

Fig. 8.1. Schematic outline of the representation of three adjacent skin areas at four successively higher levels of the neuraxis in a normal animal (*left*), and the influence of one of these areas at each level following denervation of the other two skin areas (*right*). Quantitatively, the areas of skin are scaled to the central representation of the body surface, such that, irrespective of their location on the body, they would be represented, in a primate, over about 1.0–2.0 mm² of SI cortex (see Figs. 7.1, 7.2 and 7.14). The *thick arrows* represent effective connections that existed before, and survived after, partial deafferentation. *Thin arrows* represent newly effective connections that develop, at *each* level, after partial deafferentation. The *dotted arrows* represent pathways that, after partial deafferentation, no longer receive an input from the skin. Note that after partial deafferentation, innervated skin establishes an influence over regions of level *1* (*central box, level 1*) that were previously activated by the *nearby* skin within the central (*stippled*) patch. This influence can thus be relayed, via two pathways, directly to those regions of level *2* that were previously activated by the *nearby* skin within the central patch. These pathways are the pre-existing effective connections between the two areas (*thick arrow* between *central boxes* at *levels 1 and 2*) and the newly effective connections (*thin arrows* to *central box, level 2*) from regions of level 1 that normally respond to the innervated skin (*left box, level 1*). This pattern of change is repeated at levels *3*, and *4* such that at levels *2*, *3*, and *4*, regions previously activated by denervated skin, receive inputs from any *nearby* innervated skin via two pathways – pre-existing effective connections and newly effective connections. On present experimental evidence there is no reason to reject the hypothesis that, at levels above the trigeminal nuclear complex and the spinal cord, denervation-induced reorganization of somatotopy is the result of *both* newly formed effective connections and pre-existing connections from reorganized centres lower in the neuraxis. See text for further discussion

results in some compensatory shifts of inputs to the damaged area into nearby intact cortex (Sect. 7.2.10). In conclusion, it is clear that the spinal cord and higher centres have the machinery that permits somatotopic reorganization in response to partial deafferentation.

Although empirically each level of the somatosensory system has the potential for the redistribution of somatosensory input, more data on this matter are required. It would be useful, for instance, to determine the effects of digit amputation in the cat at the level of the DCN, VB complex and the SI cortex. In the absence of these data we might consider the model reorganizing system shown in Fig. 8.1. At the lowest level (level 1) of the neuraxis, removal of input to a particular region will result in that region becoming responsive to nearby innervated skin. Projection neurons in that region will therefore relay this new input to the next level (level 2). If denervation has also elicited a restructuring of input to level 2 from level 1, then innervated skin might well have two pathways by which it may activate cells at level 2 – the original inputs from those cells at level 1 that have newly acquired receptive fields on nearby skin *and* new inputs from cells at level 1 that always did have receptive fields on nearby skin. This process will be repeated at higher levels of the nervous system and the restructuring of the receptive field profile will therefore be the same at each level irrespective of the underlying mechanism. If two input pathways are involved in the development of new receptive fields, a degree of security of transmission may be achieved in what might otherwise be a weak chain of synaptic linkages between the periphery and the cortex. The mechanisms by which somatotopic reorganization occurs at each level of the neuraxis are discussed in the following sections (Sects. 8.3 and 8.4.1).

8.3 The Case for Ineffective Connections

Studies in normal intact animals have shown that at all levels of the neuraxis there are connections, or opportunities for connections, that are not in keeping with skin area or modalities represented in the naturally evoked discharges of postsynaptic cells. Thus cutaneous primary afferents and thalamocortical fibres have terminals in areas of the dorsal horn and the SI cortex, respectively, that are significantly displaced from their somatotopically appropriate region (Sects. 4.3.3 and 7.3.2). Similarly, the dendrites of cells in the TNC and the DCN range far from the termination sites of afferents that supply their effective input (Sects. 3.2 and 5.4.3). In the VB complex of the thalamus there is little information on the relationship between the size of dendritic trees or the spread of afferent terminations and the receptive fields of thalamic afferents and neurons (Sect. 6.2). There is, however, much data showing that thalamic neurons receive inputs that are essentially subthreshold, at least in the presence of anaesthetics (Sect. 6.3). Finally, although there is no direct evidence for the contribution of interneurons or tract neurons to deafferentation-induced changes in receptive fields, there are, at all levels of the nervous system (with perhaps the exception of the VB complex of the rat) many excitatory interneurons and/or projection neurons that by virtue of their local axonal arborizations could relay a subthreshold influence from one point on the skin to cells that have receptive fields on another skin area (see Fig. 8.5b). The pyramidal cells of the SI cortex and the spinocervical tract neurons in the spinal cord are two well-known examples of tract cells with collateral projections that are sufficiently extensive to cross representational boundaries (Brown et al. 1977; Szentagothai 1978).

Information presented above has been derived from basic studies of the organization of the somatosensory system. In the following sections we will focus on the results of studies that have sought directly to establish the presence of ineffective connections within the somatosensory system with the specific goal of understanding more about the mechanisms that might be responsible for alterations in connectivity following peripheral and central lesions.

8.3.1 Elucidation of Sub-Threshold Inputs

Many investigators have used electrical stimulation of the skin and cutaneous nerves to demonstrate the existence of weak connections linking afferent input from one area of the body surface to somatotopically inappropriate regions of a central representation of the body. The synchronous activation of a group of primary afferent fibres by electrical stimulation of the skin or cutaneous nerves presumably leads to a summing of the subthreshold effects of weak synaptic connections and the subsequent generation of action potentials in central neurons that are not excited by naturally evoked impulses in those afferent fibres. Unfortunately, little can be said about the identity of the afferent fibres responsible for such responses except that, in terms of modality specificity and axonal conduction velocity, they would probably be a functionally heterogeneous group.

Regionally restricted responses to the electrical stimulation of wide areas of skin have been reported in the spinal cord, DCN, VB complex and the SI cortex (Sects. 4.5.1, 5.4.4, 6.3 and 7.3.3). The inputs revealed by these responses are thus clearly not in keeping with the somatotopic representation as it is revealed by natural forms of stimulation. In order to determine the potential role of weak synaptic linkages in the plasticity of somatotopic maps we would like to know whether the primary afferents being activated are mono- or polysynaptically linked to the neuron from which recordings are being made (Sects. 4.5.1 and 5.4.4). In the thalamus and SI cortex this is clearly not the case but even at lower levels the unequivocal definition of the monosynaptic effects of electrical stimulation is extremely difficult. This becomes clear if one considers that synaptic delays in the mammalian central nervous system are in the order of 0.2–0.3 ms whereas the conduction velocity range of group II cutaneous afferents varies from 30–70 m/s, giving a *minimal* conduction delay from skin to spinal cord of around 1.4–3.3 ms per 100 mm. The range of latencies covered by the monosynaptic EPSPs evoked in central neurons by electrical stimulation of a large number of cutaneous afferents is therefore likely to be many times the monosynaptic latency. In addition, determining the monosynaptic nature of weak connections is all the more difficult because of the low safety factor that links the EPSP and the action potential.

While the electrical stimulation of nerves or skin may reveal weak synaptic linkages throughout the somatosensory system, such experiments give little information about the nature of these linkages or the location of the synapses involved. In some cases these linkages are very sensitive to anaesthetics – particularly barbiturates that are known to potentiate GABAergic mechanisms. However, this observation only underscores their proximity to threshold, even following the sychronous activation of many afferents. Subthreshold inputs to some central cells have been indicated by the expansion, or *de novo* expression, of the receptive fields in response to the microiontophoretic administration of GABA antagonists, the excitatory amino acid, glutamate, or the convulsant, 4-aminopyridine (Sects. 4.5.1, 5.4.5 and 7.3.3). Perhaps if this approach were used in combination with electrical stimulation, it might be possible to determine whether the weak synaptic linkages revealed by electrical stimulation are actually at the site of recording and drug application.

In conclusion, we can say that electrical stimulation of the skin can reveal weak synaptic linkages in the somatosensory system but that to determine the identity of the synapses and neurons involved we need to turn to other techniques.

8.3.2 Somatotopically Inappropriate Projections of Afferent Axons

Somatotopically inappropriate projections constitute any axonal terminations of afferent neurons that are located in those regions of the next highest level of the neuraxis, which contain cells with receptive fields that do not encompass the receptive field of the afferent neuron. By definition these projections are *ineffective* in firing postsynaptic cells under the conditions in which somatotopic organization has been defined.

It is important to appreciate that to reveal somatotopically inappropriate projections, the area of skin supplied by the afferent(s) and the distribution of afferent terminations with respect to the somatotopic organization of the target tissue must be known. At the microanatomical level anterograde and retrograde transport of tracers, or degeneration staining, have been used in attempts to establish whether primary afferent fibres contained within particular nerves, or dorsal roots normally have somatotopically inappropriate terminations (Sects. 4.3.3 and 5.4.1). Similar techniques have been applied to the thalamocortical projections to the digit representations in the raccoon (Sect. 7.4.3). In general, such projections have not been revealed using these techniques. Clearly, this does not mean that such projections do not exist. They may, for instance, be so weak, in terms of the number of axonal arborizations and/or synaptic boutons, that they do not show up with anterograde tracing methods (Sect. 4.3.3) (Snow and Wilson 1989, 1990). Similarly, they may not have enough synapses to permit them to take up the quantities of tracer necessary for the detectable labelling of their cells of origin by retrograde axonal transport (Sects. 7.4.3 and 7.4.4).

Electrophysiological and anatomical studies of single afferent axons with a known receptive field, in combination with somatotopic mapping of single units within the target tissues has provided unequivocal evidence for the existence of somatotopically inappropriate afferent projections in the spinal cord and the SI cortex (Sects. 4.3.3 and 7.3.2). From these studies it is possible to say that the spatial extent of the somatotopically inappropriate projections is equivalent to that over which deafferentation-induced alterations of receptive fields has been measured. More important is the observation that, within the spinal cord, the somatotopically inappropriate projections of cutaneous primary afferents lack well-developed terminal arborizations and bear few synaptic boutons (Fig. 4.13; Sect. 4.3.3) (Meyers and Snow 1984; Snow and Wilson, 1990; Wilson and Snow, 1990a). Unfortunately, no comparable experiments have been conducted in the SI cortex and so it is not possible to say whether the somatotopically inappropriate projections of thalamocortical neurons bear fewer boutons than those that are appropriately placed with respect to somatotopic organization (Fig. 7.15).

It may be concluded that at the spinal and cortical levels of the somatosensory system of adult animals, afferent neurons do have somatotopically inappropriate afferent projections. There is reason to believe that such projections exist at all levels but more data on the morphology of afferent terminations within somatotopically defined areas are urgently required. The involvement of somatotopically inappropriate projections in plasticity will be discussed below.

8.4 Spatial Extent of Immediate and Long-Term Changes in Somatotopic Organization

8.4.1 Distance Limits of Somatotopic Reorganization

Following denervation of an area of skin some cells in the deprived part of the SI cortex of adult rats and primates immediately develop new receptive fields. These cells are recorded at points that are within 600 μm of the representation of skin that has intact innervation. With time after chronic deafferentation the influence of this skin spreads across the deprived cortex to a distance of 700–800 μm from its normal representation. Larger distances have been reported following digit amputation in raccoons and chronic deafferentation of the paw, or cordotomy, in kittens, suggesting that the distances involved may be both species- and age-dependent (Sect. 7.6). In the DCN, dorsal rhizotomy elicits an immediate shift of 500–600 μm in the representation of abdominal skin and a further expansion occurs over some months (Sect. 5.3.2). In the spinal cord *no* immediate expression of inputs accompanies partial deafferentation but after a number of weeks, cells in the deprived region develop receptive fields that were previously represented as much as 5–10 mm away from the recording site (Sects. 4.4.1 and 4.4.3). These maximum transneural distances over which the effectiveness of an input expands are known as the *distance limits.* That in both the cortex and the spinal cord the distance limits reflect the transneural distances over which somatotopically inappropriate afferent projections have been identified seems more than coincidental. It also seems to us more than coincidental that in the spinal cord, where the somatotopically inappropriate projections of primary afferents are sparse and lack boutons, an immediate stage of deafferentation-induced plasticity is absent, and that somatotopic reorganization occurs over a period of weeks after nerve section. Although, in the cortex, no part of the axonal arborizations of thalamocortical neurons appears to lack boutons (Fig. 7.15a), deafferentation-induced plasticity involves both an immediate release of presumably subthreshold inputs followed by a slow expression of new inputs that is more compatible with changes seen in the spinal cord. The response of the DCN to dorsal rhizotomy resembles the two-stage process seen in the SI cortex.

In summary, we would suggest that in the spinal cord local sprouting somatotopically inappropriate primary afferent projections is responsible for the development of new receptive fields (Fig. 8.2a; Sects. 4.5.5 and 8.4.2) whereas in the cortex deafferentation elicits an immediate release of subthreshold, somatopically inappropriate, thalamocortical inputs followed by proliferation of these terminals and a consolidation of their influence over the deprived region (Fig. 8.2b). At other levels of the neuraxis, it seems that reorganization follows a two-stage process similar to that observed in the SI cortex (Fig. 8.2b). Alternatively, it is tempting to suggest that at levels above the spinal cord, the immediate reorganization of somatotopy reflects a release from normal physiological and pharmacological inhibitory mechanisms that are intrinsic to those levels (Sect. 8.5.1), while the slower changes reflect *solely* the proliferation of somatotopically inappropriate afferents within the spinal cord (Sect. 4.5.5) and the transfer of this change to higher levels through previously existing, effective, ascending pathways (see Fig. 8.1; Sect. 8.2.4). However, this seems unlikely on two counts. First, there is some evidence for sprouting both within the DCN and the VB complex following the destruction of pathways afferent to these levels (Sect. 8.4.2). Second, Lamour and Dykes (1988) have suggested that a slow change in ACh receptor density in the cortex is the basis of some of the slow changes in somatotopic reorganization (Sect. 7.5.4). In considering these matters it should be remembered that we have no information concerning the involvement (if any) of excitatory interneurons (Fig. 8.5b) or the local collateral projections of tract neurons in the plasticity of the somatosensory maps seen at any level of the neuraxis.

It seems important to emphasize that not only is the spatial extent of alterations of central connectivity both species- and age-dependent, but that it also appears to be a characteristic of the target tissue (including its afferent terminals). For example, in the SI cortex of rats and primates, the distance limits of 700–800 μm are maximum values. These figures infer that, in these species, a piece of cortex will *not* adopt new inputs unless it is within 800 μm of an area that normally receives those inputs. What this means is that an island of unresponsive cortex will be left if input is removed from a region some of which lies further than 800 μm from the normal representation of intact nerves. In the primate and the cat each digit occupies approximately a 1.0 mm^2 area of cortex – an area equivalent to about four cortical columns or segregates (Figs. 7.2, 7.13a, b and 7.14). Thus removal of two digits should, and does (Fig. 7.12d), leave an area of permanently unresponsive cortex. With removal of an entire limb we would expect this unresponsive area to be much larger and one wonders whether random or spurious activity in such a region could be responsible for the phantom limb sensations which haunt the amputee (Fig. 8.6; Sect. 8.6). Perhaps the slow disappearance of a phantom limb (telescoping) is associated with the transcortical growth of thalamocortical fibres or cortical interneurons,

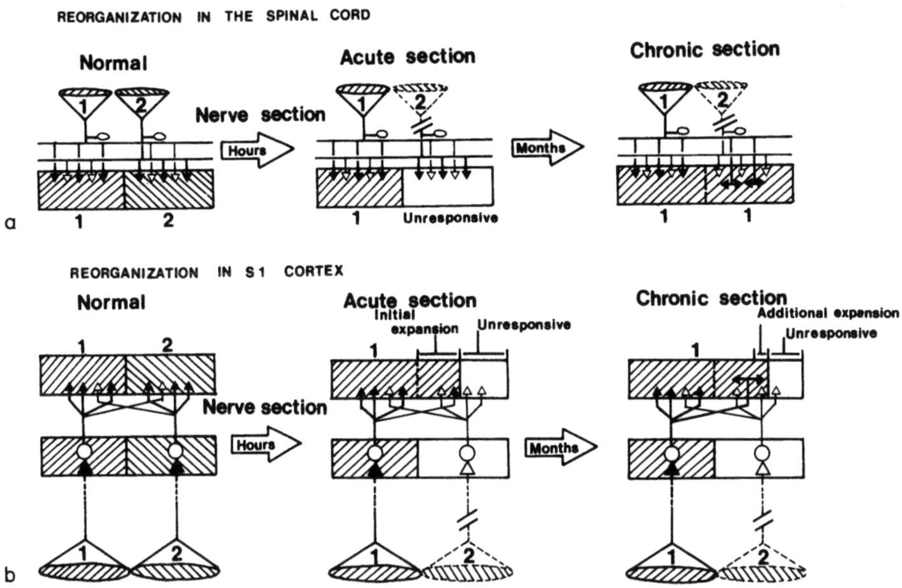

Fig. 8.2 a, b. The immediate and long-term effects of partial deafferentation in the spinal cord and the SI cortex. Quantitatively, the areas of skin are scaled to the central representation of the body surface, such that, irrespective of their location of the body, they would be represented, in a primate, over about 1.0–2.0 mm^2 of SI cortex. In the cat and the primate this area would encompass the representation of a single digit or the skin of half a limb (see Figs. 7.1, 7.12 and 7.14). **a** Organization of afferent projections from two adjacent digits in the parasagittal plane of the spinal cord. *Filled arrows* represent effective connections and *unfilled arrows* represent somatotopically inappropriate, ineffective projections. Hours after denervation of one digit (acute section) the somatotopically inappropriate projection from the adjacent digit remain ineffective and the deprived region is unresponsive to peripheral stimuli. Over the succeeding months (chronic section) the somatotopically inappropriate projections of the intact afferents proliferate, develop synapses and become effective in exciting spinal cells such that the representation of the neighbouring, innervated digit expands along the rostrocaudal axis of the spinal cord (see Figs. 4.21 and 4.37). **b** Organization in the SI cortex of the projections from thalamic neurons representing two adjacent digits. *Filled arrows* represent effective connections and *unfilled arrows* represent somatotopically inappropriate, ineffective projections. Hours after denervation of one digit (Acute Section) the somatotopically inappropriate, thalamocortical projections become effective and there is an *initial expansion* of the representation of the intact digit over the *nearby* (within 600 μm) regions of deprived cortex. Over the succeeding months (Chronic Section) the newly effective thalamocortical projections proliferate and establish an influence over a wider area of deprived cortex. There is thus an *additional expansion* of the representation of the intact digit over a distance of 700–800 μm from the border of its original representation. We suggest that the *initial expansion* represents a release from *in-field* inhibition but that the *additional expansion* results from *sprouting* of the previously somatotopically inappropriate thalamocortical projections (see Figs. 8.3 and 8.5). Of course, both phenomena might also reflect changes lower in the neuraxis (see Fig. 8.1). See text for further discussion

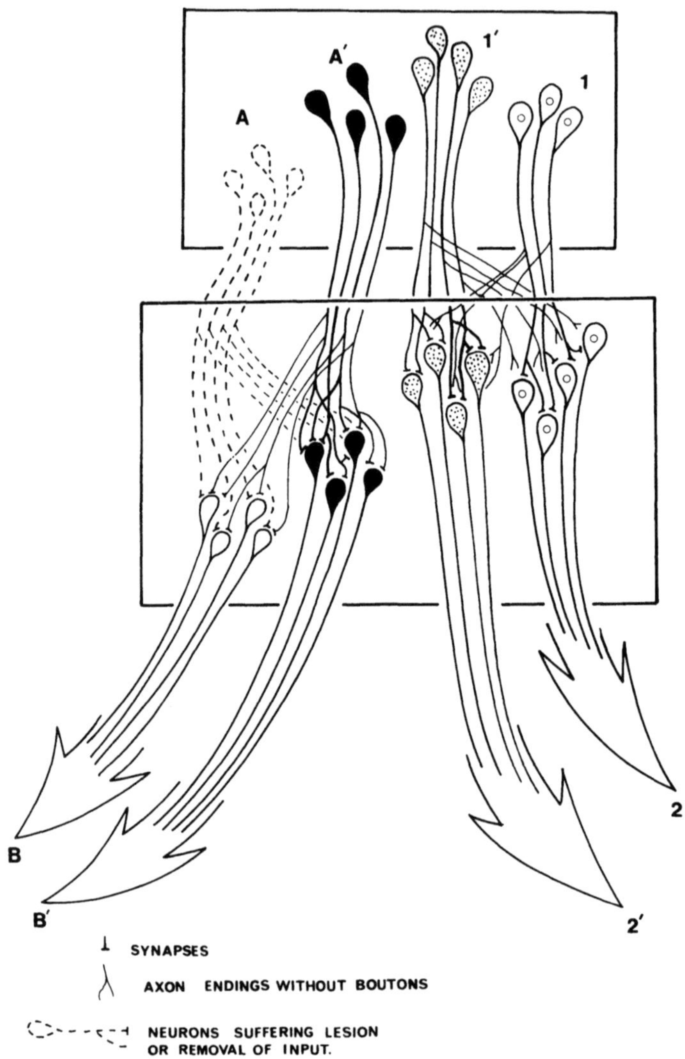

Fig. 8.3. Schematic diagram illustrating the possible compensatory role for somatotopically inappropriate, afferent projections within two parallel somatosensory pathways which relay information from quite different receptor classes. The components of the *left* pathway (*A* to *B* and *A'* to *B'*) relay information from the same receptor class but respond to nearby skin areas. The components of the *right* pathway (*1* to *2* and *1'* to *2'*) also relay information from the same class of receptors which are different from those that activate the pathway on the *left*. Thus the pathways on the *left* and the *right* serve quite different functions. Somatotopically inappropriate afferent terminations form functionally inert, anatomical links between A and B', A' and B, 1 and 2' and 1' and 2. This link does not involve synapses. No such links exist between alphabetically (*left*) and numerically labelled (*right*) sets of neurons. The removal of input to, or lesion of, A permits the ineffective axonal terminations from A' to proliferate and develop synaptic boutons, thereby enabling A' and to influence B. Thus B might, for example, remain responsive to the same class of receptors but these are now on the same area of skin as those exciting B'. The new output therefore serves a closely allied, though not identical, function to that served by B.

or similar changes at lower levels of the neuraxis (Fig. 8.2a with Fig. 8.2b). Certainly some very large transcortical shifts in somatotopy have been reported following digit amputation in raccoons, chronic denervation of the paw in kittens and adult cats and cordotomy in adult cats (Sects. 7.3.4, 7.3.6 and 7.4.4).

Finally, where they have been measured, the transneural shifts in somatotopy following local cortical damage in the adult primate are also in the order of hundreds of microns (Sect. 7.2.10). It thus seems unlikely that the plasticity that has been demonstrated in the somatosensory system could form the basis for recovery from cases of cortical damage that involve more than 1–2 mm^2 of cortex. However, the degree of plasticity may be sufficient to provide a basis for the progressive reduction of the length of a phantom limb (Fig. 1.2; Sects. 8.5 and 8.6).

8.4.2 Sprouting and Synaptogenesis in the Mature System

While alterations in a somatotopic map can reveal a change in neural circuitry they cannot tell us whether the underlying mechanisms include the growth of neuronal processes or whether they involve changes in efficacy of synaptic transmission in the absence of morphological changes. It is generally conceded that morphological alterations cannot be involved if changes in the map occur over a number of hours but this point of view may need revision in the light of recent evidence showing that the dendrites of postganglionic neurons of the mouse superior cervical ganglion are constantly changing their form (Purves and Hadley, 1985). Alterations of somatotopy that proceed over a number of weeks have been described in the spinal cord, the DCN, the VB complex and the SI cortex (Sects. 3.4, 4.4, 5.7 and 7.6). Nevertheless, in the somatosensory system of adult animals, there are surprisingly few good studies aimed at revealing synaptogenesis or axonal sprouting following either central lesions or peripheral nerve damage.

In the VPL of the thalamus, the reticular zone of the gracile nucleus and the spinal cord, partial destruction of afferent neurons results in an increase in the density of the terminations of a population of surviving afferent fibres (Sects. 3.4.2, 4.5.5, 5.3.2 and 6.5.2). These studies are interesting because they do not reveal sprouting over a long distance but instead they show a proliferation of terminations and preterminal arborizations within the territory that is already occupied by a sparse projection from the afferent fibres. It therefore seems that, in the adult nervous system, mechanisms persist that, in embryonic life, elicit the proliferation of axonal arborizations and permit the survival of correctly placed synaptic terminals. Thus the removal of an input to a region permits fibres that already have projections to that region to give rise to more axonal arborizations and to generate new synapses. These new terminations are generated, and survive, only within the *boundaries* of the

Fig. 8.3 (*continued*). No communication is established between the alphabetically labelled pathways and those labelled numerically. In this model structural plasticity does not violate the localization of function that is such a basic feature of central nervous system organization. The outcome of these alterations is different from that described in Fig. 8.6. (From Snow and Wilson 1989, p. 301) original projections even if, nearby, there exist areas that remain permanently deafferentated. We would suggest that *the position of these boundaries determines the distance limits of somatotopic reorganization* following chronic, partial deafferentation of a somatotopic representation.

In the spinal cord many cutaneous primary afferents give rise to somatotopically inappropriate collaterals. These collaterals are simple in structure and appear to bear few boutons. When the area of skin normally represented in a particular region is chronically deafferentated in neonatal animals the terminations of somatotopically inappropriate collaterals from neighbouring skin areas appear to branch and develop many new synapses and at the same time cells in the region develop receptive fields on neighbouring skin areas (Snow and Wilson 1989, 1990). It is therefore possible that the major role of somatotopically inappropriate projections is to lay a latent claim over a region of the central nervous system – a claim that is only staked if the dominant input to the region is lost (Fig. 8.3). By such a mechanism, local axonal sprouting and synaptogenesis could contribute to the restructuring of the connections within the mature central nervous system without requiring the guidance of axonal sprouts over long distances. This mechanism is compatible with the changes in local circuitry that presumably result from the ongoing modifications of dendritic trees (Purves and Hadley 1985). Gross alteration in function would not necessarily result from such changes for new connections would continue to be confined to nearby neurons by the boundaries set up in development and maintained throughout life. Ascertaining the stability and functional significance of these boundaries seems an imperative step in future studies of plasticity.

8.4.3 Recovery of Function

While it is clear that very much remains to be discovered before the neural basis of somatic sensation is understood, it is probably true to say that understanding the relationship between the short- or long-term effects of lesions on the organization of the somatosensory pathways and the ensuing loss and recovery of function will require an even greater conceptual leap. The size of the problem of relating structure to function may be highlighted by considering that arguments still exist over whether lesions of as massive an afferent somatosensory pathway as the dorsal funiculi in primates have any effect on somatosensory acuity (Sect. 5.5). Nevertheless, both recovery of function and permanent disorders of sensory perception following lesions of afferent pathways are regarded by many as bearing a direct relationship to the anatomical and physiological changes that have been described at different levels of the somatosensory system.

Goldberger and Murray (1985) have attempted to relate the recovery of locomotor function following chronic deafferentation of the cat hindlimb, to sprouting and synaptogenesis in primary afferent fibres, descending pathways or intrinsic neurons within the dorsal horn of the spinal cord. Similarly, Vierck (1982) has suggested that a relationship might be found between collateral sprouting in the DCN and recovery of somatosensory function in primates. However, in neither situation is there unequivocal evidence of collateral sprouting. Attempts to relate established somatotopic changes after lesions to behavioural observations would seem more likely to bear fruit at the present time. Indeed, several unifying hypotheses have emerged from anatomical and physiological studies of plasticity at different levels of the neuraxis and two of these are of particular interest. The first concerns the effects of peripheral nerve injury and regeneration on the somatotopic organization of somatosensory cortex and its relation to behavioural phenomena. Various well-known behavioural sequelae of peripheral nerve injury have been related to changes in cortical somatosensory maps. For example, the phenomenon of mislocalization following nerve injury has been attributed to the permanent scrambling of the somatosensory cortical representation of the reinnervated skin due to the lack of somatotopic specificity of regenerating primary afferent axons (Sects. 2.5 and 7.2.7) (Wall and Kaas 1985). However, according to Wynn-Parry (1981), it is almost always possible to retrain patients so that localization is nearly perfect (Fig. 8.4a, b). Similarly, stereognosis (the ability to recognize objects by manipulation alone) is always seriously impaired after median nerve resuture but can be considerably improved with suitable training (Wynn-Parry 1981; Dellon 1981). Wynn-Parry (1981) makes the important suggestion that such improvements in sensory perception must involve the reorganization of central connections. On the basis of this hypothesis, the degree of reorganization necessary to explain the improvement would inevitably involve changes in somatotopy over much greater distances than the distance limits that appear to hold for SI cortex in primates (Sects. 7.2.7 and 8.4.1). However, the fact that these patients take significantly longer to make correct decisions about localization (Fig. 8.4c) clearly implies that this kind of training may involve cognitive processes and rote learning, rather than a functional adaptation of the somatosensory pathways. Unfortunately, as we have no knowledge of the somatotopic organization of the SI cortex of retrained subjects we cannot eliminate the possibility that training restores, at least partially, the correct somatotopic organization. That this would not happen naturally is indicated by the need for training in human patients. It would be very instructive to map SI cortex following nerve section and regeneration in a primate and again following training to correct mislocalization, but since cortical mapping in anaesthetized primates probably reveals only the most robust of somatosensory pathways (Sect. 7.2.6) this approach may not reveal the compensatory pathways involved.

After nerve section and resuture, classical two-point discrimination is always poorer than normal. Again, however, it can be improved with practice (Wynn-Parry 1981; Dellon 1981). Improvement of two-point discrimination with practice also occurs in normal subjects and sometimes this improvement proceeds to a remarkable degree (Jones et al. 1973). We find it difficult to reject the suspicion that, during this process, the cues used by the subject for two-point discrimination change in some unknown manner. Alternatively, there may be an increase in acuity that results in an increase in the size of the central representation of the body part or increase and/or decrease in the size of the receptive fields of central neurons located on that part. Perhaps there are no changes in the cortical representation of the skin and information is simply extracted by mechanisms that serve directed attention, mechanisms that are part of what we call cognitive strategies (Sects. 8.5.2 and 8.6). The fact that the increased performance is localized and not immediately transferable to non-

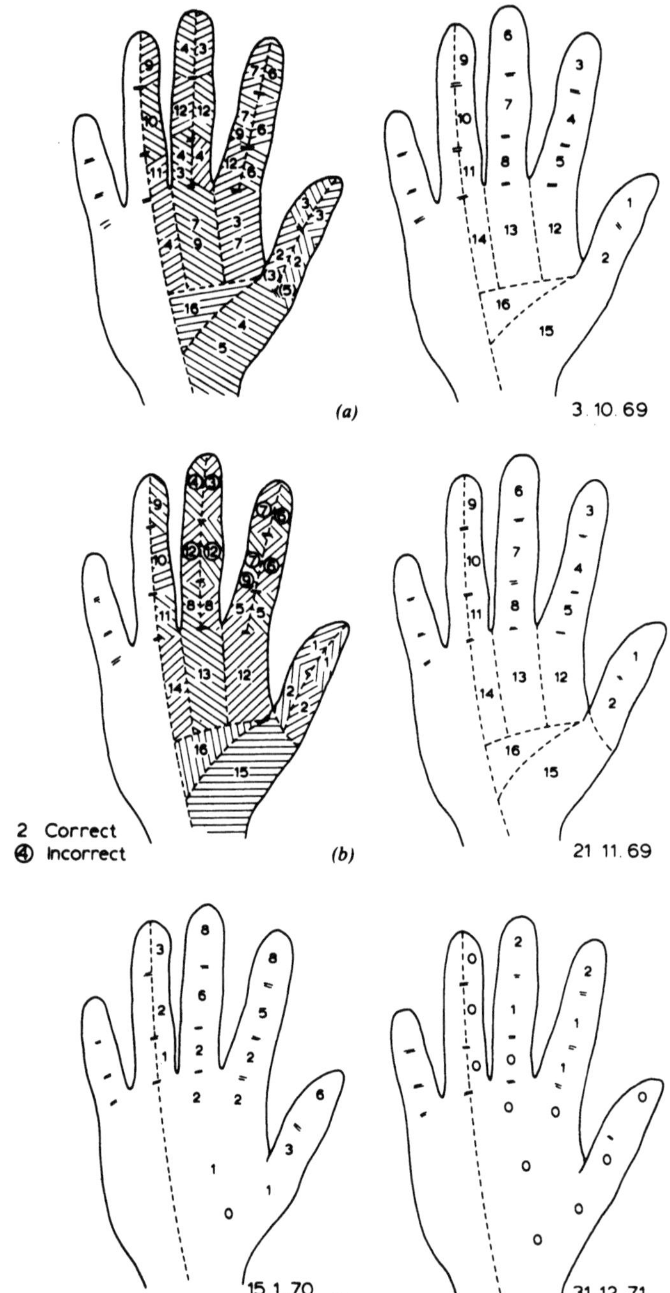

Fig. 8.4 a–c. Retraining of a human subject following section and regeneration of the median nerve. **a** The mislocalization phenomenon after nerve regeneration. The points within the median nerve territory indicated in the *right* diagram were reported by the subject to be at the positions shown on the left. Note the poor correspondence between actual position and perceived position. **b** The same patient after retraining, 7 weeks later. **c** Delay in seconds taken to answer correctly after 14 weeks of training (*left*) compared with values from normal subjects (*right*). (From Wynn-Parry 1981, p. 122)

homotypic regions of skin indicates that the underlying process, like somatotopic representation itself, is localized across the cortex (Vierck 1982).

Wall and Kaas (1985) have also pointed out a possible relation between the hypersensitivity to tactile stimulation that develops in skin regions bordering a denervated region, and the expansion of the representation of this region in SI cortex. The question here is what perceptual change is likely to result from a takeover of regions of cortex that normally contain the representation of the now denervated skin. This issue is discussed more fully below (Sect. 8.6). For the moment let us suggest that electrical activation of an area of cortex deprived of its normal input will produce sensations that are perceptually related to the missing input. If we consider that the areal extent of cortex activated by a peripheral stimulus has nothing to do with sensitivity, then what is responsible for this phenomenon? It seems to us more likely that deafferentation-induced hypersensitivity might be caused by a reduction of the tonic influences of surround inhibition on neurons responsive to the innervated skin that directly abuts the denervated area (Sect. 8.5.1). Denervation would remove part of the source of this tonic inhibition and so enhance the excitation of cells receiving input from the skin bordering the denervated area.

In conclusion, studies of plasticity still have much to tell us about changes of normal sensory function. It remains questionable as to whether neurophysiological observations following central or peripheral lesions tell us anything about recovery in the clinical or behavioural situation. In these situations recovery is usually intimately associated with reward and, consequently, a degree of attention not present in the normal postoperative routine of experimental animals. The matter of relationships between attention and reward systems and plasticity in somatic sensation is broached in Sect. 8.6.

8.5 Normal Physiological Mechanisms and Plasticity

There is an increasing amount of evidence to suggest that in the somatosensory system of normal animals, the responsiveness of neurons to a particular stimulus, the size of their receptive fields and even their modality specificity are not stable parameters. While there is more evidence for such plasticity at the thalamic and cortical levels, this probably reflects an expectation that the circuitry at higher centres will be more malleable than that in, for example, the spinal cord. At all levels, the use of anaesthetics has doubtless led to the elucidation of only the most resistant and powerful synaptic connections. In recent years an increasing appreciation of the specific effects of these agents, the study of somatosensory neurons in awake and behaving animals and the collection of many neuroanatomical and neuropharmacological data on systems that regulate synaptic transmission in normal animals had led us to re-evaluate the view that the functional connections involved in ascending pathways are stable.

8.5.1 Inhibitory Receptive Fields and Partial Deafferentation

In the somatosensory system there are numerous accounts of the inhibition of cells from regions of the skin and the deep tissues that are displaced from the excitatory receptive field. However, at all levels of the somatosensory system there is clear evidence that primary afferents within the excitatory receptive field produce a strong postexcitatory inhibition of central neurons. This has been studied using intracellular recordings, postexcitatory suppression of spontaneous activity and the inhibition of the response to a stimulus to the receptive field by a preceding stimulus. In general, we may refer to this inhibition as *in-field* inhibition (Figs. 6.7 and 8.5; Sects. 4.2.5, 5.2.3, 6.3 and 7.3.1). Across the excitatory receptive field the distribution of the magnitude of excitatory influence of afferent input matches that of the inhibitory influence such that the point that yields maximum excitation also yields maximum inhibition. In a disputed number of cases the distribution of inhibitory influence extends beyond the boundary of the excitatory receptive field forming an inhibitory fringe that is considered to be the basis of surround inhibition. This fringe is characteristically narrow when compared with the diameter of the excitatory receptive field.

It seems, to us, very likely that release from this in-field inhibition is responsible for the *immediate* expression of new receptive fields following deafferentation. The evidence for this, though somewhat circumstantial, is nevertheless compelling. First, there are considerable data to suggest that along the somatosensory axis there are numerous weak or ineffective connections that link central cells with skin areas that are near to, or abut, their normal receptive fields (Sect. 8.3). Second, there is little argument that anaesthetics, particularly those that potentiate GABAergic mechanisms, reduce to subthreshold levels all but the strongest excitatory synaptic inputs and, in doing so, cause a reduction in the size of the receptive fields of central neurons. Barbiturates, in particular, potentiate in-field inhibition while GABA antagonists can elicit expansions of receptive fields and block the post-stimulus inhibition of cortical neurons that follows thalamic stimulation. Third, following partial deafferentation, the immediate expression of new receptive fields is not seen in the presence of anaesthetics that potentiate GABAergic mechanism. Fourth, many slowly adapting cutaneous primary afferents have a resting discharge and many rapidly adapting cutaneous afferents are so sensitive that they would regularly fire action potentials in response to very small, and arguably insignificant (to survival), environmental perturbations (Fig. 8.5).

It follows that even under stable environmental conditions, central cells would be receiving an ongoing barrage of inhibitory and excitatory inputs that are derived from afferents within the excitatory receptive field. *Natural stimulation* of skin nearby, or abutting, the excitatory receptive field will produce weak excitatory inputs to these cells. It is easy to imagine that, in the presence of a tonic level of in-field inhibition, weak excitatory inputs from nearby skin will not be expressed in the discharges of central cells. However, denervation of the excitatory receptive field will release these cells from this ongoing inhibition and may permit them to become immediately responsive to the pre-existing weak inputs from nearby skin (Fig. 8.5). How many cells will respond in this way is debatable but it should be kept in mind

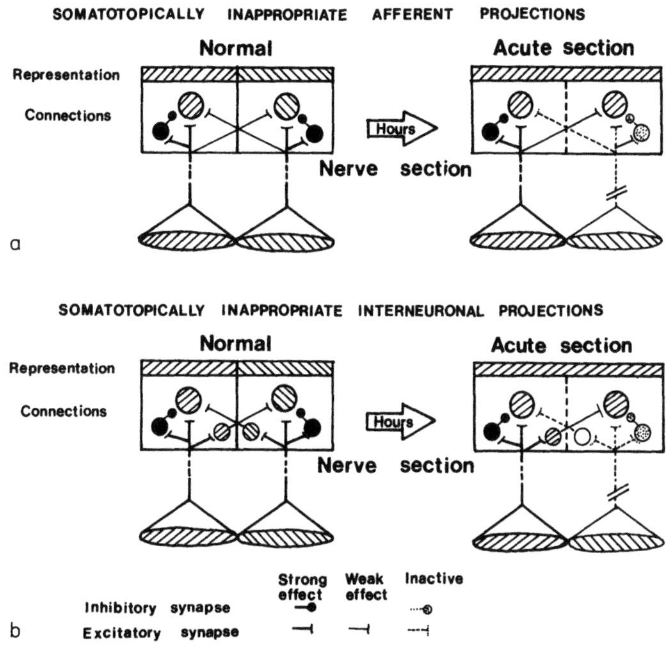

Fig. 8.5 a, b. Schematic diagram illustrating how release from *in-field* inhibition could lead to reorganization of somatotopy following partial deafferentation. Quantitatively, the areas of skin are scaled to the central representation of the body surface, such that, irrespective of their location on the body, they would be represented, in a primate or a cat, over about 1.0–2.0 mm^2 of SI cortex (see Figs. 7.1 and 7.14). In the normal animal (*left*) weak somatotopically inappropriate projections either from afferent fibres (**a**) or from excitatory interneurons (**b**) are rendered ineffective by inhibitory interneurons that are tonically excited by spontaneous afferent discharge. This inhibition is not sufficient to suppress the stronger excitatory effects to the *hatched* cells from direct afferent input although it is responsible for the postexcitatory inhibition seen after stimulation of the receptive field (see Fig. 6.7). Each skin area is thus faithfully represented in each region of the central nervous system. Immediately (hours) following denervation of one skin area (*right*) the level of spontaneous afferent activity and thus tonic inhibition falls dramatically, permitting the weak somatotopically inappropriate pathways to become effective in exciting cells. Thus some cells (e.g. the *hatched* cell) in the deprived area become newly responsive to stimulation of nearby innervated skin. See text for further discussion.

that neurophysiological experiments can only tell us that *some* cells develop new receptive fields.

It is interesting to note that, at least in the thalamus, in-field inhibition reflects the general principles of peripheral innervation density in that it is most intense in cells with distal receptive fields (Sect. 6.3). It is these distal parts and their central representations that have almost always been selected for studies of plasticity in adult animals – an observation that perhaps further emphasizes the potential importance of in-field inhibition in explaining the many existing reports of the immediate effects of nerve section.

In systems in which immediate changes in receptive fields have been documented (e.g. SI cortex of rats and monkeys), it would be most interesting to test this hypothesis by temporarily blocking the afferent input from the excitatory receptive field. At present, wherever reversible block of a group of afferents has been applied, the observed alterations in receptive fields are compatible with this theory (Sects. 6.5.1 and 7.3.5). As pointed out by Wall (1987), local anaesthesia in man does not simply produce the perception of a "hole" in the sensorium but instead is accompanied by a distorted perception of the *related parts*.

8.5.2 Neurotransmitters and Neural Systems That Regulate Sensory Input

The relationship of known transmitter substances to deafferentation-induced reorganization of receptive fields in the somatosensory system is poorly understood. To date there are few studies, but of particular interest are those led by Dykes (Sects. 7.3.3 and 7.5.4) on the effects of GABA and ACh on cells in the SI cortex and by Waterhouse, Woodhouse and Hand on the effects of norepinephrine (Sects. 3.4.1 and 7.5.4). Given its adaptability to well-controlled experiments, work on the dorsal horn of the spinal cord has lagged behind (Sect. 4.5.1), and in the DCN and the VB complex of the thalamus no such experiments have been conducted.

The principal effects of glutamate and the GABA antagonist, bicuculline, are to produce an expansion of the receptive fields of some neurons in SI cortex and a *de novo* expression of a receptive field in neurons that were previously unresponsive to cutaneous stimuli. In the anaesthetized animals used for these studies, these agents thus reveal subthreshold, cutaneous inputs to cortical neurons – a phenomenon for which there is some independent evidence (Sect. 8.3). Both glutamate and bicuculline also increase the responsiveness of neurons to stimulation of their receptive fields and so their effects are in keeping with a model in which subthreshold inputs may arise from stimulation of skin, both within and abutting the normal receptive field. The fact that GABA antagonists produce these effects indicates the presence of tonic inhibitory mechanisms that, at least in the anaesthetized animals, limit receptive field size or expression. Furthermore, Lamour and Dykes (1988) have provided evidence that tonic inhibition is reduced following nerve section thereby permitting the emergence of new receptive fields in neurons deprived of their normal input (see Fig. 8.5a, b; Sect. 7.5.4).

One potential origin of such tonic inhibition has been discussed (Sect. 8.5.1) in relation to in-field inhibition. There are, however, certainly other sources of inhibition that could be affected by nerve section. Amongst these, particularly worthy of mention are the GABAergic projections of the reticular nucleus of the thalamus to the nuclei of the dorsal thalamus (Jones 1986). This system itself receives somatosensory input that is to some extent somatotopically organized. It follows that local deafferentation will regionally affect its output and thus, potentially, the levels of inhibition within specific regions of the dorsal thalamus. Of such interactions we know nothing.

In the SI cortex the iontophoretic application of ACh has been shown to produce an increase in the responsiveness of many neurons to glutamate and to natural stimulation that may last for several hours. However, ACh alone produces either some excitation or no change in spike activity (Sects. 7.3.3 and 7.5.4). While ACh thus fulfils the role of a regulator of cortical input, its relationship to denervation-induced changes in receptive fields is currently dependent on the suggestions of Lamour and Dykes (1988) and Metherate et al. (1987) that deafferentation might induce an increase in ACh receptors or the release of ACh. If such changes could be confirmed they would, like those slow changes in somatotopic organization, take place over a period of many days – a time-course traditionally regarded as representative of anatomical changes.

In some very recent experiments, Rasmusson and Dykes (1988) have sought, with some success, to link the regulation of input to cortical neurons by ACh with the widespread cortical projections of cholinergic cells in the basal forebrain (Sect. 7.3.3). This is indeed interesting for these cells probably belong to the nucleus basalis of Meynert which is considered to be vitally involved with the acquisition of information during learning and arousal (Richardson and DeLong 1988). This nucleus receives inputs from many subcortical and cortical sites and it projects in a coarsely topographic manner to wide areas of the cortex (Marsel et al. 1984; Husschen et al. 1985; Foote and Morrison 1987). Perhaps partial deafferentation acts via this nucleus to increase the responsiveness of cells in the deprived area of cortex to previously subthreshold inputs from nearby skin with an intact innervation (Sect. 8.6). Alternatively, this increase might result from the enhancement of the effects of released ACh by deafferentation-induced increases in the density of ACh receptors within deprived cortex.

Norepinephrine also has distinct and complex effects on the responses of neurons in the SI cortex (Waterhouse et al. 1980, 1981; Woodward et al. 1979; Foote and Morrison 1987). Functionally, its principal effect is to reduce the background activity of cortical neurons relative to the evoked response. In addition, however, norepinephrine increases the responsiveness of some cells and/or increases the postexcitatory suppression of activity. These observations are compatible with the enhancement of the effects of iontophoretically applied GABA and ACh by norepinephrine. For instance, the enhancement of GABA would increase the effects of ongoing inhibition and result in the suppression of spontaneous activity while at the same time enhancing the level of in-field inhibition – a phenomenon that appears to be responsible for the postexcitatory suppression of excitability (Sect. 8.5.1). Furthermore, the enhancement of the effects of ACh by norepinephrine could in turn account for increases in the responsiveness of some neurons – an effect that can also be produced by the application of ACh or stimulation of the nucleus basalis of Meynert (see above).

The source of all norepinephrine in the neocortex is the locus coeruleus and stimulation of this structure causes a reduction in the background activity of cortical neurons. Although the locus coeruleus contains a relatively small number of neurons it projects to every major area of the neuraxis including the nucleus basalis of Meynert (Russchen et al. 1985). Single axons project extensively throughout the cortex (Foote and Morrison 1987) and so it is difficult to see how these direct connections could exert a focal influence over the receptive fields of cortical neurons, such as that seen following deafferentation. However, it is not so difficult to see how norepinephrine released from the terminals of locus coeruleus axons might modulate GABAergic and cholinergic mechanisms in the cortex of intact animals, or modulate any alterations of these mechanisms produced by partial deafferentation. Perhaps these interactions explain how the locus coeruleus can regulate the responsiveness of cortical neurons following deafferentation of a nearby region of cortex and how, in an otherwise intact animal, lesion of the locus coeruleus defocusses the cortical expression of a localized stimulus (Levin et al. 1988) (Sect. 3.4.1).

In conclusion, it is not known how these neural systems, that clearly influence the receptive fields and responses of somatosensory neurons, might be involved in mediating the short- or long-term effects of peripheral nerve section. Their influences seem to be functionally related to the internal state, or the level of arousal, of an animal and so we might expect their regulatory functions to be particularly sensitive to anaesthetics. Some discussion of their role in what we consider to be demonstrations of plasticity in awake animals and humans appears in Sect. 8.6.

8.6 Role of Plasticity in the Mature Somatosensory System

Plasticity is a normal feature in the developing somatosensory system. Thus while in developing animals the responses to a specific lesion might help us derive the rules of developmental neurobiology, it is doubtful whether they demonstrate any fundamental adaptive strategy (Sect. 8.1). In contrast, it seems reasonable to consider that the plasticity that persists in the mature somatosensory system might have some important adaptive significance. In discussing this tissue it seems wise to consider plasticity that is induced experimentally by peripheral or central damage separately from that which involves the repeated activation of afferent pathways, or training, in man or in awake, behaving animals. As will become apparent below, consideration of both of these problems quickly raises the question of whether the phenomenon of somatotopy has any functional significance or whether it is just a convenient way of organizing the brain inside a number of sheets of receptors. What we can learn from developmental studies emphasizes the latter view (Sect. 8.1).

Lesion-Induced Plasticity. Although somatotopic organization is found at all levels of the neuraxis it is unlikely that its disruption at each level will have an identical outcome. For instance, sectioning the nerves to a single digit will alter not only the somatotopic organization of dorsal horn neurons, but also the direct effects of those neurons on a variety of spinal circuits which may be involved in things as unrelated to stimulus perception as spinal reflexes. In comparison, if we could lesion, within the VB complex, all neurons receiving inputs from the same digit, local spinal circuitry (and spinal reflexes) might be little affected but the SI cortex would still be deprived of that input. It follows that changes in functional connectivity following partial deafferentation or central damage might fulfil different functions at different levels of the neuraxis. Thus, as suggested above (Sects. 8.2.3 and 8.5.2), we might anticipate that different underlying mechanisms prevail at various levels.

The SI cortex presents the most intriguing basis for consideration of the functional implications of plasticity within the somatosensory system. At lower levels of the neuraxis it is very difficult to estimate what activity in a set of neurons means to the perceptions or behaviour of an animal. In contrast, direct stimulation of the cortex of man, the neurological deficits associated with focal lesions and very recent results of monitoring cortical activity in awake humans all indicate that activity in at least some neurons at a particular cortical locus is a corollary of *stimulus awareness* (Luria 1962; Penfield 1967; Roland 1981). Thus when partial deafferentation induces the expansion of the representation of nearby innervated skin into the deprived area of cortex (Fig. 7.12), we can ask what this means in terms of the conscious appreciation of stimuli applied to that skin. What it does mean should tell us something about the function of denervation-induced plasticity at the level of the cortex and perhaps something about the function of cortical tissue in conscious experience.

In the adult somatosensory system partial deafferentation usually involves small positional changes in the representation of nearby innervated skin. The absolute magnitude of these shifts seems to be species-dependent. Furthermore, although the area of cortex devoted to the representation of a single digit might vary widely between species such as the raccoon and the primate (compare Fig. 7.2 with Fig. 7.19), the observed representation shift following digit amputation is sufficient to permit cortex once excited by skin on a single digit to develop a responsiveness to the skin of adjacent digits (Figs. 7.12 and 7.23). Deafferentation of larger parts of a representation result in silent areas or areas in which few cells are responsive to peripheral input (Fig. 8.2b; Sects. 7.2.8, 7.3.5, 7.4.4, 7.5.4 and 7.6).

There are two mutually exclusive proposals that we might make regarding the function of deafferentation-induced, somatotopic reorganization. Neither is related to the underlying mechanisms but both have profound implications regarding the link between cortical localization and conscious experience. The first regards the deprived part of a central representation, at any level of the neuraxis, as a functionally valuable entity that can be adapted to serve the purposes of intact inputs from nearby skin. This principle is implied by Fig. 8.3 and has been previously discussed by us elsewhere (Snow and Wilson 1989). It can be argued that, functionally speaking, skin adjacent to deafferentated skin serves a closely allied purpose in the sensing of the environment. There are three observations that support this proposal: (a) the receptive fields of cortical neurons in reorganized areas of cortex are smaller than the receptive fields of cells in the normal representation of that skin, an observation that might indicate an increased spatial acuity across skin adjacent to the denervated skin (Sect. 7.2.7); (b) skin adjacent to chronically denervated skin becomes hypersensitive (Sunderland 1978) – an observation that *might* be interpreted as meaning that more cortical tissue is activated by a point stimulus (but see Sect. 8.4.3); (c) Wall and Kaas (1985) have pointed out that denervation-induced shifts occur only across the boundaries between the adjacent representations of skin areas that are adjacent on the body surface (e.g. adjacent digits) and not across those boundaries between skin areas that though represented in adjacent regions of cortex are not adjacent to one another on the body surface (e.g.

the face and the hand; but see Sects. 3.4.1, 7.2.8, 9). Although each of these observations relates to the cortex, this first proposal seems to us more easily acceptable at subcortical levels. What seems difficult to understand in the SI cortex is, given the localization of sensory perception across the cortex, how does a new input to a deprived region not merely elicit an illusion, or a phantom sensation, of the lost input or appendage. Is sensory cortex so malleable that the relationship between the locus at which a stimulus is represented and the conscious experience of stimulus position can be changed? While an appraisal of this difficult question appears below, it should be emphasized that it is completely negated by consideration of the second proposal relating to the function of deafferentation-induced, somatotopic reorganization.

As we have no idea what spatio-temporal pattern of cortical, neuronal activity constitutes a conscious experience we can safely say that linking the responses of single neurons to a stimulus cannot tell us that the stimulus reached consciousness. For instance, in deeply anaesthetized animals many neurons respond to natural stimulation of the skin but presumable no conscious experience results. Perhaps, under anaesthetic, too few neurons are responsive to elicit a conscious experience, or perhaps the anaesthetic selectively silences *subsets of neurons* that are the substrate for the conscious experience of that stimulus (Edelman 1978) while those neurons that remain responsive are relevant only to the subconscious (Libet 1965). Perhaps the activity that remains is totally meaningless to either the conscious or the subconscious realms.

In the light of these possibilities it is interesting to ask what happens in the cortex when a nerve is sectioned. We may ask, for example, which neurons within the deprived region of cortex receive newly effective inputs from intacts nerves? We do not know this but herein lies an important aspect of our alternative and second proposal (Fig. 8.6). This proposal is that the cells that are reactivated by a foreign input serve no role in conscious experience. Instead they may constitute a group of inhibitory neurons that inhibit many or all inhibitory and excitatory neurons that receive excitatory inputs from the denervated part of the body (Fig. 8.6). Activity in this group of inhibitory neurons does not evoke sensation but it does prevent activity in surrounding neurons that are part of a subset for stimulus awareness of what are now lost, or phantom, sensory inputs. When limb amputation deprives a large area of sensory cortex of input (compare Fig. 1.2 with Fig. 7.1) perhaps the subsequent and progressive reduction in the length of the phantom limb reflects the slow transcortical spread in the disynaptic inhibitory influence of surviving inputs rather than a transcortical extension of their monosynaptic excitatory effects on subsets of neurons that subserve awareness (Figs. 1.2, 7.11, 8.2b and 8.6). Implicit in this proposal is the view that consciousness of the location of a stimulus is permanently linked to a particular, small area of cortex, but that conscious perception requires activation of a minimum number of neurons within a particular subset of neurons at that locus (Edelman 1978). When this happens in the absence of the normal sensory input, or even in the absence of a peripheral stimulus (sensory deprivation), the experience is that of a phantom – be it of the digits alone or a whole limb (Figs. 1.1, 1.2, 7.11 and 8.6). Crudely speaking, we might predict that both the rate of reduction of a phantom (telescoping) and the final size might reflect the area

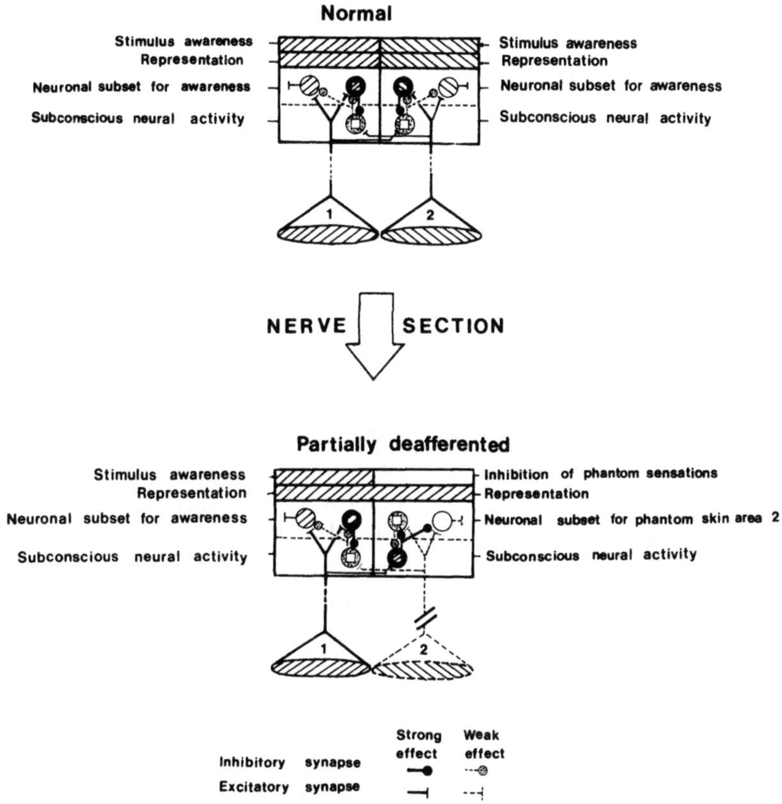

Fig. 8.6. Schematic illustration of how somatotopic reorganization following partial deafferentation may function to inhibit spurious sensations of the denervated area of the body. The theory assumes that activity in a subset of neurons at a particular cortical locus is responsible for conscious perception of a stimulus (*stimulus awareness*). Other neurons may be activated by the stimulus but this activity will not necessarily be correlated with stimulus awareness and it is therefore considered here to be *subconscious neural activity*. The *neuronal subset for awareness* consists of excitatory and inhibitory neurons that are strongly excited by afferent input. The neurons *not* involved in this subset are inhibitory. They can inhibit neurons within the subset and they are weakly excited by somatotopically inappropriate projections. These weak inputs are rendered ineffective by inhibitory connections from neurons within the subset that are excited by spontaneous, or stimulus-related, afferent activity (see Fig. 8.5). Denervation of one skin area (*area 2*) removes this source of inhibition and the suppressed somatotopically inappropriate projections become newly effective in activating those neurons *not* involved in the subset (*lower neuron, right box, partial deafferentation*). Electrophysiological examination reveals an extension of the representation of nearby innervated skin (*area 1*) but within the region of this extension neuronal subsets that lead to stimulus awareness are held repressed. Consequently, there is an *inhibition of phantom sensations*. If a large area of skin is denervated, as in the amputation of a limb, the transcortical influence of somatotopically inappropriate projections will be insufficient to repress subsets throughout the deprived area of cortex and phantom limb sensations may occur. The progressive reduction of the length of the phantom may reflect long-term spread of the influence of surviving afferents over deprived cortex in the manner outlined in Fig. 8.2b. See text for further discussion

of cortex deprived of its normal input. The question of the modality of the phantom sensation is an extension, rather than a requirement, of this proposal but it is one that can be most easily satisfied by the allocation of specific subsets of neurons to the conscious experience of input from particularly receptor groups. It would be interesting to know whether direct stimulation of a chronically or acutely deprived cortical region is perceived as stimulation of the denervated area of the skin or of a nearby, innervated area.

In the above theory, phantom limb sensations are elaborated by, and within, that area of cortex that once translated sensory messages from the missing appendage into discrete sensory experiences. Implicit in this view is that a necessary condition of phantom sensations is the survival of a *deprived* area(s) of sensory cortex following limb amputation. Should this not be the case, we would argue that no phantom sensations will be elaborated. Recently, Melzack (1990) has cited the papers of Weinstein et al. (1964) and Poeck (1964) as "convincing evidence" that children born without all or part of a limb feel a "vivid phantom". However, scrutiny of these papers does not, in our view, reveal such a conclusive case for phantom sensations in cases of congenital aplasia. Primarily this is because it is difficult to rule out the involvement of imagination, which, in association of sensory experiences from intact body parts, might be responsible for a relatively vivid, yet *conceptual*, image of a missing appendage. It would be fascinating to know whether the descriptions of phantoms by individuals born without a limb, vary in accordance with the survival through embryonic development of a chronically deprived part of sensory cortex.

Use, Training and Plasticity. There is considerable neurophysiological, neuroanatomical and clinical evidence that the size of a representation of a given skin area in SI cortex and the size of the receptive fields of single units are influenced by the frequency of use of a skin surface (Sects. 3.3.6, 3.4.1, 7.2.9, 7.3.6 and 8.4.3). While the somatotopic alterations following partial deafferentation may not be appropriate models for analysing the bases of use-related changes, they may nevertheless reveal some underlying mechanisms and principles that are fundamental to somatosensory plasticity as it is seen both clinically and in intact, behaving animals (Sects. 8.4.3 and 8.5.2).

When the effects of repeated stimulation are considered it is necessary to separate the results of experiments using frequent but random stimulation of a skin area from stimulation that is associated with reward and, as a consequence, directed attention. In relation to the former circumstance, stimulus deprivation (clipping) to many mystacial vibrissae has been shown to increase the influence of an intact stimulated vibrissa (Sects. 3.3.5 and 3.4.1). No other data are available on this matter but Merzenich et al. (1987) have suggested that inter-animal differences in the representation of the primate hand might reflect differences in use of component skin surfaces (Fig. 7.2; Sect. 7.2.9). The latter circumstance (increased stimulation in relation to reward and attention) has been studied in a preliminary manner in the monkey where, in the anaesthetized animal, an expanded representation of the stimulated skin areas was reported (Jenkins et al. 1984, 1990) (Sect. 7.2.9). Consideration of how such an expansion could come about brings to mind the somatotopically inappropriate projections of thalamocortical neurons and the cholinergic, GABAergic and adrenergic regulation of receptive fields in SI cortex (Sects. 8.3.2, 8.4.1, 8.5.1 and 8.5.2). Can activity in the postcentral gyrus be related to these parameters, as well as to directed attention and the neural systems that subserve this phenomenon?

Much evidence has implicated the prefrontal cortex in attention to a stimulus (Stuss and Benson 1986). In human subjects trained to attend to stimulation of a finger, activity could be recorded in the finger representation and in the prefrontal cortex, even when the stimulus was not actually delivered (Roland 1981). Whether a similar phenomenon is involved in retraining after nerve injury or improvements in two-point discrimination is unknown (Sect. 8.4.3) but atypical small receptive fields have been recorded within the training-induced expansions of the monkey finger representation in the SI cortex (Sect. 7.2.9). It is also unknown whether prefrontal cortex can be implicated in activation of deprived areas of the SI body representation in a manner that would potentially involve it in the mental elaboration and control of phantom limb sensations (Fig. 1.1; Chap. 1). Certainly, the pathways linking prefrontal cortex with SI cortex are diverse and probably consist of corticocortical links (Roland 1981; Mesulan 1981) and preferential projections to the nucleus basalis of Meynert and the locus coeruleus (Russchen et al. 1985; Stuss and Benson 1986; Sara 1985) – structures which are known to have the potential to modify tonically the responsiveness, and improve the single-to-noise ratio, of cortical neurons. Foote and Morrison (1987) have concluded that the general function of the nucleus basalis of Meynert is to convey information concerning relationships between *complex environmental events and the internal milieu* to cortical areas involved in many different operations, whereas they suggest that the locus coeruleus is involved in *arousal and alerting, in relation to internally generated modes and states.* A potential role for those nuclei in modulation of the receptive field size, sensitivity and signal-to-noise ratio of neurons in the SI cortex is in keeping with these broad functions. Many further studies are needed in this area but as they progress we may expect to gain an insight into the interplay between so-called higher cortical functions, reward systems and perception – an interplay that probably gave rise to man's earliest predictions of plasticity in the central nervous system.

References

Adeymo O, Wyburn GM (1957) Innervation of skin grafts. Transplant Bull 4:152–153

Ainsworth A, Hall P, Wall PD, Allt G, Mackenzie M, Lynn B, Gibson S, Polak JM (1981) Effects of capsaicin applied locally to adult peripheral nerve. II. Anatomy and enzyme and peptide chemistry of peripheral nerve and spinal cord. Pain 11:379–388

Akers RM, Killackey HP (1979) Segregation of cortical and trigeminal afferents to the ventrobasal complex of the neonatal rat. Brain Res 161:527–532

Aldskogius H, Arvidsson J (1978) Nerve cell degeneration and death in the trigeminal ganglion of the adult rat following peripheral nerve transection. J Neurophysiol 7, 229–250

Aldskogius H, Risling M (1981) Effect of sciatic neurectomy on neuronal number and size distribution in the L7 ganglion of kittens. Exp Neurol 74:597–604

Aldskogius H, Risling M (1983) Preferential loss of unmyelinated L7 dorsal root axons following sciatic nerve resection in kittens. Brain Res 289:358–361

Aldskogius H, Cerne H, Holmberg A (1985) The effect of sciatic nerve transection on myelinated fibres in the L5 dorsal root and lumbar dorsal column. Anat Embryol (Berl) 171:181–186

Alloway KD, Burton H (1985a) Homotypical ipsilateral cortical projections between somatosensory areas I and II in the cat. Neuroscience 14:15–35

Alloway KD, Burton H (1985) Submodality and columnar organization of the second somatic sensory area in cats. Exp Brain Res 61:128–140

Aloe L, Cozzari C, Calissano P, Levi-Montalcini R (1981) Somatic and behavioural postnatal effects of fetal injection of nerve growth factor antibodies in the rat. Nature 291:413–415

Altman J, Bayer S (1982) Development of the cranial nerve ganglia and related nuclei in the rat. Adv Anat Embryol Cell Biol 74:1–90

Altman J, Bayer S (1984) The development of the rat spinal cord. Adv Anat Embryol Cell Biol 85:1–164

Amassian VE, de Vito JL (1957) La transmission dans le noyau de Burdach (nucleus cuneatus). Etude analytique par unites isolees d'un relais somato-sensoriel primaire. Coll Int Cent Nat Rech Sci Paris 67:353–393

Andersen P, Eccles JC, Oshima T, Schmidt RF (1964) Mechanisms of synaptic transmission in the cuneate nucleus. J Neurophysiol 27:1096–1116

Andersen P, Etholm B, Gordon G (1970) Presynaptic and post-synaptic inhibition elicited in the cat's dorsal column nuclei by mechanical stimulation of skin. J Physiol (Lond.) 210:433–455

Andres FL, Van der Loos H (1982) Whisker patterns form in cultured non-innervated muzzle skin from mouse embryos. Neurosci Lett 30:37–41

Andres FL, Van der Loos H (1983) Cultured embryonic non-innervated mouse muzzle is capable of generating a whisker pattern. Int J Dev Neurosci 1:319–338

Andres FL, Van der Loos H (1985) From sensory periphery to cortex: the architecture of the barrel field as modified by various early manipulations of the mouse whisker pad. Anat Embryol (Berl) 172:11–20

Andres FL, Van der Loos H, Anders D (1981) A simple device for microdissection in the SEM. Scanning. 4:83–90.

Angaut-Petit D (1975a) The dorsal column system: I. Existence of long ascending postsynaptic fibres in the cat's fasciculus gracilis. Exp Brain Res 22:457–470
Angaut-Petit D (1975b) The dorsal column system: II. Functional properties and bulbar relay of the postsynaptic fibres of the cat's fasciculus gracilis. Exp Brain Res 22:471–493
Angel A, Clarke KA (1975) An analysis of the representation of the forelimb in the ventrobasal thalamic complex of albino rats. J Physiol (Lond) 249:399–423
Angulo-y-Gonzalez AW (1932) The prenatal development of behaviour in the albino rat. J Comp Neurol 55:395–442
Armstrong-James M (1975) The functional status and columnar organization of single cells responding to cutaneous stimulation in the neonatal rat somatosensory cortex. J Physiol (Lond) 246:501–538
Arvidsson J (1979) An ultrastructural study of transganglionic degeneration in the main sensory trigeminal nucleus of the rat. J Neurocytol 8:31–45
Arvidsson J (1982) Somatotopic organization of vibrissae afferents in the trigeminal sensory nuclei of the rat studied by transganglionic transport of HRP. J Comp Neurol 211:84–92
Arvidsson J, Grant G (1979) Further observations on transganglionic degeneration in trigeminal primary sensory neurons. Brain Res 162:1–12
Arvidsson J, Johansson K (1988) Changes in the central projection pattern of vibrissae innervating primary sensory neurons after peripheral (nerve injury in the rat. Neurosci Lett 84:120–124
Avenarius H-J, Gerstenbrand F (1967) Phantomerlebnisse bei Ruckenmarks-Verletzung. Wien Klin Wochenschr 24:450–453
Axelrad H, Verley R, Farkas E (1976) Responses evoked in mouse and rat S1 cortex by vibrissa stimulation. Neurosci Lett 3:265–274
Baker MA (1971) Spontaneous and evoked activity in neurones in the somatosensory thalamus of the waking cat. J Physiol (Lond) 217:359–380
Baker MA, Tyner CF, Towe AL (1971) Observations on single neurons recorded in the sigmoid gyri of anala, non-pigmented cats. Exp Neurol 32:388–403
Baker HJ, Lindsey JR Weisbroth SH (1979) The laboratory rat: Biology and Diseases, vol 1. Academic, New York
Barbut D, Polak JM, Wall PD (1981) Substance P in the spinal dorsal horn decreases following peripheral nerve injury. Brain Res 205:289–298
Basbaum AI, Hand PJ (1973) Projections of cervicothoracic dorsal roots to the cuneate nucleus of the rat, with observations on cellular "bricks". J Comp Neurol 148:347–360
Basbaum A, Wall P (1976) Chronic changes in the response of cells in adult cat dorsal horn following partial deafferentation: the appearance of responding cells in a previously non-responsive region. Brain Res 116:181–204
Bates CA, Killackey HP (1985) The organization of the neonatal rat's brainstem trigeminal complex and its role in the formation of central trigeminal patterns. J Comp Neurol 240:265–287
Bates CA, Erzurumlu RH, Killackey HP (1982) Central correlates of peripheral pattern alterations in the trigeminal system of the rat III. Neurons of the principal sensory nucleus. Dev Brain Res 5:108–113
Batuev AS, Alexandrov AA, Scheynikov NA (1982) Picrotoxin action on the receptive fields of the cat sensorimotor cortex neurons. J Neurosci Res 7:49–55
Beal JA (1979) The ventral dendritic arbor of marginal (lamina I) neurons in the adult primate spinal cord. Neurosci Lett 14:201–206
Beal JA, Knight DS (1987) Classification of aberrant primary afferents in the substantia gelatinosa of the rat following neonatal capsaicin treatment. Neurosci Letts 74:139–144
Beattie M, Bresnahan J, Liuzzi F (1986) Aberrant axons from the dorsal roots after chronic spinal hemisection in the cat: the dorsal horn. J Comp Neurol 232:456–465

Beck CHM (1976) Dual Dorsal Columns: A Review. Can J Neurol Sci 3:1–7
Beck CHM (1981) Mapping of forelimb afferents to the cuneate nucleus fo the rat. Brain Res Bull 6:503–516
Belford GR, Killackey HP (1979a) The development of vibrissae representation in subcortical trigeminal centers of the neonatal rat. J Comp Neurol 188:63–74
Belford GR, Killackey HP (1979b) Vibrissae representation is subcortical trigeminal centers of the neonatal rat. J Comp Neurol 183:305–322
Belford GR, Killackey HP (1980) The sensitive period in the development of the trigeminal system of the neonatal rat. J Comp Neurol 193:335–350
Benjamin RM, Thompson RF (1959) Differential effects of cortical lesions in infant and adult cats on roughness discrimination. Exp Neurol 1:305–321
Benkenstein M (1979) Veranderungen der ultrastruktur der merkelschen nervenendigungen an sinushaaren von ratten nach denervation. Acta Anat (Basel) 105:409–422
Bennett GJ, Abdelmoumene M, Hayashi H, Dubner R (1980) Physiology and morphology of substantia gelatinosa neurons intracellularly stained with horseradish peroxidase. J Comp Neurol 194:809–827
Bennett GJ, Seltzer Z, Lu G-W, Nishikawa N, Dubner R (1983) The cells of origin of the dorsal column postsynaptic projection in the lumbosacral enlargements of cats and monkeys. Somatosens Res 1:131–149
Bennett GJ, Nishikawa N, Lu GW, Hoffert MJ, Dubner R (1984) The morphology of dorsal column postsynaptic spinomedullary neurons in the cat. J Comp Neurol 224:568–578
Bennett RE, Ferrington DG, Rowe M (1980) Tactile neuron classes within second somatosensory area (SII) of cat cerebral cortex. J Neurophysiol 43:292–309
Berger M, Gerstenbrand F (1981) Phantom illusions in spinal cord lesions, In: Siegfried J, Zimmermann M (eds) Phantom and stump pain. Springer, Berlin Heidelberg New York, pp 66–73
Berkley KJ (1973) Response properties of cells in ventrobasal and posterior group nuclei of the cat. J Neurophysiol 36:940–952
Berkley KJ (1975) Different targets of different neurons in nucleus gracilis of the cat. J Comp Neurol 163:285–304
Berkley KJ, Hand PJ (1978) Efferent projections of the gracile nucleus in the cat. Brain Res 153:263–283
Berkley KJ, Vierk CJ Jr (1987) Transient reduction in retrograde labelling of diencephalic-projecting neurons in monkey dorsal column nuclei following removal of their dorsal column input. In: Pubols LM, Sessle BJ (eds) Effects of injury on trigeminal and spinal somatosensory systems. Liss, New York, pp 429–436
Berkley KJ, Budell RJ, Blomqvist A, Bull M (1986) Output systems of the dorsal column nuclei in the cat. Brain Res Rev 11:199–225
Berridge DA (1973) Information transfer in the dorsal column sensory pathway of the rat. PhD Thesis, University of Sheffield
Berry MS, Pentreath VW (1976) Criteria for distinguishing between monosynaptic and polysynaptic transmission. Brain Res 105:1–20
Besson JM, Abdelmoumene M, Aleonard P, Conseiller C (1968) Effects des anesthesiques sur les differentes composantes des potentiels de racine dorsale. J Physiol (Paris) 60:217
Bessou P, Perl ER (1969) Response of cutaneous sensory units with unmyelinated fibres to noxious stimuli. J Neurophysiol 32:1025–1043
Beuerman RW, Rozsa AJ (1984) Collateral sprouts are replaced by regenerating neurites in the wounded corneal epithelium. Neurosci Lett 44:99–104
Bicknell HR, Beal JA (1984) Axonal and dendritic development of substantia gelatinosa neurons in the lumbosacral spinal cord of the rat. J Comp Neurol 226:508–522

Birren JE, Wall PD (1956) Age changes in conduction velocity, refractory period, number of fibres, connective tissue space and blood vessels in the sciatic nerve of rats. J Comp Neurol 10:1–16

Blomqvist A (1980) Gracilo-diencephalic relay cells: a quantitative study in the cat using retrograde transport of horseradish peroxidase. J Comp Neurol 193:1097–1125

Blomqvist A, Westman J (1976) Interneurons and initial axon collaterals in the feline gracile nucleus demonstrated with the rapid Golgi technique. Brain Res 111:407–430

Blumberg H, Janig W (1981) Neurophysiological analysis of efferent sympathetic and afferent fibres in skin nerves with experimentally produced neuromata. In: Siegfried J, Zimmermann M (eds.) Phantom and stump pain. Springer Berlin Heidelberg New York, pp 15–31

Boeke J (1922) On the regeneration of sensitive end-corpuscles after section of the nerve. K Acad van Wentenschoppen (Amsterdam) 25:319–323

Boeke J, Dijkstra C (1932) De- and regeneration of sensible end-corpuscles in the duck's bill (corpuscles of Grandry and Herbst) after the cutting of the nerve, the removing of the entire skin or the transplantation of the skin in another region. K. Acad. van Wentenschoppen (Amsterdam) 35:1114–1119

Boivie J, Perl ER (1975) Neural substrates of somatic sensation. In: MTP International Review of Science, Physiology Series I. vol. 3, Neurophysiology. University Park Press, Baltimore, pp 303–411

Bondok AA, Sansone FM (1984) Retrograde and transganglionic degeneration of sensory neurons after a peripheral nerve lesion at birth. Exp Neurol 86:322–330

Boring EG (1916) Cutaneous sensation after nerve-division. Q J Exp Physiol 10:1–95

Brenan A (1983) Collateral reinnervation of skin by C-fibres following nerve injury in the rat. J Physiol (Lond) 338:59P

Brenan A (1986) Collateral reinnervation of skin by C-fibres following nerve injury in the rat. Brain Res 385:152–155

Brenowitz G, Pubols L (1981) Increased receptive field size of dorsal horn neurons following chronic spinal cord hemisection in cats. Brain Res 216:45–59

Brenowitz GL, Devor M (1981) Reinnervation of rat glabrous hindpaw skin by collateral sprouts following denervation by sciatic nerve section. Anat Rec 199:37A

Brinkhus M, Zimmerman M (1983) Characteristics of spinal dorsal horn neurons after partial chronic deafferentation by dorsal root transection. Pain 15:221–236

Bromberg MB, Whitehorn D (1974) Myelinated fiber types in the superficial radial nerve of the cat and their central projections. Brain Res 78:157–163

Brooks VB, Rudomin P, Slayman CL (1961a) Sensory activation of neurons in the cat's cerebral cortex. J Neurophysiol 24:286–300

Brooks VB, Rudomin P, Slayman CL (1961b) Peripheral receptive fields of neurons in the cat's cerebral cortex. J Neurophysiol 24:301–325

Brown AG (1968) Cutaneous afferent fibre collaterals in the dorsal columns of the cat. Exp Brain Res 5:293–305

Brown AG (1970) Descending control of the spinocervical tract in decerebrate cats. Brain Res 17:152–155

Brown AG (1971) Effects of descending impulses on transmission through the spinocervical tract. J Physiol (Lond) 219:103–125

Brown AG (1973) Ascending and long spinal pathways: dorsal columns, spinocervical tract and spinothalamic tract. In: Iggo A (ed) Handbook of Sensory Physiology, vol. 2, Somatosensory system. Springer, Berlin Heidelberg New York, pp 315–338

Brown AG (1981) Organization in the Spinal Cord. The Anatomy and Physiology of Identified Neurones. Springer-Verlag, Berlin Heidelberg New York

Brown AG, Franz DN (1969) Responses of spinocervical tract neurones to natural stimulation of identified cutaneous receptors. Exp Brain Res 7:231–249

Brown AG, Fyffe REW (1981) Form and function of dorsal horn neurones with axons ascending the dorsal columns in cat. J Physiol (Lond) 321:31–47
Brown AG, Gordon G (1977) Subcortical mechanisms concerned in somatic sensation. Br Med Bull 33:121–128
Brown AG, Iggo A (1963) The structure and function of cutaneous touch corpuscles after nerve crush. J Physiol (Lond) 165:28–29
Brown AG, Noble R (1982) Connexions between hair follicle afferent fibres and spinocervical tract neurons in the cat: the synthesis of receptive fields. J Physiol (Lond) 323:77–91
Brown AG, Short AD (1974) Effects from the somatic sensory cortex on transmission through the spinocervical tract. Brain Res 74:338–341
Brown AG, Hamann K-F, Martin HF III (1973a) Descending influences on spinocervical tract cell discharges evoked by non-myelinated cutaneous afferent nerve fibres. Brain Res 53:218–221
Brown AG, Hamann K-F, Martin HF III (1973b) Interactions of cutaneous myelinated (A) and non-myelinated (C) fibres on transmission through the spinocervical tract. Brain Res 53:222–226
Brown AG, Kirk EJ, Martin HF III (1973c) Descending and segmental inhibition of transmission through the spinocervical tract. J Physiol (Lond) 230:689–705
Brown AG, Gordon G, Kay RH (1974) A study of single axons in the cat's medial lemniscus. J Physiol (Lond) 236:225–246
Brown AG, Hamann K-F, Martin HF III (1975) Effects of activity in non-myelinated afferent fibres on the spinocervical tract. Brain Res 98:243–259
Brown AG, Rose PK, Snow PJ (1977a) The morphology of spinocervical tract neurones revealed by intracellular injection of horseradish peroxidase. J Physiol (Lond) 270:747–764
Brown AG, Rose P, Snow PJ (1977b) The morphology of hair follicle afferent fibre collaterals in the spinal cord of the cat. J Physiol (Lond) 272:779–797
Brown AG, Coulter JD, Rose PK, Short AD, Snow PJ (1977a) Inhibition of spinocervical tract discharges from localized areas of the sensorimotor cortex in the cat. J Physiol (Lond) 264:1–16
Brown AG, Rose P, Snow P (1978) Morphology and organization of axon collaterals from afferent fibres of slowly adapting type I units in cat spinal cord. J Physiol (Lond) 277:15–27
Brown AG, Fyffe REW, Noble R (1980a) Projections from Pacinian corpuscles and rapidly adapting mechanoreceptors of glabrous skin to the cats spinal cord. J Physiol (Lond) 307:385–400
Brown AG, Fyffe REW, Noble R, Rose PK, Snow PJ (1980b) The density, distribution and topographical organization of spinocervical tract neurones in the cat. J Physiol (Lond) 300:409–428
Brown AG, Rose P, Snow P (1980c) Dendritic trees and cutaneous receptive fields of adjacent spinocervical tract neurons in the cat. J Physiol (Lond) 300:429–440
Brown AG, Brown PB, Fyffe REW, Pubols LM (1981a) Effects of deafferentation on spinocervical tract neurons in the cat. J Physiol (Lond) 105–106
Brown AG, Fyffe REW, Rose P, Snow P (1981b) Spinal cord collaterals from axons of type II slowly adapting units in the cat. J Physiol (Lond) 316:469–480
Brown AG, Brown PB, Fyffe REW, Pubols LM (1983a) Effects of dorsal root section on spinocervical tract neurones in the cat. J Physiol (Lond) 337:589–608
Brown AG, Brown PB, Fyffe REW, Pubols LM (1983b) Receptive field organization and response properties of spinal neurones with axons ascending the dorsal columns in the cat. J Physiol (Lond) 337:575–588
Brown AG, Fyffe REW, Noble R, Rowe MJ (1984) Effects of hindlimb nerve section on lumbosacral dorsal horn neurons in the cat. J Physiol (Lond) 354:375–394

Brown AG, Koerber HR, Noble R (1987) An intracellular study of spinocervical tract cell responses to natural stimuli and single hair afferent fibres in cats. J Physiol (Lond) 382:331–354

Brown PB (1987) A reassessment of evidence for primary afferent sprouting in the dorsal horn. In: Pubols LM, Sessle BJ (eds) Effects of injury on trigeminal and spinal somatosensory systems Liss, New York, pp 273–280

Brown PB, Culberson J (1981) Somatotopic organization of hindlimb cutaneous dorsal root projections to cat dorsal horn. J Neurophysiol 45:137–143

Brown PB, Fuchs J (1975) Somatotopic representation of hindlimb skin in cat dorsal horn. J Neurophysiol 38:1–9

Brown PB, Koerber HR (1978) Cat hindlimb tactile dermatomes determined with single-unit recordings. J Neurophysiol 41:260–267

Brown PB, Moraff H, Tapper DN (1973) Functional organization of the cat's dorsal horn: spontaneous activity and central cell response to single impulses in single type I fibres. J Neurophysiol 36:827–839

Brown PB, Fuchs JL, Tapper DN (1975) Parametric studies of dorsal horn neurons responding to tactile stimulation. J Neurophysiol 38:19–25

Brown PB, Busch G, Whittington J (1979) Anatomical changes in cat dorsal horn cells after transection of a single dorsal root. Exp Neurol 64:453–468

Brown PB, Koerber HR, Culberson JL (1981) Somatotopic organization of cutaneous projections to cat dorsal horn. In: Brown AG, Rethelyi M (eds) Spinal Cord Sensation. Sensory Processing in the Dorsal Horn. Scottish Academic Edingburgh, pp 45–54

Bryan R, Trevino D, Coulter J, Willis W (1973) Location and somatotopic organization of the cells of origin of the spino-cervical tract. Exp Brain Res 17:177–189

Burchiel KJ, Russell LC (1987) Has the amount of spontaneous electrical activity in experimental neuromas been overestimated? In: Pubols LM, Sessle BJ (eds) Effects of injury on trigeminal and spinal somatosensory systems. Liss, New York, pp 77–83

Burgess PR, Clark FJ (1969) Dorsal column projection of fibres from the cat knee joint. J Physiol (Lond) 203:301–315

Burgess PR, Horch KW (1973) Specific regeneration of cutaneous fibers in the cat. J Neurophysiol 36:101–114

Burgess PR, Perl ER (1973) Cutaneous mechanoreceptors and nociceptors. In: Iggo A (ed) Somatosensory system. Springer, Berlin Heidelberg New York, pp 29–78 (Handbook of sensory physiology, vol. 2)

Burgess PR, Petit D, Warren RM (1968) Receptor types in cat hairy skin supplied by myelinated fibres. J Neurophysiol 31:833–848

Burgess PR, English KB, Horch KW, Stensaas LJ (1974) Patterning in the regeneration of type I cutaneous receptors. J Physiol (Lond) 236:57–82

Burton H, Carlson M, Warren S (1987) Development of tactile function following damage to somatic sensory cortex in infant Macaca. In: Pubols LM, Sessle BJ (eds) Effects of injury on spinal and trigeminal somatosensory systems. Liss, New York, pp 437–450

Bystrzycka E, Nail BS, Rowe MJ (1977) Inhibition of cuneate neurones: its afferent source and influence on dynamically sensitive 'tactile' neurones. J Physiol (Lond) 268:251–270

Cabaud HE, Rodkey WG, McCarroll HR (1976) Epineurial and perineurial nerve repairs: a critical comparison. J Hand Surg 1:131–137

Cabral RJ, Johnson JL (1971) The organization of mechanoreceptive projections in the ventrobasal thalamus of sheep. J Comp Neurol 141:17–36

Carlson JA, Lais AC, Dyck PJ (1979) Axonal atrophy from permanent peripheral axotomy in adult cat. J Neuropath Exp Neurol 38:579–585

Carlson M (1981) Characteristics of sensory deficits following lesions of Brodman's areas 1 and 2 in the postcentral gyrus of *Macaca mulatta*. Brain Res 204:424–430

Carlson M (1984a) Development of tactile discrimination capacity in *Macaca mulatta*. I. Normal infants. Dev Brain Res 16:69–82
Carlson M (1984b) Development of tactile discrimination capacity in *Macaca mulatta*. II. Effects of partial removal of primary somatic sensory cortex (Sm1) in infants and juveniles. Dev Brain Res 16:83–101
Carlson M and Fitzpatrick KA (1982) Organization of the hand area in the primary somatic sensory cortex (SmI) of the prosimian primate, *Nycticebus coucang*. J Comp Neurol 204:280–295
Carlson M, Welt C (1980) Somatic sensory cortex (SmI) of the prosimian primate *Galago crassicaudatus*: organization of mechanoreceptive input from the hand in relation to cytoarchitecture. J Comp Neurol 189:249–271
Carpenter SE and Lynn B (1982) Responses of cutaneous nociceptor units in rats treated with capsaicin. Adv Pain Res Ther 5:83–86
Carson LV, Kelahan AM, Ray RH, Massey CE, Doetsch GS (1981) Effects of early peripheral lesions on the somatotopic organization of the cerebral cortex. In: Carmel PW (ed) Clinical neurosurgery, vol. 28. Williams and Wilkins, Baltimore, pp 532–546
Cavanaugh MW (1951) Quantitative effects of the peripheral innervation area on nerves and spinal ganglion cells. J Comp Neurol 94:181–219
Caviness VS Jr, Frost DO, Hayes NL (1976) Barrels in the somatosensory cortex of normal and reeler mutant mice. Neurosci Lett 3:7–14
Cervero F (1986) Dorsal horn neurons and their sensory inputs. In: Yaksh TL (ed) Spinal afferent processing. Plenum, New York, pp 197–218
Cervero F, Iggo A (1980) The substantia gelatinosa of the spinal cord: a critical review. Brain 103:717–772
Cervero F, Plenderleith MB (1984) Dorsal root potentials are unchanged in adult rats treated at birth with capsaicin. J Physiol (Lond) 357:357–368
Cervero F, Plenderleith MB (1985) C-fibre excitation and tonic descending inhibition of dorsal horn neurons in adult rats treated at birth with capsaicin. J Physiol (Lond) 365:223–237
Cervero F, Plenderleith MB (1987) Spinal cord sensory systems after neonatal capsaicin. Acta Physiol Hung 69:393–401
Cervero F, Iggo A, Ogawa H (1976) Nociceptor-driven dorsal horn neurones in the lumbar spinal cord of the cat. Pain 2:5–24
Cervero F, Iggo A, Molony U (1977) Responses of spincervical tract neurones to noxious stimulation of the skin. J Physiol (Lond) 267:537–558
Cervero F, Iggo A, Molony U (1979) An electrophysiological study of neurones in the substantia gelatinosa Rolandi of the cat's spinal cord. J Exp Physiol 64:297–314
Cervero F, Shouenborg J, Sjolund BH, Waddell PJ (1984) Cutaneous input to dorsal horn neurones in adult rats treated at birth with capsaicin. Brain Res 301:47–57
Chambers MR, Andres KH, Duering M, von Iggo A (1972) The structure and function of the slowly adapting type II mechanoreceptor in hairy skin. J Exp Physiol 57:417–445
Chang HT, Ruch TC (1947) Organization of the dorsal columns of the spinal cord and their nuclei in the spider monkey. J Anat 81:140–149
Chapin JK (1986) Laminar differences in sizes, shapes and response profiles of cutaneous receptive fields in the rat SI cortex. Exp Brain Res 62:549–559
Chapin JK, Lin CS (1984) Mapping the body representation in the SI cortex of anaesthetized and awake rats. J Comp Neurol 229:199–213
Cheema S, Whitsel BL, Rustioni A (1983) The corticocuneate pathway in the cat: relations among terminal distribution patterns, cytoarchitecture, and single neuron functional properties. Somatosen Res 1:169–205

Cheema SS, Darian-Smith I, Darian-Smith C, Goodwin AW (1986) Postnatal regressive changes in thalamic projections to somatic sensorimotor cortex of the Macaque: a quadruple fluorescent dye tracer study. Soc Neurosci Abstr 12:1430

Chiaia NL, Hess PR, Rhoades RW (1986) Long-term consequences of neonatal transection of the infraorbital nerve in rat are not influenced by reestablishment of peripheral connections. Soc Neurosci Abstr 12:1576

Cho HJ, Basbaum AI (1987) Light and electron microscopic analysis of immunoreactive dynorphin in the spinal dorsal horn of the rat. Soc Neurosci Abstr 13:295

Chmielowska J, Carvell GE, Simons DJ (1989) Spatial organization of thalamocortical and corticocortical projection systems in the rat SmI barrel cortex. J Comp Neurol 285:325–338

Christensen BN, Perl ER (1970) Spinal neurons specifically excited by noxious or thermal stimuli: marginal zone of the dorsal horn. J Neurophysiol 33:293–307

Chung JM, Fang ZR, Cargill CL, Willis WD (1983) Prolonged naloxone-reversible inhibition of the flexion reflex in the cat. Pain 15:35–53

Chung JM, Fang ZR, Hori Y, Lee KN, Wilis WD (1984a) Prolonged inhibition of primate spinothalamic tract cells by peripheral nerve stimulation. Pain 19:259–275

Chung JM, Lee KN, Hori Y, Endo K, Willis WD (1984b) Factors influencing peripheral nerve stimulation produced inhibition of primate spinothalamic tract cells. Pain 19:277–293

Chung K, Coggeshall RE (1985) Unmyelinated primary afferent fibers in dorsal funiculi of cat sacral spinal cord. J Comp Neurol 238:365–369

Clark D, Hughes J, Gasser HS (1935) Afferent function in the group of nerve fibres of slowest conduction velocity. Am J Physiol 114:69–76

Clark HC (1979) A light and electron microscopic analysis of the development of the main sensory nucleus of the rat. Anat Rec 187:553–554

Clark SA, Allard TT, Jenkins WM, Merzenich MM (1986) Cortical map reorganization following neurovascular island skin transfers on the hands of adult owl monkeys. Soc Neurosci Abst 12:391

Clark SA, Allard TT, Jenkins WM, Merzenich MM (1988) Syndactyly results in the emergence of double digit receptive fields in somatosensory cortex in adult owl monkeys. Nature 332:444–445

Cliffer KD, Giesler GJ (1989) Postsynaptic dorsal column pathway of the rat. III. Distribution of ascending afferent fibers. J Neurosci 9:3146–3158

Coggeshall RE (1986) Nonclassical features of dorsal root ganglion cell organization. In: Yaksh TL (ed) Spinal afferent processing. Plenum, New York, pp 83–96

Coimbra A, Magalhaes MM, Sodre-Borges BP (1970) Ultrastructural localization of acid phosphatase in synaptic terminals of the rat substantia gelatinosa. Brain Res 22:142–146

Cole AE, Nicoll RA (1983) Acetylcholine mediates a slow synaptic potential in hippocampal pyramidal cells. Science 221:1299–1301

Cole AE, Nicoll RA (1984) Characterization of a slow cholinergic post synaptic potential recorded in vitro from rat hippocampal pyramidal cells. J Physiol (Lond) 352:173–188

Collins JG, Ren Ke (1986) WDR response profiles of spinal dorsal horn neurons may be unmasked by barbiturate anesthesia. Pain 28:369–378

Collins WF, Nulsen FE, Randt CT (1960) Relation of peripheral nerve fibre size and sensation in man. Arch Neurol 3:381–385

Comans PE, Hendry IA (1983) Initial neurite growth into the rat hindlimb. Neurosci Letts 11 Suppl S37

Cook AJ, Woolf CJ (1985) Cutaneous receptive field and morphological properties of hamstring flexor alpha-motoneurones in the rat. J Physiol (Lond) 364:249–263

Cook AJ, Woolf C, Wall P, McMahon S (1987) Dynamic receptive field plasticity in rat spinal cord dorsal horn following C-primary afferent input. Nature 325:151–153

Costanzo RM, Gardner EP (1980) A quantitative analysis of responses of direction-sensitive neurons of somatosensory cortex of awake monkeys. J Neurophysiol 43:1319–1341
Cotman CW, Nieto-Sampedro M, Harris EW (1984) Synapse replacement in the nervous system of adult vertebrates. Physiol Revs 61:684–784
Cragg BG (1975) Absence of barrels and disorganization of thalamic afferent distributions in the sensory cortex of reeler mice. Exp Neurol 49:858–862
Craig AD, Kniffki K (1985) Spinothalamic lumbosacral lamina I cells responsive to skin and muscle stimulation in the cat. J Physiol (Lond) 365:197–221
Crain SM, Peterson ER (1974) Development of neural connections in culture. Ann N Y Acad Sci 228:6–34
Crain SM, Peterson ER, Leibman M, Schulman H (1980) Dependence on nerve growth factor of early human fetal dorsal root ganglion neurons in organotypic cultures. Exp Neurol 67:205–214
Crandell JE, Cariness VS Jr (1984) Axon strata of the cerebral wall in embryonic mice. Dev Brain Res 14:185–195
Csillik B, Knyihar E (1975) Degenerative atrophy and regenerative proliferation in the rat spinal cord. Z mikrosk anat Forsch 89:1099–1103
Csillik B, Knyihar E (1978) Biodynamic plasticity in the Rolando substance. Prog Neurobiol 10:203–230
Csillik B, Knyihar-Csillik E (1981) Reactions of the substantia gelatinosa Rolandi to injury of peripheral axons. In: Brown AG, Rethelyi M (ed) Spinal cord sensation. Scottish Academic, Edinburgh, pp 309–318
Csillik B, Knihyar E, Elshiekh AA (1977) Degenerative atrophy of central terminals of primary sensory neurons induced by blockade of axoplasmic transport in peripheral nerves. Experientia 33:656–657
Culberson J, Brown P (1984) Projections of hindlimb dorsal roots to lumbosacral spinal cord of cat. J Neurophysiol 51(3):516–528
Curtis DR (1979) A method for continuously monitoring the electrical threshold of single intraspinal nerve fibres. Electroencephalog Clin Neurophysiol 47:503–506
Darian-Smith I, Proctor R and Ryan RD (1963) A single-neurone investigation of somatotopic organization within the cat's trigeminal brainstem nuclei. J Physiol (Lond) 168:147–157
Darian-Smith I (1973) The trigeminal system. In: Iggo A (ed) Somatosensory system. Springer Heidelberg New York Berlin 271–314. Handbook of sensory physiology vol. II
Dart AM, Gordon G (1973) Some properties of spinal connections of the cat's dorsal column nuclei which do not involve the dorsal columns. Brain Res 58:61–68
Davenport HA, Chor H, Dolkart RE (1937) The ratio of myelinated to unmyelinated fibers in regenerated sciatic nerves of *Macacus rhesus.* J Comp Neurol 67:483–491
Davenport HA, Chor H, Cleveland DA (1939) Fiber ratios in regenerating nerves, II: the status of regrowth of the sciatic and nerves to the gastrocnemius in *Macacus* 3 months after section. J Comp Neurol 70:153–159
Davidoff RA (1989) The dorsal columns Neurol 39:1377–1385
Davis AM, Lumsden AGS, Slavkin HC, Burnstock G (1981) Influence of nerve growth factor on the embryonic mouse trigeminal ganglion in culture. Dev Neurosci 4:150–156
Davis LA, Gordon T, Hoffer JA, Jhamandas J, Stein RB (1978) Compound action potentials recorded from mammalian peripheral nerves following ligation or resuturing. J Physiol (Lond) 285:543–559
Dawson DR, Killackey HP (1985) Distinguishing topography and somatotopy in the thalamocortical projections of the development rat. Dev Brain Res 17:309–313

Dawson DR, Killackey HP (1986) Morphological changes in rat somatosensory cortex following prenatal limb removal. Soc Neurosci Abst 12:1436

Dawson DR, Killackey HP (1987) The organization and mutability of the forepaw and hindpaw representations in the somatosensory cortex of the neonatal rat. J Comp Neurol 256:246–256

De Jong RH, Wagman IH (1968) Block of afferent impulses in the dorsal horn of monkey. A possible mechanism of anesthesia. Exp Neurol 20:352–358

Dellon AL (1976) Reinnervation of denervated meissner corpuscles: a sequential histological study in the monkey following fascicular nerve repair. J Hand Surg 1:98–109

Dellon AL (1981) Evaluation of sensitivity and re-education of sensation in the hand. Williams and Wilkins, Baltimore

Dellon AL, Curtis RM, Edgerton MT (1972) Evaluating recovery of sensation in the hand following nerve injury. Johns Hopkins Med J 130:235–243

Dellon AL, Witebsky FG, Terrill RE (1975) The denervated Meissner corpuscle: a sequential histologic study following nerve division in the rhesus monkey. J Plast Reconstr Surg 56:182–193

De Olmos JS, Ebbesson SO, Heimer L (1981) Silver methods for the impregnation of degenerating axoplasm. In: Heimer L, Robards MJ (eds) Neuroanatomical tract-tracing methods. Plenum, New York, pp 117–170

De Santis M, Duckworth JW (1982) Properties of primary afferent neurons from muscles which are spontaneously active after a lesion of their peripheral process. Exp Neurol 75:261–274

Devor M (1984a) The pathophysiology and anatomy of damaged nerve. In: Wall PD, Melzack R (eds) Textbook of pain Churchill Livingstone, Edinburgh, pp 49–64

Devor M (1984b) Spinal cord plasticity induced by peripheral nerve injury and some consequences for sensory dysfunction. In: Von Euler C, Franzen O, Lindblom U and Ottoson D (eds) Wenner-Gren center international symposium 40, somatosensory mechanisms. Macmillan, London, pp 339–351

Devor M (1987) On mechanisms of somatotopic plasticity. In: Pubols LM, Sessle BJ (eds) Effects of injury on trigeminal and spinal somatosensory systems, Liss, New York, pp 215–225

Devor M, Claman D (1980) Mapping and plasticity of acid phosphat afferents in the rat dorsal horn. Brain Res 190:17–28

Devor M, Govrin-Lippmann R (1979) Maturation of axonal sprouts after nerve crush. Exp Neurol 64:260–270

Devor M, Janig W (1981) Activation of myelinated afferents ending in a neuroma by stimulation of the sympathetic supply in the rat. Neurosci Letts 24:43–47

Devor M, Raber P (1983) Autotomy after nerve injury and its relation to spontaneous discharge originating in nerve-end neuromas. Behav neural Biol 37:276–283

Devor M, Wall PD (1976a) Type of sensory nerve fibre sprouting to form a neuroma. Nature 262:705–708

Devor M, Wall PD (1976b) Dorsal horn cells with proximal cutaneous receptive fields. Brain Res 118:325–328

Devor M, Wall P (1978) Reorganization of spinal cord sensory map after peripheral nerve injury. Nature 276:75–76

Devor M, Wall PD (1981a) Plasticity in the spinal cord sensory map following peripheral nerve injury in rats. J Neurosci 1:679–684

Devor M, Wall PD (1981b) Effect of peripheral nerve injury on receptive fields of cells in the cat spinal cord. J Comp Neurol 199:277–291

Devor M, Wall PD (1986) Spinal plasticity after nerve injury: mediolateral localization of rewired cells. Exp Brain Res [Suppl. 13]:143–149

Devor M, Merrill EG, Wall PD (1977) Dorsal horn cells that respond to stimulation of distant dorsal roots. J Physiol (Lond) 270:519–531

Devor M, Schonfeld D, Seltzer Z, Wall PD (1979) Two modes of cutaneous reinnervation following peripheral nerve injury. J Comp Neurol 185:211–220

Devor M, Govrin-Lippmann R, Frank I, Raber P (1985) Proliferation of primary sensory neurons in adult rat dorsal root ganglion and the kinetics of retrograde cell loss after sciatic nerve section. Somatosens Res 3:139–167

Devor M, Basbaum AI, Seltzer Z (1986) Spinal somatotopic plasticity: possible anatomical basis for somatopically inappropriate connections. In: Goldberger ME, Gorio A, Murray M (eds) Development and plasticity of the mammalian spinal cord. Liviana, Padova, pp 239–253 (Fidia Research Series, vol. 3)

Diamond J, Cooper E, Turner C, Macintyre L (1976) Trophic regulation of nerve sprouting. Science 193:371–377

Diamond J, Coughlin M, Macintyre L, Holmes M, Visheau B (1987) Evidence that endogenous ß nerve growth factor is responsible for the collateral sprouting, but not the regeneration, of nociceptive axons in adult rats. Proc Natl Acad Sci USA 84:6596–6600

Dickhaus H, Zimmermann M, Zottermann Y (1976) The development in regenerating cutaneous nerves of C fiber receptors responding to noxious heating of the skin. In: Zottermann Y (ed) Sensory functions of the skin in primates. Pergamon, Oxford, pp 415–425

Dijkstra C (1933) Die De- und Regeneration der sensiblen Endkorperchen des Entenschnabels (Grandry-und Herbst-Korperchen) nach Durchschneidung des Nerven, nach Fortnahme der ganzen Haut und nach Transplantation des Hautstuckchens. Z Mikrosk Anat Forsch 34:75–158

Doetsch, Towe AL (1976) Response properties of distinct neuronal subsets in hind limb sensorimotor cerebral cortex of the domestic cat. Exp Neurol 53:520–547

Donaldson L, Hand PJ, Morrison AR (1975) Cortical-thalamic relationship in the rat. Exp Neurol 47:448–458

Donoghue JP, Kerman KL, Ebner FF (1979) Evidence for two organizational plans within the somatic sensory-motor cortex of the rat. J comp Neurol 183:647–664

Dorfl J (1982) The musculature of the mystacial vibrissae of the white mouse. J Anat 135:147–154

Dorfl J (1985) The innervation of the mystacial region of the white mouse. A topographical study. J Anat 142:173–184

Dorfman LJ, Bosley TM (1979) Age related changes in peripheral and central nerve conduction in man. Neurology 29:38–44

Dostrovsky JO, Millar J (1977) Receptive fields of gracile neurons after transection of the dorsal columns. Exp Neurol 56:610–621

Dostrovsky JO, Millar J, Wall PD (1976) The immediate shift of afferent drive of dorsal column nucleus cells following deafferentation: a comparison of acute and chronic deafferentation in gracile nucleus and spinal cord. Exp Neurol 52:480–495

Dostrovsky JO, Jabbur S, Millar J (1978) Neurones in cat gracile nucleus with both local and widefield inputs. J Physiol (Lond) 278:365–375

Doucette R, Diamond J (1987) Normal and precocious sprouting of heat nociceptors in the skin of adult rats. J Comp Neurol 261:592–603

Douglas PR, Ferrington DG, Rowe M (1978) Coding of information about tactile stimuli by neurones of the cuneate nucleus. J Physiol (Lond) 285:493–513

Dreyer DA, Schneider RJ, Metz CB, Whitsel BL (1974) Differential contributions of spinal pathways to body representation in postcentral gyrus of *Macaca mulatta*. J Neurophysiol 37:119–145

Dubner R, Hylden JLK, Nahin RL, Traub R (1989) Superficial dorsal horn changes after sciatic nerve transection and cutaneous inflammation. In: Cervero F, Bennett GJ and Headley PM (eds) Processing of sensory information in the superficial dorsal horn of the spinal cord. Plenum, New York

Dubner R, Sessle BJ, Storey AT (1978) The neural basis of oral and facial function. Plenum, New York

Dubuisson D, Wall PD (1980) Descending influences on receptive fields and activity of single units recorded in laminae 1,2 and 3 of the cat spinal cord. Brain Res 199:283–298

Dubuisson D, Fitzgerald M, Wall PD (1979) Ameoboid receptive fields of cells in laminae 1, 2 and 3. Brain Res 177:376–378

Duncan GL, Dryer DA, McKenna TM, Whitsel BL (1982) Dose- and time-dependent effects of ketamine on SI neurons with cutaneous receptive fields. J Neurophysiol 47:677–699

Durham D, Woolsey TA (1978) Acute whisker removal reduces neuronal activity in barrels of mouse SmI cortex. J Comp Neurol 178:629–644

Durham D, Woolsey TA (1984) Effects of neunatal whisker lesions on mouse central trigeminal pathways. J Comp Neurol 223:424–447

Durham D, Woolsey TA (1985) Functional organization in cortical barrels of normal and vibrissae-damaged mice: a (3H) 2-Deoxyglucose study. J Comp Neurol 235:97–110

Dykes RW (1983) Parallel processing of somatosensory information: a theory. Brain Res Rev 6:47–115

Dykes RW, Lamour Y (1988) An electrophysiological laminar analysis of single somatosensory neurons in partially deafferented rat hindlimb granular cortex subsequent to transection of the sciatic nerve. Brain Res 449:1–17

Dykes RW, Terzis JK (1979) Reinnervation of glabrous skin in baboons: properties of cutaneous mechanoreceptors subsequent to nerve crush. J Neurophysiol 42:1461–1478

Dykes RW, Rasmusson DD, Hoeltzell PB (1980) Organization of primary somatosensory cortex in the cat. J Neurophysiol 43:1527–1546

Dykes RW, Sur M, Merzenich MM, Kaas JH, Nelson RJ (1981) Regional segregation of neurons responding to quickly-adapting, slowly adopting deep and pacinian receptors within thalamic ventroposterior lateral and ventroposterior inferior nuclei in the squirrel monkey (*Saimiri sciuriens*). Neurosci 6:1687–1692

Dykes RW, Rasmusson DD, Stretavan D, Rehman NB (1982) Submodality segregation and receptive-field sequences in cuneate, gracile and external cuneate nuclei of the cat. J Neurophysiol 47:389–416

Dykes RW, Landry P, Metherate R, Hicks TP (1984) Functional role of GABA in primary somatosensory cortex shaping receptive fields of cortical neurons. J Neurophysiol 52:1066–1093

Dykes RW, Herron P, Lin C-S (1986) Ventroposterior thalamic regions projecting to cytoarchitectonic areas 3a and 3b in the cat. J Neurophysiol 56:1521–1541

Eccles JC, Schmidt RF, Willis WD (1963) Pharmacological studies on presynaptic inhibition. J Physiol (Lond) 168:500–530

Edelman GM (1977) Group selection and phasic reentrant signalling: a theory of higher brain function. In: Schmitt FO (ed) The mindful brain. MIT, Cambridge, Massachusetts, pp 51–100

Egger MD, Freeman NCG, Jacquin M, Proshansky E, Semba K (1986) Dorsal horn cells in the cat responding to stimulation of the plantar cushion. Brain Res 383:68–82

Ellis LC, Rustioni A (1981) A correlative HRP, Golgi, and EM study of the intrinsic organization of the feline dorsal column nuclei. J Comp Neurol 197:341–367

English KB (1977) The ultrastructure of cutaneous type I mechanoreceptors (Haarscheiben) in cats following denervation. J Comp Neurol 172:137–163

Erlanger J, Schoepfle GM (1946) A study of nerve degeneration and regeneration. Am J Physiol 147:550–581
Erzurumlu RS, Bates CA, Killackey HP (1980) Differential organization of thalamic projection cells in the brain stem trigeminal complex of the rat. Brain Res 198:427–433
Erzurumlu RS, Killackey HP (1982) Order in the developing rat trigeminal nerve. Brain Res 255:305–310
Erzurumlu RS, Killackey HP (1983) Development of order in the rat trigeminal system. J Comp Neurol 213:365–380
Fadiga E, Haimann E, Margnelli M, Sotgui ML (1978) Variability of peripheral representation in ventrobasal thalamic nuclei of the cat; effects of acute reversible blockade of the dorsal column nuclei. Exp Neurol 60:484–498
Falls W, Gobel S (1979) Golgi and EM studies of the formation of dendritic and axonal arbors: the interneurons of the substantia gelatinosa of rolando in newborn kittens. J Comp Neurol 187:1–18
Farkas-Bargeton E, Sary C, Verley R (1986) L'organisation du systeme nerveux apres coagulation des follicules des vibrisses mystaciales chez la souris noveau-nee: un example de plasticite neuronale. Rev Neurol (Paris) 142:215–225
Favorov OV, Diamond ME, Whitsel BL (1987) Evidence for a mosaic representation of the body surface in area 3b of the somatic cortex of cat. Proc Natl Acad Sci 84:6606–6610
Feldman SG, Kruger L (1980) An axonal transport study of the ascending projection of medial lemniscal neurons in the rat. J Comp Neurol 192:427–454
Felleman DJ, Nelson RJ, Sur M, Kaas JH (1983a) Representations of the body surface in areas 3b and 1 of the postcentral parietal cortex of Cebus monkeys. Brain Res 268:15–26
Felleman DJ, Wall JT, Cusick CG, Kaas JH (1983b) The representation of the body surface in SI of cats. J Neurosci 3:1648–1669
Feringa ER, Lee GW, Vahlsing HL, Gilbertie WJ (1985) Cell death in the adult rat dorsal root ganglion after hindlimb amputation, spinal cord transaction, or both operations. Exp Neurol 87:349–357
Ferraro A, Barrera SE (1934) Effects of experimental lesions of the posterior columns in Macacus rhesus monkeys. Brain 57:307–332
Ferraro A, Barrera SE (1935) Posterior column fibres and their termination in Macacus rhesus. J Comp Neurol 62:507–530
Ferrington DG, Rowe MJ (1980a) Functional capacities of tactile afferent fibres in neonatal kittens. J Physiol (Lond) 307:335–353
Ferrington DG, Rowe MJ (1980b) Differential contributions to coding of cutaneous vibratory information by cortical somatosensory area I and II. J Neurophysiol 43:310–331
Ferrington DG, Rowe MJ (1982) Specificity of connections and tactile coding capacities in cuneate nucleus of the neonatal kitten. J Neurophysiol 47:622–640
Finger S, Reyes R (1975) Long-term deficits after somatosensory cortical lesions in rats. Physiol Behav 16:289–293
Finger S, Simons D (1976) Effects of serial lesions of somatosensory cortex and further neodecortication on performance of a rough-smooth discrimination in rats
Finger S, Cohen M, Alongi R (1972) The roles of somatosensory cortical areas 1 and 2 in tactile discrimination in the rat. Int J Psychobiol 2:93–102
Finger S, Simons DJ, Posner R (1978) Anatomical, physiological and behavioural effects of neonatal sensorimotor cortex abalation in the rat. Exp Neurol 60:347–373
Fitzgerald MJT, Martin F, Paletta FX (1967) Innervation of skin grafts. Surg Gynecol Obstet 124:808–812
Fitzgerald M (1981) A study of the cutaneous afferent input to substantia gelatinosa. Neurosci 6:2229–2237

Fitzgerald M (1982) Alteration in the ipsi- and contralateral afferent inputs of dorsal horn cells produced by capsaicin treatment of one sciatic nerve in the rat. Brain Res 248:97–107
Fitzgerald M (1983) Capsaicin and sensory neurones – a review. Pain 15:109–130
Fitzgerald M (1984) The course and termination of primary afferent fibres. In: Wall PD, Melzack R (eds) Textbook of pain. Churchill-Livingstone, Edinburgh, pp 34–48
Fitzgerald M (1985a) The sprouting of saphenous nerve terminals in the spinal cord following early postnatal sciatic nerve section in the rat. J Comp Neurol 240:407–413
Fitzgerald M (1985b) The post-natal development of cutaneous afferent fibre input and receptive field organization in the rat dorsal horn. J Physiol (Lond) 364:1–18
Fitzgerald M (1987a) Prenatal growth of fine-diameter primary afferents into the rat spinal cord: a transganglionic tracer study. J Comp Neurol 261:98–104
Fitzgerald M (1987b) Spontaneous and evoked activity of fetal primary afferents *in vivo*. Nature 326:603–605
Fitzgerald M, Gibson S (1984) The postnatal physiological and neurochemical development of peripheral sensory C-fibres. Neurosci 13:933–944
Fitzgerald M, Swett J (1983) The termination pattern of sciatic nerve afferents in the substantia gelatinosa of neonatal rats. Neurosci Letters 43:149–154
Fitzgerald M, Vrbova G (1985) Plasticity of acid phosphatase (FRAP) afferent terminal fields and of dorsal horn cell growth in the neonatal rat. J Comp Neurol 240:414–422
Fitzgerald M, Wall PD (1980) The laminar organization of dorsal horn cells responding to peripheral C fibre stimulation. Exp Brain Res 41:36–44
Fitzgerald M, Wall PD, Goedert M, Emson PC (1985) Nerve growth factory conteracts the neurophysiological and neurochemical effects of chromic sciatic nerve section. Brain Res 332:131–141
Florence SL, Wall JT, Kaas JH (1988) The somatotopic pattern of afferent projections from the digits to the spinal cord and cuneate nucleus in macaque monkeys. Brain Res 452: 388–392
Fodstad O, Johannessen JV, Schjerven L, Pihl A (1979) Toxicity of abrin and ricin in mice and dogs. J Toxicol Environ Health 5:1073–1084
Foote SL, Morrison JH (1987) Extrathalamic modulation of cortical function. Ann Rev Neurosci 10:67–95
Forbes DJ, Welt C (1982) Neurogenesis in the trigeminal ganglion of the albino rat: a quantitive autoradiographic study. J Comp Neurol 199:133–147
Ford FR, Woodhall B (1938) Phenomena due to misdirection of regenerating fibres of cranial spinal and autonomic nerves. Arch Surg 36:480–499
Franck JI (1980) Functional reorganization of cat somatic sensory-motor cortex (SmI) after selective dorsal root rhizotomies. Brain Res 186:458–462
Fried K (1982) Development, degeneration and regeneration of nerve fibres in the feline inferior alveolar nerve and mandibular incisor pulps. Acta Physiol Scand Suppl 504:1–28
Frommer GP, Tref BR, Casey KL (1977) Somatosensory function and cortical unit activity in cats with only dorsal column fibres. Exp Brain Res 27:113–129
Fukushima T, Kerr FW (1979) Organization of trigeminothalamic tracts and other thalamic afferent systems of the brainstem in the rat: presence of gelatinosa neurons with thalamic connections. J Comp Neurol 183:169–184
Fullerton PM, Gilliat RW (1965) Axon reflexes in human motor nerve fibres. J Neurol Neurosurg Psychiat 28:1–11
Fyffe REW (1984) Afferent fibres. In: Davidoff RA (ed) Anatomy and physiology. Dekker, New York, pp 76–136. Handbook of the spinal cord, vols 2, 3
Fyffe REW, Cheema SS, Rustioni A (1986a) Intracellular staining study of the feline cuneate nucleus. I. Terminal patterns of primary afferent fibres. J Neurophysiol 56:1268–1283

Fyffe REW, Cheema SS, Light AR, Rustioni A (1986b) Intracellular staining study of the feline cuneate nucleus. II. Thalamic Projecting neurons. J Neurophysiol 56:1284–1296

Ganchrow D, Bernstein JJ (1981) Bouton renewal patterns in rat hindlimb cortex after thoracic dorsal funicular lesions. J Neurosci Res 6:525–537

Ganchrow D, Margolin J, Perez L, Bernstein JJ (1980) Component analysis of hind limb behavioural alterations after damage to the rat dorsal funiculus. Exp Neurol 70:339–355

Ganchrow D, Margolin J, Perez L, Bernstein JJ (1981) Patterns of reafferentation in rat nucleus gracilis after thoracic dorsal column lesions. Exp Neurol 71:437–451

Giesler, GJ, Cannon JT, Urca G, Liebeskind JC (1978) Long ascending projections of the substantia gelatinosa and the subjacent dorsal horn in the rat. Science 202:984–986

Giesler GJ, Cliffer KD (1985) Postsynaptic dorsal column pathway of the rat. II. Evidence against an important role in nociception. Brain Res 326:347–356

Giesler GJ, Nahin RL, Masden AM (1984) Postsynaptic dorsal column pathway of the rat. I. Anatomical studies. J Neurophysiol 51:260–275

Gilbert M, Stelzner DJ (1979) The development of descending and dorsal root connections in the lumbosacral spinal cord of the postnatal rat. J Comp Neurol 184:821–838

Gilliatt RW (1982) Paresthesine. In: Culp WJ, Ochoa J (eds) Abnormal nerves and muscles as impulse generators. Oxford University Press, New York, pp 477–489

Gilman S, Denny-Brown D (1966) Disorders of movement and behaviour following dorsal column lesions. Brain 89:397–418

Giolli RA, Karamanlidis AN (1978) The study of degenerating nerve fibres using silver-impregnation methods. In: Robertson RT (ed) Neuroanatomical Research Techniques. Academic, London, pp 212–240

Glees P, Soler J (1951) Fibre content of the posterior column and synaptic connections of nucleus gracilis. Z Zellforsch 36:381–400

Gobel S (1984) An electron microscopic analysis of the trans-synaptic effects of peripheral nerve injury subsequent to tooth pulp extirpations on neurons in laminae I and II of the medullary dorsal horn. J Neurosci 4:2281–2290

Gobel S, Binck JM (1977) Degenerative changes in primary trigeminal axons and in neurons in nucleus caudalis following tooth pulp extirpations in the cat. Brain Res 132:347–354

Goldberger ME, Murray M (1974) Restitution of function and collateral sprouting in the cat Cotman CW (ed) spinal cord: the deafferented animal. J Comp Neurol 158:37–54

Goldberger ME, Murray M (1978) Recovery of movement and axonal sprouting may obey some of the same laws. In: Cotman CW (ed) Neuronal Plasticity. Raven Press, New York, pp 73–96

Goldberger ME, Murray M (1982) Lack of sprouting and its presence after lesions of the cat spinal cord. Brain Res 241:227–239

Goldberger ME, Murray M (1985) Recovery of function and anatomical plasticity after damage to the adult and neonatal spinal cord. In: Cotman CW (ed) Synaptic plasticity. Guildford, New York, pp 77–110

Goldberger ME, Paige E (1986) Anatomical plasticity of serotonin (5HT) projections after deafferentation of cat spinal cord. Soc Neurosci Abstr 12:512

Golovchinsky V (1980) Patterns of responses of neurons in cuneate nucleus to controlled mechanical stimulation of cutaneous velocity receptors in the cat. J Neurophysio 43:1673–1699

Golovchinsky V, Kruger L, Saporta SA, Stein BE, Young DW (1981) Properties of velocity-mechanosensitive neurons in the cat ventrobasal thalamic nucleus with special reference to the concept of convergence. Brain Res 209:355–374

Gordon G, Grant G (1982) Dorsolateral spinal afferents to some medullary sensory nuclei: an anatomical study in the cat. Exp Brain Res 46:12–23

Gorin PD, Johnson EM (1979) Experimental autoimmune model of nerve growth factor deprivation: effects on developing peripheral sympathetic and sensory neurons. Proc Natl Acad Sci USA 76:5382–5386

Gordon G, Jukes MGM (1964) Dual organization of the exteroceptive components of the cat's gracile nucleus. J Physiol (Lond) 173:291–319

Gordon G, Kuypers HGJM (1964) The functioning of the dorsal column nuclei. IBRO Bull 3:71

Gordon G, Manson JR (1967) Cutaneous receptive fields of single nerve cells in the thalamus of the cat. Nature 215:597–599

Gordon G, Paine CH (1960) Functional organization in nucleus gracilis of the cat. J Physiol (Lond) 153:331–349

Gordon G, Seed WA (1961) An investigation of the nucleus gracilis of the cat by antidromic stimulation. J Physiol (Lond) 155:589–601

Goshgarian HG, Koistinen JM, Schmidt ER (1983) Cell death and changes in the retrograde transport of horseradish peroxidase in rubrospinal neurons following spinal cord hemisection in the adult rat. J Comp Neurol 214:251–257

Govri-Lippmann R, Devor M (1978) Ongoing activity in severed nerves: source and variation with time. Brain Res 159:406–410

Grant G, Arvidsson J (1975) Transganglionic degeneration in trigeminal primary sensory neurons. Brain Res 95:265–279

Grant G, Ygge J (1981) Somatotopic organization of the thoracic spinal nerve in the dorsal horn of the spinal cord demonstrated with transganglionic degeneration. J Comp Neurol 202:357–364

Green EL (1966) Biology of the laboratory mouse. McGraw-Hill, New York

Greenfield Z, Devor M (1981) Collateral reinnervation of rat hindlimb skin does not depend on repeated sensory testing. Neurosci Letts 25:305–309

Greenman MJ (1913) Studies on the regeneration of the peroneal nerve of the albino rat: number and sectional areas of fibers; area relation to axis of sheath. J Comp Neurol, 23:479–513

Gregg JM, Dixon AD (1973) Somatotopic organization of the trigeminal ganglion in the rat. Arch Oral Biol 18:487–498

Guilbaud G, Peschanski M, Gautron M (1981) Functional changes in VB thalamic neurons responsive to noxious and non-noxious cutaneous stimuli after chloralose treatment: New evidence for the presence of preexisting, "silent connections" in the adult nervous system. Pain 11:9–19

Gulley RL (1973) Golgi studies of the nucleus gracilis in the rat. Anat Rec 177:325–342

Gutmann E, Sanders FK (1943) Recovery of fibre numbers and diameters in the regeneration of peripheral nerves. J Physiol (Lond) 101:498–518

Haddad RK, Rabe A, Laqueur L, Spatz M, Valsamis MP (1969) Intellectual deficit associated with transplancentally induced microencephaly in the rat. Science 163:88–90

Haimann C, Marguelli M, Solgin ML (1978) Variability of peripheral representation in ventrobasal thalamic nuclei of the cat: effects of chloralose treatment. Exp Neurol 60:469–483

Hall RD, Lindholm EP Organization of motor and somatosensory neocortex in the albino rat. Brain Res 66:23–38

Hallas BH, Renehan WE, Klein BG, Jacquin MF (1987) Absence of intramedullary sprouting in trigeminal primary afferent axons caudal and rostral to a partial medullary hemisection. In: Pubols LM, Sessle BJ (eds) Effects of injury on trigeminal and spinal somatosensory systems Liss, New York, pp 205–214

Hallin RG, Wiesenfeld Z, Lindblom U (1981) Neurophysiological studies on patients with sutured median nerves: faulty sensory localization after nerve generation and its physiological correlates. Exp Neurol 73:90–106

Hamburger V, Levi-Montalcini R (1949) Proliferation, differentiation and degeneration in the spinal ganglia of the chick embryo under normal and experimental conditions. J Exp Zool 111:457–501

Hamilton TC, Johnson JI (1973) Somatotopic organization related to nuclear morphology in the cuneate-gracile complex of opossums *Didelphis marsupialis virginiana.* Brain Res 51:125–140

Hamori J, Savy C, Madarasz M, Somogyi J, Takacs J, Verley R, Farkas-Bargeton E (1986) Morphological alterations in subcortical vibrissal relays following vibrissal follicle destruction at birth in the mouse. J Comp Neurol 254:166–183

Hand PJ (1966) Lumbosacral dorsal root terminations in the nucleus gracilis of the cat: some observations on terminal degenerations in other medullary sensory nuclei. J Comp Neurol 126:137–156

Hand PJ (1982) Plasticity of the rat barrel system. In: Changing concepts of the nervous system. Academic, New York, 49–68

Hand PJ, Morrison AR (1970) Thalamocortical projections from the ventrobasal complex to somatic sensory areas I and II. Exp Neurol 26:291–308

Hand PJ, Van Winkle T (1977) The efferent connections of the feline nucleus cuneatus. J Comp Neurol 171:83–110

Hardin WB, Arumugasamy H, Jameson HD (1968) Pattern of localization in "precentral" motor cortex of raccoon. Brain Res 11:611–627

Harding GW, Stogsdill RM, Towe AL (1979) Neuroscience 4:369–378

Harper AA, Lawson SN (1985) Conduction velocity is related to morphological cell type in rat dorsal root ganglion neurones. J Physiol (Lond) 359:31–46

Harris AJ (1943) An experimental analysis of the inferior mesenteric plexus. J Comp Neurol 79:1–13

Harris FA (1970) Population analysis of somatosensory thalamus in the cat. Nature 225:559–562

Harris FA (1980) Wide-field neurons in somatosensory thalamus of domestic cats under barbiturate anaesthesia. Exptl Neurol 68(1):27–49

Harris RM, Woolsey TA (1979) Morphology of Golgi-impregnated neurons in mouse cortical barrels following vibrissal damage at different postnatal ages. Brain Res 161:143–149

Hartman CG, Strauss WL Jr (1961) The Anatomy of the Rhesus Monkey. Hafner, New York

Hassler R, Muhs-Clements K (1964) Architektonischer aufbau des sensomotorischen und parietalen cortex der katze. J Hirnforsch 6:377–420

Hawkins GL (1948) Faulty sensory localization in nerve regeneration. An index of functional recovery following suture. J Neurosurg 5:11–18

Hayashi H (1980) Distributions of vibrissae afferent fibre collaterals in the trigeminal nuclei as revealed by intraaxonal injection of horseradish peroxidase. Brain Res 183:442–446

Hayashi H (1985a) Morphology of central terminations of intraaxonally stained, large, myelinated primary afferent fibres from facial skin in the rat. J Comp Neurol 237:195–215

Hayashi M (1985b) Morphology of terminations of small and large myelinated trigeminal primary afferent fibres in the cat. J Comp Neurol 240:71–89

Haycock JW, Levy WB, Cotman CW (1977) Pentobarbital depression of stimulus-secretion coupling in brain-selective inhibition of depolarization-induced calcium-dependent release. Biochem Pharmacol 26:159–161

Hazlett JC, Dom R, Martin GF (1972) Spino-bulbar, spino-thalamic and medial lemniscal connections in the American opossum, *Didelphis marsupialis virginiana.* J Comp Neurol 146:95–118

Head H, Rivers WH, Sherren J (1905) The afferent nervous system from a new aspect. Brain 28:99–115

Heimer L, Wall PD (1968) The dorsal root distribution to the substantia gelatinosa of the rat with a note on the distribution in the cat. Exp Brain Res 6:89–99

Hendry SCH, Jones EG (1981) Sizes and distributions of intrinsic neurons incorporating tritiated GABA in monkey sensory-motor cortex. J Neurosci 1:390–408

Hendry SCH, Houser CR, Jones EG, Vaughin JE (1983) Synaptic organization of immunocytochemically identified GABA neurons in the monkey sensory-motor cortex. J Neurocytol 12:639–660

Hentall ID, Fields HL (1979) Segmental and descending influences on single intraspinal C fibres. J Neurophysiol 42:1527–1537

Hentall ID, Fields HL (1980) Potentiation of transmission from C-fibres to dorsal horn neurons after tetanus of peripheral nerve. Brain Res 189:540–543

Herron P (1983) The connections of cortical somatosensory areas I and II with separate nuclei in the ventroposterior thalamus in the raccoon. Neuroscience 8:243–257

Herron P, Dykes RW (1986) The ventroposterior inferior (VPI) nucleus in the thalamus of cats: a relay nucleus in the pacinian pathway to somatosensory cortex. J Neurophysiol 56:1475–1497

Hess A (1956) Fine structure and morphological organization of non-myelinated nerve fibres. Proc R Soc Lond B 144:496–506

Hicks TP, Dykes RW (1983) Receptive field size for certain neurons in primary somatosensory cortex is determined by GABA-mediated intracortical inhibition. Brain Res 274:160–164

Hillman P, Wall PD (1969) Inhibitory and excitatory factors influencing the receptive fields of lamina 5 spinal cord cells. Exp Brain Res 9:284–306

Hodes R, Larrabee MG, German W (1948) The human electromyogram in response to nerve stimulation and the conduction velocity of motor axons. Studies on normal and on injured peripheral nerves. Arch Neurol Psychiat Chicago 60:340–365

Hokfelt T, Elde R, Johansson O, Luft R, Nilsson G, Arimura A (1976) Immunohistochemical evidence for separate populations of somatostatin-containing and substance P-containing primary afferent neurons in the rat. Neurosci 1:131–136

Holland GR, Robinson PP (1984) Evidence for the persistence of axons at the apex of the cat's lower canine tooth after section of the inferior alveolar nerve. Anat Rec 208:175–183

Holland GR, Robinson PP (1987) A morphological study of the collateral reinnervation of the cat's canine tooth. Exp Neurol 98:489–498

Holland GR, Matthews B, Robinson PP (1987) An electrophysiological and morphological study of the innervation of cat dentine. J Physiol (Lond) 386:31–43

Hongo T, Koike H (1975) Some aspects of synaptic organization in the spinocervical tract cell in the cat. In: Kornhuber HH (ed) The somatosensory system Thieme, Stuttgart, pp 218–226

Horch KW (1976) Ascending collaterals of cutaneous neurons in the fasciculus gracilis of the cat during peripheral nerve regeneration. Brain Res 117:19–32

Horch K (1979) Guidance of regrowing sensory axons after cutaneous nerve lesions in the cat. J Neurophysiol 42:1437–1449

Horch K (1981) Absence of functional collateral sprouting of mechanoreceptor axons into denervated areas of mammalian skin. Exp Neurol 74:313–317

Horch KW, Whitehorn D, Burgess PR (1974) Impulse generation in type I cutaneous mechanoreceptors. J Neurophysiol 37:267–281

Horch KW, Burgess PR, Whitehorn D (1976) Ascending collaterals of cutaneous neurons in the fasciculus gracilis of the cat. Brain Res 117:1–17

Horch KW, Lisney SJW (1981a) On the number and nature of regenerating myelinated axons after lesions of cutaneous nerves in the cat. J Physiol (Lond) 313:275–286

Horch KW, Lisney SJW (1981b) Changes in primary afferent depolarization of sensory neurons during peripheral nerve regeneration in the cat. J Physiol (Lond) 313:287–299

Hori Y, Lee KH, Chung JM, Endo K, Willis WD (1984) The effects of small doses of barbiturate on the activity of primate nociceptive tract cells. Brain Res 307:9–15

Houser CR, Vaughan JE, Barber RP, Roberts E (1980) GABA neurons are the major cell type of the nucleus reticularis thalami. Brain Res 200:341–341

Houser CR, Hendry SHC, Jones EG, Vaughan JE (1983) Morphological diversity of immunocytochemically identified GABA neurons in the monkey sensory-motor cortex. J Neurocytol 12:617–638

Howe JF, Loeser JD, Calvin WH (1977) Mechanosensitivity of dorsal root ganglia and chronically injured axons: a physiological basis for radicular pain of nerve root compression. Pain 3:25–41

Howe JR, Yaksh TL, Go VLW (1987) The effect of unilateral dorsal root ganglionectomies or ventral rhizotomies on alpha-adrenoceptor binding to, and the substance P, enkephalin and neurotensin content of, the cat lumbar spinal cord. Neurosci 21:385–394

Hu JW Dostrovsky JO, Lenz YE, Ball GJ, Sessle BJ (1986) Tooth pulp deafferentation is associated with functional alterations in the properties of neurons in the trigeminal spinal tract nucleus. J Neurophysiol 56:1650–1668

Hu JW, Broton JG, Lenz Y, Shohara E, Sessle BJ (1987) Brainstem effects of naturally occurring dental deafferentation in kittens. In: Pubols LM, Sessle BJ (eds) Effects of injury on trigeminal and spinal somatosensory systems. Liss, New York, pp 189–196

Huerta MF, Wall JT, Kass JH (1986) Changes in topography of somatosensory cortex after nerve loss in adult and neonatal marmoset monkeys. Soc Nerurosci Abst 12:954

Hulsebosch CE, Coggeshall RE (1981a) Quantitation of sprouting of dorsal root axons. Science 213:1020–1021

Hulsebosch CE, Coggeshall RE (1981b) Sprouting of dorsal root axons. Brain Res 224:170–174

Hulsebosch CE, Coggeshall RE (1983a) Age related sprouting of dorsal root axons after sensory denervation. Brain Res 288:77–83

Hulsebosch, CE, Coggeshall RE (1983b) A comparison of axonal numbers in dorsal roots following spinal cord hemisection in neonate and adult rats. Brain Res 265:187–197

Humbertson A (1963) A chronological study of the degenerative phenomena of dorsal root ganglia cells following section of the sciatic nerve. Anat Rec 145:244

Hunt CC, McIntyre AK (1960) An analysis of fibre diameter and receptor characteristics of myelinated cutaneous afferent fibres in cat. J Physiol (Lond) 153:99–112

Hylden JL, Nahin RL, Dubner R (1987) Altered responses of nociceptive cat lamina I spinal dorsal horn neurons after chronic sciatic neuroma formation. Brain Res 411:341–350

Hyvarinen J, Poranen A (1978a) Movement-sensitive and direction- and orientation-selective cutaneous receptive fields in the hand area of the post-central gyrus in monkeys. J Physiol (Lond) 283:523–537

Hyvarinen J, Poranen A (1978b) Receptive field integration and submodality convergence in the hand area of the post-central gyrus of the alert monkey J Physiol (Lond) 283:539–556

Hyvarinen J, Poranen A, Jokinen Y (1980) Influence of attentive behaviour on neuronal responses to vibration in primary somatosensory cortex of the monkey. J Neurophysiol 43:870–882

Inbal R, Rousso M, Ashur H, Wall PD, Devor M (1987) Collateral sprouting in skin and sensory recovery after nerve injury in man. Pain 28:141–154

Ito M (1981) Some quantitative aspects of vibrissa-driven responses in rat neocortex. J Neurophysiol 46:705–715

Ito M, Seo ML, (1983) Avoidance of neonatal cortical lesions by developing somatosensory barrels. Nature 301:600–602

Ivy GO, Killackey HP (1982) Ephemeral cellular segmentation in the thalamus of the neonatal rat. Dev Brain Res 2:1–17

Ivy GO, Killackey HP (1982) Ontogenetic changes in the projections of neocortical neurons. J Neurosci 2:735–743

Ivy GO, Akers RM, Killackey HP (1979) Differential distribution of callosal projection neurons in the neonatal and adult rat. Brain Res 173:532–537

Iwamura, Y, Inubushi S (1974) Regional diversity in excitatory and inhibitory receptive field organization of cat thalamic ventrobasal neurons. J Neurophysiol 37:910–919

Iwamura Y, Tanaka M (1978a) Functional organization of receptive fields in the cat somatosensory cortex. I. Integration within the coronal region. Brain Res 151:49–60

Iwamura Y, Tanaka M (1978b) Functional organization of receptive fields in the cat somatosensory cortex. II. Second representation of the forepaw in the ansate region. Brain Res 151:61–72

Iwamura Y, Gardner EP, Spencer WA (1972) Geometry of the ventrobasal complex: functional significance in skin sensations. Brain Beh Evol 6:185–202

Iwamura Y, Tanaka M, Hikosaka O (1980) Overlapping representation of fingers in the somatosensory cortex (area 2) of the conscious monkey. Brain Res 197:516–520

Iwamura Y, Tanaka M, Hikosaka O (1985) Comparison of hand and finger representation in areas 3, 1 and 2 of the monkey somatosensory cortex. In: Development, organization and processing in somatosensory pathways. Liss, New York, pp 239–245

Jabaley ME Burns JE Orcutt BS and Bryant WM (1976) Comparison of histologic and functional recovery after peripheral nerve repair. J Hand Surg 1:119–130

Jabbur SJ Baker MA, Towe AL (1972) Wide-field neurons in thalamic nucleus ventralis posteriolateralis of the cat. Exp Neurol 36:213–238

Jackson PC, Diamond J (1981) Regenerating axons reclaim sensory targets from collateral nerve sprouts. Science 214:926–928

Jackson PC, Diamond J (1983) Failure of intact cutaneous mechanosensory axons to sprout functional collaterals in skin of adult rabbits. Brain Res 273:277–283

Jackson PC, Diamond J (1984) Temporal and spatial constraints on the collateral sprouting of low-threshold mechanosensory nerves in the skin of rats. J Comp Neurol 226:336–345

Jacquin MF, Rhoades RW (1983) Central projections of normal and "regenerate" infraorbital nerve in adult rats subjected to neonatal unilateral lesions: a transganglionic horseradish peroxidase study. Brain Res 269:137–144

Jacquin MF, Rhoades RW (1985) Effects of neonatal infraorbital lesions upon central trigeminal primary afferent projections in rat and hamster. J Comp Neurol 235:129–143

Jacquin MF, Rhoades RW (1987) Development and plasticity in hamster trigeminal primary afferent projections Dev Brain Res 31:161–175

Jacquin MF, Hess A, Yaing G, Adamo P, Math M, Brown A, Rhoades RW (1984) Organization of the infraorbital nerve in rat: a quantitive electron microscopic study. Brain Res 290:131–135

Jacquin MF, Mooney RD, Rhoades RW (1986a) Morphology, response properties and collateral projections of trigeminothalamic neurons in brainstem subnucleus of rat. Expt Brain Res 61:457–468

Jacquin MF, Renehan WE, Klein BG, Hallas BH (1986b) Fetal cortical transplants into partially hemisected medulla of adult rat. Soc Neurosci Abst 12

Jacquin MF, Rehehan WE, Klein BG, Mooney RD, Rhoades RW (1986c) Functional consequences of neonatal infraorbital nerve section in rat trigeminal ganglion J Neurosci 6:3706–3720

Jacquin MF, Renehan WE, Mooney RD, Rhoades RW (1986d) Structure-function relationships in rat medullary and cervical dorsal horns. I. Trigeminal primary afferents. J Neurophysiol 55:1153–1186

Jacquin MF, Woerner D, Szczepanik AM, Rieter V, Mooney RD, Rhoades RW (1986e) Structure-function relationships in rat brainstem subnucleus interpolaris. I. Vibrissa primary afferents. J Comp Neurol 243:266–279

Jalowy B (1935) Über die De- und Regeneration der Nevenendigungen in den Figerbeeren der oberen Extremitaten der Affen (Macacus rhesus). Z Zellforsch mikrosk Anat 23:84–116

Jänig W, McLachlan EM (1984) On the fate of sympathetic and sensory neurons projecting into a neuroma of the superficial peroneal nerve in the cat. J Comp Neurol 225:302–311

Jänig W, Schoultz T, Spencer WA (1977) Temporal and spatial parameters of excitation and afferent inhibition in cuneothalamic relay neurons. J Neurophysiol 40:822–835

Jänig W, Spencer WA, Younkin SG (1979) Spatial and temporal features of the afferent inhibition in thalamocortical relay cells. J Neurophysiol 42:1450–1460

Jancso G, Kiraly E (1978) Distribution of chemosensitive primary sensory afferents in the central nervous system of the rat. J Comp Neurol 190:781–792

Jancso G, Kiraly E (1983) Cutaneous nerve regeneration in the rat: reinnervation of the denervated skin by regenerative but not by collateral sprouting. Neurosci Letts 36:133–137

Jeanmonod R, Rise FL, Van der Loos H (1977) Mouse somatosensory cortex: development of the alterations in the barrel field which are caused by injury to the vibrissal follicles. Neurosci Lett 6:151–156

Jeanmonod D, Rise FL, Van der Loos H (1981) Mouse somatosensory cortex: alterations in the barrel field following receptor injury at different early postnatal ages. Neuroscience 6:1503–1535

Jenkins WA, Merzenich MM, Zook JM, Fowler BC, Stryker MP (1982) The area 3b representation of the hand in owl monkeys reorganizes after induction of restricted cortical lesions. Soc Neurosci Abstr 8:141

Jenkins WM, Merzenich MM, Ochs MT (1984) Behaviourally controlled differential use of restricted hand surfaces induced changes in the cortical representation of the hand in area 3b of adult owl monkeys. Soc Neurosci Abstr 10:665

Jenkins WM, Merzenich MM, Ochs MT, Allard T, Guíc-Robles E (1990) Functional reorganization of primary somatosensory cortex in adult owl monkeys after behaviourally controlled tactile stimulation. Nature 63:82–104

Jenq C-B, Coggeshall RE (1984) Effects of sciatic nerve regeneration of axonal populations in tributary nerves. Brain Res 295:91–100

Jensen TS, Rasmussen P (1984) Amputation. In: Wall PD, Melzack R The Textbook of Pain. (eds) Churchill-Livingstone, Edinburgh, pp 402–412

Jessell TM, Iversen LL, Cuello AC (1978) Capsaicin-induced depletion of substance P from primary sensory neurons. Brain Res 152:183–188

Jessell T, Tsunoo A, Kanazawa I, Otsuka M (1979) Substance P: Depletion in the dorsal horn of rat spinal cord after section of the peripheral processes of primary sensory neurons. Brain Res 168:247–260

Johnson EM, Jr, Gorin PD, Brandeis LD, Pearson J (1980) Dorsal root ganglion neurons are destroyed by exposure *in utero* to maternal antibody to nerve growth factor. Science 210:916–918

Johnson FH (1952) Microelectrode studies on the cuneate and gracile nuclei of the cat. Amer J Physiol 171:737

Johnson JI, Welker WI, Pubols BH (1968) Somatotopic organization of raccoon dorsal column nuclei. J Comp Neurol 132:1–44

Johnson JI, Ostapoff EM, Warach S (1982) The anterior border zones of primary somatic sensory (SI) neocortex and their relation to cerebral convolutions, shown by micro mapping of peripheral projections to the region of the 4th forepaw digit representation in raccoons. Neurosci 7:915–936

Johnson JL, Hamilton TC, Hsung JC, Ulinski PS (1972) Gracile nucleus absent in adult opossums after leg removal in infancy. Brain Res 38:421–424

Johnson LR, Westrum LE (1980) Brain stem degeneration patterns following tooth extractions: visualization of dental and periodontal afferents. Brain Res 194:489–493

Jones EG (1967) Pattern of cortical and thalamic connections of the somatic sensory cortex. Nature 216:704–705

Jones EG (1975) Possible determinants of the degree of retrograde neuronal labelling with horseradish peroxidase. Brain Res 85:249–253

Jones EG (1983a) Distribution patterns of individual medial lemniscal axons in the ventrobasal complex of the monkey thalamus. J Comp Neurol 215:1–16

Jones EG (1983b) Lack of collateral thalamocortical projection to fields of the first somatic sensory cortex in monkeys. Exp Brain Res 52:375–384

Jones EG (1986) The thalamus. Plenum, New York

Jones EG, Friedman DP (1982) Projection pattern of functional components of thalamic ventrobasal complex on monkey somatosensory cortex. J. Neurophysiol 48:521–544

Jones EG, Leavitt RY (1974) Retrograde axonal transport and the demonstration of nonspecific projections to the cerebral cortex and striatum from thalamic intralaminar nuclei in the rat, cat and rhesus monkey. J Comp Neurol 154:349–378

Jones EG, Porter R (1980) What is area 3a? Brain Res Rev 2:1–43

Jones EG, Powell TPS (1969) The cortical projection of the ventroposterior nucleus of the thalamus in the cat. Brain Res 13:298–318

Jones EG, Coulter JD, Hendry SHC (1978) Intracortical connectivity of architecture fields in the somatic sensory, motor and parietal cortex of monkeys

Jones EG, Wise SP, Coulter JD (1979) Differential thalamic relationships of sensory-motor and parietal cortical fields in monkeys. J Comp Neurol 183:833–882

Jones EG, Friedman DP, Hendry SHC (1982) Thalamic basis of place- and modality specific columns in monkey somatosensory cortex: a correlative anatomical and physiological study. J Neurophysiol 48:545–568

Jones EG, Hendry SHC, Brandon C (1986) Cytochrome oxidase staining reveals functional organization of monkey somatosensory thalamus. Exp Brain Res 62:438–442

Jones MB, Vierck CJ, Graham RB (1973) Line-gap discrimination of the skin. Percept Mot Skills 36:563–570

Kaas J (1983) What, if anything, is SI? Organization of first somatosensory area of cortex. Physiol Rev 63:206–231

Kaas JH, Nelson RJ, Sur M, Lin C-S, Merzenich MM (1979) Multiple representations of the body within the primary somatosensory cortex of primates. Science 204:521–523

Kaas JH, Nelson RJ, Sur M, Dykes RW, Merzenich MM (1984) The somatotopic organization of the ventroposterior thalamus of the squirrel monkey (*Saimiri sciureus*). J Comp Neurol 226:111–140

Kadanoff D (1925) Untersuchungen uber die Regeneration der sensiblen Nervenendigungen nach Vertauschung verschieden innervierten Hautstucke. Arch Entw Mech Organ 106:249–278

Kalaska J, Pomeranz B (1979) Chronic paw denervation causes an age-dependent appearance of novel responses from forearm in "paw cortex" of kittens and adult cats. J Neurophysiol 42:618–633

Kalaska J, Pomeranz B (1982) Chronic peripheral nerve injuries alter the somatotopic organization of the cuneate nucleus in kittens. Brain Res 236:35–47

Kaneko T, Hicks TP (1988) Baclofen and gamma-aminobutyric acid differentially suppress the cutaneous responsiveness of primary somatosensory cortical neurons. Brain Res 443:360–366

Kelahan AM, Doetsch GS (1984) Time dependent changes in the functional organization of somatosensory cerebral cortex following digit amputation in adult raccoons. Somatosen Res 2:49–81

Kelahan AM, Ray RH, Carson LV, Massey CE, Doetsch G (1981) Functional reorganization of adult raccoon somatosensory cerebral cortex following neonatal digit amputation. Brain Res 223:152–159

Keller JH, Hand PJ (1970) Dorsal root projections to nucleus cuneatus of the cat. Brain Res 20:1–17

Kelly JG, Krnjevic K (1969) The action of glycine on cortical neurons. Exp Brain Res 9:155–163

Kenins P (1981) Identification of the unmyelinated sensory nerves which evoke plasma extravasation in response to antidromic stimulation. Neurosci Lett 25:137–141

Kenshalo DR Jr, Leonard RB, Chung JM, Willis WD (1979) Responses of primate spinothalamic neurons to graded and to repeated noxious heat stimuli. J Neurophysiol 42:1370–1389

Kenshalo DR, Leonard RB Chung JM, Willis WD (1982) Facilitation of the responses of primate spinothalamic cells to cold and to tactile stimuli by noxious heating of the skin. Pain 12:141–152

Kerr FWL (1972) The potential of cervical primary afferents to sprout in the spinal nucleus of V following long term trigeminal denervation. Brain Res 43:527–546

Kerr FWL (1975) Neuroplasticity of primary afferents in the neo-natal cat and some results of early deafferentation of the trigeminal spinal nucleus. J Comp Neurol 163:305–328

Kerr FWL, Kruger L, Schwassman HO, Stern R (1968) Somatotopic organization of mechanoreceptor units in the trigeminal nuclear complex of the macacque. J Comp Neurol 134:127–144

Killackey HP (1973) Anatomical evidence for cortical subdivisions based on vertically discrete thalamic projections from the ventral posterior nucleus to cortical barrels in the rat. Brain Res 51:326–331

Killackey HP (1980) Pattern formation in the trigeminal system of the rat. Trends Neuroscience 3:303–305

Killackey HP (1985) Intrinsic order in the developing rat trigeminal system. In: Development, organization and processing in somatosensory pathways. Liss, New York, pp 43–51

Killackey HP, Belford GR (1979) The formation of afferent patterns in the somatosensory cortex of the neonatal rat. J Comp Neurol 183:285–304

Killackey HP, Belford GR (1980) Central correlates of peripheral alterations in the trigeminal system of the rat. Brain Res 183:205–210

Killackey HP, Dawson DR (1989) Expansion of the central hindpaw representation following fetal forelimb removal in the rat. European J. Neuroscience 1:210–221

Killackey HP, Leshin S (1975) The organization of specific thalamocortical projections to the posteriomedial barrel subfield of the rat somatosensory cortex. Brain Res 86:469–472

Killackey HP, Shinder A (1981) Central correlates of peripheral pattern alterations in the trigeminal system of the rat. II. The effect of nerve section. Dev Brain Res 1:121–126

Killackey HP, Belford G, Ryugo R, Ryugo DK (1976) Anomalous organization of thalamocortical projections consequent to vibrissae removal in the newborn rat and mouse. Brain Res 104:09–315

Killackey HP, Ivy GO, Cunningham TJ (1978) Anomalous organization of SmI somatotopic map consequent to vibrissae removal in the newborn rat. Brain Res 155:136–140, J Comp Neurol 219:384–419

Kinnman E (1987) Collateral sprouting of sensory axons in the hairy skin of the trunk: a morphological study in adult rats. Brain Res 414:385–389

Kinnman E, Aldskogius H (1986) Collateral sprouting of sensory axons in the glabrous skin of the hindpaw after chronic sciatic nerve lesion in adult and neonatal rats: a morphological study. Brain Res 377:73–82

Kirk EJ (1974) Impulses in dorsal spinal nerve rootlets in cats and rabbits arising from dorsal root ganglia isolated from the periphery. J Comp Neurol 155:165–176

Kjerulf TD, O'Neal JT, Calvin WH, Loeser JD, Westrum LE (1973) Deafferentation effects in lateral cuneate nucleus of the cat: correlation of structural alternations with firing pattern changes. Exp Neurol 39:86–102

Knyihar E, Csillik B (1976) Effect of peripheral axotomy on the fine structure and histochemistry of the Rolando substance: degenerative atrophy of central processes of pseudounipolar cells. Exp Brain Res 26:73–87

Knyihar E, Gerebtzoff MA (1973) Extra-lysosomal localization of acid phosphatase in the spinal cord of the rat. Exp Brain Res 18:383–395

Knyihar E, Csillik B, Rakic P (1978) Transient synapses in the embryonic primate spinal cord. Science 202:1206–1209

Knyihar-Csillik, Bezzegh A, Boti S, Csillik B (1986) Thiamine monophosphatase: a genuine marker for transganglionic regulation of primary sensory neurons. J Histochem Cytochem 34:363–371

Koerber HR, Brown PB (1980) Projections of two hindlimb cutaneous nerves to cat dorsal horn. J Neurophysiol 44:259–269

Koerber HR, Brown PB (1982) Somatotopic organization of hindlimb cutaneous nerve projections to cat dorsal horn. J Neurophysiol 48:481–489

Koerber HR, Mendell LM (1988) Functional specialization of central projections from identified primary afferent fibers. J Neurophysiol 60:1597–1614

Koerber HR, Seymour AW, Mendell LM (1989) Mismatches between peripheral receptor type and central projections after peripheral nerve regeneration. Neurosci Lett 99:67–72

Korenman EMD, Devor M (1981) Ectopic adrenergic sensitivity in damaged peripheral nerve axons in the rat. Exp Neurol 72:63–81

Korthals JK, Wisniewsky HM, Ghetti B, Cook RD (1974) The fate of the axon and its terminal in the Pacinian corpuscle following sciatic nerve section. J Neurocytol 3:385–403

Kosar E, Hand PJ (1981) First somatosensory cortical column and associated neuronal clusters of nucleus ventralis posteriolateralis of the cat: an anatomical demonstration. J Comp Neurol 198:515–539

Kossut M, Hand P (1984a) Early development changes in cortical representation of C3 vibrissa following neonatal denervation of surrounding vibrissa receptors: a 2-deoxyglucose study in the rat. Neurosci Lett 46:7–12

Kossut M, Hand P (1984b) The development of the vibrissal cortical column: a 2-deoxyglucose study in the rat. Neurosci Lett 46:1–6

Krishnamurti A, Kanasungtheram R, Vij S (1973) Failure of reinnervation of Pacinian corpuscle after nerve crush: an electron microscopic study. Acta Neuropathol (Berlin) 23:338–341

Krnjevic K (1974) Chemical nature of synaptic transmission in vertebrates. Physiol Rev 54:418–540

Krnjevic K, Phillis JW (1963a) Iontophoretic studies of neurons in the mammalian cerebral cortex. J Physiol (Lond) 165:274–304

Krnjevic K, Phillis JW (1963b) Acetylcholine sensitive cells in the cerebral cortex. J Physiol (Lond) 166:296–327

Krnjevic J and Phillis JW (1963c) Pharmacological properties of acetylcholine sensitive cells in the cerebral cortex. J. Physiol (Lond) 166:328–350

Krnjevic K, Randic M, Straughan DW (1966) Pharmacology of cortical inhibition. J Physiol (Lond) 184:78–105

Krnjevic K, Pumain R, Renaud L (1971) The mechanism of excitation by acetylcholine in the cerebral cortex. J Physiol (Lond) 215:247–268

Kruger L, Siminoff R, Witkovsky P (1961) Single neuron analysis of dorsal column nuclei and spinal nucleus of trigeminal in cat. J Neurophysiol 24:333–349

Kuhn RA (1949) Topographical pattern of cutaneous sensibility in the dorsal column nuclei of the cat. Trans Am Neurol Assoc 74:227–230

Kunze WAA, Wilson P, Snow PJ (1987) Response of lumbar spinocervical tract cells to natural and electrical stimulation of the hindlimb footpads in cats. Neurosci Lett 75:253–258

Kuypers HGJM, Tuerk JD (1964) The distribution of the cortical fibres within the nuclei cuneatus and gracilis in the cat. J Anat 98:143–162

LaMotte CL, Kapadia SE (1987) Deafferentation-induced alterations in the rat dorsal horn: II. Effects of selective poisoning by pronase of the central processes of a peripheral nerve. J Comp Neurol 266:198–208

LaMotte CL, Kapadia SE, Kocol CM (1989) Deafferentation-induced expansion of saphenous terminal field labelling in the adult rat dorsal horn following pronase injection of the sciatic nerve. J Comp Neurol 288:311–325

Lamour Y and Dykes RW (1988) Somatosensory neurons in partially deafferented rat hindlimb granular cortex subsequent to transection of the sciatic nerve: effects of glutamate and acetylcholine. Brain Res. 449:18–33

Lamour Y, Dutar P, Jobert A (1982) Excitatory effect of acetylcholine on different types of neurons in the first somatosensory neocortex of the rat laminar distribution and pharmacological characteristics. Neuroscience 7:1483–1494

Lamour Y, Willer JC, Guilbaud G (1983) Rat somatosensory (SmI) cortex: I. Characteristics of neuronal responses to noxious stimulation and comparison with responses to non-noxious stimulation. Exp Brain Res 49:35–45

Land PW, Simons DL (1985a) Cytochrome oxidase staining in the rat SmI barrel cortex. J Comp Neurol 238:225–235

Land PW, Simons DJ (1985b) Metabolic activity in SmI cortical barrels of adult rats is dependent on patterned sensory stimulation of the mystacial vibrissae. Brain Res 341:189–194

Land PW, Simons DJ, Buffer SA Jr (1986) Specificity of thalamocortical connections in the rat somatosensory system. Soc Neurosci Abst 12:1434

Landry P, Deschenes M (1981) Intracortical arborizations and receptive fields of identified ventrobasal thalamocortical afferents to the primary somàtic sensory cortex in the cat. J Comp Neurol 199:345–371

Landry P, Villemure J and Deschenes M (1982) Geometry and orientation of thalamocortial arborizations in the cat somatosensory cortex as revealed by computer reconstruction. Brain Res 237:222–226

Landry P, Diadori P, Leclerc S and Dykes RW (1987) Morphological and electrophysiological characteristics of somatosensory thalamocortical axons studied with intra-axonal staining and recording in the cat. Exp Brain Res 65:317–330

Langford LA, Coggeshall RE (1981) Branching of sensory axons in the peripheral nerve of the rat. J Comp Neurol 203:745–750

Laskin SE, Spencer WA (1979a) Cutaneous masking I. Psychophysical observations on interactions of multipoint stimuli in man. J Neurophysiol 42:1048–1060

Laskin SE, Spencer WA (1979b) Cutaneous masking II. Geometry of excitatory and inhibitory receptive fields of single units in somatosensory cortex of the cat. J Neurophysiol 42:1061–1082

La Velle A, La Velle FW (1958) The nucleolar apparatus and neuronal reactivity to injury during development. J Exp Zool 137:285–315

Lawson SN (1979) Dorsal root ganglion neurones and dorsal roots: effects of neonatal capsaicin. In: Brown AG, Rethelyi M (eds) Spinal cord sensation. Scottish Academic, Edinburgh, pp 57–58

Lawson SN, Caddy KWT, Biscoe, TJ (1974) Development of rat dorsal root ganglion neurons: studies of cell birthdays and changes in mean cell diameter. Cell Tiss Res 153:399–413

Lee FC (1936) A study of the Pacinian corpuscle. J Comp Neurol 64:497–522

Lee KJ, Woolsey TA (1975) A proportional relationship between peripheral innervation density and cortical neuron number in the somatosensory system of the mouse. Brain Res 99:349–353

Leonard MH (1973) Return of skin sensation in children without repair of nerve. Clin Orthop Rel Res 95:273–277

Levi-Montalcini R, Angeletti PU (1963) Essential role of the nerve growth factor in the survival and maintenance of dissociated sensory and sympathetic embryonic nerve cells in vitro. Dev Biol 7:653–659

Levin BE, Craik RL, Hand PJ (1988) The role of norepinephrine in adult rat somatosensory (SmI) cortical metabolism and plasticity. Brain Res 443:261–271

Levitt M, Levitt J (1968a) Sensory hindlimb representation in SmI cortex of the cat. A unit analysis. Exp Neurol 22:259–275

Levitt M, Levitt J (1986b) Sensory hindlimb representation in SmI cortex of the cat after spinal tractotomies. Exp Neurol 22:276–302

Libet B (1965) Cortical activation in conscious and unconscious experience. In: Perspectives in biology and medicine, pp 77–86

Lieberman AR (1971) The axon reaction: a review of the pricipal features of perikaryal responses to axon injury. Int Rev Neurobiol 14:49–124

Lieberman AR (1974) Some factors affecting retrograde neuronal responses to axonal lesions. In: Bellairs R and Gray EG (eds) Essays on the Nervous System. Clarendon Press, Oxford pp 71–105

Light AR, Durkovic RG (1984) Features of laminar and somatotopic organization of lumbar spinal cord units receiving cutaneous inputs from hindlimb receptive fields. J Physiol (Lond) 52:449–457

Light AR, Kavookjian AM (1988) Morphology and ultrastructure of physiologically identified substantia gelatinosa (lamina II) neurons with axons that terminate in deeper dorsal horn laminae (III–V). J Comp Neurol 267:172–189

Light AR, Perl ER (1979) Spinal termination of functionally identified primary afferent neurons with slowly conducting myelinated fibres. J Comp Neurol 186:133–150

Light AR, Trevino DL, Perl ER (1979) Morphological features of functionally defined neurons in the marginal zone and substantia gelatinosa of the spinal dorsal horn. J Comp Neurol 186:151–172

Lin, C-S, Merzenich MM, Sur M, Kaas JH (1979) Connections of areas 3b and 1 of the parietal somatosensory strip with the ventroposterior nucleus in the owl monkey (*Aotus trivirgatus*). J Comp Neurol 180:355–372

Lisney SJ (1982) Receptive fields of spinal cord dorsal horn neurons in cats shortly after peripheral nerve injury. J Physiol (Lond) 325:76P

Lisney SJ (1983) Changes in the somatotopic organization of the cat lumbar spinal cord following peripheral nerve transection and regeneration. Brain Res 259:31–39
Liu CN and Chambers WW (1958) Intraspinal sprouting of dorsal root axons. Arch Neurol Psychiat 79:46–61
Livingston SK (1947) Evidence of active invasion of denervated areas by sensory fibres from neighbouring nerves in man. J Neurosurg 4:140–145
Loe PR, Whitsel BL, Dreyer DA, Metz CB (1977) Body representation in ventrobasal thalamus of macaque: A single unit analysis. J Neurophysiol 40:1339–1355
Loeser JD, Ward AA Jr (1967) Some effects of deafferentation on neurons of the cat spinal cord. Arch Neurol 17:629–636
Loewy AD (1973) Transneuronal changes in the gracile nucleus. J Comp Neurol 147:497–510
Lund RD, Webster KE (1967a) Thalamic afferents from the dorsal column nuclei: an experimental anatomical study in the rat. J Comp Neurol 130:301–312
Lund RD, Webster KE (1967b) Thalamic afferents from the spinal cord and trigeminal nuclei: an experimental anatomical study in the rat. J Comp Neurol 130:313–328
Lundberg JM, Rokaeus A, Hokfelt T, Rosell S, Brown M, Goldstein M (1982) Neurotensin-like immunoreactivity in the preganglionic sympathetic nerves and in the adrenal medulla of the cat. Acta Physiol Scand 114:153–155
Luria AM (1962) Higher cortical functions in man, 2nd edn, Basic, New York
Lynn B (1984) Effect of neonatal treatment with capsaicin on the numbers and properties of cutaneous afferent units from the hairy skin of the rat. Brain Res 322:255–260
Lynn B, Carpenter SE (1982) Primary afferent units from the hairy skin of the rat hindlimb. Brain res 238:29–43
Ma PM, Woolsey TA (1984) Cytoarchitectonic correlates of the vibrissae in the medullary trigeminal complex of the mouse. Brain Res 306:374–379
MacConnell WP, Eurman DW, Kaplan NO (1982). The action of ricin A-chain on ribosomes. Biochem Biophys Res Comm 108:809–814
Machne X, Unna KRW (1963) Actions at the central nervous system. In: Eichler O, Farah A (eds) Cholinesterase and anticholinesterase agents. Springer Berlin Heidelberg New York, pp 697–700 (Handbook of experimental pharmacology)
Mackel R, Kunesch E, Waldohr F, Struppler A (1983) Reinnervation of mechanoreceptors in the human glabrous skin following peripheral nerve repair. Brain Res 268:49–65
MacKenzie RA, Burke D, Skuse NF, Lethlean AK (1975) Fibre function and perception during cutaneous nerve block. J Neurol Neurosurg Psychiatry 38:865–873
Mann MD (1979) Sets of neurons in somatic cerebral cortex of the cat and their ontogeny. Brain Res Rev 1:3–45
Mannen H (1975) Reconstruction of axonal trajectory of individual neurons in the spinal cord using Golgi-stained serial sections. J Comp Neurol 159:357–374
Markus M, Pomeranz B, Krushelnycky D (1984) Spread of saphenous somatotopic projection map in spinal cord and hypersensitivity of the foot after chronic sciatic denervation in adult rat. Brain Res 296:27–39
Markus H, Pomeranz B (1987) Saphenous has weak ineffective synapses in sciatic territory of rat spinal cord: electrical stimulation of the saphenous or application of drugs reveal these somatotopically inappropriate synapses. Brain Res 416:315–321
Matsushita M (1969) Some aspects of the interneuronal connections in cat's spinal gray matter. J Comp Neurol 136:57–80
Matthews MA, Kruger L (1973) Electron microscopy of non-neuronal cellular changes accompanying neural degeneration in thalamic nuclei of the rabbit. II. Reactive elements within the neuropil. J Comp Neurol 148:313–346

Mattio TG, Rosenquist TH, Kirby ML (1981) Appearance of acid phosphatase in neonatal rat substantia gelatinosa. Exp Brain Res 41:411–413

Maxwell DJ, Rethelyi M (1987) Ultrastructure and synaptic connection of cutaneous afferent fibres in the spinal cord. Neurosci 10:117–123

Maxwell DJ, Fyffe REW, Rethelyi M (1983) Morphological properties of physiologically characterized lamina III neurones in the cat spinal cord. Neurosci 10:1–22

McAllister JP, Das GD (1977) Neurogenesis in the epithalamus, dorsal thalamus and ventral thalamus of the rat: an autoradiographic and cytological study. J Comp Neurol 172:647–686

McAllister JP, Wells J (1981) The structural organization of the ventral posteriolateral nucleus in the rat. J Comp Neurol 197:271–301

McComas AJ (1963) Responses of the rat dorsal column system to mechanical stimulation of the hindpaw. J Physiol (Lond) 166:435–448

McComas AJ, Sica REP, Banerjee S (1978) Long-term effects of partial limb amputation in man. J Neurol Neurosurg Psychiat 41:425–432

McCormice DA, Prince DA (1985) Two types of muscarinic response to acetylcholine in mammalian cortical neurons. Proc Natl Acad Sci USA 82:6344–6348

McCouch GP, Austin GM, Liu CN, Liu CY (1958) Sprouting as a cause of spasticity J Neurophysiol 21:205–216

McKenna TM, Whitsel BL, Dreyer DA, Metz CB (1981) Organization of cat anterior parietal cortex: relations among cytoarchitecture, single neuron functional properties and interhemispheric connectivity. J Neurophysiol 45:667–697

McKenna TM, Whitsel BL, Dreyer DA (1982) Anterior parietal cortical topographic organization in macaque monkey: a reevaluation. J Neurophysiol 48:289–318

McKinley PA, Swyter L (1986) Progressive change in SI cortex of kittens after spinal transection. Soc Neurosci Abstr 12:1050

McKinley PA, Jenkins WT, Smith JL, Merzenich MM (1987) Age dependent capacity for somatosensory cortex reorganization in chronic spinal cats. Dev Brain Res 31:136–139

McKinley PA, Kruger L (1988) Nonoverlapping thalamocortical connections to normal and deprived primary somatosensory cortex for similar forelimb receptive fields in chronic spinal cats. Somatosens Res 5:311–323

McKinley, PA, Smith, JL (1990) Age-dependent differences in reorganization of primary somatosensory cortex following low thoracic (T12) Spinal transection in cats. J Neurosci 10, 1429–1443

McMahon SB, Wall PD (1983a) A system of rat spinal cord lamina I cells projecting through the contralateral dorsolateral funiculus. J Comp Neurol 214:217–223

McMahon SB, Wall PD (1983b) Plasticity in the nucleus gracilis of the rat. Exp Neurol 80:195–207

McMahon SB, Wall PD (1984) Receptive fields of rat lamina I projection cells move to incorporate a nearby region of injury. Pain 19:235–247

McMahon SB, Wall PD (1987) Physiological evidence for branching of peripheral unmyelinated sensory afferent fibers in the rat. J Comp Neurol 261:130–136

McMahon SB, Wall PD, Granum SL, Webster KE (1984) The effects of capsaicin applied to peripheral nerves on responses of a group of lamina I cells in adult rats. J Comp Neurol 227:393–400

McNeill DI, Hulsebosch CE (1987) Intraspinal sprouting of rat primary afferents after deafferentation. Soc Neurosci Abstr 13:1392

Melzack R (1990) Phantom limbs and the concept of a neuromatrix. Trends Neurosci 13:88–92

Melzack R, Bridges JA (1971) Dorsal column contributions to motor behaviour. Exp Neurol 33:53–68

Melzer P, Welker E, Dorfl J, Van der Loos H (1986) Development of order in the whisker-to-barrel pathway of mouse: neuronal metabolic responses to whisker stimulation. Soc Neurosci Abstr 12:953
Mendell LM (1966) Physiological properties of unmyelinated fibre projection to the spinal cord. Exp Neurol 16:316–332
Mendell LM (1984) Modifiability of spinal synapses. Physiol Rev 64:260–324
Mendell LM, Wall PD (1965) Responses of single dorsal cord cells to peripheral cutaneous unmyelinated fibres. Nature 206:97–99
Mendell LM, Sassoon EM, Wall PD (1978) Properties of synaptic linkage from long ranging afferents onto dorsal horn neurones in normal and deafferented cats. J Physiol (Lond) 285:299–310
Menetrey D, Besson JM (1982) Electrophysiological characteristics of dorsal horn cells in rats with cutaneous inflammation resulting from chronic arthritis. Pain 13:343–364
Menetrey D, Giesler GJ, Besson JM (1977) An analysis of response properties of spinal cord dorsal horn neurones to non-noxious and noxious stimuli in the spinal cord. Exp Brain Res 27:15–33
Merrill EG, Wall PD (1972) Factors forming the edge of a receptive field: the presence of relatively ineffective afferent terminals. J Physiol (Lond) 226:825–846
Merrill EG, Wall PD (1978) Plasticity of connection in adult nervous system. In: Cotman CW (ed) Neuronal plasticity. Raven, New York, pp 97–111
Merzenich MM, Kaas JH, Sur M, Lin CS (1978) Double representation of the body surface within cytoarchitectonic areas 3b and 1 in "SI" in the owl monkey (Aotus trivirgatus). J Comp Neurol 181:41–74
Merzenich MM, Kaas JH, Wall J, Nelson RJ, Sur M, Felleman D (1983a) Topographic reorganization of somatosensory cortical areas 3b and 1 in adult monkeys following restricted deafferentation. Neuroscience 8:33–55
Merzenich MM, Kaas JH, Wall JT, Sur M, Nelson RJ, Felleman DJ (1983b) Progression of change following median nerve section in the cortical representation of the hand in areas 3b and 1 in adult owl and squirrel monkeys. Neuroscience 10:639–665
Merzenich MM, Nelson RJ, Stryker MP, Cynader MS, Schoppman A, Zook JM (1984) Somatosensory cortical map changes following digit amputation in adult monkeys. J Comp Neurol 224:591–605
Merzenich MM, Nelson RJ, Kaas JH, Stryker P, Jenkins WM, Zook JM, Cynader MS, Schoppman A (1987) Variability in hand surface representations in area 36 and 1 in adult own and squirrel monkeys. J Comp Neurol 258:281–296
Mesulam MM (1981) A cortical network for directed attention and unilateral neglect. Ann Neurol 10:309–325
Mesulam MM, Mufson EJ (1984) Neural inputs into the nucleus basalis of the substantia innominata in the Rhesus monkey. Brain 107:253–274
Metherate RS, Da Costa DCN, Herron P, Dykes RW (1986) A thalamic terminus of the lateral cervical nucleus: the lateral division of the posterior nuclear group. J Neurophysiol 56:1498–1520
Metherate RS, Tremblay N, Dykes RW (1987) Acetylcholine permits long term enhancement of neuronal responsiveness in cat primary somatosensory cortex. Neuroscience 22:75–81
Metzler J, Marks PS (1979). Functional changes in cat somatic sensory-motor cortex during short-term reversible epidural blocks. Brain Res 177:379–383
Meyer RA, Campbell JN, Raja SN (1984) Peripheral neural mechanisms of cutaneous hyperalgesia. In: Fields HL, Dubner R, Cervero F (eds) Advances in pain research and therapy, vol 9. Raven New York, pp 53–71

Meyers DER, Snow PJ (1984) Somatotopically inappropriate projections of single hair follicle afferent fibres to the cat spinal cord. J Physiol (Lond) 347:59–73

Meyers DER, Snow PJ (1986) Distribution of activity in the spinal terminations of single hair follicle afferent fibres to somatotopically identified regions of the cat spinal cord. J Neurophysiol 56:1022–1038

Meyers DER, Wilson P, Snow PJ (1984) Distribution of the central terminals of cutaneous primary afferents innervating a small skin patch: the existence of somatotopically inappropriate projections. Neurosci Letts 44:179–185

Micevych PE, Yaksh TL, Szolcsanyi J (1983). Effect of intrathecal capsaicin analogues on the immunofluorescence of peptides and serotonin in the dorsal horn in rats. Neurosci 8:123–131

Micevych PE, Rodin BE, Kruger L (1986) The controversial nature of the evidence for neuroplasticity of afferent axons in the spinal cord. In: Yaksh TK (ed) Spinal afferent processing. Plenum, New York, pp 417–443

Micevych PE, Stroink A, Yaksh T, Go UL (1986) Immunochemical studies of substance P and cholecystokinin octapeptide recovery in dorsal horn following unilateral lumbosacral ganglionectomy. Somatosens Res 3:239–260

Millar J, Basbaum AI (1975) Topography of the projection of the body surface of the cat to cuneate and gracile nuclei. Exp Neurol 49:281–290

Millar J, Basbaum AI, Wall PD (1976) Restructuring of the somatotopic map and appearance of abnormal neuronal activity in the gracile nucleus after partial deafferentation. Exp Neurol 50:658–672

Miller MS, Buck SH, Sipes IG, Yamamura HI, Burks TF (1982) Brain Res 250:193–196

Miyahara JT, Esplin DW, Zablocka B (1966) Differential effects of depressant drugs on presynaptic inhibition. J Pharmacol Exp Ther 154:118–127

Molander C, Grant G (1986) Laminar distribution and somatotopic organization of primary afferent fibres from hindlimb nerves in the dorsal horn. A study by transganglionic transport of horseradish peroxidase in the rat. Neurosci 19:297–312

Molander C, Kinnman E, Aldskogius H (1988) Expansion of spinal cord primary sensory afferent projection following combined sciatic nerve resection and saphenous nerve crush: a horseradish peroxidase study in the adult rat. J Comp Neurol 276:436–441

Molinari HH (1982) The cutaneous sensitivity of units in laminae VII and VIII of the cat. Brain Res 234:165–169

Moradian GP, Rustioni A (1977) Transganglionic degeneration in the dorsal horn and dorsal column nuclei of adult rats. Anat Rec 187:660

Morgan C, Sharp FR (1986) Mystacial vibrissae and common fur brainstem sensory pathways in the rat. Soc Neurosci Abstr 12:954

Morris JH, Hudson AR, Weddell G (1972) A study of degeneration and regeneration in the divided rat sciatic nerve based on electron microscopy, I–IV. Z Zellforsch 124:76–203

Morse RW, Adkins RJ, Towe AL (1965) Population and modality characteristics in the coronal region of somatosensory area I of the cat. Exp Neurol 11:419–440

Mountcastle VB (1957) Modality and topographic properties of single neurons of cat's somatic sensory cortex. J Neurophysiol 20:408–434

Mountcastle VB, Darian-Smith I (1968) Neural mechanisms in somethesia. In: Mountcastle VB (ed) Medical physiology vol 2, 12th Ed. Moseby, St. Louis, pp 1372–1423

Mountcastle VB, Henneman E (1949) Pattern of tactile representation in thalamus of cat. J Neurophysiol 12:85–100

Mountcastle VB, Davies RW, Berman AL (1957) Response properties of neurons of cat's somatic sensory cortex to peripheral stimuli. J Neurophysiol 20:374–407

Murray EA, Mishkin M (1984) Relative contributions of SmII and area 5 to tactile discrimination of monkeys. Behav Brain Res 11:67–84

Murray M, Edwards MAE (1982) A quantitative study of the reinnervation of the goldfish optic tectum following optic nerve crush. J Comp Neurol 209:363–373
Murray M, Goldberger ME (1974) Restitution of function and collateral sprouting in the cat spinal cord: the partially hemisected animal. J Comp Neurol 158:19–36
Murray M, Goldberger ME (1986) Replacement of synaptic terminals in lamina II and Clarke's Nucleus after unilateral lumbosacral dorsal rhizotomy in adult cats. J Neurosci 6:3205–3217
Nagy JI, Hunt SP (1982) Fluoride-resistant acid phosphatase-containing neurons in dorsal root ganglia are separate from those containing substance P or somatostatin. Neuroscience 7:89–97
Nagy JI, Hunt SP (1983) The termination of primary afferents within the rat dorsal horn: evidence for rearrangement following capsaicin treatment. J Comp Neurol 218:145–158
Nagy JI, Iversen LL, Goedert, Chapman D, Hunt SP (1983) Dose-dependent effects of capsaicin on primary sensory neurons in the neonatal rat. J Neurosci 3:399–406
Nakahama H, Nishioka S, Otsuka T (1966) Excitation and inhibition in ventrobasal thalamic neurons before and after cutaneous input deprivation. Prog Brain Res 21A:180–196
Narayanan CH, Fox MW, Hamburger V (1971) Prenatal development of spontaneous and evoked activity in the rat. Behaviour 40:100–134
Nelson RJ, Kaas JH (1981) Connections of the ventroposterior nucleus of the thalamus with the body surface representations in cortical areas 3b and 1 of the cynomologous macaque (Macaca fascicularis). J Comp Neurol 199:29–64
Nelson RJ, Sur M, Felleman DJ, Kaas J (1980) Representations of the body surface in post central parietal cortex of Macaca fascicularis. J Comp Neurol 192:611–643
Nijensohn DE, Kerr FWL (1975) The ascending projections of the dorsolateral funiculus of the spinal cord in the primate. J Comp Neurol 161:459–470
Nixon BJ, Doucette R, Jackson PC, Diamond J (1984) Impulse activity evokes precocious sprouting of nociceptive nerves into denervated skin. Somatosens Res 2:97–126
Noble R, Short AD (1989) Spatial spread of in-field afferent inhibition in the cat's spinocervical tract. J Physiol (Lond) 413:107–118
Nord SG (1967) Somatotopic organization in the spinal trigeminal nucleus, the dorsal column nuclei and related structures in the rat. J Comp Neurol 130:343–355
Nord SG (1968) Receptor field characteristics of single cells in the rat spinal trigeminal complex. Exp Neurol 21:236–243
Nornes HO, Das GD (1972) Temporal pattern of neurogenesis in spinal cord: cytoarchitecture and directed growth of axons. Proc Natl Acad Sci USA 69:1962–1966
Nornes HO, Das GD (1974) Temporal pattern of neurogenesis in spinal cord of rat. I. An autoradiographic study – time and sites of origin and migration and settling patterns of neuroblasts. Brain Res 73:121–138
Nornes HO, Morita M (1979) Time of origin of the neurone in the caudal brain stem of rat: an autoradiographic study. Dev Neurosci 2:101–114
Nussbaumer JC, Van der Loos H (1985) An electrophysiological and anatomical study of projections to the mouse cortical barrelfield and its surroundings. J Neurophysiol 53:686–698
Nussbaumer JC, Wall PD (1985) Expansion of receptive fields in the mouse cortical barrelfield after administration of capsaicin to neonates or local application on the infraorbital nerve in adults. Brain Res 360:1–9
Nyberg G, Blomqvist A (1982) The termination of forelimb nerves in the feline cuneate nucleus demonstrated by the transganglionic transport method. Brain Res 248:209–222
Nyberg G, Blomqvist A (1985) The somatotopic organization of forelimb cutaneous nerves in the brachial dorsal horn: an anatomical study in the cat. J Comp Neurol 242:28–39

Ochoa J (1982) Pain in local nerve lesions. In: Culp WJ, Ochoa J (eds) Abnormal nerves and muscles as impulse generators. Oxford University Press, New York, pp 568–587

Ochoa JL, Torebjork E, Marchettini P, Sivak M (1985) Mechanisms of neuropathic pain: cumulative observations, new experiments, and further speculation. In: Fields HL, Dubner R, Cervero F (eds) Advances in pain research and therapy, vol 9. Raven, New York, pp 431–450

Odutola AB (1977) Patterns and fields of dorsal column fiber terminals in the cuneogracile nuclei of the rat. Exp Neurol 57:112–120

Oester YT, Davis L (1956) Recovery of sensory function. In: Woodhall B, Beebe GW (eds) peripheral nerve regeneration: a follow up study of 3,656 World War II injuries. US Govt. Printing Office, Washington DC, pp 241–347

Ohara PT, Liberman AR, Hunt SP, Wu J-Y (1983) Neural elements containing glutamic decarboxylase (GAD) in the dorsal lateral geniculate nucleus of the rat: immunohistochemical studies by light and electron microscopy. Neuroscience 8:189–212

Olavarria J, Van Sluyters RC, Killackey HP (1984) Evidence for the complementary organization of callosal and thalamic connections within rat somatosensory cortex. Brain Res 291:354–368

Onne L (1962) Recovery of sensibility and sudomotor activity in the hand after nerve suture. Acta Chir Scand [Suppl] 300:1–69

Osborne WA, Kilvington B (1909) Axon bifurcation in regenerated nerves. II. J Physiol (Lond) 38:268–275

Ostapoff EM, Johnson JI, Albright BC (1983) Mechanosensory projections to cuneate, gracile, and external cuneate nuclei in a tree squirrel (fox squirrel, *Sciurus niger*). Neuroscience 9:107–127

Otten U, Lorez HP, Businger F (1983) Nerve Growth Factor antagonizes the neurotoxic action of capsaicin on primary sensory neurones. Nature 301:515–517

Palmer P (1965) Ultrastructural alterations of Merkel cells following denervation. Anat Rec 151:396–397

Pasternak JF, Woolsey TA (1975) The number, size and spatial distribution of neurons in lamina IV of the mouse SmI neocortex. J Comp Neurol 160:291–306

Patel N, Poo MM (1982) Orientation of neurite growth by extracellular electric fields. J Neurosci 2:483–496

Patrizi G, Munger BL (1966) The ultrastructure and innervation of rat vibrissae. J Comp Neurol 126:423–436

Patterson JT, Head PA, McNeill DL, Chung K, Coggeshall RE (1989) Ascending unmyelinated primary afferent fibres in the dorsal funiculus. J Comp Neurol 290:384–390

Paul RL, Goodman H, Merzenich MM (1972a) Alterations in mechanoreceptor inputs to Brodmann's areas 1 and 3 of the post-central hand area of Macaca mulatta after nerve section and regeneration. Brain Res 39:1–19

Paul RL, Merzenich MM, Goodman H (1972b) Representation of slowly and rapidly adapting cutaneous mechanoreceptors of the hand in Brodman's area 3 and 1 of Macaca mulatta. Brain Res 36:229–249

Penfield W, Boldney E (1937) Somatic motor and sensory representation in the cerebral cortex of man as studied by electrical stimulation. Brain 60:389–443

Penfield W, Rasmussen T (1950) The cerebral cortex of man: a clinical study of localization of function. MacMillan, New York

Penfield W (1967) The excitable cortex in conscious man. The Sherrington lectures V. Liverpool University Press, Liverpool

Perl ER (1968) Myelinated afferent fibres innervating the primate skin and their response to noxious stimuli. J Physiol (Lond) 197:593–615

Perl ER (1985) Unraveling the story of pain. In: Fields HL, Dubner R, Cervero F (eds) Advances in pain research and theraphy, vol 9. Raven, New York, pp 1–30
Perl ER, Whitlock DG, Gentry JR (1962) Cutaneous projection to second order neurons of the dorsal column system. J Neurophysiol 25:337–358
Peschanski M (1984) Trigeminal afferents to the diencephalon in the rat. Neurosci 12:465–487
Peschanski M, Besson JM (1986) Structural alteration and possible growth of afferents after kainate lesion in the adult rat thalamus. J Comp Neurol 257:1–20
Peschanski M, Mantyh PW, Besson JM (1983) Spinal afferents to the ventrobasal thalamic complex in the rat: an anatomical study using wheat-germ agglutinin conjugated to horseradish peroxidase. Brain Res 278:240–244
Peschanski M, Lee CL, Ralston HJ (1984) The structural organization of the ventrobasal complex of the rat as revealed by the analysis of physiologically characterized neurons injected intracellularly with horseradish peroxidase. Brain Res 297:63–74
Peters A, Jones EG (1984) Classification of cortical neurons. In: Peters A, Jones EG (eds) Cellular components of the cerebral cortex. Plenum, New York, pp 107–121 (Cerebral Cortex, vol 1)
Petit D (1972) Postsynaptic fibres in the dorsal columns and their relay in the nucleus gracilis. Brain Res 48:380–384
Petit D, Burgess PR (1968) Dorsal column projection of receptors in cat hairy skin supplied by myelinated fibres. J Neurophysiol 31:849–855
Petrovicky P, Druga R (1972) Peculiarities of the cytoarchitectonics and some afferent systems of the parietal cortex. Folia Morphol 20:161–163
Peyronnard JM, Charron L (1983) Decreased horseradish peroxidase labelling in deafferented spinal motoneurons of the rat. Brain Res 275:203–214
Peyronnard JM, Charron L, Lavoie J, Messier JP (1986) Differences in horseradish peroxidase labelling of sensory, motor, and sympathetic neurons following chronic axotomy of the rat sural nerve. Brain Res 364:137–150
Pickel VM, Sumal KK, Miller RJ (1982) Early prenatal development of substance P and enkephalin-containing neurons in the rat. J Comp Neurol 210:411–422
Pidoux B, Verley R (1978) Projections on the cortical somatic SI barrel subfield from ipsilateral vibrissae in adult rodents. EEG Clin Neurophysiol 46:715–726
Pidoux B, Verley R, Farkas E, Scherrer J (1979) Projections of the common fur of the muzzle upon the cortical area for mystacial vibrissae in rats dewhiskered since birth. Neurosci Lett 11:301–306
Pidoux B, Diebler MF, Sary C, Farkas E, Verley R (1980) Cortical organization of the postero-medial barrel-subfield in mice and its reorganization after destruction of vibrissal follicles after birth. Neuropath Appl Neurobiol 6:93–107
Pierau FK, Taylor DC, Abel W, Friedrich B (1982) Dichotomizing peripheral fibres revealed by intracellular recording from rat sensory neurones. Neurosci Lett 31:123–128
Poeck K (1964) Phantoms following amputation in early childhood and in congenital absence of limbs. Cortex 1:269–275
Poggio GF, Mountcastle VB (1960) A study of the functional contributions of the lemniscal and spinothalamic systems to somatic sensibility. Central nervous mechanisms in pain. Johns Hopkins Med J 106:266–316
Poggio GF, Mountcastle VB (1963) The functional properties of ventrobasal thalamic neurons studied in unanaesthetized monkeys. J Neurophysiol 26:775–806
Pollin B, Albe-Fessard D (1979) Organization of somatic thalamus in monkeys with and without section of dorsal spinal tracts. Brain Res 173:431–449
Pomeranz B, Wall PD, Weber WV (1968) Cord cells responding to fine myelinated afferents from viscera, muscle and skin. J Physiol (Lond) 199:511–523

Pomeranz B, Mullen M, Markus H (1984) Effect of applied electrical fields on sprouting of intact saphenous nerve in adult rat. Brain Res 303:331–336

Pons TP, Kaas JH (1985) Connections of area 2 of somatosensory cortex with the anterior pulvinar and subdivisions of the ventroposterior complex in macaque monkeys. J Comp Neurol 240:16–36

Pons TP, Garraghty PE, Cusick CG, Kaas JH (1985a) The somatotopic organization of area 2 in macaque monkeys. J Comp Neurol 241:445–466

Pons TP, Garraghty PE, Cusick CG, Kass JH (1985b) A sequential representation of the occiput, arm, forearm and hand across the rostrocaudal dimension of areas 1, 2 and 5 in Macaque monkeys. Brain Res 335:350–353

Powell TPS, Mountcastle VB (1959) The cytoarchitecture of the postcentral gyrus of the monkey Macaca mulatta. Johns Hopkins Med J 105:108–120

Price DD (1984) Dorsal horn mechanisms of pain. In: Davidoff RA (ed) Anatomy and physiology. Dekker, New York, pp 751–777 (Handbook of the Spinal Cord, vols 2, 3)

Price DD, Mayer DJ (1974) Physiological laminar organization of the dorsal horn of *M. Mulatta*. Brain Res 79:321–325

Priestley JV, Somogyi P, Cuello AC (1982) Immunocytochemical localization of substance P in the spinal trigeminal nucleus of the rat: a light and electron microscopic study. J Comp Neurol 211:31–49

Princince TJ, Donoghue JP (1986) Receptive field architecture in the whisker representation of rat somatosensory cortex. Soc Neurosci Abstr 12:1433

Proshansky E, Egger MD (1977) Dendritic spread of dorsal horn neurons in cats. Exp Brain Res 28:153–166

Pubols BH, Warren S (1985) Information processing within the raccoon somatosensory system. In: Rowe M, Willis WD (eds) Development, organization and processing in somatosensory pathways. Liss, New York, pp 183–190

Pubols LM (1984) The boundary of proximal hindlimb representation in the dorsal horn following peripheral nerve lesion in cats: a reevaluation of plasticity in the somatotopic map. Somatosen Res 2:19–32

Pubols LM, Brenowitz GL (1982) Maintenance of dorsal horn somatotopic organization and increased high threshold response after single root or spared root deafferentation in cats. J Neurophysiol 47:103–112

Pubols LM, Goldberger ME (1980) Recovery of function in dorsal horn following partial deafferentation. J Neurophysiol 43:102–117

Pubols LM, Leroy RF (1977) Orientation detectors in the primary somatosensory neocortex of the raccoon. Brain Res 129:61–74

Pubols LM, Foglesong ME, Vahle-Hinz C (1986) Electrical stimulation reveals relatively ineffective sural nerve projections to dorsal horn neurons in the cat. Brain Res 371:109–122

Purves D, Hadley RD (1985) Changes in the dendritic branching of adult mammalian neurons revealed by repeated imaging in situ. Nature 315:404–406

Rainey WT, Jones EG (1983) Spatial distribution of individual medial lemniscal axons in the thalamic ventrobasal complex of the cat. Exp Brain Res 49:229–246

Raisman G (1977) Formation of synapses in the adult rat after injury: similarities and differences between a peripheral and a central nervous site. Philos Trans R Soc Lond [Biol] 278:349–359

Raja SN, Campbell JN, Meyer RA (1984) Evidence for different mechanisms of primary and secondary hyperalgesia following heat injury to the glabrous skin. Brain 107:1179–1188

Rakic P, Bourgeois J-P, Eckenhoff F, Zecevic N, Goldman-Rakic PS (1986) Concurrent overproduction of synapses in diverse regions of the primate cerebral cortex. Science 232:232–235

Ralston HJ (1968) Dorsal root projections to dorsal horn neurons in the cat spinal cord. J Comp Neurol 132:303–329

Ralston HJ, Ralston DD (1979) The distribution of dorsal root axons in laminae I, II and III of the macaque spinal cord: A quantitative electron microscope study. J Comp Neurol 184:643–644

Ramon y Cajal S (1928) Degeneration and regeneration of the nervous system, vol 1. Oxford University Press, London

Ramon y Cajal S (1952) Histologie du systeme nerveus de l'homme et des vertebres, vol 1. Inst Cajal, Madrid

Randolph M, Semmes J (1974) Behavioral consequences of selective subtotal ablations in the postcentral gyrus of *Macaca mulatta*. Brain Res 70:55–70

Randic M, Siminoff R, Straughan DW (1964) Acetylcholine depression of cortical neurons. Exp Neurol 9:236–242

Ranson SW (1906) Retrograde degeneration in the spinal nerves. J Comp Neurol 16:265–293

Ranson SW (1909) Alterations in the spinal ganglion cells following neurotomy. J Comp Neurol 19:125–149

Rasmusson DD (1982) Reorganization of raccoon somatosensory cortex following removal of the fifth digit. J Comp Neurol 205:313–326

Rasmusson DD (1988) Projections of digit afferents to the cuneate nucleus in the raccoon before and after partial deafferentation. J Comp Neurol 277:549–556

Rasmusson DD, Dykes RW (1988) Long-term enhancement of evoked potentials in cat somatosensory cortex produced by co-activation of the basal forebrain and cutaneous receptors. Brain Res 70:276–286

Rasmusson DD, Nance DM (1986) Non-overlapping thalamocortical projections for separate forepaw digits before and after cortical reorganization in the raccoon. Brain Res Bull 16:399–406

Rasmusson DD, Turnbull BG (1983) Immediate effects of digit amputation on SI cortex in the raccoon: unmasking of inhibitory fields. Brain Res 228:368–370

Rasmusson DD, Dykes RW, Hoeltzell PB (1979) Segregation of modality and submodality formation in SI cortex of the cat. Brain Res 166:409–412

Rasmusson DD, Turnbull BG, Leech CK (1985) Unexpected reorganization of somatosensory cortex in a raccoon with extensive forelimb loss. Neurosci Letts 55:167–172

Recanzone GH, Allard TT, Jenkins WM, Merzenich MM (1986) Electrical stimulation of a peripheral cutaneous nerve results in widespread changes in cortical receptive fields. Soc Neurosci Abstr 12:1436

Renehan WE, Munger BL (1986) Degeneration and regeneration of peripheral nerve in the rat trigeminal system. I. Identification and characterization of the multiple afferent innervation of the mystacial vibrissae. J Comp Neurol 246:127–145

Renehan WE, Rhoades RW (1984) A quantitative electron microscopic analysis of the infraorbital nerve in the new born rat. Brain Res 322:369–373

Renehan WE, Jacquin MF, Mooney RD, Rhoades RW (1986a) Structure-function relationships in rat medullary and cervical dorshal horns. II. Medullary dorsal horn cells. J Neurophysiol 55:1187–1201

Renehan WE, Rhoades RW, Mooney RD, Hallas BH, Jacquin MF (1986b) Central arbor morphology of trigeminal primary afferent neurons peripherally axotomized on the day of birth. Soc Neurosci Abstr 12:698

Rethelyi M, Light AR, Perl ER (1982) Synaptic complexes formed by functionally defined primary afferent units with fine myelinated fibres. J Comp Neurol 207:381–393

Rethelyi M, Salim MZ, Jancso G (1986) Altered distribution of dorsal root fibres in the rat following neonatal capsaicin treatment. Neurosci 18:749–761

Rexed B (1952) The cytoarchitectural organization of the spinal cord in the cat. J Comp Neurol 96:415–466

Rhoades RW, Fiore JM, Math MF, Jacquin MF (1983) Reorganisation of trigeminal primary afferents following neonatal infraorbital nerve section in hamster. Dev Brain Res 7:337–342

Rhoades RW, Belford GR, Killackey HP (1987a) Receptive-field properties of rat ventral posterior medial neurons before and after selective kainic and lesions of the trigeminal brain stem complex. J Neurophysiol 57:1577–1600

Rhoades RW, Mooney RD, Chiaia NL, Szczepanik AM, Hess PR, Renehan WW, Klein BG, Nagele RG, Jacquin MF (1987b) Development and lesion induced reorganisation of the peripheral projections of the trigeminal ganglion in rat. In: Pubols LM, Sessle BJ (eds) Effects of injury on trigeminal and spinal somatosensory systems. Liss, New York, pp 49–61

Rhoades RW, Chiaia NL, Mooney RD, Klein BG, Renehan WE, Jacquin MF (1987c) Reorganization of the peripheral projections of the trigeminal ganglion following neonatal transection of the infraorbital nerve. Somatosens Res 5:35–62

Ribeiro da Silva A, Coimbra A (1984) Capsaicin causes selsective damage to type I synaptic glomeruli in rat substantia gelatinosa. Brain Res 290:380–383

Rice FL (1985) Gradual changes in the structure of the barrels during maturation of the primary somatosensory cortex in rats. J Comp Neurol 236:496–503

Rice FL, Van der Loos H (1977) Development of the barrels and barrel field in the somatosensory cortex of the mouse. J Comp Neurol 171:545–560

Rice FL, Gomez C, Barstow C, Burnet A, Sands P (1985) A comparative analysis of the development of primary somatosensory cortex: interspecies similarities during barrel and laminar development. J Comp Neurol 236:477–495

Rich KM, Luszczynski JR, Osborne PA, Johnson EM (1987) Nerve growth factor protects adult sensory neurons from cell death and atrophy caused by nerve injury. J Neurocytol 16:261–268

Richardson RT, De Long MR (1988) A reappraisal of the functions of the nucleus basalis of Meynert. Neuroscience 11:264–267

Richter JA, Jackson SK (1980) 3H-acetylcholine and 3H-5-hydroxy-tryptamine release from rat midbrain slices and the effects of calcium and pentobarbital. Neurochem Res 5:719–730

Risling M, Remahl S, Hildebrand C, Aldskogius H (1980) Structural changes in kittens' ventral and dorsal roots L7 after early postnatal sciatic nerve transection. Exp Neurol 67:265–279

Risling M, Aldskogius H, Hildebrand C (1983a) Effects of sciatic nerve crush on the L7 spinal roots and dorsal root ganglia in kittens. Exp Neurol 79:176–187

Risling M, Aldskogius H, Hildebrand C, Remahl S (1983b) Effects of sciatic nerve resection on L7 spinal roots and dorsal root ganglia in adult cats. Exp Neurol 82:568–580

Ritz LA, Greenspan JD (1985) Morphological features of lamina V neurons receiving nociceptive input in cat sacrocaudal spinal cord. J Comp Neurol 238:440–452

Rivers WHR, Head H (1908) A human experience in nerve division. Brain 31:323–450

Robinson PP (1980) An electrophysiological study of pathways of pulpal nerves from mandibular teeth in the cat. Arch Oral Biol 25:825–829

Robinson PP (1981) Reinnervation of teeth, mucuous membrane and skin following section of the inferior alveolar nerve in the cat. Brain Res 220:241–253

Rodin BE, Kruger L (1984) Absence of evidence for intraspinal sprouting in dorsal root axons caudal to a partial spinal hemisection: a horseradish peroxidase study. Somatosens Res 2:171–192

Rodin BE, Sampogna SL, Kruger L (1983) An examination of intraspinal sprouting in dorsal root axons with the tracer horseradish peroxidase. J Comp Neurol 215:187–198

Roland PE (1981) Somatotopical tuning of postcentral gyrus during focal attention in man. A regional cerebral blood flow study. J Neurophysiol 46:744–754

Rose RD, Koerber HR, Sedivec MJ, Mendell LM (1986) Somal action potential duration differs in identified primary afferents. Neurosci Letts 63:259–263

Rosen I (1967) Functional organization of group I activated neurones in the cuneate nucleus of the cat. Brain Res 6:770–772

Rosen I (1969a) Localization in caudal brain stem and cervical spinal cord of neurons activated from forelimb group I afferents in the cat. Brain Res 16:55–71

Rosen I (1969b) Afferent connexions to group I activated cells in the main cuneate nucleus of the cat. J Physiol (Lond) 205:209–236

Rosen I, Sjolund B (1973a) Organization of group I activated cells in the main and external cuneate nuclei of the cat: identification of muscle receptors. Exp Brain Res 16:221–237

Rosen I, Sjolund B (1973b) Organization of group I activated cells in the main and external cuneate nuclei of the cat: convergence patterns demonstrated by natural stimulation. Exp Brain Res 16:238–246

Roth G (1978) Intranervous regeneration of lower motor neuron. 1. Study of 1153 motor axon reflexes. Electromyogr clin Neurophysiol 18:225–288

Roth G (1979) Intranervous regeneration. The study of motor axon reflexes. J Neurol Sci 41:139–148

Rubel EW (1971) A comparison of somatotopic organization in sensory neocortex of newborn kittens and adult cats. J Comp Neurol 143:447–480

Russchen F, Amaral D, Price J (1985) The afferent connections of the substantia innominata in the monkey, *Macaca fascicularis*. J Comp Neurol 242:1–27

Rustioni A (1973) Non-primary afferents to the nucleus gracilis from the lumbar cord of the cat. Brain Res 51:81–95

Rustioni A (1974) Non-primary afferents to the cuneate nucleus in the brachial dorsal funiculus of the cat. Brain Res 75:247–259

Rustioni A (1977) Spinal neurons project to the dorsal column nuclei of Rhesus monkeys. Science 196:656–658

Rustioni A, Dekker JJ (1974) Non-primary afferents to the dorsal column nuclei of cat: distribution pattern and cells of origin. Anat Rec 178:454–455

Rustioni A, Kaufman AB (1977) Identification of cells of origin of non-primary afferents to the dorsal column nuclei of the cat. Exp Brain Res 27:1–14

Rustioni A, Macchi G (1968) Distribution of dorsal root fibers in the medulla oblongata of the cat. J Comp Neurol 134:113–126

Rustioni A, Molenaar I (1975) Dorsal column nuclei afferents in the lateral funiculus of the cat: distribution pattern and absence of sprouting after chronic deafferentation. Exp Brain Res 23:1–12

Rustioni A, Sotelo C (1974) Synaptic organization of the nucleus gracilis of the cat. Experimental identification of dorsal root fibres and cortical afferents. J Comp Neurol 155:441–468

Rustioni A, Sanyal S, Kuypers HGJM (1971) A histochemical study of the distribution of the trigeminal divisions in the substantial gelatinosa of the rat. Brain Res 32:45–52

Rustioni A, Hayes NL, O'Neill S (1979) Dorsal column nuclei and ascending spinal afferents in macaques. Brain 102:95–125

Saade NE, Banna NR, Khouri A, Jabbur SJ, Wall PD (1982) Cutaneous receptive field alterations induced by 4-aminopyridine. Brain Res 232:177–180

Saade N, Jabbur SJ, Wall PD (1985) Effects of 4-aminopyridine, GABA and bicuculline on cutaneous receptive fields of cat dorsal horn neurons. Brain Res 344:356–359

Sakai ST (1982) The thalamic connectivity of primary motor cortex (MI) in the raccoon. J Comp Neurol 204:238–252

Salt TE (1989) Gamma-aminobutyric acid and afferent inhibition in the cat and rat ventrobasal thalamus. Neuroscience 28:17–26

Salt TE, Hill RG (1983) Neurotransmitter candidates of somatosensory primary afferent fibres. Neuroscience 10:1083–1103

Sanders FK (1948) The thickness of the myelin sheaths of normal and regenerating peripheral nerve fibres. Proc R Soc [Biol] 135:323–357

Sanders FK, Whitteridge D (1946) Conduction velocity and myelin thickness in regenerating nerve fibres. J Physiol (Lond) 105:152–174

Sanders FK, Young JZ (1944) Role of peripheral stump in control of fibre diameter in regenerating nerves. J Physiol (Lond) 103:119–136

Sanders KH, Zimmermann M (1986) Mechanoreceptors in rat glabrous skin: redevelopment of function after nerve crush. J Neurophysiol 55:644–659

Saporta S, Kruger L (1977) The organization of thalamocortical relay neurons in the rat ventrobasal complex studied by retrograde transport of horseradish peroxidase. J Comp Neurol 174:187–208

Sara SJ (1985) Selective attention, memory, and the locus coeruleus. Adv Behav Biol 28:211–217

Savy C, Margules S, Farkas-Bargeton E, Verley R (1981) A morphometric study of mouse trigeminal ganglion after unilateral destruction of vibrissae follicles at birth. Brain Res 217:265–277

Savy C, Margules S, Solari A, Wehrle R, Verley R, Farkas-Bargeton E (1986) A morphometric study of mouse trigeminal roots after unilateral destruction of vibrissae follicles at birth. Int J Dev Neurosci 4:1–12

Scadding JW (1981) Development of ongoing activity, mechanosensitivity, and adrenalin sensitivity in severed peripheral nerve axons. Exp Neurol 73:345–364

Scheibel AB (1984) Organization of the spinal cord. In: Davidoff RA (ed) Anatomy and physiology. Dekker, New York, pp 47–77 (Handbook of the spinal cord, vols 2, 3)

Scheibel ME, Scheibel AB (1968a) Terminal patterns in cat spinal cord. III. Primary afferent collaterals. Brain Res 13:417–443

Scheibel ME, Scheibel AB (1968b) Terminal axonal patterns in cat spinal cord. II. The dorsal horn. Brain Res 9:32–58

Schiff J, Lowenstein WR (1972) Development of a receptor on a foreign nerve fibre in a Pacinian corpuscle. Science 177:712–715

Schoenen J (1982) The dendritic organization of the human spinal cord: the dorsal horn. Neuroscience 7:2057–2088

Schouenborg J, Sjouland BH (1983) Activity evoked by A- and C-afferent fibres in rat dorsal horn neurons and its relation to a flexion reflex. J Neurophysiol 150:1108–1121

Schroder JM (1972) Altered ration between axon diameter and myelin sheath thickness in regenerated nerve fibres. Brain Res 45:49–65

Schultz W, Wiesendanger R, Hess B, Ruffieux A, Wiesendanger M (1981) The somatotopy of the gracile nucleus in cats with agenesis of a hindfoot. Exp Brain Res 43:413–418

Schwartz AS, Eidelberg E, Marchok P, Azulay A (1972) Tactile discrimination in the monkey after section of the dorsal funiculus and lateral lemniscus. Exp Neurol 37:582–596

Scott SA (1982) The development of the segmental pattern of skin sensory innervation in embryonic chick hindlimb. J Physiol (Lond) 330:203–220

Seddon HJ (1943) Peripheral nerve injuries. Glasgow Med J 21:61–75

Sedivec MJ, Ovelmen-Levitt J, Karp R, Mendell LM (1983) Increase in nociceptive input to spinocervical tract neurons following chronic partial deafferentation. J Neurosci 3:1511–1519

Semba K, Geller H, Egger MD (1985) 4-aminopyridine induces expansion of cutaneous receptive fields of dorsal horn cells. Brain Res 343:398–402

Semmes J, Porter LA (1972) A comparison of precentral and postcentral cortical lesions on somatosensory discrimination in the monkey. Cortex 8:249–264

Semmes J, Porter LA, Randolph MC (1974) Further studies of anterior postcentral lesions in monkeys. Cortex 10:55–68

Senba E, Shiosaka S, Hara Y, Inagaki S, Sakanaka M, Takatsuki K, Kawai Y, Tohyama M (1982) Ontogeny of the peptidergic system in the rat spinal cord: immunohistochemical analysis. J Comp Neurol 208:54–66

Seo ML, Ito M (1987) Reorganization of rat vibrissa barrel field as studied by cortical lesioning on different postnatal days. Exp Brain Res 65:251–260

Shanks MF, Pearson RCA, Powell TPS (1978) The intrinsic connections of the primary somatic sensory cortex of the monkey. Proc R Soc Lond [Biol] 200:95–101

Sharp FR, Gonzalez MF, Morgan CW, Morton MT, Sharp JW (1988) Common fur and mystacial vibrissae parallel sensory pathways: 14 C 2-deoxyglucose and WGA-HRP studies in the rat. J Comp Neurol 270:446–469

Sharpey-Schafer E (1928) The effects of denervation of a cutaneous area. Q J Exp Physiol 19:85–107

Sharpey-Schafer E (1930) The permanent results of denervation of a cutaneous area. Q J Exp Physiol 20:95–99

Shawe GDH (1955) On the number of branches formed by regenerating nerve-fibres. Br J Surg 42:474–488

Shea VK, Perl ER (1985) Regeneration of cutaneous afferent unmyelinated (C) fibres after transection. J Neuryphysiol 54:502–512

Shehab SA, Atkinson ME (1986a) Vasoactive intestinal polypeptide increases in areas of the dorsal horn of the spinal cord from which other neuropeptides are depleted following peripheral axotomy. Exp Brain Res 62:422–430

Shehab SA, Atkinson ME (1986b) Vasoactive intestinal polypeptide (VIP) increase in the spinal cord after peripheral axotomy of the sciatic nerve originates from primary afferent neurons. Brain Res 372:37–44

Sherman RA, Sherman CJ, Gall NG (1980) A survey of current phantom limb pain treatment in the United States. Pain 8:85–100

Shimazu H, Yamagisawa N, Garoutte B (1965) Cortico-pyramidal influences on thalamic somatosensory transmission in the cat. Jap J Physiol 15:101–124

Shipley MT (1974) Response characteristics of single units in the rat's trigeminal nuclei to vibrissa displacements. J Neurophysiol 37:73–90

Shortland P, Molander C, Woolf CJ, Fitzgerald M (1990) Neonatal capsaicin treatment induces invasion of the substantia gelatinosa by the terminal arborizations of hair follicle afferents in the rat dorsal horn. J Comp Neurol 296:23–31

Shortland P, Woolf CJ, Fitzgerald M (1989) Morphology and somatotopic organization of the central terminals of hindlimb hair follicle afferents in the rat lumbar spinal cord. J Comp Neurol 289:416–433

Siegfried J, Zimmermann M (1981) Preface. In: Siegfried J, Zimmermann M (eds) Phantom and stump pain. Springer, Berlin Heidelberg New York, pp v–vii

Sillito AM, Kemp JA (1983) Cholinergic modulation of the functional organization of the cat visual cortex. Brain Res 289:143–155

Silver A, Montagna W, Versaci A (1964) The effect of denervation on sweat glands and Meissner corpuscles of human hands. J Invest Dermatol 42:307–324

Simmel ML (1966) Developmental aspects of the body scheme. Child Dev. 37:69–78

Simons DJ (1978) Response properties of vibrissa units in rat SmI somatosensory neocortex. J Neurophysiol 41:798–820

Simons DJ, Land PW (1987) Early tactile experience influences organization of somatic sensory cortex. Nature 326:694–697

Simons DJ, Woolsey TA (1979) Functional organization in mouse barrel cortex. Brain Res 165:327–332

Simons DJ, Woolsey TA (1984) Morphology of Golgi-cox-impregnated barrel neurons in rat SmI cortex. J Comp Neurol 230:119–132

Simons DJ, Durham D, Woolsey TA (1984) Functional organization of mouse and rat SmI barrel cortex following vibrissal damage on different postnatal days. Somatosensory Res 1:207–245

Singer W, Hollander H, Vanegas H (1977) Decreased peroxidase labelling of lateral geniculate neurons following deafferentation. Brain Res 120:133–137

Smith CL (1983) The development and postnatal organization of primary afferent projections to the rat thoracic spinal cord. J Comp Neurol 220:29–43

Smith KR Jr (1967) The structure and function of the Haarscheibe. J comp Neurol 131:459–474

Smith RL (1973) The ascending fiber projections from the principal sensory trigeminal nucleus in the rat. J Comp Neurol 148:423–441

Snow PJ, Meyers DER (1985) Projections of cutaneous afferent fibres to somatotopically identified regions of the dorsal horn. In: Rowe MJ, Willis WD (eds) Development organization and processing in somatosensory pathways. Liss, New York, pp 141–148

Snow PJ, Wilson P (1989) Deafferentation induced changes in somatotopic organization and the somatotopically inappropriate projections of primary afferent fibres. In: Cervero F, Bennett GJ, Headley PM (eds) Processing of sensory information in the superficial dorsal horn of the spinal cord. Plenum, pp 285–306

Snow, PJ, Wilson P (1990) Blind axonal endings and plasticity in the mammalian CNS. News Physiol Sci (In press)

Snow PJ, Rose PK, Brown AG (1976) Tracing axons and axon collaterals of spinal neurons using intracellular injection of horseradish peroxidase. Science 191:312–313

Snow PJ, Nudo RJ, Rivers W, Jenkins WM, Merzenich MM (1988) Somatotopically inappropriate projections from thalamocortical neurons to the SI cortex of the cat demonstrated by the use of intracortical microstimulation. Somatosens Res 5:349–372

Spreafico R, Schmechel DE, Ellis LC, Rustioni A (1983) Cortical relay neurons and interneurons in the N. ventralis posterolateralis of cats: a horseradish peroxidase, electronmicroscopic, golgi and immunocytochemical study. Neuroscience 9:491–509

Sretavan D, Dykes RW (1983) The organization of two cutaneous submodalities in the forearm region of area 3b of cat somatosensory cortex. J Comp Neurol 213:381–398

Steffen H (1976) Golgi-stained barrel-neurons in the somatosensory region of the mouse cerebral cortex. Neurosci Lett 2:57–59

Steffen H, Van Der Loos H (1980) Early lesions of mouse vibrissal follicles: their influence on dendrite orientation in the cortical barrelfield. Exp Brain Res 40:419–431

Stefken M, Van der Loos H (1980) Early lesions of mouse vibrissal follicles; their influence on dendritic orientation in the cortical barrel field. Exp Brain Res 40:419–431

Stelzner DJ (1971) The normal postnatal development of synaptic end feet in the lumbosacral spinal cord and of responses in the hindlimbs of the albino rat. Exp Neurol 31:337–357

Stelzner D, Weber E, Prendergast J (1979) A comparison of the effect of midthoracic spinal hemisection in the neonatal or weanling rats on the distribution and density of dorsal root axons in the lumbosacral spinal cord of the adult. Brain Res 172:407–426

Stennett RA, Renehan WE, Rhoades RW, Jacquin MF (1986) Neonatal deafferentation effects upon structure and function of trigeminal (V) second-order neurons. Soc Neurosci Abst 12:543

Stopford JSB (1926) An explanation of the two-stage recovery of sensation during regeneration of a peripheral nerve. Brain 49:372–386

Stryker MP, Jenkins WM, Merzenich MM (1987) Anaesthetic state does not affect map of the hand representation within area 3b of somatosensory cortex in owl monkey. J Comp Neurol 258:297–303

Stuss DT, Benson DF (1986) The frontal lobes. Raven, New York

Sugimoto T, Gobel S (1982) Primary neurons retain their central axonal arbors in the spinal dorsal horn following peripheral nerve injury: an anatomical analysis using transganglionic transport of horseradish peroxidase. Brain Res 248:377–381

Sugiura Y, Terui N, Hosoya Y, Kohno K (1989) Distribution of unmyelinated primary afferent fibres in the dorsal horn. In: Cervero F, Bennett GJ, Headley PM (eds) Processing of sensory information in the superficial dorsal horn of the spinal cord. Plenum, New York, pp 15–27

Sugiura Y, Lee CL, Perl ER (1986) Central projections of identified, unmyelinated (C) afferent fibres innervating mammalian skin. Science 234:358–361

Sunderland S (1949) Observations on the course of recovery and late end results in a series of cases of peripheral nerve suture. Aust NZ J Surg 18:264–341

Sunderland S (1978) Nerves and nerve injuries. 2nd Edn. Churchill Livingstone, London

Sunderland S, Laverack JO (1953) The branching of nerve fibres. Acta Anat 17:46–61

Sur M (1980) Receptive fields of neurons in areas 3b and 1 of somatosensory cortex in monkeys. Brain Res 198:465–481

Sur M, Nelson RJ, Kaas JH (1978) The representation of the body surface in somatosensory area I of the grey squirrel. J Comp Neurol 179:425–450

Sur M, Merzenich MM, Kaas JH (1980) Magnification, receptive-field area, and "hypercolumn" size in areas 3b and 1 of somatosensory cortex in owl monkeys. J. Neurophysiol 44:295–311

Sur M, Nelson RJ, Kaas JH (1980) Representation of the body surface in somatic koniocortex in the prosimian galago. J Comp Neurol 198:381–402

Sur M, Wall JT, Kaas JH (1981) Modular segregation of functional cell classes within the precentral somatotopy cortex of monkeys. Science 212:1059–1061

Sur M, Nelson RJ, Kaas JH (1982) Representations of the body surface in cortical areas 3b and 1 of squirrel monkeys: comparisons with other primates. J Comp Neurol 211:177–192

Sur M, Wall JT, Kaas JH (1984) Modular distribution of neurons with slowly adapting and rapidly adapting responses in area 3b of somatosensory cortex in monkeys. J Neurophysiol 51:724–744

Swett JE, Woolf CJ (1985) The somatotopic organization of primary afferent terminals in the superficial dorsal horn of the rat lumbar spinal cord. J Comp Neurol 231:66–77

Szentagothai J (1978) The neuron network of the cerebral cortex: A functional interpretation. Proc R Soc Lond [Biol] 201:219–248

Taber E (1961) The cytoarchitecture of the brain stem of the cat. I. Brain stem nuclei of cat. J Comp Neurol 116:27–70

Taber-Pierce E (1970) Histogenesis of the sensory nuclei of the trigeminal nerve in the mouse. An autoradiographic study. Anat Rec 166:388

Tapper DN, Brown PB, Moraff H (1973) Functional organization of the cat's dorsal horn: connectivity of myelinated fibre systems of hairy skin. J Neurophysiol 36:817–826

Taub A (1964) Local, segmental and supraspinal interaction with a dorsolateral spinal cutaneous afferent system. Exp Neurol 10:357–374

Taylor DCM, Pierau Fr-K, Szolcsanyi J (1985) Capsaicin-induced inhibition of axoplasmic transport is prevented by nerve growth factor. Cell Tiss Res 240:569–573

Tello JF (1923) Genese des terminaisons motrices et sensitives. II. Terminations dans les poils de la souris blanche. Trav Lab Invest Biol Univ Madr 21:257–284

Terzis JK, Dykes RW (1980) Reinnervation of glabrous skin in baboons: properties of cutaneous mechanoreceptors subsequent to nerve transection. J Neurophysiol 44:1213–1225

Tessler A, Glazer E, Artymyshyn R, Murray M, Goldberger ME (1980) Recovery of substance P in the cat spinal cord after unilateral lumbosacral deafferentation. Brain Res 191:459–470

Tessler A, Himes B, Artymyshyn R, Murray M, Goldberger ME (1981) Spinal neurons mediate return of substance P following deafferentation of cat spinal cord. Brain Res 230:263–281

Tessler A, Himes B, Soper K, Murray M, Goldberger M, Reichlin S (1984) Recovery of substance P but not somatostatin in the cat spinal cord after unilateral lumbosacral dorsal rhizotomy: a quantitative study. Brain Res 305:95–102

Tessler A, Himes BT, Krieger NR, Murray M, Goldberger ME (1985) Sciatic nerve transection produces death of dorsal root ganglion cells and reversible loss of substance P in spinal cord. Brain Res 332:209–218

Therman PO (1941) Transmission of impulses through the Burdach nucleus. J Neurophysiol 4:153–166

Thesleff S (1980) Aminopyridines and synaptic transmission. Neurosci 5:1413–1419

Thoenen H, Barde Y-A (1980) Physiology of nerve growth factor. Physiol Rev 60:1284–1335

Thomas PK (1974) Nerve injury. In: Bellairs R, Gray EG (eds) Essays on the nervous system. Clarendon, Oxford, pp 44–70

Thor K, Kuo D, deGroat W, Blais D, Backes M (1982) Alterations of HRP-labelled pudendal nerve afferent projections in the sacral spinal cord transection: correlation with physiological plasticity of a spinal somatovesical reflex. Soc Neurosci Abstr 8:305

Tomasulo KC, Emmers R (1972) Activation of neurons in the gracile nucleus by two afferent pathways in the rat. Exp Neurol 36:197–206

Towe AL, Tyner CF, Nyquist JK (1976) Facilitatory and inhibitory modulation of wide-field neuron activity in postcruciate cerebral cortex of the domestic cat. Exp Neurol 50:734–756

Tracey DJ (1980) The projection of joint receptors to the cuneate nucleus in the cat. J Physiol (Lond) 305:433–449

Tremblay N, Warren R, Dykes RW (1986) Iontophoretic application of Beta-alanine, glycine, taurine and their antagonist strychnine in cat somatosensory cortex. Soc Neurosci Abstr 12:1437

Tripp LN, Wells J (1978) Formation of new synaptic terminals in the somatosensory thalamus of the rat after lesions of the dorsal column nuclei. Brain Res 155:362–367

Trotter W, Davies HM (1909) Experimental studies in the innervation of the skin. J Physiol (Lond) 38:135–246

Tsumoto T, Nakamura S (1974) Inhibitory organization of the thalamic ventrobasal neurons with different peripheral representations. Exp Brain Res 21:195–210

Uddenberg N (1968a) Differential localization in dorsal funiculus of fibres originating from different receptors. Exp Brain Res 4:367–376

Uddenberg N (1968b) Functional organization of long, second-order afferents in the dorsal funiculus. Exp Brain Res 4:377–382
Valverde F (1966) The pyramidal tract in rodents. A study of its relations with the posterior column nuclei, dorsolateral reticular formation of the medulla oblongata, and cervical spinal cord. Golgi and electron microscopic observations. Z Zellforsch 71:297–363
Van der Loos H (1976) Barreloids in mouse somatosensory thalamus. Neurosci Lett 2:1–6
Van der Loos H, Dorfl (1978) Does the skin tell the somatosensory cortex how to construct a map of the periphery? Neurosci Lett 7:23–30
Van der Loos H, Welker E (1985) Development and plasticity of somatosensory brain maps. In: Development, organization and processing in somatosensory pathways. Liss, New York, pp 52–67
Van der Loos H, Woolsey TA (1973) Somatosensory cortex: structural alterations following early injury to sense organs. Science 179:395–398
Van Exan RJ, Hardy MH (1980) A spatial relationship between innervation and early differentiation of vibrissa follicles in the embryonic mouse. J Anat 131:643–656
Van Hees J, Gybels JM (1972) Pain related to single afferent C fibres from human skin. Brain Res 48:397–400
Varon S, Raiborn C, Tyszka C (1973) In vitro studies of dissociated cells from newborn mouse dorsal root ganglia. Brain Res 54:51–63
Verley R, Axelrad H (1977a) Functional maturation of rat trigeminal nerve. Neurosci Lett 5:133–139
Verley R, Axelrad H (1977b) Organization en "barils" des cellules de la conche IV du cortex SI chez la souris: effets des lesions ou de la privation des vibrisses mystaciales. CR Acad Sci [Paris] 284:1183–1185
Verley R, Onnen I (1981) Somatotopic organization of the tactile thalamus in normal adult and developing mice and in adult mice dewhiskered since birth. Exp Neurol 72:462–474
Vierck CJ (1973) Alterations of spatio-tactile discrimination after lesions of primate spinal cord. Brain Res 58:69–79
Vierck CJ (1975) Proprioceptive deficits after dorsal column lesions in monkeys. In: Kornhuber HH (ed) The somatosensory system. Thieme, Stuttgart, pp 311–318
Vierck CJ Jr (1982) Plasticity of somatic sensations and motor capacities following lesions of the dorsal spinal columns in monkeys. In: Morison AR, Strick PL (eds) Changing concepts of the nervous system. Academic, New York, pp 151–170
Vierck CJ (1984) The spinal lemniscal pathways. In: Davidoff RA (ed) Anatomy and physiology. Dekker, New York, pp 673–750 (Handbook of the spinal cord. vols 2, 3)
Vincent SB (1913) The tactile hair of the white rat. J Comp Neurol 23:1–38
Wagman IH, Price DD (1969) Responses of dorsal horn cells of M. mulatta to cutaneous and sural nerve A and C fibre stimuli. J Neurophysiol 32:803–817
Waite PME (1973a) Somatotopic organization of vibrissal responses in the ventrobasal complex of the rat thalamus. J Physiol (Lond) 228:527–540
Waite PME (1973b) The responses of cells in the rat thalamus to mechanical movements of the whiskers. J Physiol (Lond) 228:541–561
Waite PME (1977) Normal nerve fibres in the barrel region of developing and adult mouse cortex. J Comp Neurol 173:165–174
Waite PME (1984) Rearrangement of neuronal responses in the trigeminal system of the rat following peripheral nerve section. J Physiol (Lond) 352:425–445
Waite PME, Cragg BG (1979) The effects of destroying the whisker follicles in mice on the sensory nerve, the thalamo-cortical radiation and cortical barrel formation. Proc R Soc Lond [Biol] 204:41–55

Waite PME, Cragg BG (1982) The peripheral and central changes resulting from cutting or crushing the afferent nerve supply to the whiskers. Proc R Soc Lond [Biol] 214:191–211
Waite PME, Taylor PK (1978) Removal of whiskers in young rats causes functional changes in cerebral cortex. Nature 274:600–602
Walberg F (1966) The fine structure of the cuneate nucleus in normal cats and following interruption of afferent fibres. An electron microscopical study with particular reference to findings made in Glees and Nauta sections. Exp Brain Res 2:107–128
Wall JT, Cusick CG (1984) Cutaneous responsiveness in primary somatosensory (SI) hindpaw cortex before and after partial hindpaw deafferentation in adult rats. J Neurosci 4:1499–1515
Wall JT, Cusick CG (1986) The representation of peripheral nerve inputs in the S1 hindpaw cortex of rats raised with incompletely innervated hindpaws. J Neurosci 6:1129–1147
Wall JT, Kaas JH (1985) Cortical reorganization and sensory recovery following nerve damage and regeneration. In: Cotman CW (ed) Synaptic plasticity. Guildford, New York, 231–260
Wall JT, Kaas JH (1986) Cong term cortical consequences of reinnervation errors after nerve regeneration in monkeys. Brain Res 372:400–404
Wall JT, Kaas JH (1987) A "recovery cycle" in central somatosensory circuits of primates after damage and regeneration of nerves: system level reorganization during regrowth of connections. In: Pubols LM, Sessle BJ (eds) Effects of injury on spinal and trigeminal somatosensory systems. Liss, New York, pp 347–354
Wall JT, Felleman DJ, Kaas JH (1983) Recovery of normal topography in somatosensory cortex of monkeys after nerve crush and regeneration. Science 221:771–773
Wall JT, Kaas JH, Felleman DJ (1985) Functional recovery in the somatosensory cortex of monkeys with regenerated nerves: replication versus reintegration. In: Rowe M, Willis WD (eds) Development, organization and processing in somatosensory pathways. Liss, New York, pp 277–287
Wall JT, Huerta MF, Kaas JH (1986a) Somatosensory cortex topography after nerve repair in neonatal marmoset monkeys. Soc Neurosci Abst 12:1436
Wall JT, Kaas JH, Sur M, Nelson RJ, Felleman DJ, Merzenich MM (1986b) Functional reorganization in somatosensory cortical areas 3b and 1 of adult monkey after median nerve repair: possible relationships to sensory recovery in humans. J Neurosci 6:218–233
Wall PD (1958) Excitability changes in afferent fibre terminations and their relation to slow potentials. J Physiol (Lond) 142:1–21
Wall PD (1960) Cord cells responding to touch, damage, and temperature of skin. J Neurophysiol 23:197–210
Wall PD (1962) The origin of a spinal-cord slow potential. J Physiol (Lond) 164:508–526
Wall PD (1967) The laminar organization of dorsal horn and effects of descending impulses. J Physiol (Lond) 188:403–423
Wall PD (1970) The sensory and motor role of impulses travelling in the dorsal columns towards cerebral cortex. Brain 93:505–524
Wall PD (1977) The presence of ineffective synapses and the circumstances which unmask them. Phil Trans R Soc Lond B 278:361–372
Wall PD (1981) The nature and origins of plasticity in adult spinal cord. In: Brown AG, Rethelyi M (eds) Spinal cord sensation. Sensory Processing in the dorsal horn. Scottish Academic, Edinburgh, pp 297–308
Wall PD (1982) The effect of peripheral nerve lesions and of neonatal capsaicin in the rat on primary afferent depolarization. J Physiol (Lond) 329:21–35
Wall PD (1984) The dorsal horn. In: Wall PD, Melzack R (eds) Textbook of pain. Churchill Livingstone, Edinburgh, pp 80–87

Wall PD (1987) The control of neural connections by three physiological mechanisms. Prog Brain Res 71:239–247
Wall PD, Devor M (1978) Physiology of sensation after peripheral nerve injury, regeneration, and neuroma formation. In: Waxman SG (ed) Physiology and pathobiology of axons. Raven, New York, pp 377–388
Wall PD, Devor M (1981) The effect of peripheral nerve injury on dorsal root potentials and on transmission of afferent signals into the spinal cord. Brain Res 209:95–111
Wall PD, Devor M (1982) Consequences of peripheral nerve damage in the spinal cord and in neighbouring intact peripheral nerves. In Culp WJ, Ochoa J (eds) Abnormal nerves and muscles as impulse generators. Oxford University Press, Oxford, pp 588–603
Wall PD, Devor M (1984) Sensory afferent impulses originate from dorsal root ganglia as well as from the periphery in normal and nerve-injured rats. Pain 17:321–339
Wall PD, Dubner R (1972) Somatosensory pathways. Ann Rev Physiol 34:315–336
Wall PD, Egger MD (1971) Formation of new connexions in adult rat brains after partial deafferentation. Nature 232:542–545
Wall PD, Fitzgerald M (1981) Effects of capsaicin applied locally to adult peripheral nerve. I. Physiology of peripheral nerve and spinal cord. Pain 11:363–377
Wall PD, Fitzgerald M (1982) If substance P fails to fulfil the criteria as a neurotransmitter in somatosensory afferents, what might be its function? In: Substance P in the nervous system. Pitman, London, pp 249–266 (Ciba Foundation Symposium 91)
Wall PD, Gutnick M (1974a) Properties of afferent nerve impulses originating from a neuroma Nature 248:740–743
Wall PD, Gutnick M (1974b) Ongoing activity in peripheral nerves: the physiology and pharmacology of impulses originating from a neuroma Exp Neurol 43:580–593
Wall PD, Noordenbos W (1977) Sensory functions which remain in man after complete transection of dorsal columns. Brain 100:641–653
Wall PD, Werman R (1976) The physiology and anatomy of long ranging afferent fibres within the spinal cord. J Physiol (Lond) 255:321–334
Wall PD, Woolf CJ (1984) Muscle but not cutaneous C-afferent input produces prolonged increases in the excitability of the flexion reflex in the rat. J Physiol (Lond) 356:443–458
Wall PD, Woolf CJ (1986) The brief and the prolonged facilitatory effects of unmyelinated afferent input on the rat spinal cord are independently influenced by peripheral nerve section. Neurosci 17:1199–1205
Wall PD, Merrill EG, Yaksh TL (1979) Responses of single units in laminae 2 and 3 of cat spinal cord. Brain Res 160:245–260
Wall PD, Fitzgerald M, Nussbaumer JC, Van der Loos H, Devor M (1982a) Somatotopic maps are disorganized in adult rodents treated neonatally with capsaicin. Nature 295:691–693
Wall PD, Fitzgerald M, Woolf CJ (1982b) Effects of capsaicin on receptive fields and on inhibitions in rat spinal cord. Exp Neurol 78:425–436
Walshe, FMR (1942) The anatomy and physiology of cutaneous sensibility: a critical review. Brain 65:48–112
Warren S, Pubols BH (1984) Somatosensory thalamocortical connections in the raccoon: an HRP study. J Comp Neurol 227:597–606
Waterhouse BD, Woodward DJ (1980) Interaction of norepinephrine with cerebrocortical activity evoked by stimulation of somatosensory afferent pathways in the rat. Exp Neurol 67:11–34
Waterhouse BD, Moises HC, Woodward DJ (1980) Noradrenergic modulation of somatosensory cortical neuronal responses to iontophoretically applied putative neurotransmitters. Exp Neurol 69:30–49

Waterhouse BD, Moises HC, Woodward DJ (1981) Alpha-receptor-mediated facilitation of somatosensory cortical neuronal responses to excitatory synaptic inputs and iontophoretically applied acetylcholine. Neuropharmacology 20:907–920

Waterhouse BD, Sessler FM, Moises HC, Grady SM (1986) Neuropeptide modulation of somatosensory cortical neuronal responses to iontophoretically applied putative neurotransmitters. Soc Neurosci Abstr 12:217

Weber ED, Stelzner DJ (1977) Behavioural effects of spinal cord transection in the developing rat. Brain Res 125:241–255

Weddell G (1942) Axonal regeneration in cutaneous nerve plexus. J Anat 77:49–65

Weddell G, Guttman L, Gutmann E (1941) The local extension of nerve fibres into denervated areas of skin. J Neurol Psychiat 4:206–225

Weinstein S, Sersen EA, Vetter RJ (1964) Phantoms and somatic sensation in cases of congenital aplasia. Cortex 1:276–290

Weisberg JA, Rustioni A (1979) Differential projections of cortical sensorimotor areas upon the dorsal column nuclei of cats. J Comp Neurol 184:401–422

Weiss P, Edds MV Jr, Cavanaugh M (1945) The effect of terminal connections on the caliber of nerve fibres. Anat Rec 92:215–233

Welk E, Fleischer E, Petsche U, Handwerker HO (1984) Afferent C-fibres in rats after neonatal capsaicin treatment. Pflugers Arch 400:66–71

Welker C (1971) Microelectrode delineation of fine grain somatotopic organization of SmI cerebral neocortex in albino rat. Brain Res 26:259–275

Welker C (1973) Organization of somatosensory cerebral neocortex in micrencephalic rats. Anat Rec 175:467–468

Welker C (1976) Receptive fields of barrels in the somatosensory neocortex of the rat. J Comp Neurol 166:173–189

Welker C, Woolsey TA (1974) Structure of layer IV in the somatosensory cortex of the rat: description and comparison with the mouse. J Comp Neurol 158:437–454

Welker E, Soriano E, Van der Loos H (1989) Plasticity in the barrel cortex of the adult mouse: effects of peripheral deprivation on GAD-immunoreactivity. Exp Brain Res 74, 441–452

Welker R, Van der Loos H (1983) Genetically determined supernumerary mystacial vibrissae in albino mice (ICR): association with peripheral innervation and with size of barrelfield. Neurosci Lett [Suppl] 14:S401

Welker WI, Johnson JI (1965) Correlation between nuclear morphology and somatotopic organization in ventrobasal complex of the raccoon's thalamus. J Anat 99:761–790

Welker WI, Seidenstein S (1959) Somatic sensory representation in the cerebral cortex of the raccoon (Procyol lotor). J Comp Neurol 111:469–501

Welker W, Sanderson KJ, Shambles G (1984) Patterns of afferent projections to transitional zones in the somatic sensorimotor cerebral cortex of albino rats. Brain Res 292:261–267

Wells J, Tripp LN (1987) Time-course of reactive synaptogenesis in the subcortical somatosensory system. J Comp Neurol 255:466–475

Westrum LE, Canfield RC (1979) Normal loss of milk teeth causes degeneration in brain stem. Exp Neurol 56:169–177

Westrum LE, Canfield RC, Black RG (1976) Transganglionic degeneration in the spinal trigeminal nucleus following removal of tooth pulps in adult cats. Brain Res 101:137–140

Westrum LE, Johnson LR, Canfield RC (1984) Ultrastructure of transganglionic degeneration of brain stem trigeminal nuclei during normal primary tooth exfoliation and permanent tooth eruption in the cat. J Comp Neurol 230:198–206

White EL, De Amicis RA (1977) Afferent and efferent projections of the region in mouse SmI cortex which contains the posteriomedial barrel subfield. J Comp Neurol 175:455–482

White JC, Selverston B (1956) Pain and related phenomena, including causalgia. In: Woodhall B, Beebe GW (eds) Peripheral nerve regeneration: a follow up study of 3,656 World War II injuries. US Govt Printing Office, Washington DC, pp 311–340

Whitsel BL, Petrucelli LM, Sapiro G (1969) Modality representation in the lumbar and cervical fasciculus gracilis of squirrel monkeys. Brain Res 15:67–78

Wiley RG (1985) Effects of retrogradely transported ricin on vagal neurons: time course of ricin transport, inhibition of neuronal protein synthesis and morphologic changes. Soc Neurosci Abstr 11:440

Wiley RG, Oeltmann TN (1986) Anatomically selective peripheral nerve ablation using intraneural ricin injection. J Neurosci Methods 17:43–53

Wiley RG, Talman WT, Reis DJ (1983) Ricin transport distinguishes between central and peripheral neurons. Brain Res 269:357–360

Willis WD (1985) The pain system. The neural basis of nociceptive transmission in the mammalian nervous system. Karger, Basel.

Willis WD, Coggeshall RE (1978) Sensory mechanisms of the spinal cord. Plenum, New York

Willis WD, Trevino DL, Coulter JD, Maunz RA (1974) Responses of primate spinothalamic tract neurons to natural stimulation of hindlimb. J Neurophysiol 37:358–372

Wilson P (1987) Absence of mediolateral reorganization of dorsal horn somatotopy after peripheral deafferentation in the cat. Exp Neurol 95:432–447

Wilson P, Snow PJ (1987a) Reorganization of the receptive fields of spinocervical tract neurons following denervation of a single digit in the cat. J Neurophysiol 57:803–817

Wilson P, Snow PJ (1987b) The time course of reorganization of spinal somatotopy following denervation of a single digit in the cat. Proc Aust Physiol Pharmacol Soc 17:98P

Wilson P, Snow PJ (1988a) Somatotopic organization of the dorsal horn in the lumbosacral enlargement of the neonatal cat. Exp Neurol 101:428–444

Wilson P, Snow PJ (1988b) Reorganization of the receptive fields of spinocervical tract cells following neonatal peripheral nerve transection in the cat. Proc Aust Physiol Pharmacol Soc 19:103P

Wilson P, Snow PJ (1988c) Alteration of dorsal horn somatotopy after neonatal peripheral nerve section in the cat may result from collateral sprouting by intact primary afferent fibres. Proc Aust Physiol Pharmacol Soc 19:181P

Wilson P, Snow PJ (1990a) Ultrastructural study of spinal arborizations of cutaneous primary afferent collaterals that lack synaptic boutons. Proc Aust Neurosci Soc 1:141

Wilson P, Snow PJ (1990b) The capacity for reorganization within spinal somatosensory systems. In: Rowe M, Aitkin L (eds) Information processing in mammalian auditory and tactile systems. Wiley-Liss, New York, pp 205–220

Wilson P, Meyers DER, Snow PJ (1986) The detailed somatotopic organization of the dorsal horn in the lumbosacral enlargement of the cat spinal cord. J Neurophysiol 55:604–617

Wilson P, Meyers DER, Snow PJ (1987) Changes of spinal circuitry in response to alterations of input; an evaluation of the development and basis of contemporary theories. In: Pubols LM, Sessle BJ (eds) Effects of injury on trigeminal and spinal somatosensory systems. Liss, New York, pp 227–238

Windle WE, Baxter RE (1936) Development of reflex mechanisms in the spinal cord of albino rat embryos. Correlation between structure and function and comparisons with the cat and the chick. J Comp Neurol 63: 189–209

Winter DL (1965) N gracilis of cat. Functional organization and corticofugal effects. J Neurophysiol 28:48–70

Wise SP, Jones EG (1976) The organization and postnatal development of the commissural projection of the rat somatic sensory cortex. J Comp Neurol 168:313–343

Wise SP, Jones EG (1978) Developmental studies of thalamocortical and commissural connections in the rat somatic sensory cortex. J Comp Neurol 178:187–208
Wise SP, Hendry SHC, Jones EG (1977) Prenatal development of sensorimotor cortical projections in cats. Brain Res 138:538–544
Woerner D, Rhoades RW, Jacquin MF, Klein BG (1986) Continuous cutting of mystacial vibrissae from birth alters the morphology of whisker primary afferents in rat subnucleus interpolaris. Soc Neurosci Abstr 12:126
Wong WC, Kanagasuntheram R (1971) Early and late effects of median nerve injury on Meissner's and Pacinian corpuscles of the hand of the macaque (M fascicularis). J Anat 109:135–142
Wong-Riley, Welt (1980) Histochemical changes in cytodrome oxidase of cortical barrels after vibrissal removal in neonatal and adult mice. Proc Natl Acad Sci USA 77:2333–2337
Woodward DJ, Moises HC, Waterhouse BD, Hoffer BJ, Friedman R (1979) Modulatory actions of norepinephrine in the central nervous system. Fed Proc 38:2109–2116
Woolf CJ (1983) C-primary afferent fibres mediated inhibitions in the dorsal horn of the decerebrate spinal cord. Exp Brain Res 51:283–290
Woolf CJ (1987) Central terminations of cutaneous mechanoreceptive afferents in the rat lumbar spinal cord. J Comp Neurol 261:105–119
Woolf CJ, Fitzgerald M (1986) Somatotopic organization of cutaneous afferent terminals and dorsal horn neuronal receptive fields in the superficial and deep laminae of the rat lumbar spinal cord. J Comp Neurol 251:517–531
Woolf CJ, King AE (1987) Physiology and morphology of multireceptive neurons with C-afferent fibres inputs in the deep dorsal horn of the rat lumbar spinal cord. J Neurophysiol 58:460–479
Woolf CJ, McMahon S (1985) Injury-induced plasticity of the flexor reflex in chronic decerebrate rats. Neuroscience 16:395–405
Woolf CJ, Wall PD (1982) Chronic peripheral nerve section diminishes the primary afferent A-fibre mediated inhibition of rat dorsal horn neurons. Brain Res 212:77–85
Woolf CJ, Wall PD (1986) Relative effectiveness of C-primary afferent fibres of different origins in evoking a prolonged facilitation of the flexor reflex in the rat. J Neurosci 6:1433–1442
Woolsey TA, Van der Loos H (1970) The structural organization of layer IV in the somatosensory region (SI) of mouse cerebral cortex. The description of a cortical field composed of discrete cytoarchitectonic units. Brain Res 17:205–242
Woolsey TA, Diertner ML, Wann DF (1975) Mouse SmI cortex: qualitative and quantitative classification of Golgi-impregnated barrel neurons. Proc Natl Acad Sci USA 72:2165–2169
Woolsey TA, Anderson JR, Wann JR, Stanfield BB (1979) Effects of early vibrissal damage on neurons in the ventrobasal (VB) thalamus of the mouse. J Comp Neurol 184:363–380
Woolston DC, La Londe JR, Gibson JM (1982) Comparison of response properties of cerebellar- and thalamic-projecting interpolaris neurons. J Neurophysiol 48:160–173
Woudenberg RA (1970) Projections of mechanoreceptive fields to cuneate-gracile and spinal trigeminal nuclear regions in sheep. Brain Res 17:417–437
Wynn Parry CP (1981) Rehabilitation of the hand, 4th ed. Butterworths, London
Yamakado M (1985) Postnatal development of barreloid neuropile in the ventrobasal complex of mouse thalamus: a histochemical study for cytodrome oxidase. Brain Nerve 37:497–506
Yamakado M, Yohro J (1979) Subdivision of mouse vibrissae on an embryological basis with descriptions of variations in the number and arrangement of sinus hairs and cortical barrels in BALB/c (au/t; nu/nu) and hairless (hr/hr) strains. Am J Anat 155:153–174

Yamamoto S, Sugihara S, Kuru M (1956) Microelectrode studies on sensory afferents in the posterior funiculus of cat. Jpn J Physiol 6:68–85

Yen C-T, Conley M, Jones EG (1985) Morphological and functional types of neurons in cat ventral posterior thalamic nucleus. J Neurosci 5:1316–1338

Ygge J, Aldskogius H (1984) Intercostal nerve transection and its effect on the dorsal root ganglion. A quantitative study on thoracic ganglion cell numbers and sizes in the rat. Exp Brain Res 55:402–408

Ygge J, Aldskogius H, Grant G (1981) Asymmetries and symmetries in the number of thoracic dorsal root ganglion cells. J Comp Neurol 202:365–372

Yip HK, Rich KM, Lampe PA, Johnson EM Jr (1984) The effects of nerve growth factor and its antiserum on the postnatal development and survival after injury of sensory neurons in rat dorsal root ganglia. J Neurosci 4:2986–2992

Yokota T, Nishikawa N (1982) Effects of picrotoxin upon response characteristics of wide dynamic range neurons in the spinal cord of the cat and monkey. Neurosci Letts 28:259–263

Yoshida S, Matsuda Y (1979) Studies on sensory neurons of the mouse with intracellular-recording and horseradish peroxidase-injection techniques. J Neurophysiol 42:1134–1145

Yurkewicz L, Valentino KL, Flocter MK, Fleshman JW Jr, Jopnes DG (1984) Effects of cytotoxic deletions of somatic sensory cortex in fetal rats

Zander E, Weddell G (1951) Reaction of corneal nerve fibres to cornea injury. Br J Ophthalmol 35:61–88

Zieglganzberger W, Herz A (1971) Changes of cutaneous receptive fields of spinocervical tract neurons and other dorsal horn neurons by microelectrophoretically administered amino acids. Exp Brain Res 13:111–126

Zucker E, Welker WI (1969) Coding of somatic sensory input by vibrissae neurons in the rat's trigeminal ganglion. Brain Res 12:138–156

Subject Index

MIX
Papier aus verantwortungsvollen Quellen
Paper from responsible sources
FSC® C105338

If you have any concerns about our products,
you can contact us on
ProductSafety@springernature.com

In case Publisher is established outside the EU,
the EU authorized representative is:
Springer Nature Customer Service Center GmbH
Europaplatz 3, 69115 Heidelberg, Germany

Printed by Libri Plureos GmbH
in Hamburg, Germany